Climate and Food Security

papers presented at the
International Symposium on
Climate Variability and Food Security
in Developing Countries
5-9 February 1987
New Delhi, India

organized by
American Association for the Advancement of Science
Indian National Science Academy
International Rice Research Institute

with support from
Indian Council for Agricultural Research
Indian Meteorological Department
Indian Council of Scientific and Industrial Research
India Department of Environment, Forest, and Wildlife

1989
International Rice Research Institute
P.O. Box 933, 1099 Manila, Philipppines
in collaboration with
American Association for the Advancement of Science
1333 H. Street, N.W., Washington, D.C., 20005 U.S.A.

The International Rice Research Institute (IRRI) was established in 1960 by the Ford and Rockefeller Foundations with the help and approval of the Government of the Philippines. Today IRRI is one of the 13 nonprofit international research and training centers supported by the Consultative Group on International Agricultural Research (CGIAR). The CGIAR is sponsored by the Food and Agriculture Organization (FAO) of the United Nations, the International Bank for Reconstruction and Development (World Bank), and the United Nations Development Programme (UNDP). The CGIAR consists of 50 donor countries, international and regional organizations, and private foundations.

IRRI receives support, through the CGIAR, from a number of donors including the Asian Development Bank, the European Economic Community, the Ford Foundation, the International Development Research Centre, the International Fund for Agricultural Development, the OPEC Special Fund, the Rockefeller Foundation, the United Nations Development Programme, the World Bank, and the international aid agencies of the following governments: Australia, Belgium, Brazil, Canada, China, Denmark, Finland, France, Federal Republic of Germany, India, Italy, Japan, Mexico, The Netherlands, New Zealand, Norway, the Philippines, Saudi Arabia, Spain, Sweden, Switzerland, United Kingdom, and United States.

The responsibility for this publication rests with the International Rice Research Institute.

ISBN 971-104-210-X

Contents

Foreword

Increasing evidence and scientific consensus indicate that the world may be experiencing a significant climate change. Whatever the eventual scenario, climate change raises a number of urgent questions and poses additional challenges for scientists. Research is needed to identify the important factors involved and to design the technology needed to increase and sustain the productivity of basic food crops. Climate-defensive food systems require the fullest possible understanding of the subtle and complex relationships between crops and climate.

IRRI's cosponsorship of the International Symposium on Climate and Food Security was particularly appropriate, and its responsibility for publication of this proceedings book particularly relevant. Rice is the basic food for more than half of the world's population. It is one of the oldest cultivated crops on earth, and probably the world's most versatile crop. Over millennia, in different parts of the world, rice has been adapted to very different climatic areas. Most of today's improved rices are designed for particular water, temperature, and photoperiod regimes.

The conveners of this tri-part symposium of climatologists, meteorologists, and agricultural scientists were Roger Revelle, chairman of the Committee on Climate of the American Association for the Advancement of Science (AAAS); M.S. Swaminathan, then Director General of IRRI, and S.K. Sinha, Professor of Eminence of the Indian Agricultural Research Institute and the Indian National Science Academy. The book was edited by Janet Kayser, Washington, D.C., and LaRue Pollard, assisted by Emerita P. Cervantes, IRRI.

Klaus Lampe
Director General

Preface

The International Symposium on Climate and Food Security was planned and convened by Roger Revelle, Chairman of the Committee on Climate of the American Association for the Advancement of Science (AAAS); M. S. Swaminathan, Director-General of the International Rice Research Institute (IRRI); and S. K. Sinha, Professor of Eminence of the Indian Agricultural Research Institute and the Indian National Science Academy (INSA).

They recognized three critical world problems: that several billion people often lack the most basic human need—food security; that population growth and the need to improve living standards are putting severe pressure on the soil and water resources that sustain all food production; and that unfavorable weather and climate remain the most frequent cause of crop failure—sometimes leading to widespread distress and even famine.

They also recognized a new factor: the growing scientific consensus that the buildup of greenhouse gases in the atmosphere is likely to cause a global climate change—an environmental change on a scale unprecedented in human history—with the potential for great impacts, both beneficial and harmful, on food security.

The Symposium 6-9 February 1987 in New Delhi, India, brought together climatologists and agronomists, experts in irrigation and water-use efficiency, agricultural economists, and development planners from 21 countries to discuss how climatological information can be used to help reduce fluctuations in agricultural output. They presented the results of their research and discussed how to apply scientific understanding to the practical problems faced by farmers. The overriding concern was: how can scientists help farmers exploit favorable agroclimatic patterns and adapt to or protect against unfavorable climatic trends?

Papers presented dealt with food supply and demand; climatic variability and crop yields; climatic vulnerability of major food crops such as rice, maize, wheat, oilseeds, and pulses; the impact of climatic variability on factors of production such as soils, stress, diseases and insects, and postharvest losses; the role of modeling; social and economic implications including droughts and famine; and strategies for coping and adapting. Other papers discussed climate as a valuable, but under-exploited, resource—as important in its way as fertile soil and ample water. Development planners were urged to give more consideration to climatic factors in the planning and management of agriculture at national, regional, and local levels.

Dr. Swaminathan stressed that, although food stocks are currently in good supply in many areas, there is no cause for complacency. Weather and climate remain the most significant variable in agricultural production. If reliable production is to be maintained in the face of ever-growing world food demand, systems must be developed and implemented to provide farmers with practical and timely climatic information.

Dr. Revelle reviewed the impacts that might result from a greenhouse climate change. The certainty of their occurrence ranges from extremely likely to unknown, with effects on agriculture ranging from potentially beneficial to seriously disruptive.

The good news is that agroclimatic information, when used in a systematic and coordinated way, can help maintain and stabilize crop production, despite weather and climate fluctuations.

Logistics of the symposium were shared among the three supporting organizations. AAAS, with funds from the U.S. National Science Foundation, handled the secretariat, as well as travel arrangements for most participants coming from outside of India. INSA, with support from the Indian Council for Agricultural Research, the India Meteorological Department, the Indian Council of Scientific and Industrial Research, and the India Department of Environment, Forest and Wildlife, handled all arrangements in Delhi at the symposium itself. IRRI, in addition to the participation of its own scientists, brought the meeting to the attention of scientists in other international agricultural research centers (the CGIAR group—Consultative Group for International Agricultural Research) and undertook the publication of the proceedings volume.

To list the many individuals who contributed to the symposium would be unwieldy. But the convenors wish to acknowledge the critically important support of Dr. N. S. Randhawa, Director General of the Indian Council of Agricultural Research. S. K. Sahni, Assistant Executive Secretary of the Indian National Science Academy, also provided important assistance. The IRRI Communication and Publications Department handled editing and publishing of this volume. Carolina Carter, Program Associate of the AAAS Committee on Climate, handled many details with efficiency and grace, both in Washington and in Delhi.

The symposium provided an opportunity for scientists from many parts of the world to exchange ideas and present the results of studies on the role of weather and climate in food production. Many participants said that more work in this area could yield significant benefits, and expressed the hope their colleagues would be encouraged to pursue promising lines of research.

In the never-ending struggle to provide people everywhere with the assurance of food security, we certainly need to understand more. But the participants also emphasized the need to apply what we already know—by devising and testing better methods of conveying to farmers the timely and practical agroclimatic information they need.

These were the goals and vision of M. S. Swaminathan, Roger Revelle, and S. K. Sinha—the three eminent scientists who made this meeting possible.

DAVID BURNS
Director, AAAS Climate Project

An agenda for action

Climatic variability is a principal source of fluctuations in the production and price of agricultural commodities. Throughout history, many practices developed by farmers and livestock herders have been aimed at mitigating or avoiding risks to livelihood imposed by uncertain weather.

Climate-induced variability will continue to be a major problem, whether or not the global climate changes. But it may be exacerbated if the current secular warming trend leads to increases in the frequency or intensity of short-term fluctuations around the mean. Thus, although longer-term measures may have to be taken to cope with climatic change, policies to deal with climatic variability must continue to receive high priority, especially because climatic variability affects strategic world grain supplies and food security in low-income developing countries.

Improvements in long-range weather forecasting could help counter climate-induced variability, especially in tropical regions, although more research is needed before this becomes a practical possibility. Better forecasts would help agriculture adapt to changing conditions, and thus contribute to food security. Farmers in developed countries often have relatively easy access to research results and weather monitoring, but almost no farmers in less developed countries (LDCs) have access to even current-season information.

If scientific knowledge of climate is to be applied effectively, better two-way communication between farmers, policymakers, and researchers is crucial. Research into how farmers perceive and react to climatic variability has been woefully neglected. Improvements in food security require better communication between scientists and the societies they serve. Work to characterize agricultural environments in ecological and social terms needs to be reinforced so that the potential for sustainable management of agricultural systems to meet changing demands for food in the face of climatic variability and change can be better understood.

Climatic change

There is a scientific consensus that the continuing buildup of heat-absorbing gases, such as carbon dioxide and methane, will cause a global climatic change—often called a greenhouse effect. A 1985 World Meteorological Organization (WMO) United Nations Evironment Programme/International Council of Scientific Unions

(ICSU) study (The Villach Report) predicted that the world will likely experience the changes early in the next century. We must do everything possible to anticipate the possible effect on food security of an anthropogenically induced climatic change. Past experiences with climatically induced variations in food production may not be good guides to the future. Understanding the complex causes and effects of climatic change will require long-term work by physical, biological, and social scientists.

The implications of global warming are numerous and difficult to predict, especially for a particular location. However, vigilant, detailed investigation of climatic change as it occurs may make it possible to adjust agricultural systems to keep pace. We must identify ways for farmers and national and regional agricultural sectors to minimize the effects of adverse climate and maximize the advantages of favorable weather and climate. This will help achieve higher levels of food production and greater security of food production, despite climatic uncertainties.

Some changes caused by global warming may be beneficial. Plant growth is stimulated by the direct fertilizing effect of carbon dioxide, especially in the tropics, and temperate regions will have longer growing seasons. Disadvantages may include flooding of low-lying land due to higher sea levels caused by expansion of warmer ocean water, greater incidence of pests and diseases, and possible direct crop damage due to higher temperatures. Changes in the frequency or magnitude of severe events, such as floods, droughts, heat waves, and cold spells, may be more harmful than changes in average conditions.

The food security implications of climatic change are complicated by economic and political considerations. For example, increased carbon dioxide concentrations may benefit food production in LDCs at lower latitudes. These countries could possibly increase their food exports, resulting in economic gains. But on balance, climatic change might still be unfavorable to LDCs. If climatic change reduces output in major food-exporting countries, fluctuations in prices could hurt low-income food-importing countries.

Scientific evidence on atmospheric warming suggests that climate in the decades ahead is likely to be fundamentally different in many ways and in many areas. Climatologists must redouble their efforts to describe the details of future climate so that agronomists, plant physiologists, and plant breeders can begin now to develop crops and farming systems adapted to the expected conditions.

In addition to the uncertainties about global climatic change, the causes of climatic variability are not well understood. Three of the most urgent areas of research are 1) the relationships between ocean surface temperature and rainfall, 2) the impact on rainfall of fluctuations in wind stress over the ocean surface, and 3) data on rainfall at the local level for identifying different temporal and spatial scales.

International research cooperation

LDCs are especially vulnerable to the impacts of climatic variability and change. International agencies such as the Food and Agriculture Organization (FAO), UNESCO, WMO, ICSU, Consultative Group on International Agricultural

Research (CGIAR), and nongovernmental agencies can help by supporting anticipatory adaptation now, before changes occur. Virtually all of the adaptive actions are fully justified by current climatic variability, and thus make economic and policy sense no matter what climatic change may bring. Governments should consider "what if?" scenarios of future change, and begin to develop policy responses. Maintenance of global food production and food security in a period of rapid climatic change is likely to require an unprecedented level of cooperation among nations, and among providers and recipients of economic and technical assistance.

The impacts of global climatic change on food security in LDCs are potentially so serious, a special international interdisciplinary research effort is needed. This could be organized under the auspices of the WMO's World Climate Programme. It could also be a special component of the proposed International Geosphere/Biosphere Program. We should begin now to monitor and evaluate climatic change, establish strategic reserves, and carry out global and regional sensing and surveying. An effective research effort will require trained scientific personnel and super-computing power, perhaps incorporating a unified agroecological database. Because it is unlikely that any one developing nation could support such a large and diverse program, a consortium of nations and international and private donors should sponsor this work.

Recommendations to the scientific community

1. Climatic data analysis involving collaboration among agrometeorologists, ecologists, agronomists, soil scientists, and crop protection specialists should be encouraged. At present, agronomically useful syntheses of long-term climatic data are not generally available to agricultural advisers or farmers.
2. Interdisciplinary work by climatologists, geographers, plant scientists, and soil scientists needs to be expanded to characterize and classify agroclimatic zones. This will provide means of defining the prevalence and intensity of stresses that limit productivity in each zone and of identifying appropriate research and land-use strategy for sustainable agricultural growth.
3. Basic studies in plant-soil-water relationships should continue.
4. Crop-climate models should be developed for all major food crops.
5. Genetic variability should be exploited to develop stress-tolerant cultivars, genetic resources should be conserved and germplasm banks used to locate favorable genes, and ways to speed up screening for drought resistance and other climate-induced stresses should be developed.
6. Genetic material should be distributed by means of international crop nurseries to obtain data on yields in relation to weather in various environments, with uniform crop management trials of primary food crops to determine interaction or correlation with climatic factors in various locations.
7. Research on agroforestry and forestry—land use, species, establishment, and production—should be encouraged. Research on alley cropping in humid and subhumid areas of the tropics should investigate management, competition, allelopathy, and socioeconomic factors.

Recommendations to the agricultural science community

Agriculture should balance the practical benefits of growing the best-adapted crops for each region with the insurance provided by diversity. Better transport, storage, and financing can encourage crop diversification, particularly in climatically marginal areas.

Farmers should be encouraged to make their own decisions. They are likely to make better decisions if they are shown how to minimize risk and maximize available sun and soil moisture by adapting alternatives to shifting cultivation and to monoculture, such as intercropping, double cropping, or relay cropping.

There is an urgent need for regionally specific quantification of climatic constraints on agricultural production, particularly in sub-Saharan Africa. For example, farmers should de-emphasize maize in Africa's drier areas and instead emphasize millet and sorghum. This could be especially important if these crop areas become more arid. Plant breeders should also consider changes in plant architecture to help tailor the crop to its environment.

Government should provide better, more simple guidelines on adapting farming practices to different weather and climate. The basics of good agronomy for each crop are known, but may require modification as conditions change. Extension systems in Asia can carry out these recommendations now. Strategies may be as simple as using windbreaks to protect crops from mechanical damage and from desiccation by dry hot winds.

It is critically important to identify the most important climatic element affecting crop growth, for example, water stress in the semiarid tropics and temperature variability in wheat-growing areas of the subtropics. Where crop husbandry practices are unable to moderate critical climatic constraints, the potential of plant breeding should be reassessed.

Additional recommendations

1. Measure current growing season lengths in different locations.
2. Establish dew registration as a standard meteorological observation. This practice is particularly important for crop physiology in semiarid areas and in disease forecasting.
3. Develop methods of assessing crop vulnerability to adverse climate, for example by simulating the marginal effects of specified changes in temperature, precipitation, evapotranspiration, and humidity on pest and disease incidence or other important variables affecting crop yields.
4. Agricultural and climatic statistics could be made more useful than they are currently. Obstacles to better use of statistics include
    - Loss of valuable detailed information as a result of statistical aggregation for governments and policymakers. Agriculturists often need specifics to respond properly to weather and climate information.
    - Failure to present data in probabilistic terms. Data format often makes it impossible to calculate the different weights that should be given to more or less likely scenarios.
    - Lack of data access and comparability. Differences in classification

methods and incomplete collaboration among meteorologists and agriculturists often lessen data usefulness.

- Inadequacy of data handling and storage systems. The CGIAR, in collaboration with other international institutions, has recommended a database management system for climate, soil, topography, crop distribution, and farming system information.
- Absence of data in crucial areas. Rainfall intensity records are particularly spotty.

Scientists also face basic inadequacies in the statistical data available to them. Evaluation of agricultural adjustments to changes in climate requires understanding where climate fits among the many considerations that guide successful agricultural production. At present, there is little empirical data on how farmers adapt to climatic change. Because it is difficult to distinguish changes caused purely by climate from changes rooted in other factors, it is necessary to investigate how climate impinges on each aspect of agriculture.

Identifying the best cropping strategies for different regions requires new empirical statistical approaches. Analyses of key crops should be performed across major climatic zones, leading to the identification of agroecological zones that are homogeneous in respect to cropping potential. Cropping strategies could then be determined on the basis of essential data on each zone. Such strategies would incorporate all that is known about the zone's climate, soil, and socioeconomic conditions, thus improving the chances for maximum sustainable productivity.

Other work that would improve our statistical picture of the interaction between weather and agriculture includes the following:

- Adoption of a standard scale, allowing climatic factors to be analyzed temporally as well as spatially.
- Analysis of relationships between crop yield, precipitation, and soil moisture-holding capacity.
- Development of simulation models that show the effects of climate on crop phenology and growth and that indicate the management decisions necessary to bring a crop to harvest.
- Collection of more data on key nonfood crops, such as coffee, cotton, tea, rubber, cacao, and oil palm, as on well as on land under forest and pasture.

The usefulness of statistical data to both scientists and farmers depends on many factors: the quality of the data, its value for comparisons with other areas, and the statistical assumptions and techniques of analysis. Much depends on the ability of those who gather the data. At present, there is a serious shortage of staff qualified to collect, analyze, and apply agroclimatic information in developing countries. Training local staff members in appropriate disciplines should receive high priority from national governments and international donors and research institutes.

Applied research

The greatest need is to adapt technology to specific circumstances. Adaptation would be speeded up if farmers were seen as partners, to be consulted early in the process.

Location-specific research by people sensitive to local views and traditions can improve scientists' understanding of farm practices, farmers' choices of technology, their needs, and their responses to climatic risk and uncertainty.

Participatory research should concentrate on case studies and multilocation trials. Different types of agrometeorological and agroecological information will be useful in different areas. Small-scale case studies can generate new data to augment information from existing sources. Participatory research sites should be linked with local or regional research networks. Appropriate database systems need to be developed to record, store, and manipulate agroecological information and make the results available to potential users (research planners, extension staff, and farmers).

Gene pools and genetic diversity require paying attention to local cultivars and their origins before those resources are lost.

Improved varieties, appropriate cultural practices, and agrometeorological information can significantly improve crop yields. In humid and subhumid regions, the adverse effects of low soil fertility and of rainfall variability can be mitigated by mulch farming, leguminous cover crops, no-till, alley farming and agroforestry, and weed control. Where grain crops are grown in semiarid and arid regions, high-yielding practices include timely sowing, adequate crop stand, use of early-maturing varieties, mixed cropping, and tied-ridge systems for moisture conservation. Small-scale irrigation practices could potentially reduce vulnerability to erratic rainfall in the Sahel and other semiarid regions.

Soils

Restoring the productivity of eroded and degraded lands deserves high priority. Improved soil and crop management practices, including agroforestry, and the use of leguminous covers and mulching can reduce crop vulnerability in marginal areas and increase production or reduce its variability. Preserving forests and vegetation cover may reduce the adverse effects of rainfall. Laws can help ensure preservation of forests; national scientists can recommend better crop and soil management strategies for specific regions.

Researchers must examine the extent to which climate itself is affected by overgrazing and fuel gathering in the low-rainfall tropics, and by deforestation in rain forests. Experiments should be supplemented by detailed monitoring of climatic and soil variables so that soil-climate relationships can be better understood and extrapolated to other regions.

We need to establish criteria and thresholds of soil and climatic variables so as to quantify the trend line, magnitude, and severity of soil degradation. We must also evaluate alternate soil, water, and crop management systems, coupled with genotype improvement, as strategies for coping with climatic variability.

In semiarid regions, technological advances have led to better soil and water conservation methods, land use systems that augment fuel and fodder production, contingent crop planning to meet weather aberrations, and cropping systems designed to take advantage of periods of assured moisture supply. However, a more thorough understanding of soil-plant-water relationships is still needed to enhance water-use efficiencies and reduce vulnerability to climatic variability.

Drought and water management

Droughts do not begin when rainfall ends, but develop over time. Even imprecise drought prediction can help in coping with climatic variability. So far, efforts to predict have been fragmentary and have been based on disparate approaches—usually synoptical, climatological, or statistical. Work to apply all three methodologies simultaneously may make it possible to predict the occurrence of drought and to identify regional patterns. Climatological data from such predictions integrated with crop-climate models and data on current production could be used to calculate potential drought impact, allowing contingency plans for mitigation to be devised and for midseason corrections to be made.

Attaining skill in predicting and in devising strategies for adjustment will require a multifaceted research effort. Vulnerable regions must be identified by incidence, frequency, duration, and intensity of droughts and by crop sensitivities.

Developing mitigating strategies will require equal effort. In the past, mitigation efforts have relied on massive, timely inputs and on compensatory mechanisms, such as alternate crop planting and increased cropping intensity in irrigated areas. Mitigation might be further improved by developing rules-of-thumb that would enable field workers to estimate the effects of drought on crop yields and the potential for crop revival with drought-breaking rains.

In many parts of the world, enlarging and increasing the efficiency of irrigation systems is the most effective way to cope with climatic variability. However, in some areas, large-scale irrigation, even where technically feasible, may still be inappropriate. Some large development projects have come close to failure due to unforeseen pest and disease problems, including human diseases. Construction costs and the cost of energy for pumping may be prohibitively expensive. Large irrigation systems sometimes become the focus of political disputes, both within countries and among adjoining countries.

Integrated pest management (IPM)

Large-scale pest and disease outbreaks are often linked to variations in temperature and precipitation. If climatic change makes average weather conditions warmer and more humid, serious pest and disease outbreaks are likely to increase. In that case, biological pest control will become even more important. Biological control can help reduce problems arising from the overuse of insecticides, such as pest resistance, toxicity to human and animal life, and environmental pollution (including groundwater contamination). The following measures enhance biological control: 1) Minimize insecticide use which slows evolution of resistant pests and reduces toxicity to beneficial predators. 2) Promote host-plant resistance, and concurrently gauge response by monitoring pest population dynamics. 3) Explore the potentials of biocontrol agents introduced from other regions. 4) Develop ways to augment indigenous species by mass producing biocontrol agents. IPM depends crucially on correct timing and the integration of a complex array of control measures.

While IPM aims to increase the effectiveness of natural predators and reduce the need for chemicals, it is not certain whether its application will be effective in all climates. Some level of pesticide use will be necessary.

If integrated pest management is to succeed in reducing pesticide use, the following conditions are required:

1) Adequate weather forecasting systems are in place.
2) Government policies that ensure supplies of biological pest control agents (predators, parasites, etc.) are readily available to farmers.
3) Farmers receive timely forecasts and are able and willing to implement them.
4) Farmers have easy access to pesticides and application equipment, as well as credit for their purchase.
5) Farmers know how to use pesticides properly and safely, and are alert to potentials for abuse.
6) Specially-trained extension services personnel are available to assist farmers in IPM techniques.

Pest management measures persistently have been hindered by the lack of good long-term crop and weather records and by manufacturers who aggressively promote the use of inappropriate chemicals by farmers. Addressing these two problems would speed implementation of rational pest control.

Social education

The challenge is to select and adapt techniques that work, then transfer them to developing countries that lack extension services, information delivery systems, and often the financial resources to implement new methods. The importance of project follow-up has been dramatized by the unfortunate consequences of the introduction of infant feeding formulas into areas where water supplies are unreliable or polluted—situations made worse by climatic anomalies such as drought and flooding.

The crucial stage is the period after a pilot project. Follow-up activity should explore how whole-season and within-season agroclimatic information can be communicated effectively to farmers. Not enough attention is paid to how such information is used by farmers and their advisors, and to how best to communicate it. Physical and biological scientists must collaborate with experts in mass communication and in village-level field studies. The key may be improved extension systems.

Effective information could help improve nutrition. People succumb more often to diseases when they are weakened by inadequate diets. Traditional local diets and food preparation methods should be recorded and their nutritional quality assessed. Ways should be found to preserve the cultivation of local crops, many of which are more climatically tolerant than introduced crops. Crop diversity increases the odds of some crop survival of drought, and often improves diets.

Information programs could also discourage the use of dung as fuel rather than as fertilizer. Burning dung instead of using it as fertilizer limits crop yields where farmers use little chemical fertilizers, and the practice pollutes indoor air. Village woodlots are often the best alternative, and often can provide multiple benefits (fuel, fodder, timber, shelter belts). Kerosene and biogas are proving to be feasible alternatives in some areas.

Food grain research

Rice. Rice improvement programs should give more emphasis to cultivars adapted to unfavorable environments, particularly drought-prone, flood-prone, adverse soil, and adverse temperature conditions. Recent progress in improving yield and yield stability for different environments should be extended by

- Using new breeding approaches to capture genes from less compatible sources, biotechnology.
- Intensifying in situ selection in more representative adverse environments.
- Using agroclimatology to identify adverse environments for tests.

Sorghum and millet. Seventy percent of the world's sorghum and millet is grown in the less developed countries of Asia and sub-Saharan Africa. The crops are completely rainfed. Sorghum shows potential for adaptation to other semiarid tropical areas, such as Northeast Brazil.

Tropical sorghums are vulnerable to climatic change because of their highly localized adaptability and because their growth depends on seasonal rains. The millet and postrainy-season sorghum areas of India, West Asia, and sub-Saharan Africa are regions of potential coarse-grain deficit. It is possible, however, to incorporate in sorghum resistance to both low and high rainfall problems. Regional and national networks, in cooperation with well-established international programs, could reduce sorghum vulnerability by developing region-specific cultivars, with the assurance of diversity through international gene banks, breeding programs, and multilocation trials.

Unique problems of Africa

Sub-Saharan Africa is the only region in the world with a chronic food deficit. Widespread hunger and malnutrition there are caused by the world's fastest growth in population and slowest growth in food production. The result has been a significant drop in food per capita. Food deficits persist despite massive emergency food aid and crisis management efforts by the international community and by national governments.

The slow growth in food production over the last two decades is attributable to many factors, including adverse climate, unfavorable soil, unfavorable farm policies and terms of trade for agriculture, political instability, weak national research and extension systems, lack of understanding of the role of women in agriculture, underdeveloped rural infrastructure, and lack of access to essential agricultural inputs.

A high proportion of Africa's soils are easily degradable, with low inherent fertility. Climatic factors leading to soil degradation include water and wind erosion, high soil temperatures, compaction, laterization, reduction in plant-available water and nutrient reserves, and nutrient imbalances. Low and erratic rainfall contributes to soil degradation, and drought accentuates the process. Understanding Africa's climate, including the causes of recurrent droughts and whether these are a manifestation of climatic change or a normal variability, and understanding Africa's soil resources and learning to manage them are essential in developing sustainable agricultural systems for the long term.

In the meantime, significant changes in farming methods are needed, especially in tillage, fertilizer use, and adoption of improved varieties to reduce the climatic vulnerability of crops. Recent analyses show considerable potential in sub-Saharan areas of limited rainfall for weather-responsive crop management. More research could help develop farm-level strategies to capitalize on the strong relationship between the onset of rains and the length of the growing season.

Policy issues

Science and policy need to be better integrated. Research findings should be communicated to decisionmakers in a manner that suggests policy consequences.

The potential impact of climatic change is huge. Climatic change could affect investments in water and irrigation projects with lifetimes of 50-100 yr, as well as investments in agricultural research systems and in measures to improve storage and transportation to ensure that food is available when and where it is needed. So far, funding agencies and governments have been influenced more by current and past climatic variations than by potential dangers or by opportunities.

Local weather observers often are not skilled in the effective use and maintenance of their instruments. Relatively modest investments in equipment and training would pay high dividends in improved agroclimatic data. Remote sensing and geographic information systems could facilitate gathering, storing, and processing data; they often are cost-effective. Again, there is a need for training in the use and interpretation of the information, including surveys to verify ground truth.

Under various weather scenarios and management levels, agroclimatology can help determine crop yield and production indices. It can also determine the population dynamics for semiarid and arid regions under different drought and deforestation scenarios.

Food security

Food security is defined as physical and economic access to adequate food for all people at all times. Recent technological advances have increased worldwide food security, but the outlook is troubling. Population growth, diminishing land available for agriculture, and damage to the life support systems of soil, water, and air suggest there cannot be any slackening in measures to promote agricultural growth on an ecologically sustainable basis. Although self-sufficiency has been achieved in several less developed countries, nutritional security remains elusive.

The great increase in wheat production that occurred in India and elsewhere in the late 1960s was due largely to improvements in cultivars developed by means of traditional breeding techniques. This increase could not have occurred without concurrent improvements in irrigation systems, fertilizers, and pesticides. However, yields in farmers' fields with the new crop varieties are still far less than their biological potential. Food production remains too dependent on favorable or unfavorable rainfall situations.

Food security can be improved in the following ways:

- Increase reforestation and prevent further deforestation, to allow forests to act as sponges in areas of heavy rainfall.
- Increase emphasis on forecasts of global and national food stocks and the need to evolve more efficient food distribution systems.
- Introduce social and fiscal measures to mitigate producer losses caused by adverse climate, to encourage investment in inputs that will increase production, or at least reduce losses.
- Increase food storage safety. Even if climatic change proves beneficial to food production, the cushioning effect of more secure storage might still exceed the cost.

Views on "The new global context for agricultural research: implications for policy"

P. A. Oram

Food and fertilizer prices have fallen substantially since the mid-1970s, and globally there is a relative abundance of food. The improvement in Asia is marked. While the situation in Africa remains precarious, the number of food emergency countries has declined since 1985. This gives cause for satisfaction, but not for complacency. Severe national debt problems divert funds from long-term development, and an increasing number of developing countries have food deficits. Some countries have rising food deficits despite rapid agricultural growth because general economic progress and rising incomes have increased demand, especially for livestock products. Developing-country importers need an international environment wherein food supplies are reliable and not subject to radical price or volume fluctuations. Some large low-income countries are being forced to export grain despite severe malnutrition because of lack of domestic purchasing power, yet they are underinvesting in sectors that could increase income and employment. The improvement of food production in Asia has resulted from investment in research, expansion of irrigation and fertilizer use, infrastructural development, relative peace and stability, and emphasis on building resources of trained personnel. Similar progress has not been achieved so far in Africa. The reasons for this inadequate production growth merit careful study, especially of factors that influence the generation and adoption of new technology and of yield-increasing inputs and the sustainability of crop area expansion without environmental damage. Attention is drawn to weaknesses in the assumptions on which long-range global studies are predicated, in particular those related to climate (which is usually ignored) and the natural resource base. Research and monitoring of climate and land and water resources, bases for sound policy decisions on food security, are essential. International support for national production policies that maximize ecological comparative advantage rather than simply aiming at food self-sufficiency is needed.

John Mellor's presentation to the 1986 Centers' Week meeting of the Consultative Group on International Agricultural Research (CGIAR) looked broadly at trends in world food supply and demand as a basis for its conclusions about food policy (Mellor 1986). Mellor drew substantially on the continuing review of developments in the world food situation by the International Food Policy Research Institute (IFPRI). Along with data supplied generously by the Food and Agriculture Organization of the United Nations (FAO) and the United States Department of

Agriculture (USDA), that review has been an important source of material for IFPRI's work since its founding in 1975, as well as the basis for several research reports by Paulino (1977, 1986) and others.

This recent overview of world food issues is based on the sort of long-term and carefully documented assessment involving staff in several disciplines I believe is essential to an informed discussion of such an important issue.

Recent improvements in world food supply

Twenty years ago, after two bad monsoons, South Asia was being written off as beyond saving by certain food strategists. Ten years ago, the world was just emerging from another food crisis, compounded by rocketing energy costs. Even though the world is still far from achieving Henry Kissinger's vow to the 1974 United Nations (UN) World Food Conference, that in 10 yr time, no child would go to bed hungry, I believe that substantial progress toward food security has been made, in Asia particularly. In sharp contrast to the predictions of several global studies, food prices have fallen by 30% in the last 5 yr, fertilizer prices have greatly decreased, and the focus of concern over hunger has shifted to sub-Saharan Africa (FAO 1986).

Even there, the situation has improved considerably in the last 2 yr, although it remains precarious. The massive international and bilateral resources that have been channeled to help African countries are perhaps unprecedented in the annals of development assistance, at least in terms of aid per capita. The UN Office for Relief in Africa identified only 4 emergency countries in October 1986, compared to 20 in September 1985. But to what extent the recent improvement in African food production is due to enhanced national effort and greater development assistance, or simply to better weather, has not, to my knowledge, been adequately evaluated.

The role of aid is controversial. A review of aid to agricultural research in Africa (Oram 1985b) suggested that resources were in danger of being wasted for lack of coordination among the donors. Several African countries had more than 40 donor-supported projects in fields related to research. At a recent meeting, no fewer than 50 different agricultural research networks were listed in Africa alone.

Important steps are now being taken to rectify this situation. Better exchange of information on ideas, priorities, and policies among donors and with countries should result in more effective use of resources and better evaluation of their impact. Even so, the food situation in Africa will continue to cause concern unless technical progress and the building of institutions and infrastructure can be accelerated so as to reduce vulnerability to climate.

Some causes for concern

Mellor notes that the current appearance of global food abundance is in sharp contrast to that of a decade ago. This is a cause for satisfaction, but certainly not for complacency. He also refers to the diversionary effects of debt problems on long-term development efforts; to the twofold increase in cereal imports by developing countries, as a result of consumption exceeding production; and to the growing surpluses in developed countries.

Between 1961 and 1971, 26 developing countries had annual food production growth rates under 1%/yr; 64 countries failed to exceed 3%—the average target used by both FAO's Indicative World Plan (FAO 1969) and its Agriculture Toward 2000 study (FAO 1981) as necessary for meeting food demand generated by population and income growth. Between 1971 and 1980, growth rates in 74 countries fell below 3%; 42 of those countries did not exceed 1% (Table 1). The number of sub-Saharan countries unable to achieve a 3% growth rate rose from 72% in 1961-70 to 82% in 1971-80; the proportion of all less developed countries that maintained that level of growth fell, from 54 to 42%. This is because the number with growth rates below 3% rose in both Asia and Latin America. Consequently, net food deficits of developing countries rose from 6.3 million t in 1965 to 59.6 million t in 1983 (Table 2).

Although inability to achieve a growth rate in food production adequate for domestic needs is a source of worry to the many developing countries without the economic resources to pay for food imports, rising food imports in themselves may not be a symptom of failure. A rapid rate of growth in the agricultural sector can set in motion powerful forces for growth in other sectors. The consequent rise in income and employment may generate a demand for food that exceeds the capacity even of a fast-growing food production sector. Thus, the 16 developing countries with the fastest growth rates in food production from the late 1960s to late 1970s increased their imports of basic food staples by 360% (Bachman and Paulino 1979). This may be why the largest net deficits have been in regions other than sub-Saharan Africa, where drought, lack of suitable technology, poor infrastructure, political instability, and institutional weaknesses have slowed agricultural growth.

Table 1. Food production growth rates in developing countries, 1961-70 and 1971-80.

Average annual production growth rate (%)	Countries (no.)				
	Asia	North Africa/ Middle East	Sub-Saharan Africa	Latin America	Total
		1961-70			
Less than 1.0	5	6	13	2	26
1.0-2.9	5	6	16	11	38
3.0-4.9	8	4	10	4	26
5.0 and over	4	3	1	7	15
Total	22	19	40	24	105
		1971-80			
Less than 1.0	6	10	19	7	42
1.0-2.9	7	2	14	9	32
3.0-4.9	5	3	5	4	17
5.0 and over	4	4	2	4	14
Total	22	19	40	24	105
		1961-80			
Less than 1.0	4	6	13	4	27
1.0-2.9	10	7	19	10	46
3.0-4.9	8	5	7	8	28
5.0 and over	0	1	1	2	4
Total	22	19	40	24	105

Table 2. Surpluses and deficits (million t) of basic food staples[a] in developing countries grouped by food sufficiency, 1965, 1975, and 1983 (data from FAO 1984).[b]

Developing region	1965			1975			1983		
	Gross		Net surplus	Gross		Net surplus	Gross		Net surplus
	Surplus	Deficit		Surplus	Deficit		Surplus	Deficit	
Asia	6.10	19.19[c]	–13.09	5.84	23.18	–17.33	12.52	36.41	–23.89
North Africa/ Middle East	0.65	4.02	–3.38	0.30	13.51	–13.20	1.97	31.98	–30.02
Sub-Saharan Africa	2.11	1.35	0.76	1.62	3.05	–1.44	0.90	7.77	–6.87
Latin America	13.87	4.44	9.43	9.35	11.08	–1.73	24.27	23.12	1.15
Total	22.72	29.00	–6.28	17.11	50.82	–33.70	39.66	99.28	–59.63

[a]Basic food staples here include all the cereals and the major noncereal foodcrops (roots and tubers, pulses, peanuts, and bananas and plantains). [b]Estimates of surpluses and deficits during the indicated years. Data are based on net trade of each country, and thus ignore stocks. Parts may not add to totals due to rounding. [c]Includes Bhutan's food position based on 1966 data.

One other feature of the mid-1980s merits comment. In contrast to the situation in 1974, when world grain stocks overall had declined to what many considered the dangerously low level of 11.5% of utilization, they now represent 21%, the same level as in 1961 (Table 3). In addition, there are mountains of butter, rivers of milk, and lakes of wine. The great majority of these surpluses are in the market economies of developed countries, where income elasticity of demand for conventional staple foods as well as for meat and milk is almost saturated and response of demand to price is low. The emergence of the European Economic Community as a major food exporter is a striking example of how a combination of high technology and high support prices to farmers can generate still more food in one of the most densely populated and heavily farmed areas of the world.

Nevertheless, the world as whole remains exceedingly and perhaps dangerously dependent for grain reserves on North America and, to a lesser extent, on the mid-latitude countries of the Southern Hemisphere. In 1983, North America had about 50% of world wheat and coarse grain exports (Table 4); in 1984, 55%. While there seems to be a good deal of supply elasticity there, political factors (including price changes), energy shortages, or crises due to a secular shift in climate could dramatically alter this situation. Mellor points out that the fragility of estimates concerning developed-country futures based on projections of past trends is another reason for not regarding the present situation with complacency.

Future needs and strategies for growth

Mellor makes a number of suggestions with relevance to policy arising from this situation, including the need for structural adjustments in developed countries and greater use of food surpluses for food aid-based employment growth programs in developing countries. In addition to fostering growth in those countries, this use of surplus grain might cushion the social impact of farm price adjustments in developed economies.

Table 3. World total grains supply/demand 1960-61 to 1986-87 (million t/ha).

Year	Area harvested	Yield	Production	World trade[a]	Utilization total[b]	Ending stocks[c]	Stocks as % of utilization[d]
1960-61	646.7	1.31	846.3	72.4	832.3	199.4	24.0
1961-62	641.6	1.26	806.3	83.2	833.3	171.9	20.6
1962-63	647.4	1.34	866.5	82.7	864.9	173.5	20.1
1963-64	652.6	1.33	870.7	97.7	869.9	174.5	20.1
1964-65	663.0	1.39	924.3	95.3	919.8	179.0	19.5
1965-66	659.6	1.40	921.3	110.9	955.2	142.4	14.9
1966-67	660.4	1.52	1006.4	103.7	980.0	168.9	17.2
1967-68	672.7	1.54	1037.8	97.1	1017.4	189.3	18.6
1968-69	678.4	1.59	1079.2	89.5	1046.7	221.7	21.2
1969-70	679.6	1.60	1087.2	97.3	1102.3	206.6	18.7
1970-71	671.0	1.64	1102.5	109.6	1143.8	165.2	14.4
1971-72	680.2	1.76	1196.5	110.0	1178.5	183.3	15.6
1972-73	669.5	1.73	1160.9	134.6	1201.2	142.8	11.9
1973-74	697.8	1.82	1272.6	141.6	1266.3	148.5	11.7
1974-75	699.8	1.74	1217.7	136.6	1224.2	140.4	11.5
1975-76	717.4	1.74	1246.7	150.3	1236.9	148.0	12.0
1976-77	719.3	1.90	1363.1	157.7	1310.4	201.1	15.3
1977-78	716.8	1.87	1337.2	171.2	1336.7	201.1	15.0
1978-79	716.8	2.04	1465.7	176.7	1435.7	230.9	16.1
1979-80	713.0	2.00	1426.6	197.9	1450.3	207.1	14.3
1980-81	723.9	2.00	1447.0	215.2	1252.3	190.9	15.2
1981-82	734.0	2.04	1498.9	209.7	1461.7	227.2	15.5
1982-83	717.9	2.15	1543.9	200.5	1507.9	262.1	17.4
1983-84	707.6	2.10	1484.5	206.4	1555.1	190.6	12.3
1984-85	714.3	2.30	1642.8	219.5	1594.3	239.8	15.0
1985-86[e]	715.6	2.32	1660.3	182.7	1580.0	319.4	20.2
1986-87[f]	673.2	2.44	1639.4	193.0	1619.5	339.3	20.9

[a]Trade data expressed in this table exclude intra-EC trade. Wheat and coarse grains are on a July-June basis through 1975-76. The trade year for coarse grains from 1976-77 is October-September. [b]For countries for which stocks data are not available (excluding the USSR) utilization estimates represent "apparent" utilization (i.e., include annual stock level adjustments). [c]Stocks data are based on an aggregate of differing local marketing years and should not be construed as representing world stock levels at a fixed point in time. Stocks data are not available for all countries and exclude those such as the People's Republic of China and parts of eastern Europe. World stock levels have been adjusted for estimated year-to-year changes in USSR grain stocks, but do not purport to include the absolute level of USSR grain stocks. [d]Represents the ratio of marketing year ending stocks to total utilization. [e]Preliminary. [f]Projection. Area and yield do not include rice forecast for U.S.

Looking to the future, Mellor stresses that developing countries offer the only cereal market capable of rapid growth, due to rising effective demand for food with employment growth and to changes in the structure of demand, with increasing emphasis on livestock products as incomes increase (Sarma 1986). Thus, he considers the number one food policy need for the net food importers among developing countries (more than 80% of them in 1983), to be an international environment in which food supplies are reliable and not subject to radical fluctuations in price or volume.

The few net exporters of food staples in the Third World are dominated by China, India, Indonesia, and Pakistan. But these countries still have large numbers of malnourished people, despite success in increasing food production. Mellor regards their exports as symptoms of a failure to invest adequately in agriculture and

Table 4. Surpluses and deficits (million t) of basic food staples[a] in developed countries grouped by food sufficiency: 1965, 1975, and 1983 (data from FAO).[b]

Developing region	1965			1975			1983		
	Gross		Net surplus	Gross		Net surplus	Gross		Net surplus
	Surplus	Deficit		Surplus	Deficit		Surplus	Deficit	
North America (2)	55.87	–	55.87	89.29	–	89.29	127.63	–	127.63
Eastern Europe/USSR (8)	0.86	9.96	–9.10	1.27	21.56	–20.29	1.85	38.32[c]	–36.47
Western Europe (20)	4.47	38.57	–34.10	12.95	40.26	–27.32	25.64	34.30	–8.66
EC (12)	(4.21)	(33.84)	(–29.63)	(11.71)	(37.23)	(–25.52)	(22.93)	(32.13)	(–9.21)
Others (8)	(0.26)	(4.73)	(–4.48)	(1.24)	(3.04)	(–1.80)	(2.71)	(2.17)	(0.54)
Oceania (2)	7.20	0.17	7.03	11.46	0.17	11.29	10.59	–	10.59
Others (3)	0.68	11.87	–11.19	3.81	21.63	–17.82	–	27.92	–27.92
Total (35)	69.08	60.56	8.52	118.77	83.63	35.15	165.70	100.54	65.16

[a]Basic food staples here include all the cereals and major noncereal food crops (roots and tubers, pulses, peanuts, and bananas and plantains.) [b]Data are based on net trade of each country and thus ignore stocks. Parts may not add to totals due to rounding. Figures in parentheses denote the number of countries to which estimates relate. [c]Includes Albania's food position in 1982.

other sectors that can increase demand for food by accelerating income and employment. Even so, I feel that most Asian countries, including Bangladesh, have made remarkable progress in the last 10 yr. There have been poor crops, but no major famines; despite considerable poverty, social progress has been achieved.

Several factors have made this improvement possible, especially investment in agricultural research and technology, expansion of irrigation and fertilizer use, relatively good communications and market infrastructure, and no major wars or civil conflicts. Considerable emphasis has also been given education, including higher education, thus helping provide trained personnel to support strong technical progress.

Most of these factors are cited by Mellor as conditional to the growth of food production and employment. He illustrates them mainly with examples which contrast Asia and Africa. From these comparisons, it is hard to see how sub-Saharan Africa—with a population growth exceeding that predicted in earlier global food studies and a slower expansion of food production—can avoid further deterioration in per capita food intake between 1981 and the end of the century without much wider use of fertilizer and new technology.

Wider use of fertilizer and new technology implies a big boost to African agricultural research. My recent data show that 13 Asian countries contain about 75% of the Third World's agricultural scientists. Africa has about 7%. Although its scientific strength is growing, most African countries still rely heavily on expatriate research staff and on donor funding of research. Rather surprisingly, Mellor does not mention the vastly greater importance of irrigation in Asia, where more than 25% of the total cultivated area is irrigated, compared with about 2% in Africa. Not only does this make higher rates of fertilizer and high-yielding variety use remunerative, it also allows double cropping on a good deal of the area. Clearly, sub-Saharan Africa faces a Herculean task. FAO (1981) states that an agricultural growth rate of 4.3% will be necessary there between 1980 and 2000, if severe food crises are to be avoided (the 1961-65/1980 real growth rate was 1.8%).

For developing countries as a whole, the 1980-2000 growth rate of agricultural production advocated in the FAO Agriculture Toward 2000 study is only 3.7%, even in the more optimistic scenario. This is the same as that proposed for 1961-85 in the Indicative World Plan (IWP) (FAO 1969). The actual achievement for that period was 2.8%, closer to the lower Agriculture Toward 2000 scenario of 3.1%.

Ex post analysis shows that although the suggested IWP growth rates for input use were largely achieved, especially in Asia, the resulting yields were not as high as expected. Overall, unfavorable weather was probably not the main reason. Other possible causes include planning expectancies that failed to take adequate account of the gap between experimental and actual farm yields, incorrect use of inputs by farmers, poor water or input delivery systems, and price disincentives. Recent studies (Hazell 1986) suggest that these problems may also increase yield variability. Since future expectations lean heavily on more intensive use of inputs to avoid serious shortfalls in production, these lessons merit closer study.

The critical nature of input assumptions is clear from another important piece of work by FAO/UNFPA/IIASA (1982). It looked at potential population-supporting capacities of lands in the developing world. The conclusion was that at low inputs, 64 countries would be unable to support their projected populations in 2000; 38 would not be able to support half their people, even if all cultivable land was used solely for food production. At intermediate input levels, the number of critical countries would be reduced to 36; at high levels, to 19. Some countries have reached the intermediate level of input use, but no major geographical region as a whole has yet achieved that level.

That study shows that the developing regions have great potential, but also a long way to go to achieve it. One issue on which the jury is still out is how much expansion of cultivation into new areas will contribute to increased food supply, compared to intensification in existing cultivated areas. There is clearly some concern, perhaps especially for Asia, the Near East, and Central America, as to whether technological change alone, largely concentrated in high potential areas, can bridge the gap. Africa seems to have much more scope than is widely recognized, especially if about a billion hectares of reasonably good land can be cleared of tsetse fly (Fischer et al 1984). Large areas of land also might be brought into cultivation in the humid tropical lowlands of South America, assuming that soil management problems can be solved (Sanchez 1976).

Weaknesses of long-term forecasts

This brings me to my penultimate point: the vulnerability of global studies to their assumptions. FAO's Land, Food, and People study (1984) is excellent and methodologically innovative, but I find it difficult to believe that all cultivable land will be used to produce food and most marginal land to produce livestock. Other studies make assumptions that there will be no pest or disease losses or that Australia can be irrigated from Tasmania. Hence, estimates of global carrying capacity vary from 8 to 150 billion people (IIASA 1981). The sensitivity of food deficit projections to assumptions concerning population growth is also great. The difference in world population by 2025 between the UN medium and high projections exceeds the 1987 population of India (UN 1986b).

Another source of error in past studies that never ceases to amaze me is the vast discrepancy among the various estimates of the resource base, especially with respect to potential cultivable area. The differences between some of these estimates exceed the total area under crops in the world today. Because of this, it is not surprising that the issue of availability of new land is controversial, especially as management or reclamation of currently marginal or degraded lands for sustained arable agriculture is a weak area of research.

Perhaps the most questionable feature of the majority of studies on future global food supply is their failure to build in any climatic variables. This was the case with the IWP; Agriculture Toward 2000; Land, Food, and People; The Limits to Growth (Meadows et al 1972); and IFPRI and World Bank (1982) projections of food requirements and potential deficits. It is, of course, a weakness of particular concern here.

The basic arguments for this omission seem to be threefold. The first is that climate is not likely to change dramatically within the time span of the studies, which usually seems to be about 20 yr into the future. Therefore, it can be treated as a constant. The second is that a longer time horizon, within which climatic change might be significant, would introduce another element of uncertainty and seriously complicate assessments of how technology might shift the supply curve. The third is that the schism among climatologists as to whether a warming or cooling trend will prevail offers a cast-iron excuse to leave out such an uncertain element.

The U.S. Global 2000 study (CEQ/State 1980) does include some climatic scenarios, but these do not appear to be built into its conclusions. Other studies, which look further ahead, do not do so either. Moreover, these longer term efforts tend to be highly aggregated. The studies that are more specifically oriented to future climate scenarios point to the need for disaggregation down to agroecological zones, sometimes within individual countries, if the potential influence of climate on food production is to become clearer. Unfortunately, those studies (for example, the 1980 U.S. National Defense University Study) generally fail to adequately relate their conclusions to those involving other components of the agricultural resource base—land and water, technology, and human resources. The holistic view is lacking.

The Global 2000 report notes, from a comparative analysis of other studies, that a high degree of intersector aggregation (industry, agriculture, energy, environment) leads to less optimistic scenarios in terms of human welfare than when the analysis is largely confined to one sector, such as agriculture. Most of the in-depth analyses I have mentioned that are likely to be of use to national agricultural policymakers are of the latter type, and their conclusions must be interpreted with caution.

An important and related point is that the longer term, more highly integrated studies, such as the World Models sponsored by the Club of Rome, the UN World Model, and the 21st century preview of Global 2000, even though they largely ignore climatic change, are generally more pessimistic for the first half of the next century than for the period up to 2000. They foresee increasingly tight food supplies and steeply rising food prices, resulting in massive hunger and misery, especially in South Asia and in other areas where land is scarce. In light of the situation of relative abundance as we approach 1990 (depicted in Mellor's paper), it is hard to credit some of these predictions.

Now that there is some relief from the persistent food crises in Africa, there may be a danger of complacency as to the future. This makes it all the more essential not to overlook the potential role of climate as an element of change and to strive to understand and to make contingency plans for climate-induced shifts in food and agricultural systems while there is still time.

As Mellor's paper shows, a sustainable level of world food depends not simply on research and the wider adoption of improved agricultural technology (although this is crucial), but on sound policies in many other areas—water resources, input supplies, energy, delivery systems and other infrastructure, prices, and markets. The chances of success would be greatly enhanced by a better understanding of the natural resource base for agriculture and how to manage it, both in the climatic regime we now enjoy and with possible future climatic changes, whether for better or for worse. I use the word "enjoy" advisedly, since, despite the tragedies in parts of Africa, the world as a whole has had a rather favorable period of climate during last

50 yr compared with many periods in the past. This may have led to some overly optimistic assumptions concerning the future.

In this context, Mellor's plea that the world should use its time to plan wisely while there is still a relative abundance of food has special relevance. If climate becomes more favorable, it will be a blessing; should it deteriorate, or should variability increase, the task of meeting future needs for food and other agricultural products could be immeasurably more difficult.

Maximizing comparative advantage

Mellor notes that the new environment—an apparent global abundance of food—imposes somewhat different requirements on food production research. He sees a continuing need for yield-increasing technology, with concern about rice in Asia and low labor productivity in Africa. He argues for attempting to redistribute population according to sustainable carrying capacity and comparative ecological advantage. He stresses research on maximizing linkages between agriculture and non-agricultural sectors to raise effective demand. He emphasizes smallholder livestock development. And he pleads for innovative research to help the poorest people in the poorest regions—a difficult task indeed.

I would like to make one further plea. Do not sacrifice too much on the altar of self-sufficiency. As Mellor stresses, increased food supplies and increased employment are two sides of the same coin: both are essential to human welfare. The world as a whole currently has abundant food, it is maldistribution and lack of purchasing power, even in countries where food is plentiful, that are the primary causes of individual hunger and malnutrition.

Not all land is best suited to food production. Where comparative advantage lies in producing other agricultural commodities (such as trees in the wetter tropics or livestock in semiarid and upland areas), it may make sense both environmentally and economically to maximize that, and use the returns to pay for grain imports. "Food first" is a fine slogan, but we should not overlook the power of some nonfood commodities (jute, cotton, coffee, rubber) to create employment and purchasing power, both on and off the farm. I do not disagree with Mellor's argument for a sharper CGIAR focus on a few key food crops—the CGIAR is a global food research organization. I say it also is necessary to consider national needs, which may point to other commodity or resource management priorities. These seem to be largely overlooked in global studies, and not infrequently are also overlooked in allocating donor resources to support research and development efforts.

References cited

Bachman K, Paulino L (1979) Rapid food production growth in selected developing countries: a comparative analysis of underlying trends, 1961-76. Research Report No. 11. International Food Policy Research Institute, Washington, D.C., USA.

CEQ/State—Council on Environmental Quality and the U.S. Department of State (1980) Entering the twenty-first century. The Global 2000 report to the President. Washington, D.C., USA.

FAO—Food and Agriculture Organization of the United Nations (1969) Indicative world plan for agricultural development. Rome, Italy.

FAO—Food and Agriculture Organization of the United Nations (1981) Agriculture toward 2000. Rome, Italy.
FAO—Food and Agriculture Organization of the United Nations (1984) Land, food and people. Rome, Italy.
FAO—Food and Agriculture Organization of the United Nations (1986) Food crop and shortages. Global information and early warning system on food and agriculture. Special report. Rome, Italy.
FAO/UNFPA/IIASA—Food and Agriculture Organization/United Nations Fund for Population Activities/International Institute for Applied Systems Analysis (1982) Potential population supporting capacity in the developing world. FFA/INT/513. Rome, Italy.
Fischer G, Shah M M, Rollinson D H L (1984) Agricultural perspectives in the tsetse infested areas of Africa. International Institute for Applied Systems Analysis/Food and Agriculture Organization, Laxenburg, Austria.
Hazell P (1986) Cereal yield variability: an overview of issues. *In* Proceedings of a workshop on cereal yield variability. P. Hazell, ed. International Food Policy Research Institute/Deutsche Stiftung für Internationale Entwicklung, Washington, D.C., USA.
IIASA—International Institute for Applied Systems Analysis (1981) Food for all in a sustainable world. K. Parikh and F. Rabar, eds. Laxenburg, Austria.
Meadows D H, Meadows D L, Randers J, Behrens W (1972) The limits to growth. Report for the Club of Rome. Universe Books, New York, USA.
Mellor J (1986) The new global context for agricultural research: implications for policy. Agenda Item 03, International Centers Week, 3-6 Nov 1986, Washington, D.C., USA. CGIAR Secretariat, ICW/86/03.
Mellor J (1987) Food surpluses of developing countries: policy implications for developing countries. *In* The emerging world food situation and challenges for development policy. IFPRI Policy Seminar, Royal Tropical Institute, Amsterdam, The Netherlands.
Oram P (1985a) Agriculture research and extension: issues of public expenditure. *In* Recurrent cost and agricultural development. J. Howell, ed. Overseas Development Institute, London, UK.
Oram P (1985b) Donor assistance to agricultural research. A proposal for information exchange. Paper presented at the First Plenary Meeting of the Special Program for African Agricultural Research, The World Bank, Washington, D.C., USA.
Paulino L (1977) Food needs of developing countries: projections of production and consumption to 1990. Research Report No. 3. International Food Policy Research Institute, Washington, D.C., USA.
Paulino L (1986) Food in the Third World. Past trends and projections to 2000. Research Report No. 52. International Food Policy Research Institute, Washington, D.C., USA.
Sarma J S (1986) Cereal feed use in the Third World: past trends and projections to 2000. Research Report No. 57. International Food Policy Research Institute, Washington, D.C., USA.
Sanchez P A (1976) Properties and management of soils in the tropics. Wiley Interscience, New York, USA.
UN—United Nations (1986a) Final report on the emergency situation in Africa. Report No. OEAR/3/3. Office for Emergency Operations in Africa, New York, USA.
UN—United Nations (1986b) World population prospects: estimates and projections as assessed in 1984. New York, USA.
U.S. National Defense University (1980) Crop yields and climatic change to the year 2000. Vol. 1. Washington, D.C., USA.
World Bank (1982) Agriculture and economic development. Part II. World Development Report. Washington, D.C., USA.

Notes

The paper represents the author's views on John Mellor's paper, The New Global Context for Agricultural Research: Implications for Policy, presented at the CGIAR Centers' Week Meeting, 1986.

Author's address: P. A. Oram, International Food Policy Research Institute, 1776 Massachusetts Avenue, N.W., Washington, D.C. 20036, USA.

Citation information: International Rice Research Institute (1989) Climate and food security. P.O. Box 933, Manila, Philippines.

Sensitivity of agricultural production to climatic change, an update

P. A. Oram

Agriculture is inherently vulnerable to climatic hazards, and many agricultural practices are aimed at reducing climatic risks. While modern technology has mitigated some aspects of weather variability, it has been unable to compensate for the progressive breakdown of traditional safeguards in tropical climates under population pressure, especially in Africa. Short-, medium-, and long-term policies involving both technical and nontechnical measures to offset climatic instability and to increase food security are indicated. Four temporal patterns of climatic change and variability are described, and geographic and strategic areas of special risk are identified. Climatic variability and secular climate shifts are the two most significant problems; the first always present, the second impending. Possible causes of increased variability include long-term climatic change, technological change, expansion of cultivation into new areas, and shifts in the geographical focus of major crops. The implications for research and policy of anthropogenically induced climatic warming are examined. Most attention has been focused on the possible effects on temperature, with not enough attention on precipitation. Soil quality and moisture-retaining capacity also need greater consideration. Laboratory studies of individual plant responses to enhanced CO_2 shed little light on how increases might affect crop behavior and competition in farming systems or in unmanaged ecosystems. Difficulties are foreseen in convincing policymakers of the need for current action in anticipation of long-term climatic change.

The vulnerability of world food supply to climate and weather

In an earlier paper (Oram 1985), I suggested that because agricultural systems represent changes imposed on the natural environment by human influence and because they contain a relatively narrow range of species, they may be more vulnerable to climatic or other stochastic shocks than are undisturbed ecosystems which have achieved a stable equilibrium over time. Lamb (1981) also stresses the vulnerability of agricultural systems to climatic shocks. He argues that the development of agriculture has led to concentrations of certain crops in areas where they grow best, leading to one-crop economies and increasing the risks resulting from a climatic change.

Historically, the risks have been unavoidable. What people ate in the days before food could be transported long distances essentially depended on what could be grown best locally. Hence, there is a close correlation between the staple components of a region's diet and the suitability of crops to a region (Table 1).

This does not necessarily mean that crops or livestock from one geographic area or broad ecological region cannot be grown successfully in another. Examples include wheat and rice in Latin America, maize in East Africa and Southeast Asia, potato in Europe, and cassava and rice in West Africa. Over time, these have become dietary staples occupying large areas of land. Thus, introducing new crops does not necessarily solve the dilemma posed by Lamb, although it may spread the risk.

Because agriculture is probably the most weather-dependent of all human activities, spreading risk is an important objective. The evolution of traditional agricultural practices reflects a continuing effort to ensure the survival of selected and sometimes exotic species and to reduce the vulnerability of a food producer's livelihood to natural hazards. The devices adapted include the following:

- Mobility. Nomadism, transhumance, shifting cultivation, and (as a last resort) migration.
- Spreading risks. Large holdings, stratification across agroecological zones, fragmentation of holdings, communal land sharing, farming cooperatives, crop insurance schemes.
- Manipulation of the environment. Irrigation, drainage, soil amendment, terracing, contour farming, mulching and other soil and moisture conservation practices, fallowing, fertilizer application, shelter belts.
- Diversification of production systems. Introduction of new crops or cultivars, crop rotation, mixed cropping, multiple cropping, alley cropping, agroforestry, dual-purpose livestock breeds, mixed livestock (i.e. camels, sheep, goats).
- Manipulation of components in farming systems. Plant and animal breeding, changing the crop calendar (time of sowing, harvesting, other key points), adjusting livestock practices (breeding seasons, supplementary feeding, hormonal treatments), mechanizing critical time-dependent operations, using chemicals for crop protection or as substitutes for machine or human labor (i.e. no tillage).
- Reduction of postharvest losses. Storage, drying, fumigation, processing, refrigeration.
- Trade and aid, in particular with respect to food security. (There are problems here to which I refer later.)
- Weather wisdom. In traditional societies, folklore and experience; in more recent times, meteorological recording, analysis, forecasting and early warning systems, and research on climate and factors that influence it, in a long-range attempt to improve understanding of past events and to identify future possibilities.

In the current century, reliance on science and technology to find means of buffering agriculture against climatic shocks has increased. Technology, together with several decades of relatively benign climate and declining rates of population

Table 1. Major crop-climate zones in developing countries.

Crop zone	Predominant agroclimatic condition	Geographic region		Countries (no.)	Total population (millions)	Agricultural population (millions)	Agricultural population density (persons/ha)	Av yield of primary cereals (kg/ha)	Arable land reserves	Annual growth 1961-80	
										Net area	Yield
Root crop zone	Humid tropical	Africa:	Equatorial	16	193	120	1.6	750	Abundant	0.8	1.0
		Asia:	Pacific	8							
Rice zone	Humid tropical and humid temperate	Asia:	Southeast	10	574	358	3.5	2050	Scarce	1.0	1.9
		Asia:	South	4							
		America:	Central	5							
Maize zone	Subhumid tropical	Africa:	East	4	353	161	1.1	1450	Abundant	1.0	2.2
		Africa:	Southern	8							
		America:	Central	7							
		America:	South	6							
Sorghum and millet zone	Semiarid tropical	Africa:	West	9	86	70	1.6	670	Moderately abundant	0.8	0.8
		Africa:	East	3							
		Asia:	West	3							
Wheat zone	Temperate/ Mediterranean	Africa:	North	5	395	182	1.0	1850	Scarce	0.4	2.5
		Asia:	West	10							
		America:	South	4							
Mixed	Warm temperate and arid to humid tropical	India			673	439	2.5	1310	Very scarce	0.3	2.0
Mixed	Cold temperate to subhumid tropical	China			977	572	5.8	2700	Very scarce	−0.3	3.0

[a] Because of limited data for subdivisions of countries, the borders between crop zones have been adjusted to coincide with national boundaries.
Sources: IFPRI and FAO data.

growth, has led to large surpluses of production over consumption in some temperate zone countries. But despite successes in Asia and Latin America, technology has not enabled many developing countries to avoid large deficits and widespread malnutrition.

It is a tautology that there is more than enough food for the world as a whole. But apart from foreign exchange constraints, physical problems of distribution, and lack of purchasing power to buy food in poor countries, global trade in dietary staples is basically confined to wheat and coarse grains produced primarily in the temperate regions. Even the world rice market has a very narrow base. There is little trade in sorghum, millet, pulses, roots, and tubers or vegetables. In recent years, this has led to some shift in dietary preferences toward wheat, originally supplied as emergency relief to countries (particularly in Africa) that are ecologically poorly suited to its production (Oram et al 1979).

This heavy global dependence for food reserves on grains grown in a restricted ecological spectrum poses political and nutritional problems as well as climatic hazards. Looking further into the future, only a dramatic change in the climate of the mid-latitude grain-exporting countries (or perhaps a nuclear war) would appear likely to alter this situation substantially.

In the developing countries, traditional approaches to avoid or ameliorate climatic shocks are breaking down in the face of changing demands for food and agricultural raw materials, of rising incomes, and—even more—of rapid population growth. Hence, the widespread concern about environmental degradation and long-term sustainability of production (Worldwatch 1986, World Resources 1986).

Easy answers are not apparent. Technology that works well in certain agroclimatic situations in developed and some developing countries does not work equally well in similar situations in other developing countries. This applies particularly to Africa, where new technology has been less successful or less widely adopted than it has in Asia. The reasons are not fully understood but, in addition to differences in climate and soils, to a significant extent they relate to nontechnical conditions that are often outside the control of local people: poor infrastructure; weak research, knowledge transfer, and input delivery systems; political instability; inappropriate institutional mechanisms; and inconsistent macropolicies (Blackie 1987, World Bank 1982). These domestic problems are aggravated by agricultural protection policies in developed countries and by events in world markets where individual developing countries have little power.

Policies to reduce vulnerability

Rendering agricultural production and food availability less vulnerable to climatic events presents complex problems. Consideration must be given to ecological suitability; use of and competition for land; development and adoption of new technology; producer and consumer preferences; income growth; and the availability, nature, and location of global reserve stocks. Policies in these areas that may help to increase food security can be considered in relation to their time span.

Short-term measures

Stockholding. Developing countries need national emergency food reserves to facilitate providing a timely food supply to areas suffering climatic crises. However, because stockholding is expensive, arrangements such as the International Monetary Fund food facility should be available to enable countries in situations of prolonged hardship or wide interannual fluctuations in supply to buy grain from overseas (Valdes 1981).

Trade liberalization. Increased international and intraregional cooperation in trade, both for food and nonfood commodities, could increase food security (Valdes and Zeitz 1980, Koester 1987). Many developing countries rely heavily on exports to earn foreign exchange to import food and food grains, as well as to purchase fertilizer and other inputs to boost domestic food production. Cash crops and processed derivatives contribute significantly to increased employment and purchasing power, both on and off the farm. Mellor (1985) has shown that poor people spend most additional income on food.

Intermediate-term measures

Strengthening meteorological services. A recent survey in North America (Johnson and Holt 1986) cites 18 examples of substantial financial benefits to producers from timely climatic information. Not taking climate and weather data into account can lead to failure (e.g. in irrigation design) (Glantz 1987). Rasmussen (1986) has stated that short-term weather forecasting and global climate forecasting over several months to several years is now considerably improved, particularly global forecasting relative to the El Niño/Southern Oscillation (ENSO) phenomena. He notes, however, that opportunities for improvement in empirical-statistical forecasts in developing countries, especially in the tropics, are impeded by weaknesses in agrometeorological services.

Infrastructural development. The success of weather warning systems in helping to avert crop losses depends heavily on efficient transportation. Inability to move animals to market, even excellent advance warning of the Sahelian drought, would not have prevented heavy loss of livestock (Glantz 1976). Even when food was available at ports, failure to move the grain fast enough to areas seriously affected by climatic disasters has caused loss of life. Investment in infrastructure to provide access to the areas identified by experience and/or prediction as likely to be most vulnerable is crucially important.

Infrastructure development can also improve the rural population's access to production inputs, extension, and health and other services, as well as raise the producer's share of monetary returns from product sales (Ahmed 1987, Wanmali 1986). These benefits all contribute to food security.

Medium- to long-term measures

Investment in irrigation. Irrigation can be an important safeguard against climatic risk, provided it is well planned and does not lead to salinity or flood problems, and if low temperature is not a primary constraint. It also facilitates crop diversification and higher yields. However, competition for water for nonagricultural uses is increasing. Climatic warming accompanied by decreased precipitation could further reduce irrigation supplies. Irrigation plans should consider such problems.

Research. Several international agricultural research centers have stressed that inadequate agroclimatic information hampers agricultural research and development (CAB 1987). The need is to better understand the factors underlying short- and long-term global climatic behavior and to pay explicit attention to agroecological characterization, climatic variability, and longer term climatic trends in agricultural research planning and resource allocations (Parry and Carter 1984, Rasmussen 1986). Cooperation between developed and developing countries in collecting, analyzing, and exchanging data will be necessary to achieve better knowledge of climate and weather, as well as to effectively apply the information to agriculture. Agricultural research in general is underfunded (Boyce and Evenson 1975, Oram and Bindlish 1981), and climatic research in particular has not been adequately supported. Remedying this would be an important contribution toward longer term sustainability of global agricultural production.

Characteristic patterns of climatic change and variability

Time scales

Four main time scales of climatic change and variability can be distinguished (Oram 1982).

- Short-term natural disasters beyond human control—cyclones, hurricanes, typhoons, effects of volcanic eruptions, large-scale flooding. Nuclear winter is a new threat, although it is a disaster that is within human power to prevent.
- Interannual and intraannual variability. Part of the pattern of a longer term climatic change might be increased variability.
- Cyclical patterns of climatic anomaly across several years, but within a basically unchanged long-term situation. These patterns may affect the world as a whole (the 1940-70 cooling trend) or individual geographic regions (the deteriorating rainfall in the Sahel, which began around 1950). In such circumstances, environmental damage may become cumulative.
- Long-term climatic change. This would affect the entire world or large geographical regions and stretch over several centuries or longer.

Areas of special risk

In considering the impact of these climatic patterns on agriculture and food security, it is necessary to identify geographical or agroecological regions especially vulnerable to climatic shocks. These vary with the time scale and the nature of the climatic situation.

Disaster-prone areas. Certain geographical regions, predominantly in the more humid tropics of South and Southeast Asia, the South Pacific, and the Caribbean, are particularly prone to violent climatic events (cyclones, hurricanes, typhoons, accompanying oceanic disturbances). Although these tend to be most prevalent at certain seasons of the year, it is difficult to predict exactly where they will strike.

Weather-tracking and early warning systems are particularly useful in preventing heavy loss of life and material damage. Fortunately for agriculture, the areas most affected tend to have a relatively low interannual yield variability and are potentially productive.

Climatically marginal areas. The semiarid and arid areas in the tropics and subtropics, high-altitude zones in mid-latitude areas, and cold marginal areas toward the polar extremities of the continental land masses are climatically marginal areas (Fig. 1) (Parry and Carter 1984). Although not especially liable to violent climatic disasters, they often have large intraseasonal extremes as well as high year-to-year variability.

Strategically important areas. These areas do not necessarily have high current vulnerability or instability, but they are areas where climatic change could have a serious impact on global agricultural production and food security. They include 1) the tropical rain forests, because their destruction might have catastrophic effects on future climate, especially on precipitation; 2) the main grain-exporting countries on which world food reserves depend, principally the United States, but also Canada, Australia, Brazil, Argentina, and Thailand; and 3) the major densely populated higher income importing regions, e.g. Western Europe, USSR, Near East, Japan, because they rely on other producing areas for much of their food and feed needs.

Socially vulnerable areas. Regions with a preponderance of low-income population and/or dense population, with food deficits, are socially vulnerable. These include South Asia, Southeast Asia, sub-Saharan Africa, and possibly Central America and the Caribbean.

These situations are not mutually exclusive. For example, much of Sahelian West Africa is both climatically marginal for settled agriculture and socially

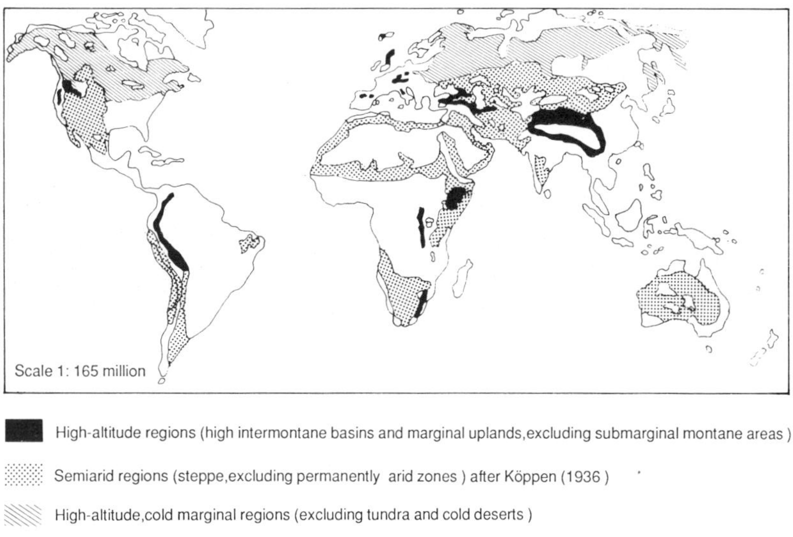

1. Some climatically marginal regions. Source: Parry and Carter (1984).

vulnerable. Canada and Australia are strategically important, with parts of both countries ecologically marginal for settled farming.

The two main sources of global vulnerability to climate are climatic variability and climatic change.

Sensitivity of production systems to climatic variability

Weather variability within and between seasons is probably the major source of uncertainty in agriculture, and is likely to remain so even if the climate changes. However, the magnitude and unpredictability of weather show wide differences among geographical regions, as does the impact of weather on major crops and farming systems.

Regional and commodity-specific patterns

Drier regions of the tropics and subtropics. High levels of variability occur in the drier regions of the tropics and subtropics because they receive unreliable and predominantly unimodal precipitation in combination with high summer temperatures. Growing seasons are short, from a minimum of 75 to a maximum of 180 d (FAO 1978). There are two main agroclimatic regions:

- The winter rainfall subtropics (Mediterranean climatic region) stretching from Turkey in West Asia to Morocco in North Africa, including much of Southern Europe. Severe winters may compound summer drought in high-altitude areas of Europe and West Asia. Wheat and barley, the main food and feed grains, have coefficients of variation (CVs) that generally exceed 25%; climatic analogs in Chile, Australia, and Southern Africa also have high CVs (Table 2,3).
- The summer rainfall semiarid tropics in Africa. The Sahelian countries grow progressively drier from south to north, the reverse of the situation in North Africa. The staples millet, sorghum, and peanuts are also important in Central India and Northeast Brazil. CVs are high. The steady downward trend in precipitation in the Sahel since the 1950s (Farmer and Wigley 1985) led to a marked southward displacement of mean annual rainfall, a parallel southward shift of some important crops such as peanuts, higher risk of crop failure, and a shorter growing season. The reasons have not been adequately explained, but it has been suggested that overgrazing and deforestation have increased albedo and reduced precipitation by decreasing water vapor in the atmosphere. Pulses (beans, chickpeas, lentils) and leguminous oilseeds (peanuts and soybeans) exhibit high yield variability where low rainfall and high evapotranspiration produce frequent drought stress.

  Low rainfall and high evapotranspiration also severely affect biomass production in the rangelands. Large fluctuations in animal numbers, as well as in animal productivity, stem from the variability of feed supply, and various systems of nomadism and transhumance have developed as coping measures. Nevertheless, these are among the most vulnerable of all human societies to climatic unpredictability.

Table 2. Variability in staple food production, 1961-76.

Region and country	Staple food production instability		Probability of actual production falling below 95% of trend (%)	Correlation coefficient between total staple food production and consumption	Correlation coefficient between cereal production and total staple food production
	Standard deviations[a] (thousand tons)	Coefficient of variation[b] (%)			
Asia					
Bangladesh	765	6.4	22	0.90	0.99
India	6 653	6.4	22	0.89	0.99
Indonesia	1 040	5.4	18	0.92	0.94
Korea, Rep. of	445	7.1	24	0.20	0.96
Philippines	346	5.7	19	0.03	0.99
Sri Lanka	107	9.3	29	0.56	0.91
North Africa/ Middle East					
Algeria	531	28.9	43	0.78	1.00
Egypt	282	4.5	13	0.29	0.96
Jordan	119	65.6	47	0.63	1.00
Libya	56	28.0	43	0.62	1.00
Morocco	1156	27.2	43	0.98	0.96
Syria	702	38.8	45	0.92	1.00
Sub-Saharan Africa					
Ghana	121	5.8	20	0.98	0.93
Nigeria	958	5.7	19	0.99	0.92
Senegal	325	18.6	39	0.99	0.81
Tanzania	430	12.7	35	0.98	0.09
Upper Volta	128	9.8	30	0.95	0.99
Zaire	190	4.9	15	0.96	−0.21
Latin America					
Brazil	1631	5.2	17	0.92	0.60
Chile	215	11.1	33	0.54	0.99
Colombia	126	4.4	13	0.51	0.85
Guatemala	56	6.5	22	0.51	0.99
Mexico	1060	7.7	26	0.53	1.00
Peru	197	9.8	30	0.37	0.97

[a]Defined as the standard deviation of the variable $Q_1 - Q_1$.
[b]Defined as the standard deviation of the variable $\frac{Q_1 - Q_1}{Q_1} \times 100$.

The subhumid tropics. With growing seasons from 180 to 270 d, the subhumid tropics are generally more favorable for annual crop production than either the very wet or the semiarid tropics. Interannual CVs of yield of the principal staple crops (maize, cassava, soybean, and cowpea) are relatively low except at the drier end of the zone. Within-year seasonal extremes are also lower than in the semiarid regions. Rainfall may be bimodal or, more commonly, unimodal, with a long but relatively cool season. At higher altitudes, potato and wheat replace maize and cassava. Frost may limit the growing season and the range of crops. Disease hot spots are common.

The humid tropics. In South and Southeast Asia, the main areas of rice cultivation, crops show relatively low variability (Barker et al 1981). Lack of rain is not usually a serious problem in the wetter lowland tropics (although too much

Table 3. Coefficients of variation of yield of major food crops in large countries for 1961-69 (1) and 1970-77 (2).

Crop	Major exporters								Other large producers					
	Argentina		Australia		Canada		USA		USSR		China		India	
	(1)	(2)	(1)	(2)	(1)	(2)	(1)	(2)	(1)	(2)	(1)	(2)	(1)	(2)
Wheat	26.4	26.7	15.4	16.5	13.0	9.3	3.4	5.5	13.1	13.0	7.3	5.6	10.8	5.6
Rice	17.1	13.8	10.6	8.9	—	—	4.2	3.2	20.4	2.4	1.9	1.2	8.6	6.8
Maize	14.7	19.1	11.2	6.3	17.1	8.5	3.8	10.2	24.4	8.4	4.3	7.2	4.8	12.1
Sorghum	22.3	19.2	19.9	16.5	—	—	6.4	10.1	12.8	22.1	21.1	3.0	7.5	6.3
Millet	22.1	17.1	19.5	10.8	—	—	—	—	17.8	37.7	6.4	2.9	9.0	15.1
Soybeans/peanuts	22.2	24.5	24.5	16.2	—	—	5.8	4.2	81.6	56.5	4.6	5.9	11.6	13.3
Pulses (total)	19.3	11.9	18.9	24.0	9.6	6.2	5.3	4.6	28.3	16.2	0.9	3.1	10.6	8.2
Roots and tubers (total)	8.9	14.5	8.0	4.2	4.1	3.6	2.6	1.8	5.0	10.0	5.2	3.6	9.7	4.0

Source: FAO and USDA production data compiled by IFPRI.

sometimes is). Temperature is not limiting. Diseases, insects, inadequate fertilizer use, and uncontrolled flooding are probably the main causes of low yields. Where rice is grown in more marginal conditions without good water control (upland, swamp, and floodplains), as in much of sub-Saharan Africa and Brazil, yields are more variable because of a complex of soil factors (including iron and aluminum toxicity), diseases, and fluctuating water supply.

The high humidity and low light intensity that prevail in much of the humid tropics do not suit most cereals and grain legumes. They favor cassava, yams, and aroids, as well as bananas and plantains. Yield variability is generally low. Tree crops become increasingly important.

Temperate climates. Although most industrialized countries have the majority of their cultivated area outside the tropics and subtropics (Australia is an exception), the annual variability of yield of staple crops is not always lower, especially at higher latitudes (Table 3). Low autumn and spring temperatures, killing frosts, fluctuating snow cover, and variable rainfall are the main causes of loss in Canada, the North American Great Plains, and parts of the USSR. In Australia and Argentina, unreliable rainfall and high summer temperatures, as well as frost in some areas, tend to limit yields. The low variability of yield in the United States compared with other major developed countries is noteworthy.

Causes of increased variability

Climatic change. An important issue is whether intra- or interannual variability of agricultural production in the world's major agricultural regions is increasing and, if so, what are the causes. The question of climatic variability and its relationship to climatic change to 2000 was addressed (but not necessarily resolved) by a U.S. National Defense University (NDU) study (1980). Some climatologists believe that any significant shift in climate will be accompanied by greater variability. This needs to be carefully monitored.

In any case, as the total volume of world production increases, the amplitude of seasonal fluctuations around the mean due to weather variability can also be expected to increase, even if there is no major shift in climate. Where coefficients of variation are relatively low, as in the U.S., China, and India, the size of the standard deviation, measured in tons, may still be very large simply because of the size of the country and the volume of production.

Technological change. The NDU study concludes that, at least for the next two decades, technology is likely to have a greater impact than climatic change on yields, and therefore on total production. If we accept this conclusion, then the influence of technology on product variability will be crucial to the stability of food supplies, and thus to food prices.

It is disturbing that recent studies (Barker et al 1981) with respect to Southeast Asia; Mehra (1981) concerning India; and Hazell (1982, 1986) suggest that production variability has increased as a result of the wider use of new technology. It appears that increased yield variance and a simultaneous loss of offsetting patterns of variation in yields between crops and regions (increased covariance) are the predominant sources of the increase in overall production variability. In general, this is attributed less to greater susceptibility of new varieties to climatic stress than to a lack of genetic diversity that increases their vulnerability to pests and diseases and to such nontechnical contributory factors as unstable prices and inefficient fertilizer and water delivery systems that lead to erratic use of inputs by producers, compounded by their unfamiliarity with new technology and consequent poor input management.

An important issue is whether these apparent increases in yield variability represent persistent trends or whether they are temporary phenomena. Increased yield variability could arise from the initial unfamiliarity of farmers with the use of new technology and from the inability of agricultural planners and administrators to cope with new demands being made on government policies and services by needs related to the successful and widespread adoption of technology. Experiences in the industrialized countries (Austin and Arnold 1986, Fischbeck 1986, Klatt et al 1986), which have had more time to adjust to such changes, and an analysis of the Indian situation based on a longer time series (Jain 1986) suggest that these may be the underlying reasons. Jain's study concludes that strong breeding programs capable of supplying new varieties suited to specific regional needs within countries, good germplasm collection and storage facilities, and efficient and flexible seed production and distribution services to back them up, are essential investments if higher and less variable yields are to be achieved. More research is recommended on soil, water, and pest management and on other production techniques that increase yield stability, as well as on nontechnical factors affecting successful adoption and use of technology.

Whether and how technology has affected yield variability in different climatic situations and over different space and time scales seems open to question. Weather-technology interactions are often difficult to disentangle. Some increases in variability and risk might be acceptable if the technology that causes it also leads to higher average yields and greater total production. Because of the great importance attached to more widespread use of yield-increasing technology in FAO's *Agriculture toward 2000* (1981) and other global studies, further ex post analysis of this crucial issue, over longer periods, in well-specified situations, is clearly necessary. So is careful monitoring of trends, particularly with respect to covariance and trade-offs between yield growth and stability (Evans 1986).

Area expansion. Most recent studies on variability have focused on relationships between climate, weather, and crop yield. Increased yield has become the main component of increases in agricultural production.

Nevertheless, in sub-Saharan Africa, where levels of technology are still low and the proportion of irrigated to total cultivated area very small, the impact of new technology on yield variability is almost a non-issue. Expansion of the land area farmed under rainfed conditions has been the main source of additions to food supply. That will continue to be important in Africa and in South America, which have large reserves of potentially cultivatable land, often in remote areas.

When production is extended into new areas, climate can affect yields or human activity could affect climate. Examples of the climate-yield relationship include increased incidence of frost damage to coffee when the area planted expanded southward in Brazil; increased yield variability of wheat in South Australia due to penetration into drier regions; damage to rice grown near the northern limits of its tolerance to cold in Japan (Carter and Parry 1986); and much greater weather-induced variability of grain yields in the Soviet Union as the area planted extended south and east into regions where temperature extremes are greater and precipitation lower than in the Ukraine (Desai 1986). Although all these climate effects on yield were somewhat adverse due to ecological marginality (which tends to dilute national averages), overall national production was sometimes larger due to increased area.

The impact of area expansion on climate is much more speculative, but it is widely assumed that the accelerated environmental degradation consequent on spontaneous movements of people, mainly cultivators, into rain forests, mountain forests and grazings, and semiarid rangelands will have adverse effects on global climate in the long term, especially on precipitation.

It is absolutely critical to find ways to manage land and water resources to conserve moisture, control erosion, and feed a rapidly growing population in new areas where the dominant influences are the interaction of production with weather in the short run and with climate in the longer run. This need is exacerbated by the lack of suitable technology, infrastructure, research, and other government services in such areas. Nor are the consequences of expansion of agriculture into new areas (whether spontaneous, commercial, or government-sponsored) being adequately controlled or evaluated by national governments or by donor agencies.

These issues have not been adequately addressed by national or international research programs. More attention is needed on agroecological characterization of new land, identification of its current and potential use, the nature and magnitude of limiting constraints (particularly with respect to soils), what research can do to remove them, and crop phenological requirements for critical growth periods in relation to the local climate and its variability. Understanding people, their motives, and their responses to incentives becomes increasingly crucial as population pressures increase. Research needs to be specific both to commodity and to ecology, within a systems framework that includes a strong social science component.

Agricultural implications of climatic change

The probability of a global climatic warming

Although climate has been receiving more public as well as scientific attention over the last 10 yr, what might happen in the longer term is still somewhat speculative. It is difficult to specify contingency plans. However, the probability and direction of climatic change seem to be less an issue of controversy and more one of degree than at the end of the 1970s. At that time, there had been a moderate cooling trend in the Northern Hemisphere for some 40 yr, with considerable uncertainty as to its probable direction (NDU 1980, Thompson 1979). The scientific evidence now points to progressive global warming, largely as a result of human influence (industrial and urban emissions and increased woodsmoke in low-income countries). The magnitude and spatial and temporal parameters are debatable. Although it has been suggested that stronger countervailing measures, especially by industrialized countries, could substantially reduce this threat (Bach 1987), the probability of lag effects and other imponderables has led to doubts that human intervention now can do much to reverse the trend, although it might delay or ameliorate it (Wigley 1987, Schelling 1984).

Implications for research

Research is needed to discover more about the causes, timing, magnitude, and geographical impact of climatic warming and to determine how the new situation may affect agriculture. Progress is being made toward learning more about climatic warming, through comparative analysis of historical records of climate and empirical studies of crop growth, simulation, and predictive modeling.

A good deal of research also has been directed toward assessing which areas, crops, and populations might be most seriously affected by climatic cooling or warming. It is believed that the largest temperature response to climatic warming will be in geographical regions at the cold margins. They should be closely monitored. But because of low populations, they are not the most vulnerable socially, nor are they agriculturally important.

Effects on precipitation. More attention should be directed to the possible effects of climatic change on precipitation. The magnitude of a warming trend is unlikely to be as great in the tropics or subtropics as in the higher latitudes, but in the drier areas of West and South Asia, China, North Africa, the Sahel, Northeastern Brazil, Mexico, and Southern Australia, a secular decrease in precipitation, greater variability, or a poorer pattern of distribution could be catastrophic. Water, rather than temperature, is the main determinant of crop yields in those areas. The effects of a cumulative decline in water availability for irrigation systems could compound the effects of decreased rainfall.

On the other hand, a modest decrease in precipitation might be beneficial in the subhumid tsetse fly-infested areas of the Soudanian grasslands, as well as in deltaic zones of Asia that are subject to uncontrollable flooding. It is probable that reduced rainfall in the rain forest zone would be undesirable. Locally, it could open new areas for settled farming; globally, any resultant deterioration of forest cover might seriously reduce precipitation and increase atmospheric pollution and carbon dioxide levels.

Increased precipitation could be crucial in offsetting some of the reductions in yield postulated to result from higher temperatures in mid-latitudes, especially in the U.S. corn belt (Thompson 1979). However, as physiological and phenological studies show, it is not so much total precipitation but its distribution that is critical to plant growth. Predicting rainfall distribution far in advance is difficult.

Soil characteristics. The first indication of climatic change may be a shift in area, as in the recent southward displacement of key crops in Sahelian Africa. In attempting to simulate or predict the spatial impact of climatic change and its possible impact on production, it is important to incorporate knowledge about soil quality and moisture-holding capacity. This does not always receive adequate weight in research on climate impact (the approach followed by FAO in its agroecological zone studies is to be commended). Studies in North America show that climatic cooling might shift crops such as barley in Canada out of areas of poor soils

(Williams and Oakes 1978). With climatic warming, the reverse might happen to maize in the U.S., thus reducing yields.

Teleconnections. Research on the causes and effects of linked climatic events in regions that are geographically distant is a relatively new and important field. Much attention has been directed toward understanding the ENSO phenomenon, whereby oceanic and atmospheric influences are interconnected. Major rain-generating patterns over tropical Southeast Asia, Amazonia, and equatorial Africa may be affected by oscillations in atmospheric pressure (Glantz et al 1987, Rasmussen 1986, Rosenberg 1981). Further understanding of this phenomenon may help scientists understand climatic variability and predict how adverse weather in one region of the globe may be offset by more favorable weather elsewhere. The causes of the blocking of the circulation of prevailing westerly winds in the Northern Hemisphere, which results in periods of extreme weather in Europe and the USSR, also require further study. Lamb (1981) has suggested that the frequency of days with general westerly winds over the British Isles may be an index of wide significance; it is positively correlated with global average temperature, with precipitation in many other mid-latitude regions, with monsoon rainfall in the Sahel-Ethiopian zone of Africa, and perhaps with the Indian summer monsoon. How these important phenomena might be affected by a global warming trend is unclear and merits close attention from climatologists.

Crop responses to elevated CO_2. Research on the probable response of major crops to climatic change is essential to provide rationale for breeding programs or other technological measures. Because of the predicted greenhouse effects of human activity, in the last few years attention seems to have been directed to understanding more the influence of higher levels of atmospheric carbon dioxide on plant growth.

This research indicates the possibility of beneficial increases in net photosynthetic productivity and water-use efficiency, although plants having a C_3 assimilation pathway might gain more than C_4 species. However, the experiments on which these conclusions are predicated were largely conducted under artificially controlled conditions on single plants. Usually they measure only plant response to varying levels of CO_2 and do not simulate any changes in temperature or precipitation that might result. Nor do they identify whether the influence of elevated CO_2 would transcend that of other limiting factors, such as moisture deficiency or low soil fertility.

Morphological, physiological, phenological, and genetic changes as a result of CO_2 enrichment have been postulated. Some of these could reduce economic yield (i.e. increased vegetative growth at the expense of grain formation). In certain species, the carbon-nitrogen balance could be upset, possibly leading to lower leaf protein and compensatory increases in consumption by pests (Kimball 1985). Or an aggressive response of weed species to CO_2 could eventually cause undesirable changes in species composition.

Because of methodological difficulties, little is known about the effects of elevated CO_2 on interspecific competition in whole farm systems or unmanaged ecosystems. These might be particularly severe in rangeland, where the buffering effects of human influence are lower than in arable land and where natural selection is not held in check by agricultural management.

Acock and Allen (1985) and Bazzazz et al (1985) attempted to synthesize the state of knowledge on a number of these important plant-CO_2 relationships. Their summaries highlight both the gaps and the complexities.

Policy implications

Planning for future climatic contingencies requires long-term vision. But a clear-sighted perspective is obscured by many uncertainties. Although considerable progress has been made toward understanding some aspects of crop-weather relationships, underlying climatic influences, and effects of CO_2 on plants, there are important gaps in our knowledge and difficulties in isolating single-factor effects from interactions. The impact of CO_2 on climate is still obscured by noise. Strain and Cure (1985) argue that the observed increase in atmospheric CO_2 during the 20th century must already be exerting a stimulating effect on the biosphere. If so, its influence on agricultural production cannot yet be separated from that of climatic variability or of production technology. These complexities act as disincentives to action by policymakers, who are normally driven by relatively short-term imperatives.

Nevertheless, timeliness will be critical in developing policies to cope with the possible effects of climatic change on human welfare. Feeding a rapidly growing population, especially in Africa, would become even more difficult if climate changed rapidly and adversely, especially if the change were accompanied by increased variability at the drier margins. The severe Ethiopian crisis, even though of relatively short duration, led to the desertion of barren farms by large numbers of people. A less sudden climatic anomaly might have allowed a carefully planned redistribution of population.

Time will be needed to facilitate the major adjustments in agricultural technology, shifts in land use and cropping patterns, and investments in irrigation and infrastructure likely be required in response to climatic change. On the basis of European Economic Community experience, Mackenzie (1981) considers that at least 30 yr after a consistent climatic shift had been established would be required before action initiated at the political level could have a major impact on the agricultural industry. In the absence of conviction that such a climatic shift has really occurred, he feels that policymakers and farmers will continue to react in an ad hoc fashion. This emphasizes the need for accurate and clear predictions.

As a first precaution against accelerated climatic change and associated increases in variability, measures to improve food security could be stepped up. However, maintaining adequate national food reserves in efficient, strategically situated storage facilities linked to effective distribution networks is costly. It will likely be difficult to persuade national planners to invest heavily in such facilities, or to persuade taxpayers to foot the bill, unless they feel under urgent pressure. Many policymakers seem to have a blind spot about climate. In this sense, a crisis climatic situation (or expectations of one) might be more likely to generate action than a gradual transition.

Enough sense of urgency and interest is needed among policymakers and the public to ensure adequate and consistent funding of national weather recording systems, early warning of climate-related shocks, research on climate-agriculture relationships, global climatic monitoring, and interdisciplinary analysis of the results and their implications. One would expect a well-planned monitoring program to give timely signals of emerging risks extending beyond agriculture to include energy, health, water resources, and other key areas affecting human activity and the global environment. These should be sufficiently clear to be heeded by policymakers, nationally and globally.

Continuing, and indeed enhanced, research is clearly essential. The needs have been articulated at several major meetings. The U.S. National Program on CO_2, Environment, and Society (US-DOE 1980) defines 6 research categories, 18 subcategories, and 50 research issues. Parry (1984) discusses approaches to estimating impacts of a CO_2-induced climatic change on the basis of data from general circulation models (GCM), instrumental data, and palaeoclimatic data. Gates (1984) deals with the use of GCM models. A U.S. Department of Energy report (White 1985) attempts to characterize information requirements for studies of CO_2 effects related to agriculture. A review of the various reports indicates the following:

- Closer linkages are needed between global (GCM) and other predictive and explanatory models.
- Although there is much ongoing research on varying aspects of climate, there are relatively few linkages between this work and agricultural research, especially in developing countries. Even the Consultative Group on International Agricultural Research institutes are only marginally and recently involved.
- To fill some of these gaps, both in research and in perception, information exchange as well as research should be strengthened. The initiatives taken by the U.S. National Center for Atmospheric Research and the World Meteorological Organization (1985) are praiseworthy, but I wonder whether its worldwide coverage is sufficient and if the information is analyzed and presented in a way intelligible to agricultural research and development planners.
- Although policymakers are becoming concerned about the more immediate problems arising from weather variability, acid rain, and environmental degradation, it will likely be difficult to interest them in longer-term possibilities requiring a sacrifice of present benefits for a more speculative future, especially where that sacrifice is to prevent externalities affecting other countries. Parry suggests that if these were expressed in terms of increasing risk, particularly with respect to extreme events, the impact on decisionmakers might be greater. An even more convincing approach might be to attempt to link the costs of enhanced research and monitoring on short- and long-term aspects of climatic variability and climatic change more closely to the potential benefits, and to specify more precisely what those might be and who might

gain or lose from national or international action to reduce human vulnerability to climatic shocks. The literature is most patchy on this important issue.

It seems apt to conclude my paper by quoting Mackenzie: "When my commission colleagues say to me, 'Thank you for all the climatological data and for your seminar report. How can I use it?', what do I reply?"

References cited

Acock B, Allen L H Jr. (1985) Crop responses to elevated carbon dioxide concentrations. Pages 53-97 *in* Direct effects of increasing carbon dioxide on vegetation. B. R. Strain and J. D. Cure, eds. U.S. Department of Energy, Washington, D.C., USA.

Ahmed R (1987) Strategic role of infrastructure in development. *In* The emerging world food situation and challenges for development policy. International Food Policy Research Institute Briefs 1:1.

Austin R, Arnold M (1986) Variability of wheat yields in the UK: analysis, future prospects and implications for plant breeding strategies. Page 236 *in* Proceedings of a workshop on cereal yield variability. P. Hazell, ed. International Food Policy Research Institute/Deutsche Stiftung für Internationale Entwicklung, Washington, D.C., USA.

Bach W (1987) The growing consensus and challenges regarding a greenhouse climate. International symposium on climate variability and food security. Indian National Science Academy, New Delhi.

Barker R, Gabler E, Winkelmann D (1981) Long-term consequences of technological change in crop yield: the case of cereal grain. Pages 53-78 *in* Food security for developing countries. A. Valdes, ed. Westview Press, Boulder, Colorado, USA.

Bazzazz F A, Garbutt K, Williams W E (1985) Effect of increased atmospheric carbon dioxide concentration on plant communities. Pages 155-170 *in* Direct effects of increasing carbon dioxide on vegetation. B. R. Strain and J. D. Cure, eds. U.S. Department of Energy, Washington, D.C., USA.

Blackie M J (1987) Restructuring marketing systems for smallholders: cases in Zimbabwe. Pages 187-198 *in* Accelerating food production in sub-Saharan Africa. J. W. Mellor, C. L. Delgado, and M. J. Blackie, eds. The Johns Hopkins University Press, Baltimore, Maryland, USA.

Boyce J, Evenson R (1975) National and international agricultural research and extension programs. Agriculture Development Council, New York, USA.

Carter T, Parry M (1986) Climatic changes and yield variability. Pages 47-70 *in* Proceedings of a workshop on cereal yield variability. P. Hazell, ed. International Food Policy Research Institute/Deutsche Stiftung für Internationale Entwicklung, Washington, D.C., USA.

CAB—Commonwealth Agricultural Bureaux (1987) Agricultural environments: characterization, classification and mapping. Proceedings of CGIAR Inter-Center Workshop, Rome, 1986. A. H. Bunting, ed. London, UK.

Desai P (1986) Weather and grain yields in the Soviet Union. Research Report No. 54. International Food Policy Research Institute, Washington, D.C., USA.

Evans L (1986) Yield variability in cereals: concluding assessment. Pages 1-12 *in* Proceedings of a workshop on cereal yield variability. P. Hazell, ed. International Food Policy Research Institute/Deutsche Stiftung für Internationale Entwicklung, Washington, D.C., USA.

Farmer G F, Wigley T M L (1985) Climatic trends for tropical Africa. Overseas Development Administration, UK.

Fischbeck G (1986) Trends in yield increase and yield variability of winter wheat and spring barley in Bavaria 1950-84. Page 185 *in* Proceedings of a workshop on cereal yield variability. P. Hazell, ed. International Food Policy Research Institute/Deutsche Stiftung für Internationale Entwicklung, Washington, D.C., USA.

FAO—Food and Agriculture Organization (1978) Report on the Agro-Ecological Zones Project. World Soil Resources Reports 48/1 (Africa). Rome, Italy.

FAO—Food and Agriculture Organization (1981) Agriculture toward 2000. Rome, Italy.

Gates W L (1984) The use of general circulation models in the analysis of the ecosystem impacts of climatic change. *In* Report of the study conference on sensitivity of ecosystems and society to climatic change (WCP-83). World Climate Programme, Geneva, Switzerland.

Glantz M D, ed. (1976) The politics of natural disaster: the case of the Sahel drought. Praeger, New York, USA.

Glantz M D (1987) Politics, forecasts and forecasting. Forecasts are the answer, but what was the question? Page 81 *in* Policy aspects of climatic forecasting. R. Krasnow, ed. Resources for the Future, Washington, D.C., USA.

Glantz M, Katz R, Krenz M, eds. (1987) The societal impacts associated with the 1982-83 worldwide climate anomalies. United Nations Environmental Programme and U.S. National Center for Atmospheric Research.

Hazell P (1982) Instability in Indian foodgrain production. Research Report No. 30. International Food Policy Research Institute, Washington, D.C., USA.

Hazell P (1986) Cereal yield variability: overview of issues, introduction. Page 15 *in* Proceedings of a workshop on cereal yield variability. P. Hazell, ed. International Food Policy Research Institute/Deutsche Stiftung fr Internationale Entwicklung, Washington, D.C., USA.

Jain H K, Dags M, Taylor T A (1986) Yield variability and the transition of the new technology. Page 77 *in* Proceedings of a workshop on cereal yield variability. P. Hazell, ed. International Food Policy Research Institute/Deutsche Stiftung fr Internationale Entwicklung, Washington, D.C., USA.

Johnson S, Holt M (1986) The value of climate information. Page 53 *in* Policy aspects of climate forecasting. R. Krasnow, ed. Resources for the Future, Washington, D.C., USA.

Kimball B A (1985) Adaptation of vegetation and management practices to a higher carbon dioxide world. Pages 185-204 *in* Direct effects of increasing carbon dioxide on vegetation. B. R. Strain and J. D. Cure, eds. U.S. Department of Energy, Washington, D.C., USA.

Koester U (1986) Regional cooperation to improve food security in southern and eastern African countries. Research Report No. 53. International Food Policy Research Institute, Washington, D.C., USA.

Lamb H H (1981) Effects of climatic change on the people of Europe in the past. Page 1 *in* Climatic change and European agriculture. Seminar Paper No. 12. S. W. Burrage and M. K. V. Carr, eds. Wye College, University of London, UK.

Klatt A, Pfeiffer W H, Braun H J (1986) Yield stability in bread wheat. Page 115 *in* Proceedings of a workshop on cereal yield variability. P. Hazell, ed. International Food Policy Research Institute/Deutsche Stiftung fr Internationale Entwicklung, Washington, D.C., USA.

Mackenzie L (1981) Interpretation of climatological data by agricultural policy-makers in an EEC context. *In* Climatic change and European agriculture. Seminar Paper No. 12. S. W. Burrage and M. K. V. Carr, eds. Wye College, University of London, UK.

Mehra S (1981) Instability in Indian agriculture in the context of the new technology. Research Report No. 25. International Food Policy Research Institute, Washington, D.C., USA.

Mellor J (1985) The role of government and new agricultural technologies. Food Policy Statement No. 4. International Food Policy Research Institute, Washington, D.C., USA.

NDU—U.S. National Defense University (1980) Crop yields and climatic change to the year 2000. Vol. 1. Washington, D.C., USA.

Oram P (1982) The economic cost of climatic variation. Page 355 *in* Food, nutrition and climate. Sir K. Blaxter and L. Fowden, eds. Applied Science Publishers, London, UK.

Oram P (1985) The sensitivity of agricultural production to climatic change. Climatic Change 7(1):129-152.

Oram P, Bindlish B (1981) Resource allocations to national agricultural research: trends in the 1970s. International Service for National Agricultural Research (ISNAR), The Hague. International Food Policy Research Institute, Washington, D.C., USA.

Oram P, Zepata J, Alibaruho G, Roy S (1979) Investment and input requirements for accelerating food production in low-income countries by 1990. Research Report No. 10. International Food Policy Research Institute, Washington, D.C., USA.

Parry M (1984) Climate impact assessment in cold regions. *In* Report of the study conference on sensitivity of ecosystems and society to climate change (WCP-83). World Climate Programme, Geneva, Switzerland.

Parry M L, Carter T R, eds. (1984) Assessing the impact of climatic change in cold regions. Summary Report SR-84-1. International Institute for Applied Systems Analysis.

Rasmussen E (1986) Understanding and predicting year-to-year climate variability. Page 1 *in* Policy aspects of climatic forecasting. R. Krasnow, ed. Resources for the Future, Washington, D.C., USA.

Rosenberg N (1981) The increasing CO_2 concentration in the atmosphere and its implication on agricultural production. Climatic Change 3:265-279.

Schelling T (1984) Anticipating climatic change. Environment 28(8):6.

Strain B R, Cure J D (1985) Status of knowledge and recommendations for future work. Pages 205-213 *in* Direct effects of increasing carbon dioxide on vegetation. B. R. Strain and J. D. Cure, eds. U.S. Department of Energy, Washington, D.C., USA.

Thompson L (1979) Climate change and world grain production. The Chicago Council on Foreign Relations, Chicago, USA.

US-DOE (1981) Research issues of the national program on CO_2, environment and society, fiscal year 1980. National Climate Program Office, U.S. Department of Energy, EV1366TN, Washington, D.C., USA.

Valdes A (1981) Food security for developing countries. A. Valdes, ed. Westview Press, Boulder, Colorado, USA.

Valdes A, Zeitz J (1980) Agricultural protection in OECD countries: its cost to less-developed countries. Research Report No. 21. International Food Policy Research Institute, Washington, D.C., USA.

Wanmali S (1986) Rural infrastructure and agricultural development in Zambia. Pages 203-218 *in* Planning and operating rural centres in developing countries. H. U. Thimm, D. A. G. Green, M. Leupolt, and R. M. Mkandawire, eds. Verlag Weltarchiv GMBH., Hamburg, W. Germany.

White M R (1985) Characterization of information requirements for studies of CO_2 effects: water resources, agriculture, fisheries, forestry and human health. U.S. Department of Energy, Washington, D.C., USA.

Wigley T M L (1987) The impact of climate on resource use and availability. Page 79 *in* Resources and world development. Energy, minerals and land. D. J. McLaren and B. J. Skinner, Dahlem Conferenzen, eds. John Wiley and Sons, Ltd., Chichester, UK.

Williams G D V, Oakes W T (1978) Climatic resources for maturing barley and wheat in Canada. Page 367 *in* Essays on meteorology and climatology. Monograph 3. K. D. Hage and E. R. Reinelt, eds. Department of Geography, University of Alberta, Edmonton, Canada.

World Bank (1982) Agriculture and economic development. World Development Report, Part II. The World Bank, Washington, D.C., USA.

WMO—World Meteorological Organization (1985) Infoclima: catalogue of climate system data sets. WCP-101. World Climate Programme, Geneva, Switzerland.

World Resources (1986) An assessment of the resource base that supports the global economy. A report by the World Resources Institute and the International Institutes for Environment and Development. Basic Books, Inc., New York, USA.

Worldwatch (1986) State of the world 1986. A Worldwatch Institute report on progress towards a sustainable society. L. Starke et al, eds. W. W. Norton and Company, New York, USA.

Notes

Author's address: P. A. Oram, International Food Policy Research Institute, 1776 Massachusetts Avenue, N.W., Washington, D.C. 20036, USA.

Citation information: International Rice Research Institute (1989) Climate and food security. P.O. Box 933, Manila, Philippines.

Climatic variability and agricultural production in temperate regions: six case studies

W. J. Maunder

Climate variability can be evaluated in ways useful to both political and economic planning and management. Six case studies are used to discuss climate variability and agricultural production in temperate regions: 1) the climate/wool scene in one temperate climate country; 2) the significance of climate variations to investment in livestock farming in New Zealand; 3) the development and use of national climate productivity indices; 4) the development and use of international climate productivity indices; 5) the impact of a specific climate event on a commodity price; and 6) the weather and the U.S. commodity futures market.

One of the major problems in considering the atmosphere as a resource is recognition that it is not only an elite resource, but also a variable resource. Of prime importance is the impact of short-term variations, particularly variations related to water and temperature. Clearly this impact will continue to increase in importance as the demand for food and water and the relative cost of energy increase. Hence it is important to understand the effect that variations in weather and climate can have on economic, social, and political activities (Maunder 1978).

Particular responses of agricultural production to climatic variability in temperate regions can be examined through case studies. Key factors in the six case studies discussed here include 1) the increasing impact that weather and climate variations will have on economic, social, and political activities, irrespective of any long-term trend in climate; 2) the importance of the climate/agricultural production mix at both local and national levels; and 3) the independence, competitiveness, and political strength of farmers in many temperate regions.

Climate and wool production

An example of the sensitivity to weather and climate of pastoral production in different parts of the world is New Zealand. That country has a significant climatic advantage for grassland agriculture, which has made it the world's leading exporter of grassland products (Table 1). A comparison of the world's wool climates (Table 2) shows the climatic advantage that New Zealand has over her competitors (Maunder 1963, 1977). (Similarly, the Australian climate could be compared with the climates of the chief wheat production areas of the world, and the climate advantages that Australia may have over her agricultural competitors in wheat assessed.)

Table 1. Comparison of New Zealand and world food production and exports, 1977.

Product	N. Z. exports as % of world exports	N. Z. exports as % of N. Z. production	World exports as % of world production	N. Z. production as % of world production
Lamb and mutton	50.0	80.9	14.4	8.9
Beef and veal	8.9	46.7	6.3	1.2
Butter	20.8	76.8	15.0	4.1
Cheese	6.6	97.5	11.7	0.8
Dried milk	10.6	85.4	39.5	4.9

Source: Maunder (1980). Data from New Zealand Monthly Abstracts of Statistics, FAO Production/Trade Yearbooks. Original table was published in *New Zealand in the future world,* a booklet on 'sustainability' prepared by Diane Hunt for the NZ Commission of the Future, Government Printer, Wellington, 1979.

Table 2. Selected climate data of wool-producing areas in the world.

Station	No. of months with	
	Av rainfall ⩾ 38 mm	Av temp ⩾ 13°C
New Zealand		
Whatawhata	12	12
Masterton	12	12
Lake Coleridge	12	9
Omarama	11	8
Mid Dome	12	8
Argentina		
Buenos Aires	12	12
Bahia Blanca	8	12
Santa Cruz	0	7
Australia		
Hay	1	12
Dubbo	12	12
Bourke	1	12
Port Augusta	0	12
China		
Hanchow	3	7
Tai-Yuan	4	7
South Africa		
Grootfontein	5	12
Victoria West	2	12
USA		
Abilene	7	12
Austin	12	12
Cheyenne	5	6
USSR		
Ordzhonkikdze	7	6
Astrakhan	0	7
Semipalatinsk	1	5
Chernovtsy	6	7
Irkutsk	4	5

Sources: Maunder 1963, 1977.

The wool-producing areas of New Zealand receive more rain than other wool-producing areas in the world. For example, Omarama (in one of New Zealand's driest areas) has only 1 mo with an average rainfall less than 38 mm: that is a shorter dry season than is experienced in most wool-producing areas outside New Zealand. Also average midsummer temperatures in New Zealand lie at the cool end of the 15-30 °C temperature range that is characteristic of most wool areas. Average daily maximum midsummer temperatures in New Zealand are 21-23 °C, compared with 24-34 °C in most other wool-producing areas. Average daily minimum midwinter temperatures at New Zealand stations are nearly as high as in the milder wool-producing areas of northern Argentina and Australia. In all other major wool-producing areas of the world, nighttime midwinter temperatures are 4-22 °C colder than those experienced at Omarama (which is also one of New Zealand's colder wool-producing areas).

Climate variation and livestock investment

In all countries, production, whether agriculture-based or not, varies considerably. This variability is caused by economic, cultural, political, and environmental factors. If production is not agriculture-based, the variations usually are economically manageable. Where production is agriculture-based, the variations can assume considerable economic and political significance.

Of New Zealand's total 1984-85 gross domestic product, 11.4% was from the agricultural production group and 6.1% from the food production group. Moreover, nearly 70% of New Zealand's exports are agriculture-based.

Agrometeorological research traditionally has focused on marketing, production, animal health, crop protection, and crop quality. An additional factor is farmers' investment response; Walsh (1981) examined this aspect in detail. In his analysis, the factors influencing farmers' livestock investment response in New Zealand are, in order of importance, 1) monetary terms of exchange, 2) climatic conditions and biological factors, 3) technological change and innovation, and 4) the availability of land, labor, and credit.

Walsh considered that climatic conditions operate as a major constraint on growth, in that livestock feed is grown largely on the farm and it usually is not economically feasible to import large amounts of feed in times of widespread drought. There are significant fluctuations in rainfall from season to season in New Zealand.

The effect of climate as a constraint on livestock investment by farmers is largely a constraint on pasture production and consequent carrying capacity, especially during winter. The effect on reproductive rates is also significant, since feed intake just before conception and birth strongly influences lambing and calving percentages. Aberrations from normal climatic conditions, especially drought, reduce carrying capacity and performance; an effect that may carry through to the following season or even longer.

Table 3 shows data on three main factors that affect investment in livestock for 20 yr of New Zealand sheep and beef production. The year-by-year changes indicate

Table 3. Sheep and beef production of the New Zealand sheep/beef farm, as affected by agroeconomic and climatic factors.

Season	Sheep and beef stock units[a] (million)	Terms of exchange[b] (1975-76=1000)	Real expenditure per stock unit[b] (in 1975-76 $)	Days of soil water deficit[c]
1960-61	60.1	1079	7.77	18.9
1961-62	61.0	940	7.54	34.5
1962-63	62.8	1049	7.90	29.4
1963-64	63.8	1214	8.94	40.5
1964-65	66.3	1064	9.27	17.5
1965-66	70.7	998	9.46	18.1
1966-67	74.8	918	8.80	33.9
1967-68	76.8	899	7.83	33.0
1968-69	77.5	951	8.32	30.4
1969-70	78.8	1075	8.54	34.2
1970-71	78.6	963	8.23	35.5
1971-72	79.2	884	8.29	28.7
1972-73	78.4	1362	9.68	57.1
1973-74	79.5	1194	10.04	39.4
1974-75	80.8	750	8.10	28.5
1975-76	82.6	1000	9.10	32.8
1976-77	83.9	1063	8.92	25.7
1977-78	85.9	877	8.48	54.1
1978-79	84.4	993	9.06	27.3
1979-80	89.2	964	9.16	9.9

Source: Maunder (1986), from Walsh (1981).
[a]Estimated from national census figures, sheep at 30 Jun of second year, cattle at 31 Jan. [b]N. Z. Meat and Wool Boards' Economic Service data. [c]Weighted average of the New Zealand Meteorological Service indices for sheep and beef cattle populations.

the nature of livestock investment responses. A clearer picture can be obtained by aggregating the data into periods with similar monetary exchange and input volumes, and examining the resulting growth rates of livestock numbers. This simple procedure to some extent overcomes the effects of lag in livestock investment response.

For purposes of this analysis, it is convenient to look at three 5-yr periods: 1962-63 to 1966-67, 1967-68 to 1971-72, and 1972-73 to 1976-77 (Table 4). Walsh (1981) noted that the early periods correspond to conventional economic notions about the effect of relative prices on investment. For example, in the period 1962-67, growth of livestock numbers (approximately 4%/yr) was associated with relatively high and stable levels of both monetary terms of exchange and expenditure per stock unit, as well as generally favorable climatic conditions. There was some carry-over of the higher growth rate of stock numbers into 1967-72, but in general this was a period of slow growth associated with significantly lower terms of exchange and real expenditure; climatic conditions remained favorable, but perhaps slightly less so than in 1962-67.

In 1972-77 there was a definite recovery in the rate of growth of stock numbers, but the growth was much less than might be expected from the high average levels of

Table 4. Agroeconomic and climatic factors: The New Zealand sheep/beef farm, selected periods.

	Annual average		
	1962-63 to 1966-67	1967-68 to 1971-72	1972-73 to 1976-77
Sheep and beef stock units (Average annual % change)	+4.1	+0.4	+1.4
Terms of exchange index	1049	954	1074
Real expenditure per stock unit (1975-76 $)	$8.87	$8.24	$9.17
Weighted no. of days of soil water deficit	28	32	37

Source: Maunder (1986b), from Walsh (1981).

relative prices and input volumes. The reasons for this lower productivity of expenditure included 1) climatic constraints in the form of drought, especially in 1972-73 and in some regions in 1973-74, which affected stock-carrying capacity and reproductive rates in both those and subsequent seasons; and 2) relatively low levels of expenditure in 1967-72, which had a damaging effect on livestock physiology and, therefore, on reproductive rates. This effect carried over into 1971-77. It appears that growth requires stable as well as favorable terms of exchange and real expenditure.

In general, the overall weather and climatic conditions in New Zealand during 1970-80 did not foster confidence among farmers. Table 3 shows many days with soil water deficit during 1972-73, 1973-74, and 1977-78. The impact of these droughts, coupled with rapidly increasing costs of farming, needs to be considered if livestock farming in New Zealand is to remain a viable economic activity.

Development of climate productivity indices

In many countries, factors other than weather and climate are readily cited to explain the vagaries of national economic conditions. Some reasons for omitting climate as an economic determinant include the ready availability of nationwide economic indicators, the absence of suitable nationwide climate indicators, and the difficulty of understanding climate indicators when they are available. For example, in the United States, it is difficult to ascertain from the *Weekly weather and crop bulletin* the effect of the overall state of the weather on national grain production or meat prices. This publication is excellent in its station-by-station and state-by-state analyses, but is difficult to use as an indicator of national climatic conditions.

Because of the impact of weather and climate on a nation, on international trade, on commodity prices, and on food shortages and surpluses, a climate productivity index could be useful as a single indicator of influence of weather on the economy. The availability of real-time weather and climate information from around the world, and the availability of a few reasonably credible agro-weather-economy models would also mean that, for the first time, a world and/or hemispheric weather-productivity or climate-productivity index could be a reality. Two aspects of any national econoclimatic indicator need to be considered. First, for the information to be useful, it must be available in real time or near real time; second, the information must be available on a national basis. If it is to be used at a political level, it should be a single index.

Consider the development of a climate productivity index for New Zealand. The importance of agriculture to New Zealand's economy has long been stressed by numerous observers, scientists, and politicians. Nevertheless, factors other than weather and climate have usually been advanced to explain New Zealand's economic conditions.

In the belief that real-time information on the agricultural climate of New Zealand is essential if the business community and the farmer are to maximize the country's agricultural potential, an Agroclimatic Productivity Index was devised. The index compares past weather with present weather to give a single-figure indicator of what could be called the true economic state of the nation's agriculture.

The Agroclimatic Productivity Index computed for New Zealand is essentially an economic weighting of the standard commodity-weighted weather and climate indices computed by the New Zealand Meteorological Service. The commodity weightings of sheep, dairy cows, beef cattle, and lambs shorn (the main components of New Zealand's agricultural production) are economically weighted according to their contribution to total national agricultural production. In reality, this means giving a weight of 30% to the weighted weather and climate indices for wool (sheep/hoggets shorn 25%, lambs shorn 5%), plus weights of 25% for mutton and lamb production (sheep population 5%, breeding ewe population 20%) and 20% for beef and veal production (beef cattle population 15%, calves born population 5%). The combination of the economic weightings and the normal commodity weightings (i.e. the weighting of weather and climate data based on the distribution of a commodity such as crop production or animal numbers) in effect produces a double weighting—one for the significance of the commodity production and one for the significance of that commodity to the nation as a whole.

Two aspects of weather are considered in computing the Agroclimatic Productivity Index: days of soil water deficit (more specifically, days below wilting point) and growth limiting degree days (base 5 °C). Each factor is in turn weighted according to the significance of the months being considered. (For example, the number of days of water deficit is much more significant in midsummer than in midwinter.) These weighted values are then further adjusted by the number of months being considered in the composite index. The final index should provide a single value indicator of the overall state of New Zealand's agroclimate.

The index is constructed so that a value of 1,000 or greater implies confidence in New Zealand's pastoral farming sector (that is, overall it should be good for farmers), while an index of less than 1,000 implies concern. It is not suggested that weather is the only factor to be considered in evaluating the general level of national farm productivity or confidence, but as a purely climate-based indicator, the Agroclimatic Productivity Index is a useful and simple guide.

New Zealand Agroclimatic Productivity Indices for consecutive months from June 1977 to May 1985 are given in Figure 1. Each index is a cumulative weighted

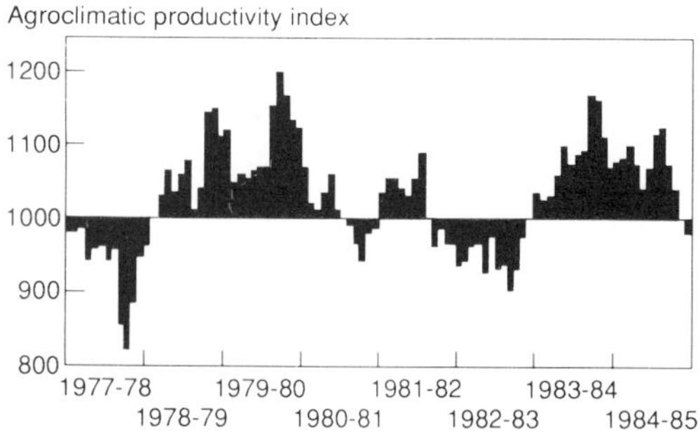

1. New Zealand agroclimatic productivity indices. Compiled from data supplied by the New Zealand Meteorological Service.

Table 5. International commodity-weighted agroclimatic indices, May 1981.

Country	Commodity	Rainfall (percentage probability[a])	Temperature (percentage probability[a])
USA	Maize	72	17
	Soybean	84	2
	Winter wheat	96	49
	Spring wheat	23	56
Canada	Wheat	17	82
USSR	Winter wheat	36	8
	Spring wheat	28	16
	Sunflower seed	16	4
	Maize	24	4

Source: Maunder (1984), from McQuigg Consultants Inc. (1981). [a]100 = very wet/very warm, 0 = very dry/very cold.

value of the previous 6-mo period. The index during the 8 seasons shown varied from 815 for the 6 mo ending March 1978 (the peak of the 1977-78 drought) to 1200 for the 6 mo ending February 1980. Comparing the average values for whole seasons shows that 1979-80 and 1983-84 were the best of the six seasons shown, and 1977-78 and 1982-83 the worst. Changes in the seasonal average of the Agroclimatic Productivity Index, from 947 in 1982-83 to 1085 in 1983-84, were dramatic.

International climate productivity indices

Agroclimatic productivity indices could be developed for international commodities, for nations as a whole, for commodities for selected nations, and for commodities for combinations of nations. A global (or at least a hemispheric) agroclimatic productivity index for maize, wheat, soybean, wool, meat, and coffee would have obvious political and international trade significance. Table 5 shows a sample of climate values for various national commodities in May 1981; those could easily be developed into hemispheric commodity indices.

Such econoclimatic developments could fit neatly into videotext systems, which make detailed information available to individual farmers with visual display units. Alternatively, the data could be interpreted for farmers, with appropriate advice and recommendations, on a regional basis.

At the international level, commodity-weighted weather information could be used by a country to monitor and, in some cases, to predict the agricultural production of another country. The advantages (economic, strategic, or political) such information could give a country are wide ranging, and involve decisionmakers at government, producer board, producer, and processing levels.

Climate events and commodity prices

A specific climate event can have a considerable effect on commodity prices. Commodity price changes affect export markets and a country's ability to import, as well as consumer prices. For example, a dramatic price change in one key commodity, soybean, occurred in the United States between the comparatively normal 1979 season and the very hot and dry 1980 season (Fig. 2).

The price of soybeans in the United States varied during the 1979 season from a little more than $239/t in January to a peak of about $272 in midsummer, before decreasing to about $231 in December. In marked contrast, the price of soybeans in 1980 reached a low of less than $217/t in May, and then—primarily due to an anticipated weather-related shortfall in production—increased rapidly to a high of more than $294 in November. Such weather-induced prices, often coupled with similar price increases in other key food commodities, can play a major role in overall price structures, both nationally and internationally.

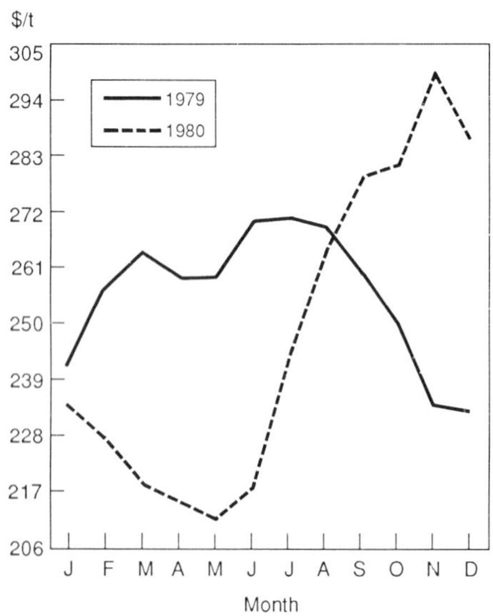

2. Changes in the U.S. price of soybean, 1979 and 1980 (Maunder 1985, Liebhardt 1981).

Weather and the futures market

The commodity market is the exchange where commodities such as wheat, maize, soybean, sugar, wool, cocoa, and frozen orange juice are bought and sold for delivery at a stated time in the future. The prices for commodity futures are related to the anticipated demand for, and supply of, a particular commodity at some future date. A close relationship exists between actual and expected weather conditions in the various producing areas and the price at which futures in a commodity will be bought and sold. Reports of the weather that has occurred and forecasts of the expected weather thus have a real dollar value as far as the commodity markets are concerned, and consequently influence the prices that the consumers will eventually pay.

The commodity market is perhaps the best example of sensitivity to weather and climate in the business world, particularly because the market reflects not only the sensitivity of commodity prices to the actual weather and climate, but in particular to information about weather and climate. For example, reports of frosts in Brazil or freezing in Florida can affect commodity prices before the actual weather conditions are known. (Although weather and climate data are collected, analyzed, and assessed in real time, the data are at best sample surveys.)

Unofficial reports of weather conditions from secondary networks also are part of the information package available to commodity market dealers and their customers, and are acted upon according to the sensitivity of the market. In addition to the weather that has occurred, or is reported to have occurred, there is the weather and climate expected to occur tomorrow, next week, next month, or next season. The commodity market is, therefore, a melting pot of past, present, and future weather and climate information, and it is in this melting pot that the real sensitivity to weather and climate may be measured.

The role of the meteorological community in supplying weather and climate information (which may or may not affect the market) is an important aspect of the commodity market. In a few countries, including the United States, such information is supplied by both the government-funded national weather service and private meteorological companies. Almost daily, in the commodities pages of the (New York) *Wall street journal*, a meteorologist (usually from the private sector) comments on the future prices of maize, wheat, or orange juice, with a weather-related explanation of what has happened or is likely to happen on the commodity market. Two typical examples of such comments, from the *Wall street journal* during Jan-Mar 1981, are as follows:

> Soybeans rose more than 21 cents a bushel amid reports of dry weather in South America and U.S. growing areas. Hot, dry weather in Brazil and Argentina during the last month is leading traders to re-assess the outlook for the soybean crops there.
>
> News of weekend rains in the Grain Belt pushed prices lower, with May delivery wheat dropping 7.5 cents, to $4.33 a bushel. Generally improved weather is prompting some analysts to look for larger crops.

Such examples are common, and are acted upon daily in the U.S. and some other countries by consumers and producers, as well as by investors and speculators. However, despite the obvious sensitivity of the commodity market to weather information, even in the sophisticated U.S. market, the overall emphasis is on providing and assessing fairly basic weather information, not specific commodity-weighted weather information. In the U.S. and presumably in other futures markets, there is, therefore, a largely untapped market for the more imaginative meteorological consultant to assist those people who want to know whether to buy or sell wheat, wool, coffee, or soybeans.

Conclusion

Both weather and climatic fluctuations have major economic, social, and political consequences. Indeed, our vulnerability to such fluctuations has undoubtedly grown as world population has increased and the use of available resources has become more intense. Furthermore, it is now well recognized that, particularly in regard to food, energy, and commodity flows, the monitoring and prediction of local, regional, national, and international agricultural productivity are ultimately associated with the availability, use, and interpretation of data on past, present, and future weather and climate.

The multidisciplinary nature of the climate system calls for a mechanism to make concise, condensed, easily readable information available to research scientists and technical/managerial/governmental personnel. To this end, the innovative work of the Climate Analysis Center of the (U.S.) National Weather Service should be commended, as well as the associated monthly climatic summaries distributed by the World Meteorological Organization (Geneva) through its World Climate Programme. Nevertheless, to be really useful for decisionmaking, such global and national climate analyses must be overlain with an appropriate economic and social framework. Such an overlay is essential if the value of last week's weather or the current season's climate is to be evaluated in terms of food shortages, food surpluses, profit for the farmer, or political advantage.

If we are to live within the limits of our "climatic income" or our "elite atmospheric resource," appropriate meteorological and climatological planning must occur. The politician and the planner must become more weather- and climate-oriented, for only then will optimum use be made of the climate resources of the 1990s. Central to this planning is the need for much more comprehensive monitoring and analysis of the world's climate, both to detect and to predict changes. The consequences of such changes must also be better understood. Indeed, the need for more relevant and more timely information about the weather and the climate offers an important challenge to the meteorological community.

A key factor in all the case studies discussed here is the significance of weather and climate in the marketplace. Meteorologists and climatologists have an important role to play in educating and influencing governments and the community, as well as the key weather and climate-sensitive sectors within each community, on the importance of the elite resource we call the atmosphere. The need for appropriate climatic guidance to politicians and decisionmakers during times of major climatic aberrations—such as the recent droughts in the Sahel, Ethiopia, Brazil, and Australia, as well as the 1982-83 El Niño event (Hare 1985)—is now apparent. However, the need for climatic guidance during normal times must also be given due recognition.

References cited

Center for Environmental Assessment Services (1981) Retrospective evaluation of the 1980 heat wave and drought. United States Department of Commerce, Washington, D.C.

Hare F K (1985) Climate variations, drought and desertification. WMO Publ. 653. World Meteorological Organization, Geneva. 35 p.

Liebhardt K (1981) The socioeconomic impact of climate. Enquiry 1(3).

McQuigg Consultants Inc. (1981) The weighted weather index. McQuigg Crop/Weather News, June 5.

Maunder W J (1963) The climates of the pastoral production areas of the world. Proc. N. Z. Inst. Agric. Sci. 9:25-40.

Maunder W J (1977) Climate constraints to agricultural production: prediction and planning in the New Zealand setting. N. Z. Agric. Sci. 11:110-119.

Maunder W J (1978) Economic and political issues. Pages 327-334 *in* Climatic change and variability: a southern perspective. A. B. Pittock et al, eds. Cambridge University Press, Cambridge.

Maunder W J (1979) New Zealand's real economic climate - evaluation and prospects. Pages 203-245 *in* Symposium on the value of meteorology in economic planning. New Zealand Meteorological Service, Wellington.

Maunder W J (1980) The use of econoclimatic models in national supply forecasting: New Zealand wool production. Pages 22-28 *in* Proceedings of the tenth New Zealand geography conference.

Maunder W J (1984) Climatology: past, present, future... a personal view. Weather Climate 4:2-10.

Maunder W J (1985) Climate and socio-economics. Pages 33-59 *in* Scientific lectures presented at the ninth world meteorological congress, 1983. WMO Publ. 614. World Meteorological Organization, Geneva.

Maunder W J (1986) Economic aspects of agricultural climatology. *In* Agricultural meteorology. J. L. Monteith, J. F. Elston, and L. E. Mount, eds. Academic Press. (forthcoming)

Maunder W J (1987) The uncertainty business: risks and opportunities in weather and climate. Methuen and Co., London. 420 p.

Tarpley J D, Schneider S R, Money R L (1984) Global vegetation indices from the NOAA-7 meteorological satellite. J. Climatol. Appl. Meteorol. 23:491-494.

Taylor B F, Dini P W, Kidson J W (1985) Determination of seasonal and interannual variation in New Zealand pasture growth from NOAA-7 data. Remote Sensing Environ. 18:177-192.

Walsh M J (1981) Farm income production and exports. Part II. Pages 299-327 *in* External economic structure and policy - an analysis of New Zealand's balance of payments. R. S. Deane, P. W. E. Nicholl, and M. J. Walsh, eds. Reserve Bank of New Zealand, Wellington.

Yates H W, Tarpley J D, Schneider S R, McGinnis D F, Schofield R A (1984) The role of meteorological satellites in agricultural remote sensing. Remote Sensing Environ. 14:219-233.

Notes

Acknowledgment: Paper based in part on extracts from Maunder (1987).
Author's address: W. J. Maunder, New Zealand Meteorological Service, P.O. Box 722, Wellington, New Zealand..
Citation information: International Rice Research Institute (1989) Climate and food security. P.O. Box 933, Manila, Philippines.

Climatic variability and sustainability of crop yields in the humid tropics

J. W. Pendleton and T. L. Lawson

Food demand predictably increases as populations increase. Food supplies fluctuate strongly with weather and climate changes. Of the global land mass, 28% lies within the humid tropics, in areas characterized by high temperatures, variable rainfall, and fragile soils which deteriorate rapidly when cleared. High rainfall and humidity often adversely affect reproduction, ripening, drying, and storage, and increase pest and disease problems of cereals and grain legumes. Shifting cultivation—a widely used, low-input, energy-efficient system—does not meet the food needs of rapidly increasing populations. Crop yields are limited by the climate. Constraints are high temperature and low solar radiation combined with relatively poor, highly erosive soils, which are often acidic, with low CEC. Fluctuating rainfall patterns combined with soils of low moisture-holding capacity often lead to periodic drought. Effective soil moisture management, crop and variety selection, minimum tillage, organic matter preservation, and integrated pest management offer hope for sustainable yields. With such new technologies as alley cropping, agroforestry offers a large opportunity for increased production. Climatological analyses that predict the probability of events such as the onset or cessation of the rainy season can be useful in selecting crops and in land preparation and seeding.

Demographers and other social scientists are alarmed at the rate populations in the developing world are increasing. The increases must be checked, they say, for there to be any hope of providing enough food (Table 1). The demographic situation in Africa is acute (Barker and Dorosh 1986, Davidson 1986). The densest populations there are in the wetter areas—the humid tropics (Goldman 1986).

Agricultural production in sub-Saharan Africa needs to increase by about 3% annually over the next 2 decades to maintain even the currently substandard nutritional levels of the growing population. But the production growth rate has been decreasing, from 2.5% in the 1960s to 1.9% in the 1970s (Davidson 1986). The problem is compounded by urban population increases, with population migrating away from the agricultural production sectors.

Comparing averages of population growth and food needs masks an important agricultural fact: population growth tends to be smooth, without large deviations; agricultural production fluctuates widely, in response to variations in weather. Long-term variations in climate may further change production (Bunting et al 1982, Hare 1981). Negative swings away from the long-term norm cause production havoc in marginal situations, as was so drastically manifest with the recent Sahel drought. Total production must increase and yields must stabilize.

Table 1. Population projections (thousands) and growth rates (%) 1980-2025, for coastal West Africa and Central Africa countries (adapted from Davidson 1986).

Country	1980	Growth rate	1985	Growth rate	1990	Growth rate	2000	Growth rate	2025
Benin	3,464	3.1	4,043	3.3	4,757	3.1	6,472	2.7	11,562
Cameroon	8,701	3.2	10,191	3.5	12,128	3.2	16,645	2.8	30,081
Ghana	10,828	3.2	12,710	3.1	14,850	2.9	19,821	2.4	33,510
Guinea	5,488	2.0	6,049	2.1	6,703	2.2	8,318	2.0	13,197
Guinea Bissau	809	1.8	886	2.0	981	2.2	1,226	2.1	1,964
Ivory Coast	8,358	4.2	10,300	3.9	12,500	3.3	17,300	2.5	29,099
Liberia	1,871	3.2	2,196	3.3	2,590	3.1	3,542	2.8	6,426
Nigeria	84,732	3.3	99,669	3.4	118,254	3.2	163,481	2.8	297,936
Sierra Leone	3,358	2.2	3,745	2.3	4,205	2.4	5,360	2.3	8,891
Togo	2,578	3.3	3,038	3.4	3,605	3.2	4,961	2.8	9,072
Central African Republic	2,286	2.5	2,583	2.8	2,966	2.9	3,962	2.5	6,841
Congo	1,605	3.1	1,872	3.6	2,238	3.5	3,170	2.8	5,610
Equatorial Guinea	341	1.8	373	2.3	418	2.5	537	2.2	879
Gabon	755	1.8	827	2.6	941	2.8	1,242	2.4	2,126
Zaire	26,379	2.9	30,557	3.1	35,633	2.8	47,391	2.4	79,881
Burundi	4,114	2.6	4,696	2.9	5,431	3.0	7,367	2.7	13,236
Rwanda	5,139	3.2	6,026	3.4	7,151	3.6	10,239	3.1	20,098
Total	170,806		199,761		235,351		321,034		570,409

The humid tropics

The geographical mandate of the International Institute of Tropical Agriculture (IITA) includes all areas where rainfall equals or exceeds potential evapotranspiration at least 5 mo a year. We shall consider this analogous to the humid tropics (Fig. 1), about 28% of the global land area. Some definitions of the humid tropics are more restrictive, using 8 or more months in which rainfall exceeds evapotranspiration. That area covers about 10% of the world's land surface (Sanchez et al 1982).

Temperature

The humid tropics have moderately high and relatively stable temperatures, higher where it is drier and more variable in higher altitudes (Table 2). Soil temperatures of bare soils skyrocket at the beginning of the cropping season (Lawson et al 1979) and are deleterious to seed germination (Harrison-Murray and Lal 1979, Lal 1979). High nighttime temperatures increase respiration losses.

Solar radiation

Cloudiness moderates solar radiation in the humid tropics. Insolation improves with distance from the wetter areas (distance from the equator), particularly in the sub-Saharan region (Table 2).

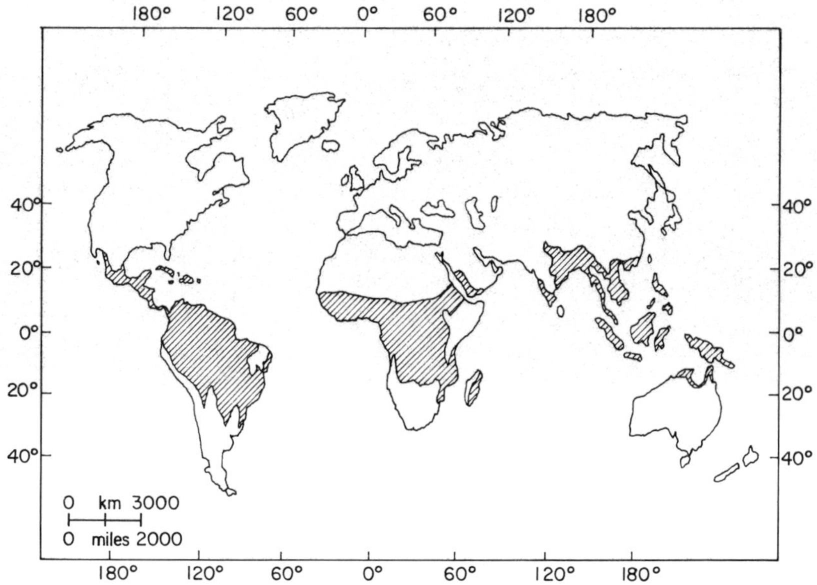

1. Areas of the tropics with five or more humid months (Source: Map, Seasonal climates of the earth, by C. Troll and K. H. Paften, *in* H. E. Landsberg et al 1963).

Table 2. Mean maximum and minimum temperatures and monthly global radiation at representative stations for selected months, West Africa.[a]

Station	Location		Max temp (°C)			Min temp (°C)			Global radiation (cal/cm² per d)				
	Latitude N	Longitude	Apr	Jul	Oct	Apr	Jul	Oct	Apr	Jul	Aug	Sep	Oct
Daru	8°00′	10°51′W	33.3	28.9	31.7	22.2	21.7	21.7	490	394	304	358	415
Ibadan	7°24′	3°53′E	32.8	27.8	30.0	22.8	21.1	22.2	434	384	335	367	401
Monrovia	6°18′	10°48′W	30.6	26.3	28.3	22.8	22.2	22.2	443	354	298	332	395
Port Harcourt	4°46′	7°01′E	31.7	28.9	29.4	22.8	21.7	21.7	397	314	294	322	357

[a]Data on temperatures are from Lebedes (1970), data on radiation are adapted from Table 1 and Lawson (1980).

Table 3. Incidence of extreme rainfall and resulting floods in Ibadan, Nigeria (7°23′N, 3°54′E), 1955-80 (Oguntala and Oguntoyinbo 1982).

Date	Rainfall		Resulting floods
	Amount per incidence (mm)	Annual total (mm)	
17 Jun 1955	166.1	1502.1	Disastrous
17 Aug 1960	86.6	1371.3	Mild
28 Aug 1963	128.5	1510.5	Disastrous
5 Aug 1973	108.2	1367.5	Disastrous
20 Apr 1978	112.7	1518.4	Disastrous
31 Aug 1980	274.1	1967.0	Most disastrous

Moisture

Available moisture determines crop growth in the tropics, and rainfall is the most relevant agricultural factor. The seasonal positive moisture balance ranges from 12 mo to only 5 mo, with rainfall variability increasing as the dry season lengthens. The irregularities stem primarily from the variable start and cessation of the rains (Lawson 1980) or from intraseasonal dry spells (Fig. 2).

Total water yield and the intensity of individual storms can be rather high, with total water yields of 150 mm per incidence and with rates of fall exceeding 100 mm/h (Charreau 1974, Lawson 1980) (Table 3). At Ibadan, Nigeria, the mean annual rainfall over the past 30 yr was 1,294 mm, but the range was 790-1,940 mm.

Soils

The apparent luxuriant growth that is so striking in tropical forests gives a deceptive impression of the predominantly fragile soils. These soils deteriorate rapidly when cleared and cultivated. Alfisols cover 15%; Ultisols, 20%; and Oxisols, 23% of the area (Lal 1985). Inceptisols occur across the entire zone and Aridisols in the drier areas.

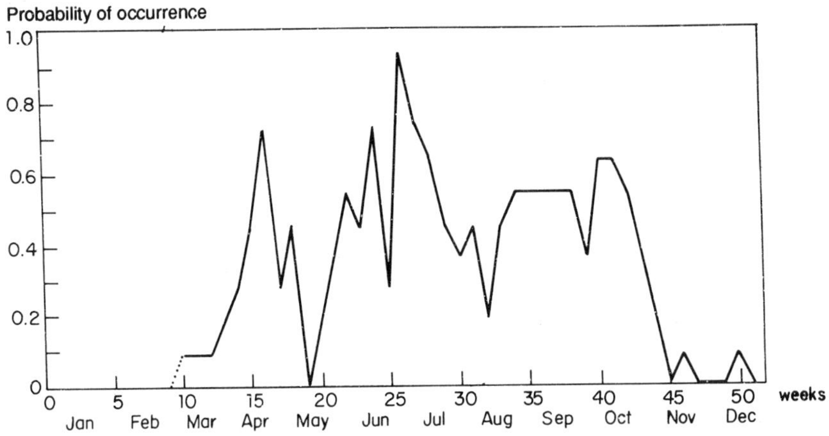

2. Probability of the occurrence of positive potential water balance in Ibadan (precipitation: pan evaporation (Base period: 1952-75) (Lawson 1980).

The soils generally have low effective cation exchange capacity (CEC) and low phosphorus and potassium reserves; are largely acidic, with or without Al toxicity; have rapidly declining or low organic matter when cultivated; and are prone to compaction (Cochrane and Sanchez 1982, Lal 1979). They are intrinsically not productive, or only marginally productive compared with most agricultural regions. The Oxisols and Alfisols that make up 75% of the Amazon basin soils are acid and infertile (Sanchez et al 1982). But unlike sub-Saharan and other forest areas, these South American soils have good physical properties, except for poor drainage in about one-fourth of the basin (Cochrane and Sanchez 1982). In West and Central Africa, gravel layers at relatively shallow depths severely limit moisture storage and root development in some areas.

Principal food crops

Natural selection, adaptation, and history have determined the crops now grown in the humid tropics. Only a few crops are dietary mainstays (Bunting et al 1982, Kawano and Jennings 1983). The primary crops are rice, maize, cassava, yam, sweet potato, cocoyam, and plantain, supplemented with field beans, cowpeas, peanuts, soybeans, and an assortment of vegetables and fruits.

Root crops are by far the best adapted to the humid lowlands. They are subject to fewer diseases and pests than cereals and grain legumes, are more efficient biologically, and are more tolerant of poor edaphic conditions (Bunting et al 1982, Coursey and Booth 1977, Tanaka 1983). High rainfall and humidity often adversely affect reproduction, ripening, drying, and storage, and increase pest and disease problems of cereals and grain legumes.

Food production systems

Shifting cultivation is the dominant production system in the humid tropics. Land is cleared, cropped a few seasons, then left fallow several years to restore its productive

Table 4. West and Central Africa: soil groups, locations, and rest period requirements (adapted from Goldman 1986).

Soil group	General characteristics	Location[a]	Rest period requirements (*R* value[b])		
			Low inputs	Intermediate inputs	High inputs
1. Oxisols and Ultisols [Ferralsols and Acrisols]	High acidity and aluminum; low nutrient availability; rapid leaching and organic matter loss. Moderate fertilizer use helpful but heavy use wasteful due to leaching.	Mainly in forests and transition zones. Constitute 50% of perhumid and 39% of humid areas.	15-20	35	65-70
2. Alfisols and some Inceptisols [Luvisols and Cambisols]	Moderate nutrient and organic matter content; good response to fertilization.	Mainly in moist and dry savanna. Constitute 5% of perhumid, 17% of humid, and 30% of subhumid areas.	30-40 (25 in forest)	50-60	75-85
3. Vertisols and some Alfisols [Vertisols and Nitosols]	Good nutrient and organic matter levels. Vertisols have physical problems both when wet and dry. Alfisols have erosion hazard in humid areas.	Mainly in savanna areas. Constitute 10% of humid, 11% of subhumid, and 32% of dry subhumid areas.	50	75	90
4. Entisols and Inceptisols [Fluvisols and Gleysols]	Good fertility. Occur on level land with high water table. Fallowing mainly for weed and pest control.	Throughout zones. Constitute 17% of perhumid, 9% of humid, and 12% of subhumid areas.	60-70	80	90

Sources: Young and Wright (1980); Le Houerou and Popov (1981).
[a]Soil distribution percentages from Le Houerou & Popov. Using their ecological terms, "perhumid" areas include rainforest; "humid" includes derived and southern Guinea savanna; "subhumid" includes northern Guinea and southern Sudan savanna; "dry subhumid" include northern Sudan savanna in West Africa. [b]The *R* value is the percentage of years of cultivation-fallow rotation.

capacity (Greenland and Okigbo 1983, Nicholaides et al 1985). Table 4 shows rest periods for typical soil groups in West and Central Africa's humid tropics. Productivity in terms of labor is low, but productivity in terms of total energy required per unit of food produced is favorable.

Cropping systems vary in intensity and complexity. Stability, a major consideration on small farms, dictates that different species or varieties, with different biophysical requirements, be used. Permanent cultivation is practiced in limited areas, on household compound plots where household refuse ensures a supply of organic matter to maintain productivity (Okigbo 1983).

The climate factor in crop production

Climate affects crop production by influencing physiological processes and by indirectly influencing soil characteristics, pest and disease incidence, and the timing of farm activities. Because man is unable to change or substantially modify climate, it has received little attention in agricultural production beyond defining the choices of particular species and varieties for given localities (Bunting et al 1982). Climate's role in agriculture has been seen as passive rather than active, with "normal" values used to characterize climate. Historically, this appears to have been appropriate, particularly during the relatively mild, stable 1940s and 1950s (Oram 1982).

Climate, however, is not static. It is subject to fluctuations of different scales in time and space (Hare 1981). Those fluctuations can cause variations in crop yields that can be devastating, as occurred during the 1930s dust bowl in the United States and the famine associated with the Sahel drought in the late 1960s and 1970s. Sad as those events were, they forced attention to the role of climate in agricultural production and the need to minimize the impact of climatic variations on yields through "climate-defensive" production practices. The success of such approaches depends on the nature and extent of the variability.

Extent of climate variability

Climate variations usually are variations in temperature or precipitation, or both (manifested by ice ages and wet and dry geological episodes). Such episodes may be brought about by changes in the primary forcing function, incident radiation. The direct primary role of solar radiation in photosynthesis is important to agriculture. Whatever mechanism triggers fluctuations in weather and climate, we should recognize the following basic categories in considering the implications to crop production:

- Long-term wide-scale variations that last for centuries.
- Shorter term cyclical changes that last for several years or decades. These fluctuations within a basically unchanged regime have a more confined geographical impact.
- Year-to-year and within-year variabilities more limited in area.
- Sudden occurrences, such as unusual storms, destructive winds, and flash floods, more "local" in nature.

The importance of the first two major classes is that they determine crop species distribution and the genesis or evolution of general soil classes. The fragility of tropical soils (particularly in the humid tropics) and the management problems they pose derive in part from major climatic changes in the past. The year-to-year, intraseasonal variations and the sudden occurrences of unusual weather more directly and immediately influence crop production. Fortunately, they are also more amenable to designs and practices that reduce their impact.

To provide yield stability, we need to evaluate existing technologies and practices and develop improvements that will significantly reduce fluctuations in crop production.

Variation in crop yields

Variations in crop yields in response to fluctuations in climate differ in space and time, but responses to fluctuations are by no means linear (Hare 1981). The responses tend to be more pronounced in the transition zones, where the variations may indicate changes from one climate regime to another (Lawson 1978). Figure 3 shows variations in rice yields in four West African countries. The drop in production in 1972-73, which coincided with the drought, is clearly more pronounced in the drier areas (Senegal, Niger) than in the wetter coastal region (Sierra Leone, Ivory Coast) (Lawson 1981). Figure 3 also reflects spatial variation in the anomalies, as do changes in average yields of five principal crops of the moist tropical region between 1974-76 and 1982-84 (Table 5). The improvement in cereal crops, particularly rice, probably reflects the improved moisture regime after a 1974-77 drought in many regions (Abel et al 1981). Responses of crops differ, which indicates the possibility of spreading risk and obtaining stability in overall yields.

Although the variation in yields appears to be ascribable primarily to the fluctuations in rainfall or moisture supply, changes in temperature and solar radiation also contributed, as did extreme weather on the local scale. The complexity of the relationship between weather and climate and crop yields (Monteith and Scott 1982) underlines the difficulties involved in making production technologies broadly weatherproof. Individual aspects of the environment can, however, be readily studied and the results used to enhance yield stability.

Increasing and sustaining crop yields

Research on food crop production, particularly in recent years, has helped us to understand the moist tropical environment and to develop technologies to help

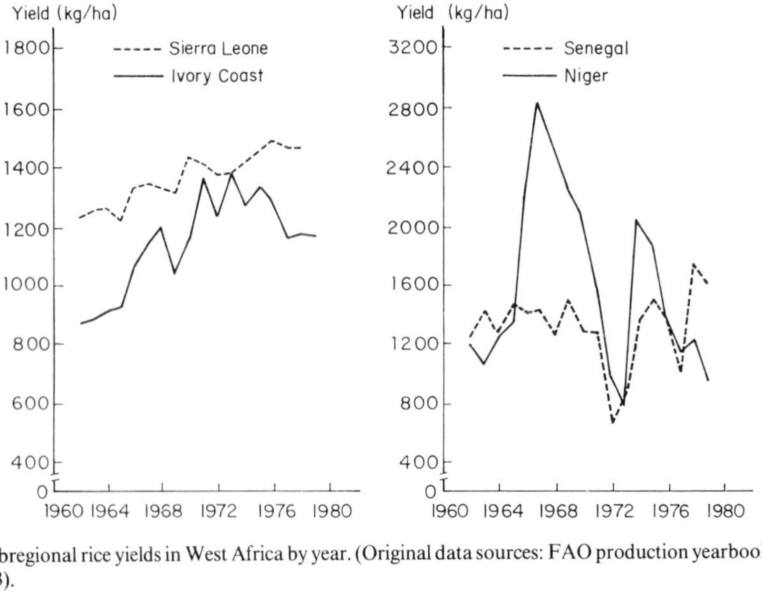

3. Subregional rice yields in West Africa by year. (Original data sources: FAO production yearbook, Vols. 21-33).

Table 5. Change (%) in average yields between 1974-76 and 1982-84 (adapted from FAO Annual Reports).

	Rice	Maize	Cassava	Yam	Sweet potato
World	+25.6	+22.8	+3.0	+2.6	+8.7
Africa	−0.3	−3.3	+0.7	+3.0	+6.1
North and Central America	+13.5	+15.1	−1.6	−13.9	−4.3
South America	+27.4	+34.6	−6.4	−3.5	−8.3
Oceania	+15.8	+9.1	−5.3	−2.4	+1.3
Far East	+25.3	+19.0	+21	−20.1	−4.5

overcome limitations induced or enhanced by bringing land under cultivation: soil erosion and general physiochemical degradation of the soil, acidification, and loss of organic matter (Kang and Juo 1982, Lal and Greenland 1979, Nicholaides et al 1985). To increase and sustain yields, land must be managed with minimum disturbance of the soil at clearing and postclearing, so that the soil structure and other physiohydrological properties are not disturbed in ways that cause erosion and loss of organic matter (Cochrane and Sanchez 1982, Lal and Greenland 1979, Lal and Russel 1982, Oram 1982).

Relatively high infiltration rates in forest soils initially high in stored moisture have been maintained by minimally disturbing the soil, using a combination of manual clearing and a crawler tractor/shear blade and no-till crop management after clearing (Lal 1985, Nicholaides et al 1985).

Infiltration rates and capacity to store moisture in the soil can be significantly improved by using cover crops such as *Mucuna utilis* and other in situ mulches. In all cases, the preserved or improved moisture storage capacity of the soil provides a buffer against the adverse effects on crops of short-term rainfall fluctuations (Fig. 1), and thus enhances yield stability (Cooper 1982; IITA 1982, 1983).

Crop residue and in situ mulches also improve soil nutrient status and productivity (IITA 1982, 1983). They can play important roles in sustaining crop yields on a range of soils in the humid tropics (IITA 1982, Lal 1979, Lawson and Lal 1979).

Alley cropping also effectively conserves and improves soils and increases land productivity (IITA 1982, 1983, 1984). Using Leucaena prunings from alley hedges as mulch substantially improved maize yields, in response to both the nutrient contribution of the Leucaena and the improvement of soil moisture (Kang et al 1981, 1984).

Effective soil moisture management is the most effective and direct means of enhancing crop yield stability in the humid tropics, where rainfall variability is the main factor in year-to-year variability. Where a significant portion of the productive land is irrigated (an estimated 30% in Asia), total crop production fluctuates much less than in West and Central Africa, where there is little or no irrigation.

Using appropriate genetic resources also can significantly improve yield stability (Slater 1981). High pest and disease incidences make choosing appropriate resistant varieties even more important in the humid tropics than elsewhere (Lawson and Terry 1984).

Mixed cropping systems designed to optimize use of sunlight, moisture, and soil nutrients also may reduce the spread and severity of certain pests. Their inherent yield stability affords food security to subsistence farmers and provides maximum returns from relatively small farms. An additional degree of stability can be achieved by cropping systems that involve agroforestry or harvestable trees, because they show less response to weather and climate fluctuations.

Use of agroclimatological information

Forecasting does not yet provide a basis for a farmer's day-to-day decisions during the cropping season. However, meteorological analyses that predict the probability of the occurrence of events, such as the onset or cessation of the rainy season, can be useful in selecting crops and in land preparation and seeding. Matching crop attributes to the environment allows the selection of appropriate varieties and practices for local areas, substantially reducing the vulnerability of marginal situations where yield fluctuations are most pronounced.

Basing the design of cropping systems and the choice of crop combinations on data about a crop's microclimate and the physiological attributes of species/varieties will help optimize production. Similarly, basic information on weather and pest or disease interaction should make possible an integrated approach to pest and disease control that affords better crop protection.

Estimating yields after specific weather incidents during the cropping season may provide adequate lead time for planning storage, distribution, and marketing policies.

Other factors

Many other factors related to institutional policy, socioeconomic situations, credit, and infrastructure obviously play roles, directly or indirectly, in determining food production stability. These also are sometimes subject to the vagaries of weather. Roads cut off by streams or floods during the cropping season may prevent the movement of certain farm inputs, just as badly eroded roads may impede or prevent movement of produce at the end of the season.

Summary and conclusions

Food demand in the humid tropics, as elsewhere in the world, is relatively stable, with a predictable increase as populations increase. Food supplies, however, fluctuate with weather and climate changes. Recent research in land management, cropping systems, and crop species or varieties provides a basis of hope for stable yields. Additional broadbased, interdisciplinary, and sustained research on the complex interrelationships involved and more accurate forecasting capabilities will contribute greatly to ensuring more stable crop yields.

References cited

Abel M E, Schnittker J A, Brown D C (1981) Climate-defensive policies to assure food supplies. Pages 45-60 *in* Climate's impact on food supplies. Strategies and technologies for climate-defensive food production. AAAS selected symposium 62. L. E. Slater and S. K. Levin, eds. Westview Press, Boulder, Colorado.

Barker R, Dorosh P (1986) The changing agricultural economy of West and Central Africa; implications for IITA. Paper prepared for the IITA Program Strategy Planning Review, International Institute of Tropical Agriculture, Ibadan, Nigeria.

Bunting A H, Dennett M D, Elston J, Speed C B (1982) Climate and crop distribution. Pages 43-74 *in* Food, nutrition and climate. Sir K. Blaxter and L. Fowden, eds. Applied Science Publishers, London.

Charreau C (1974) Soils of the tropical dry and dry-wet climatic areas and their use and management. Cornell University, Ithaca, New York.

Cochrane T T, Sanchez P A (1982) Land resources, soils, and their management in the Amazon Region: a state of knowledge report. Pages 137-209 *in* Amazonia agriculture and land use research. S. B. Hecht, ed. Centro Internacional de Agricultura Tropical, Cali, Colombia.

Cooper J P (1982) Plant breeding for different climates. Pages 79-99 *in* Food, nutrition and climate. Sir K. Blaxter and L. Fowden, eds. Applied Science Publishers, London.

Coursey D G, Booth R H (1977) Root and tuber crops. Pages 75-96 *in* Food crops of the lowland tropics. C. L. A. Leakey and J. B. Wills, eds. Oxford University Press, London.

Davidson R K (1986) Population growth and agriculture in the IITA primary focus countries. Paper prepared for the IITA Program Strategy Planning Review, International Institute of Tropical Agriculture, Ibadan, Nigeria.

FAO—Food and Agriculture Organization of the United Nations (1968-80) FAO production yearbook. Vol. 21-33. Rome, Italy. (various issues)

FAO—Food and Agriculture Organization of the United Nations (1974-84) Annual report. Rome, Italy. (various issues)

Goldman A (1986) Agroecological characterization and its relation to research issues in West and Central Africa. Paper prepared for the IITA Program Strategy Planning Review, International Institute of Tropical Agriculture, Ibadan, Nigeria.

Greenland D J, Okigbo B N (1983) Crop production under shifting cultivation and maintenance of soil fertility. Pages 505-524 *in* Potential productivity of field crops under different environments. International Rice Research Institute, P.O. Box 933, Manila, Philippines.

Hare K F (1981) Climate's impact on food supplies: Can it be identified? Pages 9-21 *in* Climate's impact on food supplies. Strategies and technologies for climate-defensive food production. AAAS selected symposium 62. L. E. Slater and S. K. Levin, eds. Westview Press, Boulder, Colorado.

Harrison-Murray, Lal R (1979) High soil temperature and the response of maize to mulching in the lowland humid tropics. Pages 285-304 *in* Soil physical properties and crop production in the tropics. R. Lal and D. J. Greenland, eds. John Wiley and Sons, Chichester, Great Britain.

IITA—International Institute of Tropical Agriculture (1982) Annual report for 1981. Ibadan, Nigeria.

IITA—International Institute of Tropical Agriculture (1983) Annual report for 1982. Ibadan, Nigeria.

IITA—International Institute of Tropical Agriculture (1984) Annual report for 1983. Ibadan, Nigeria.

Kang B T, Juo A S R (1982) Effect of forest clearing on soil chemical properties and crop performance. Paper presented at the Conference on Land Clearing and Development in the Tropics, 23-26 Nov 1982, International Institute of Tropical Agriculture, Ibadan, Nigeria.

Kang B T, Wilson G F, Lawson T L (1984) Alley cropping. International Institute of Tropical Agriculture, Ibadan, Nigeria. 22 p.

Kang B T, Wilson G F, Sipkens L (1981) Alley cropping maize (*Zea mays* L.) and Leucaena (*Leucaena leucocephala* Lam) in southern Nigeria. Plant Soil 63:165-179.

Kawano K, Jennings P R (1983) Tropical crop breeding, achievements and challenges. Pages 81-99 *in* Potential productivity of field crops under different environments. International Rice Research Institute, P.O. Box 933, Manila, Philippines.

Lal R (1979) The role of physical properties in maintaining productivity of soils in the tropics. Pages 3-5 *in* Soil physical properties and crop production in the tropics. R. Lal and D. J. Greenland, eds. John Wiley and Sons, Chichester, Great Britain.

Lal R (1985) Upland production systems: achievements, priorities and future strategies. A program document. International Institute of Tropical Agriculture, Ibadan, Nigeria.

Lal R, Greenland D J, eds. (1979) Soil physical properties and crop production in the tropics. John Wiley and Sons, Chichester, Great Britain. 551 p.

Lal R, Russel E W, eds. (1982) Tropical agricultural hydrology watershed management and land use. John Wiley and Sons, Chichester, Great Britain. 428 p.

Lawson T L (1978) Effects of agroclimatological changes in food production in Africa. Paper presented at the Association for the Advancement of Agricultural Sciences in Africa 3rd General Conference and 10th Anniversary, 9-15 Apr 1978, Ibadan, Nigeria.

Lawson T L (1980) Agroclimatic constraints to dryland rice production in West Africa. Pages 37-45 *in* Proceedings of a symposium on the agrometeorology of the rice crop. International Rice Research Institute, Los Baños, Philippines.

Lawson T L (1981) Climatic changes in the sub-Saharan region and their effects on cereal production with special emphasis on rice. Paper presented at the WARDA Annual Rice Review Meeting, 18-23 May 1981, Monrovia, Liberia.

Lawson T L, Lal R (1979) Response of maize (*Zea mays* L.) to surface and buried straw mulch on a tropical Alfisol. Pages 63-74 *in* Soil tillage and crop production. R. Lal, ed. International Institute of Tropical Agriculture, Ibadan, Nigeria.

Lawson T L, Oguntoyinbo J S, Ojo O (1979) Agroclimatic conditions of West Africa. Paper presented at the IITA Annual Research Conference on Soil and Climatic Resources and Constraints in Relation to Food Crop Production in West Africa, 15-19 Oct 1979, International Institute of Tropical Agriculture, Ibadan, Nigeria.

Lawson T L, Terry E R (1984) Weather and plant diseases in Africa. Pages 314-320 *in* Advancing agricultural production in Africa. D. L. Hawkworth, ed. Commonwealth Agricultural Bureaux, Slough, U.K.

Lebedev A N, ed. (1970) The climate of Africa. Part 1. Air temperature, precipitation (Leningrad 1968). IPST, Jerusalem.

Le Houerou H N, Popov G F (1981) An eco-climatic classification of intertropical Africa. Food and Agriculture Organization, Rome.

Monteith J L, Scott R K (1982) Weather and yield variation of crops. Pages 127-149 *in* Food, nutrition and climate. Sir K. Blaxter and L. Fowden, eds. Applied Science Publishers, London.

Nicholaides J J, Bandy D E, Sanchez P A, Benites J R, Villachica J H, Coutu A J, Valverde C S (1985) Agricultural alternatives for the Amazon Basin. Bioscience 35(5):279-285.

Oguntala A B, Oguntoyinbo J S (1982) Urban flooding in Ibadan: a diagnosis of the problem. Urban Ecology 7:39-46.

Okigbo B N (1983) Cropping systems and related research in Africa. Association for the Advancement of Agricultural Sciences in Africa (AAASA). Ogunsanya Press Publishers, Ibadan, Nigeria. 81 p.

Oram P A (1982) The economic cost of climate variation. Pages 355-392 *in* Food, nutrition and climate. Sir K. Blaxter and L. Fowden, eds. Applied Science Publishers, London.

Sanchez P A, Bandy D E, Villachica J A, Nicholaides J J (1982) Amazon Basin soils: management for continuous crop production. Science 216:821-827.

Slater L E (1981) Three approaches to reducing climate's impact on food supplies. Pages 99-130 *in* Climate's impact on food supplies strategies and technologies for climate-defensive food production. AAAS selected symposium 62. L. E. Slater and S. K. Levin, eds. Westview Press, Boulder, Colorado, USA.

Tanaka A (1983) Physiological aspects of productivity in field crop. Pages 61-80 *in* Potential productivity of field crops under different environments. International Rice Research Institute, P.O. Box 933, Manila, Philippines.

Young A, Wright A C S (1980) Rest period requirements of tropical and subtropical soils under annual crops. *In* Report on the second FAO/UNFPA consultation on land resources for populations of the future, Food and Agriculture Organization, Rome.

Notes

Authors' address: J. W. Pendleton and T. L. Lawson, International Institute of Tropical Agriculture, Ibadan, Nigeria.
Citation information: International Rice Research Institute (1989) Climate and food security. P.O. Box 933, Manila, Philippines.

Climatic variability and food production in West Asia and North Africa

P. J. M. Cooper, H. C. Harris, and W. Goebel

The major climatic factors that affect crop yield stability in North Africa and West Asia are interseasonal and intraseasonal rainfall variability and temperature extremes. These are discussed in the context of crop yield variability and systems sustainability. Crop yields are highly variable, but there are prospects for increasing food production within the climatic constraints.

Rainfed agricultural production occupies more than 80% of the total cropped area in West Asia and North Africa (Table 1). Rainfed farming systems in areas receiving 200-600 mm expected rainfall/yr occupy about 125 million ha of land (Kassam 1981).

Crop production occurs against a background of limited and often chronically deficient rainfall, characterized by substantial temporal and spatial variability. Rainfall occurs largely during the cool or cold winter months of November-March. Most cereal and pulse crops are sown at the onset of this period. During the winter months, rainfall exceeds crop evapotranspiration and moisture is stored in the soil profile. As temperatures rise during March and April, crop canopies increase, evapotranspiration exceeds rainfall, and crops become increasingly dependent on stored soil moisture. Rainfall usually ceases by May, and crops almost inevitably mature under increasing drought stress. The length of the growing season is dependent on the amount of water available. The moisture storage capacity of the soil is determined by soil texture and profile depth.

In general, rainfall is greatest near the coast, decreasing rapidly as one moves inland. Eventually, rainfed crop production systems give way to large expanses of arid grazing land or desert. Overlaid on this general pattern is a decrease in rainfall from west to east in North Africa and from the high plateau areas in the north to south in West Asia.

The principal crops in the region are cereals, mainly wheat in the areas with higher potential and barley in the drier and harsher environments, with small areas of rye, oats, and triticale in some countries. Barley is grown as livestock feed (both grain and straw), and is also a significant food crop in Morocco and Tunisia. Pulses, mainly faba bean, chickpea, and lentils, also are grown. Although they occupy only 5-10% of the area planted to cereals (Table 1), they meet most of the local demand.

Faba beans commonly are irrigated or limited to the wettest areas. Chickpea and lentils are grown in areas of medium rainfall (300-600 mm/yr).

Other rainfed crops are primarily summer crops grown in the higher-rainfall areas (>350 mm) on soils with high moisture storage capacity. They are planted in the spring after a winter fallow, to grow on stored soil moisture during the summer. Such crops include watermelon, cantaloupe, sesame, cotton, sunflower, sorghum, and maize. Permanent crops include citrus, olives, vineyards, pomegranates, and nuts. In areas of higher potential rainfall, tree crops are a major source of income for many farmers.

Close integration of livestock in the farming systems plays a key role in determining the current crop production strategy of the majority of farmers (Tully 1986). Sheep, goats, and cattle are the most important (Table 2). In general, most of the sheep and goats are found in the steppe or drier crop production areas, cattle are

Table 1. Land use (million ha) in West Asia and North Africa, 1981-83 (FAO 1984).

Country	Area (million ha)							
	Arable crops	Permanent crops	Total crops	Irrigated crops	Rainfed crops	Wheat	Barley	Pulses
Algeria	6,827	642	7,469	310	7,159	1,741	880	122
Libya	1,738	315	2,053	200	1,853	266	264	9
Morocco	7,269	430	7,699	426	7,273	1,770	2,142	347
Tunisia	3,440	1,420	4,860	123	4,737	809	490	144
Iraq	5,100	187	5,287	1,572	3,715	1,187	958	46
Jordan	357	34	391	36	355	104	52	16
Lebanon	240	108	348	85	263	18	5	10
Syria	5,356	369	5,725	547	5,178	1,256	1,485	201
Cyprus	365	67	432	94	338	14	50	6
Turkey	24,977	2,787	27,764	1,983	25,781	9,040	2,951	909

Table 2. Livestock production in West Asia and North Africa, 1981-83 (FAO 1984).

Country	Animals (thousand)			Meat (thousand t)			Milk (thousand t)	
	Sheep	Goat	Cattle	Sheep	Goat	Cattle	Sheep, Goat	Cattle
Algeria	13,730	2,763	1,389	68	13	43	316	530
Libya	4,598	1,412	175	53	2	29	54	64
Morocco	14,913	6,240	3,047	59	23	116	50	790
Tunisia	4,980	875	572	24	3	34	30	226
Iraq	11,883	3,767	3,033	43	15	55	234	1,075
Jordan	1,021	540	37	6	7	2	40	4
Lebanon	143	442	53	11	4	9	56	91
Syria	10,969	1,103	799	88	7	34	539	599
Cyprus	518	360	43	6	4	4	64	51
Turkey	49,288	18,727	17,003	294	108	211	1,865	3,613

more common in the wetter or irrigated areas and in dairies near cities. As a major source of income, particularly on the smaller farms (Campbell et al 1977, Papachristodoulou 1979), they act as a financial buffer against yearly fluctuations in crop production.

Throughout the region, integrated crop - livestock farming systems have evolved that are adapted to the prevailing climatic conditions. The current distribution of crops and cropping systems largely reflects the long-term average rainfall expected by farmers. Nevertheless, in addition to the marked spatial variability in rainfall, which determines crop distribution, any given location will experience substantial within-season and between-season variability in rainfall. This temporal variability is responsible for the very marked fluctuations in crop productivity at the farm, national, and regional levels.

Wheat forms the basic diet in West Asia and North Africa and is the most widespread crop. Barley is the second most important crop. Although it is used almost exclusively as an animal feed, it is a vital constituent of livestock feeding and as such is directly linked to meat and dairy produce supply.

Figure 1 shows the national average grain yields of wheat and barley from 1966 to 1984 for Morocco, Tunisia, Jordan, and Syria. These countries have high ratios of rainfed to irrigated agriculture. Large areas of irrigated wheat and barley tend to mask the impact of climate on yield variability. Important factors apart from climate also may contribute to the substantial variability illustrated. The relative importance of rainfall can be illustrated using the data from Syria. Seasonal total rainfall for 1966-84 in four sites representing wheat-growing areas (Aleppo, Jindiress, Idlib, and Kamishly) and four representing barley-growing areas (Hama, Hassakeh, Sweida, and Breda) were averaged to provide an estimate of crop water supply for each crop each year. Simple regression analyses of mean national yield (Y kg/ha) on total seasonal rainfall (P, mm) gave the following relationships:

$$\text{Wheat } Y = 3.215P - 278 \quad (R^2 = 0.69)$$

$$\text{Barley } Y = 3.575P - 471 \quad (R^2 = 0.70)$$

This rather crude consideration of a single climatic factor, total seasonal rainfall, accounts for about 70% of the variability in national average grain yields of wheat and barley in Syria.

Climatic variability: rainfall

The season-to-season variability in rainfall, which is a feature of Mediterranean-type environments, can be shown in several parameters of rainfall.

Total seasonal rainfall

Annual rainfall is not a meaningful agricultural statistic in the Mediterranean region, where rainfall, and hence the crop growth season, spans the turn of the year. Most statistics are reported on a seasonal basis. The length of the season varies with location, but begins in September-November and ends in April-June. Rain during July and August is rare, except in some coastal areas. Any rain that does fall is

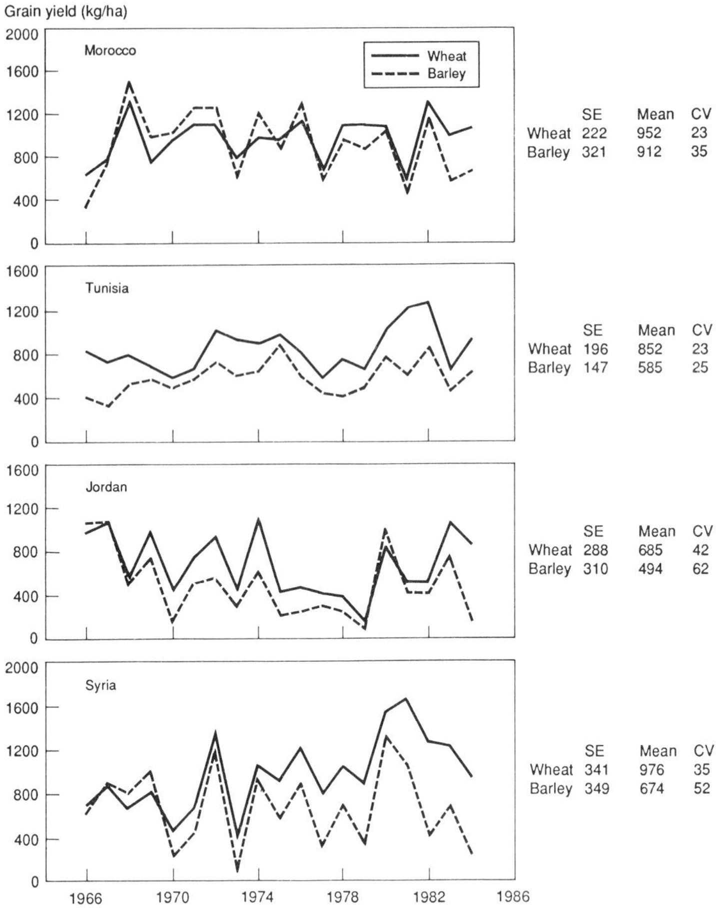

1. National average wheat and barley grain yields (kg/ha) for 4 countries in North Africa, 1966-84.

ineffective under the prevailing high evaporative conditions. Northern Syria is generally recognized as one of the more favorable agricultural areas (El-Sherbini 1985). Data for three sites illustrate the season-to-season differences, although no trends are discernible in these short records (Fig. 2).

Spatial variability in seasonal precipitation across the region deserves study. Macroscale effects are consistent, but what is less clear is whether regional trends in any one season are common to the whole area. De Brichambaut and Wallen (1963) commented on the homogeneity of rain-producing events in West Asia, but although crop yields for Syria and Jordan show some similarity, differences are also evident (Fig. 1). We echo the call by El-Sherbini (1985), for work to quantify rainfall variability to improve understanding of regional food production instability. For a center such as the International Centre for Agricultural Research in the Dry Areas (ICARDA), and for national agriculture research programs, such information is

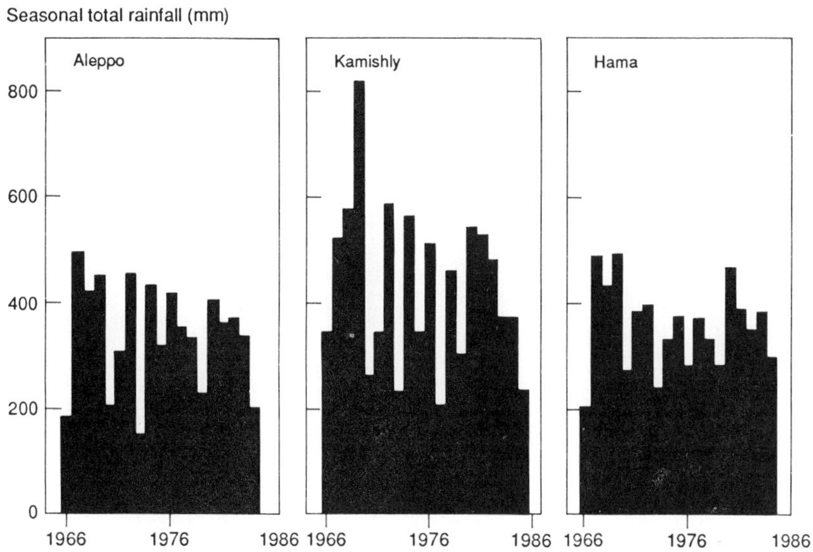

2. Total seasonal rainfall for 3 sites in northern Syria.

vital if trial results in any one season or short series of seasons are to be interpreted in terms of longer-term expectation. Such studies have been undertaken in some countries (e.g. Al-Hassani 1964, Serv. de la Meteorol. Nationale 1969), but a broader perspective is needed.

Start of the rains

Timeliness is important in crop production. Delaying the start of crop growth beyond an optimum germination date reduces yield potential; the reduction can be approximated from the length of the delay. Thus, variability in the time of the onset of rains from season to season is a significant factor in yield fluctuation. This variability has been studied in 6 sites in northwest Syria where average seasonal rainfall varies from 220 to 500 mm (ICARDA 1983). Although the condition imposed to identify the start of the season may be unduly stringent, there is clearly considerable variation between seasons and between sites (Fig. 3).

At any one site, an early start to the season may be associated with a higher seasonal rainfall (El-Sherbini 1979, Stewart 1986). However, these relationships appear somewhat tenuous; statistical significance is dependent on only a few points in the overall distribution. We suggest that much wider testing be done before this relationship is used as a seasonal predictor.

Length of rainy season

No correlation was found between the start and the end of rainfall for individual sites in Syria (Smith and Harris 1981; H. C. Harris, unpubl. data). However, some clear macroscale trends follow the trends in mean seasonal precipitation. Higher seasonal

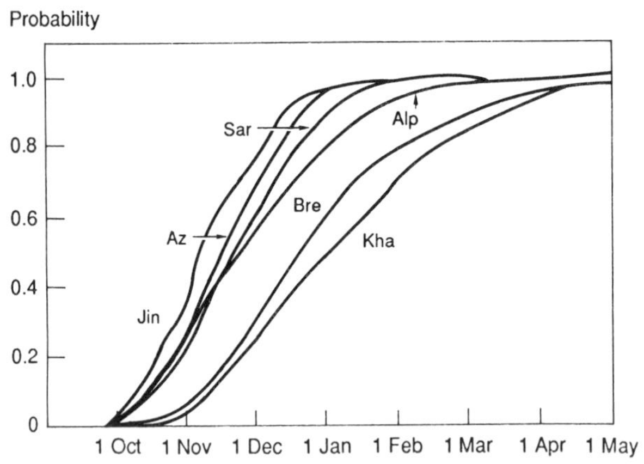

3. Probability of receiving germinating rains (20 mm in 3 d) at 6 sites in northern Syria. Jin = Jindiress, Az = Azaz, Sar = Saraqeb, Alp = Aleppo, Bre = Breda, Kha = Khanasser.

values are associated with longer duration of the rains, lower values with shorter duration. This has clear implications for research strategies, in particular for the development of crop genotypes that will maximize yields in any given environment.

Intraseasonal variability

Intraseasonal rainfall distribution can be as important as, or more important than, total precipitation in promoting crop growth and yield. Variability in rainfall distribution is illustrated by weekly data for six seasons at ICARDA's research station in Syria (Fig. 4). Characteristically, no two sets of data are alike. Crops are subjected to unique patterns of stress each season. We emphasize that if we are to understand crop-weather interactions, we need to quantify and interpret the effects of these variable stresses on crop growth and yield. Only through such understanding will we be able to identify and use the mechanisms by which plants adapt to such variations to develop more productive and stable cultivars.

Rainfall intensity

Rainfall intensity varies across the season, and it appears that there are regional differences in the manner in which it varies. In northern Syria, convective thunderstorms at the start (Oct-Nov) and end (Apr-May) of the season may yield high-intensity rain in local areas. Records from the Aleppo Airport for 1967-84 show that 5-min rainfall intensities of >75 mm/h are not uncommon (ICARDA 1986). The almost total lack of vegetative cover at the start of the season means that, without appropriate soil management, even short-duration rainfall of this intensity can cause runoff and soil erosion. Crop cover affords protection at the end of the season, but fallow land remains vulnerable.

In other areas, the highest intensities occur in midwinter (De Brichambaut and Wallen 1963). High-intensity rains in mountainous regions of Jordan, Lebanon, and Syria, and in the Zagros Mountains may lead to surface runoff and soil erosion. The

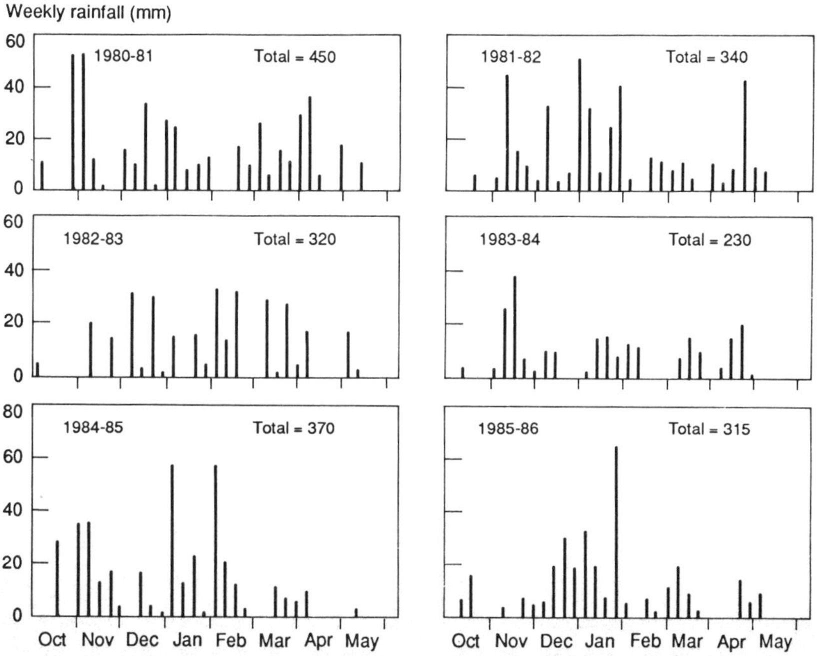

4. Weekly rainfall distribution during 6 successive seasons at Tel Hadya, Syria.

problem is not confined to West Asia. There are reports of erosion-producing rainfall events in North Africa (Floret and Pontanier 1982).

Throughout the region, average intensities for much of the season are <5 mm/d. All indicators (e.g. landscape and size and frequency of streams) support the belief that there has been little runoff throughout most of the region. Nonetheless, there is danger in complacency. With increasing pressure on land resources and encroaching cultivation on ever steeper slopes, risks of runoff and soil loss increase. We have seen evidence of current erosion in many places in the region. If food production is to be sustained, soil resources must be conserved. Problem areas should be identified and erosion control measures instituted now.

Rainfall and soil water

The water supply for plant development is buffered through the storage of water in the soil profile. Many soils of the region are clays, with good storage capacity per unit depth. But soil depth is variable.

The amount of rainfall stored is related to the amount that falls and its distribution. When, as is most frequent, rain per rainday is low, a significant proportion evaporates from the soil surface at a rate approaching that of evaporation from a free water surface. As rain per rainday exceeds the rate of evaporation, water is stored with increasing efficiency, provided runoff and drainage do not occur. On deep soils (>150 cm) where land is cropped annually, the soil is rarely wetted beyond this depth and drainage is not a problem. This also may be so on shallow soils.

The depth of profile recharge varies markedly with total rainfall and with its distribution. Profile recharge at Tel Hadya during the last 6 yr has ranged from 30 to 160 cm under the rainfall regimes shown in Figure 4. Maximum storage regularly occurs in late February or early March. After that, increasing crop water use as the canopy expands causes net depletion, and reserves are almost always exhausted at or before crop maturity.

During crop growth, stored water is either transpired or lost via evaporation from the soil surface. It is possible to increase the proportion transpired through improved crop and soil management (Cooper et al 1983). This leads to increased crop yields (in accordance with the well-recognized linear relationship between transpiration and yield [Fischer 1981]), through more efficient use of the finite quantity of water available (Cooper 1983). Similar improvements should be possible in other parts of the region as levels of agronomic management improve.

Methodology for analyzing rainfall data

Stern (1980a, 1986) makes the point that, as soon as rainfall data are summed over time, information, especially on variability, is lost. The need to summarize data in the past is understandable because of the volume of daily records. However, now that more and more records are computerized, use of daily data has become feasible.

Models designed for this purpose are available (e.g. Katz 1974, 1977; Stern 1980a). Examples using existing climate records show their usefulness in describing rainfall regimes in relation to agriculture (e.g. Garbutt et al 1980; Stern et al 1982a,b). Stern's model has been used to examine probabilities of events that affect the management of crop rotations in northern Syria (Keatinge et al 1985, 1986). Questions raised included the probabilities of rainfall during the season, probabilities of wet spells with specified rain amounts and duration, and probabilities of dry spells of prescribed lengths at particular stages of the crop cycle. These analyses suggest that, as long as some information on soils and their water-holding characteristics is available, models provide a useful first test of hypotheses relating to management strategies.

One further advantage in using models based on daily data is that only a relatively short record is required (Stern et al 1982b). They suggest that a 10-yr data record should be sufficient. However, our experience is that, when the probability of rain on any given day is low, 10 yr is not adequate and longer runs (20-50 yr, if available) are preferable.

Climatic variability: temperature

Crop productivity in the region is limited by winter temperatures and by spring temperature increases.

In West Asia and on the plateaus, winter temperatures restrict growth or, where permanent snow cover occurs, inhibit crop growth. Severe freeze damage is a problem of winter wheat in areas on the plateau where the winter snow cover is not

Table 3. Number of frosts[a], absolute minimum temperature, and date of first and last frosts during 6 cropping seasons at 4 sites in northwest Syria.

Site and mean annual rainfall		Cropping season					
		1980-81	1981-82	1982-83	1983-84	1984-85	1985-86
Tel Hadya, 330 mm	No. of frosts	23	39	52	25	44	17
	Abs. min temp (°C)	–4.0	–7.8	–9.8	–3.9	–9.5	–6.4
	First frost	24 Nov	5 Nov	27 Nov	25 Dec	30 Nov	2 Dec
	Last frost	6 Apr	28 Mar	11 Mar	20 Mar	12 Mar	3 Mar
Jindiress, 470 mm	No. of frosts	19	39	51	20	42	10
	Abs. min temp (°C)	–4.0	–7.0	–5.5	–2.3	–10.2	–5.0
	First frost	22 Nov	5 Nov	10 Nov	15 Dec	30 Nov	3 Dec
	Last frost	1 Apr	29 Mar	8 Mar	14 Mar	10 Mar	26 Jan
Breda, 280 mm	No. of frosts	37	40	62	32	42	34[b]
	Abs. min temp (°C)	–6.5	–8.0	–9.8	–5.0	–9.6	–8.0[b]
	First frost	23 Nov	5 Nov	4 Nov	14 Dec	23 Oct	2 Dec
	Last frost	6 Apr	6 Apr	18 Mar	16 Mar	10 Mar	
Khanasser, 225 mm	No. of frosts	47	47	66	31	62	27[a]
	Abs. min temp (°C)	–7.5	–7.0	–9.3	–7.0	–9.1	–5.5[a]
	First frost	22 Nov	5 Nov	1 Nov	14 Dec	19 Oct	2 Dec
	Last frost	6 Apr	30 Nov	1 Apr	16 Mar	12 Mar	3 Mar

[a]Frost = temperature of 0.0°C at 1.5 m in Stevenson Screen. [b]Incomplete.

permanent and thawing and freezing are accompanied by soil heaving. In lowland areas, except along the coasts, minimum night temperatures below about –8 °C during early growth cause severe leaf burn and, if they occur after stem elongation begins, death of tillers. Plants recover, but maturity is delayed and yield decreased (Stapper 1984). Milder frosts during late stem elongation and at heading cause unpredictable, but at times severe, damage. In North Africa, these problems are less acute and are confined predominantly to areas adjacent to the Atlas Mountains.

There is season-to-season variability in the severity and timing of frosts, illustrated by data from our main site at Tel Hadya and from three semipermanent substations in northwest Syria (Table 3). Clearly, more frosts occur in the drier areas. But, within a year, the severity is very similar across the area. This may also hold true for large parts of the region. There is evidence of synchronous and very widespread cold events in the historical records for West Asia. An event in 1984-85, when both maximum and minimum temperatures were very low for 20 d 21 Feb-12 Mar, affected both West Asia and North Africa. We think this was a once-in-50-yr event in northwest Syria.

Rapid temperature increases in spring (Fig. 5), accompanied by sharp increases in vapor pressure deficit of the atmosphere, occur at a time of maximum crop water requirement. The rate of water use rises sharply, soil water reserves are quickly exhausted, and crops nearly always suffer drought stress during grain filling. Occasional extremely high temperature, usually accompanied by strong winds, may entirely burn off crops. This appears to be a more severe problem in North Africa than in West Asia.

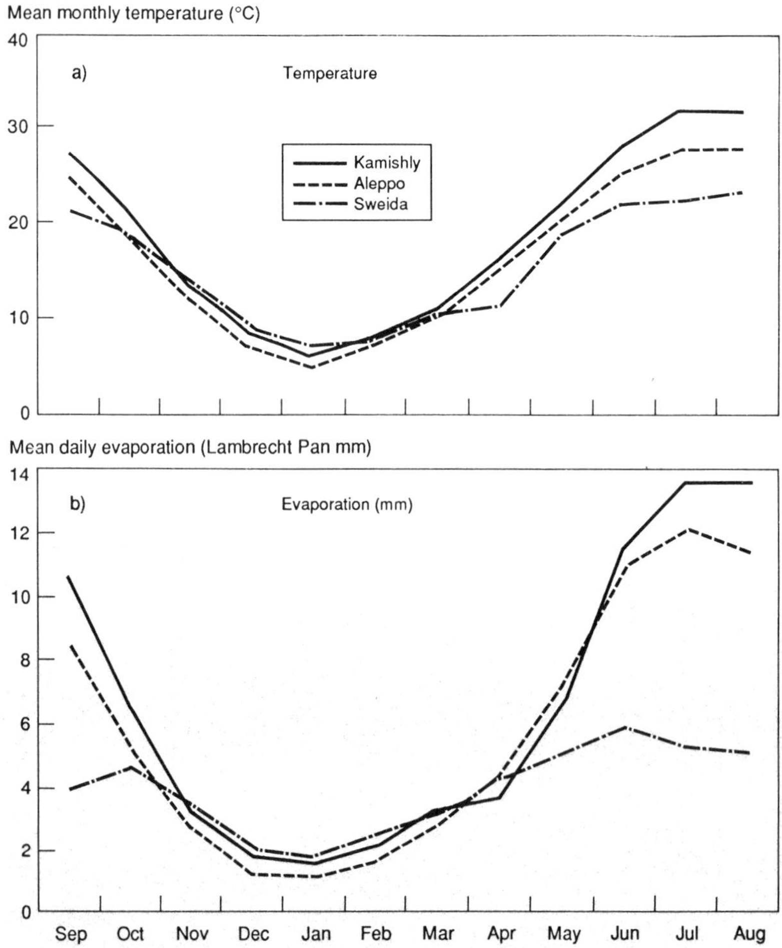

5. Mean monthly temperature and mean daily evaporation for 3 sites in Syria. Data supplied by Meteorological Department, Minsitry of Defense, Syrian Arab Republic.

Crop yields and climate

Figure 1 illustrates the instability of cereal yields in the Mediterranean region. The data also indicate that, with the exception of Syria, there is no apparent yield trend across time. This sharp contrast to recent trends in wheat yields in other parts of the world illustrates the difficulties faced by those who seek to improve production in these marginal areas.

In Syria, there is an apparent upward trend in wheat yields, especially in the last 10 yr. Some of this most likely results from increased use of supplementary irrigation for wheat production. Some undoubtedly is due to the successful introduction of improved agronomic practices by the Syrian Ministry of Agricultural and Agrarian Reform. The most important are fertilizer use and weed control (ICARDA 1985).

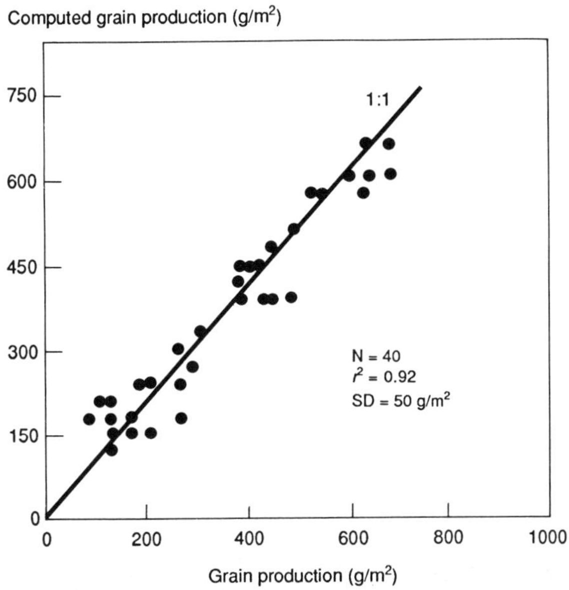

6. Actual vs computed grain yields for 26 wheat crops at 9 sites in Australia, Mexico, South Africa, and Syria. A regression line fitted to the points (Y = 27.9 + 0.92X) was not significantly different from the 1:1 line (Stapper 1984).

Supply policy has concentrated these improvements in the wetter areas (>350 mm seasonal rainfall), where wheat is grown in rotation with legumes and/or summer crops.

At the same time, yields of barley (grown principally but not exclusively in drier areas) have declined. A major factor in this decline has been an expansion of barley production into areas with <250 mm rainfall. That also may account for the greater yield fluctuations in barley than in wheat (Fig. 1). Prospects exist for a reversal of this yield trend (Jones and Harris, this volume).

Sources of variability

A wheat model (SIMTAG) has been used to explore sources of yield variability in relation to environmental factors for Syria (Stapper 1984). This model has been calibrated for semiarid regions against experimental data from diverse environments (Fig. 6). It uses a daily time step to predict growth, development, and yield from inputs of daily precipitation, maximum and minimum temperatures, and solar radiation. Assumptions are that management factors (nutrients, weed control, etc.) are nonlimiting; pests and diseases are not considered.

Grain yield variability

Grain yields, simulated for 4 sites in Syria using 25 yr of historical weather records, are shown in Figure 7 as the cumulative frequency of yield being less than the level specified. The effect of soil depth, through its influence on the water balance, also is

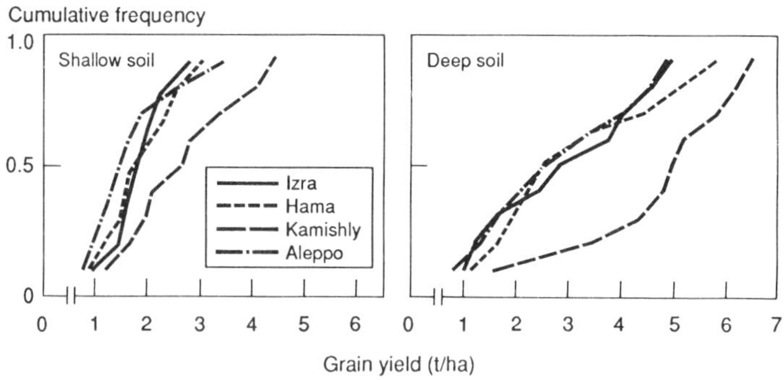

7. Cumulative frequency of simulated yields of a midseason wheat cultivar grown on shallow and deep soils at 4 sites in Syria, using 25 years of climate data.

considered. The shallow soil had a storage capacity of about 80 mm and the deep soil, about 180 mm. Three of the sites (Hama, Aleppo, and Kamishly) span a west-to-east transect across northern Syria; Izra is in the south. The cultivar simulated is a medium-duration type. Fields are assumed to be continuously cropped.

Median yields on the shallow soil ranged from 1.5 to 2.5 t/ha; on the deep soil, from 2 to 5 t/ha. These yield levels reflect not only the assumptions noted, but also the recognized yield gap between small experimental plots and large areas. Nevertheless, the yields generated are not unrealistic. In more favorable years, the best farmers, using appropriate management, achieve yields equal to or greater than these estimates.

The data indicate that 1 yr in 10, yield can be expected to be less than 1-1.5 t/ha at all sites. In the best years, yields should be 3-4 times higher. Kamishly clearly represents the most favorable environment, partly due to high seasonal rainfall throughout the season, with a high probability that rain will occur at flowering and during grain filling.

Figure 8 shows the cumulative frequency of the simulated length of growing season (i.e. the period from germination to physiological maturity for the same crops). This is governed largely by temperature, but is also strongly influenced by the date of the onset of rains that determines the germination date (Fig. 3). There is much less variability in maturity date than in germination date (Fig. 8), illustrating the effect of high spring temperatures and terminal drought which hasten development and cause premature senescence. Consistent differences of 10-15 d in length of growing season clearly suggest that the same cultivar may not be the most suitable for all sites.

Analyses of this type can be used to look at other factors not discussed here (e.g. appropriate crop maturity types [Stapper 1984], nitrogen application strategies [Godwin and Vlek 1985], and crop phasic development patterns). If we assume that historical weather records reflect current and future climate, models provide tools for quantifying the expected climatic risks. That is important to food security.

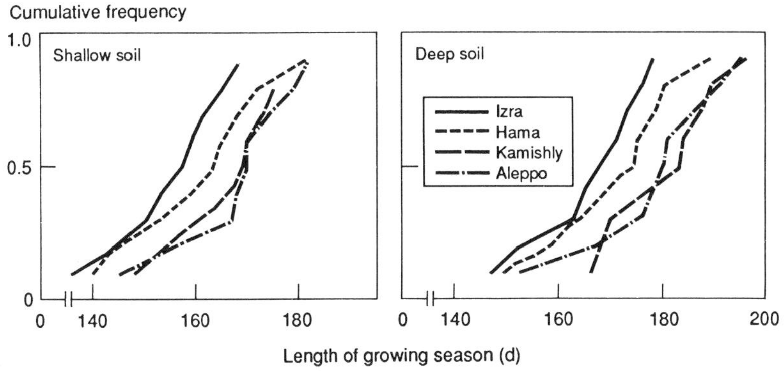

8. Cumulative probability of simulated length of growing season for crops grown on shallow and deep soil at 4 sites in Syria, using 25 years of climate data.

Future prospects

Notwithstanding the recent yield increases in Syria, mean yields there and elsewhere in the region are low (Fig. 1). Although there is room for increases, we must recognize that climatic factors dictate that low yields will always occur. We cannot expect that the yield gains achieved in other areas will be possible in the Mediterranean region. The physical principles involved, discussed in relation to the specific environments in more detail elsewhere (Cooper et al 1987, Gregory 1984), are summarized here.

Total biological yield ultimately will depend on the efficiency with which a crop can convert a finite water supply to dry matter. In this respect, the concept of water-use efficiency (WUE) is useful. Cooper (1983) has shown that

$$\text{WUE (kg/ha/mm)} = \frac{\text{TE}}{1 + \text{Es}/\text{T}} \tag{1}$$

where TE is the mean seasonal transpiration efficiency of the crop, Es is the amount of moisture lost as soil evaporation from under the crop canopy, and T is the amount of moisture transpired by the crop.

Consideration of the basic processes governing water vapor and carbon dioxide diffusion between the air of intercellular spaces within the leaf and the air surrounding the leaf indicates that

$$\text{TE} \ \alpha \ \frac{k}{e_s - e_a} \tag{2}$$

where k is a crop-specific constant of proportionality and ($e_s - e_a$) is the saturation vapor pressure deficit (SVPD) of the air, weighted for periods when crop transpiration is occurring. The relative importance of the terms TE and e_s/T in contributing to improved WUE was reviewed by Cooper et al (1987). They conclude that the TE of crops grown in the arid Mediterranean region will be lower than that of crops grown in more humid environments because of the higher SVPD experienced. Management practices that promote growth during the winter months when SVPD values are low were shown to result in larger seasonal mean TE values. Cooper et al (1987) discussed evidence for potential increases in k (see equation 2) through genetic manipulation. Nevertheless, they concluded that the greatest potential for improved WUE lies in reducing the ratio E_s/T through improved crop management.

Conclusion

Comparing our predictions of the current yield potential for wheat in the Mediterranean region with the yields now achieved by farmers using good agronomic practices has led us to conclude that there is considerable scope for increasing food production in the rainfed areas of the Mediterranean region. It remains a fact, however, that demand is also rising sharply in response to improved incomes and increasing populations (Khaldi 1984). Given the degree of climatic risk experienced and the probable limits in the extent to which yield can be increased, we are led to conclude that food security in this region is likely to remain a matter of concern into the foreseeable future.

References cited

Al-Hassani S (1964) Empirical rainfall probabilities in Iraq. Pages 3-7 *in* Studies of rainfall and temperature conditions in Iraq. N. J. Sultan, ed. Directorate of Civil Aviation, Ministry of Communications, Republic of Iraq, Baghdad.

Campbell R R, Follet R H, Howell H B, Riddle R, Stubbendieck J T, Hamway D G (1977) Applied agronomic research program for dryland research farming in 200-440 mm rainfall zone of Morocco. Publ. PN-AAP-32q. United States Agency for International Development, Washington, D.C.

Cooper P J M (1983) Crop management in rainfed agriculture with special reference to water use efficiency. Pages 19-35 *in* Proceedings of the 17th colloquium of the International Potash Institute, Rabat, Morocco, May 1983. International Potash Institute, Berne.

Cooper P J M, Gregory P J, Tully D, Harris H C (1987) Improving water use efficiency of annual crops in the rainfed farming systems of West Asia and North Africa. Exp. Agric. 23:113-158.

Cooper P J M, Keatinge J D H, Hughes G (1983) Crop evapotranspiration - a technique for calculation of its components by field measurements. Field Crop Res. 7:299-312.

De Brichambaut G P, Wallen C C (1963) A study of agroclimatology in semi-arid and arid zones of the Near East. World Meteorological Organization Tech. Note 56. Geneva.

El-Sherbini A A (1979) Food security issues in the Arab Near East. UN Economic Commission for West Asia, New York.

El-Sherbini A A (1985) Current state of agricultural development in North African and Middle-Eastern countries. Paper presented at the 1985 World Food Production Conference, 1-7 Sep 1985, Beijing, China.

FAO—Food and Agriculture Organization (1984) 1983 FAO production yearbook. Vol. 37. Rome.

Fischer R A (1981) Optimizing the use of water and nitrogen through the breeding of crops. Plant Soil 58:249-278.

Floret C, Pontanier R (1982) L'Aridite en Tunisie presaharienne. Climat, sol, vegetation et amenagement. Travaux et Documents de l'ORSTOM. Office de la Recherche Scientifique et Technique Outre-Mer, Paris. 544 p.

Garbutt D J, Stern R D, Dennett M D, Elston J (1980) A comparison of the rainfall climate of eleven places in West Africa using a two-part model for daily rainfall. Arch. Meteorol. Geophys. Bioklimatol. 29:137-155.

Godwin D C, Vlek P L G (1985) Simulation of nitrogen dynamics in wheat cropping systems. Pages 311-332 *in* Wheat growth and modelling. W. Day and R. K. Atkin, eds. Plenum Publishing Corporation, New York.

Gregory P J (1984) Water availability and crop growth in arid regions. Outlook Agric. 13:208-215.

ICARDA—International Center for Agricultural Research in the Dry Areas (1983) Annual report 1982. Aleppo, Syria.

ICARDA—International Center for Agricultural Research in the Dry Areas (1985) Annual report 1984. Aleppo, Syria.

ICARDA—International Center for Agricultural Research in the Dry Areas (1986) Annual report 1985. Aleppo, Syria.

Jones M J, Harris H C (1989) Climatic variability and agronomic management in Mediterranean barley-livestock farming systems. Pages 563-572 *in* Climate and food security. International Rice Research Institute, P.O. Box 933, Manila, Philippines.

Kassam A H (1981) Climate, soil and land resources in North Africa and West Asia. Plant Soil 58:1-29.

Katz R W (1974) Computing probabilities associated with the Markov chain model for precipitation. J. Appl. Meteorol. 13:953-954.

Katz R W (1977) Precipitation as a chain-dependent process. J. Appl. Meteorol. 16:671-675.

Keatinge J D H, Dennett M D, Rodgers J (1985) The influence of precipitation regime on the management of three-course crop rotations in northern Syria. J. Agric. Sci. (Camb.) 104:281-287.

Keatinge J D H, Dennett M D, Rodgers J (1986) The influence of precipitation regime on the crop management of dry areas in northern Syria. Field Crops Res. 13:239-249.

Khaldi N (1984) Evolving food gaps in the Middle East/North Africa: prospects and policy implications. Res. Rep. 47. International Food Policy Research Institute. 74 p.

Papachristodoulou S (1979) Socio-economic aspects of rainfed agriculture in Cyprus. Pages 327-337 *in* Proceedings of the FAO regional seminar on rainfed agriculture in the Near East and North Africa, Amman, Jordan, May 5-10, 1979. Food and Agriculture Organization, Rome.

Serv. de la Meteorol. Nationale (1969) Probabilite des pluies en Tunisie. Etudes Meteorologiques 1. Tunisie.

Smith R C G, Harris H C (1981) Environmental resources and restraints to agricultural production in a Mediterranean-type environment. Pages 31-57 *in* Soil water and nitrogen in Mediterranean-type environments. J. Monteith and C. Webb, eds. Martinus Nijhoff/Dr. W. Junk, The Hague.

Stapper M (1984) Simulations assessing the productivity of wheat maturity types in a Mediterranean climate. Ph D thesis, University of New England, Armidale, N.S.W., Australia.

Stern R D (1980) Analysis of daily rainfall at Samaru, Nigeria, using a simple two-part model. Arch. Meteorol. Geophys. Bioklimatol., Ser. B, 28:123-135.

Stern R D, Dennett M D, Dale I C (1982a) Analyzing daily rainfall measurements to give agronomically useful results. I. Direct methods. Exp. Agric. 18:223-236.

Stern R D, Dennett M D, Dale I C (1982b) Analyzing daily rainfall measurements to give agronomically useful results. II. A modelling approach. Exp. Agric. 18:237-253.

Stewart J I (1986) Formulation of response farming recommendations for variable rainfall in Jordan: an illustrative case study. Paper presented at a Workshop on Soil Water and Crop/Livestock Management Systems for Rainfed Agriculture in the Near East Region, 18-23 Jan 1986, Amman, Jordan.

Tully D (1986) Rainfed farming systems of the Near East region. Paper presented at a Workshop on Soil Water and Crop/Livestock Management Systems for Rainfed Agriculture in the Near East Region, 18-23 Jan 1986, Amman, Jordan.

Notes

Authors' address: P. J. M. Cooper, H. C. Harris, and W. Goebel, International Centre for Agricultural Research in the Dry Areas, Aleppo, Syria.

Citation information: International Rice Research Institute (1989) Climate and food security. P.O. Box 933, Manila, Philippines.

The relationship between fluctuations in food crop yields and meteorological conditions in China

Xin Naiquan and Cheng Yannian

Food crop yields in China increased by more than 200% during the last 30 yr. But the rate of growth in food production varied from year to year and from region to region. Yield fluctuations for the country as a whole and by province, city, and autonomous region were matched with meteorological events. The effect of weather and climate on yield was regional. The principal meteorological events that affected food crop yields—drought, flood and waterlogging, and low temperature—are discussed. As higher yield levels are achieved, the effects of weather and climate become more crucial. The importance of developing countermeasures is discussed.

China is a developing country with a large and increasing population. Agricultural production, especially food production, occupies a decisive and strategic position in the development of the national economy and overall social stability. Efforts to increase grain production and develop diversified agricultural practices are intense.

Since the founding of the People's Republic of China, food production has greatly increased. Total food output during the Sixth Five-Year Plan (1981-85) was 2.26 times that of 1952 (Table 1). Per capita food production increased by 26%, while the population increased by approximately 80% (Table 1). The arable land in China, only 7% of the world's total, has been able to feed 22% of the world's population. Basically, we have solved China's food problem and contributed to the world's agricultural production.

During the last 30 yr, yields of food crops increased on a quadratic curve, but the rate of increase varied from year to year and from region to region. Food production is influenced by social factors, the level of inputs, the scientific and technological level, and natural factors. When other factors are relatively stable, natural factors, especially weather and climate, become the most important influence on food production. Therefore, the relationship between yield fluctuations and meteorological factors is important in predicting and managing yields of food crops.

Methods

Yield data were analyzed in the context of climate. Changes in yield were considered in terms of two scales that could induce fluctuations: the long-term trend scale,

Table 1. Grain production in China from 1952 to 1985.

Year	Grain production (million t)	Index of increase (%)	Per capita output of grain (kg)
1952	163.9	100.0	288
1957	195.1	119.0	306
1965	194.6	118.7	272
1978	304.8	185.9	318
Av for 1981-85	370.6	226.1	363

which changed slowly across time, and the short-term fluctuation scale, which deviated from the time trend. Thus,

$$Y = Y_t + Y_w$$

where Y = yield potential output or per-unit-area yield
Y_t = the long-term yield trend, and
Y_w = the short-term yield fluctuation.

The influence of social forces was reflected by the long-term yield trend under normal conditions and the influence of other factors, especially weather and climate, were reflected in the short-term fluctuations.

We dealt with two scales, across the total time and by two periods (1949-58, 1962-81), using orthogonal polynomial analysis. The coefficient of variation (CV) was calculated by

$$C_v = \frac{1}{\overline{Y}} \sqrt{\frac{\sum_{i=1}^{n} (Y_i - \overline{Y})^2 - \sum_{i=1}^{n} (Y_{li} - \overline{Y}_l)^2}{n-1}}$$

where Y = the average yield,
Y_i = the yield of i years,
Y_{li} = the trend yield of i—years,
Y_l = the average trend yield, and
n = years.

The relationships between calculated results of yield fluctuations and weather and climate circumstances, and between primary meteorological events and areas affected by them were analyzed. In high-yield regions (e.g., the suburbs of Shanghai) yield was analyzed quantitatively.

Results

Fluctuations of food crop yields

Yield fluctuations by time. The yield data over time for the whole nation and for certain provinces, cities, or autonomous regions show both increasing yield and year-by-year fluctuations (Fig. 1).

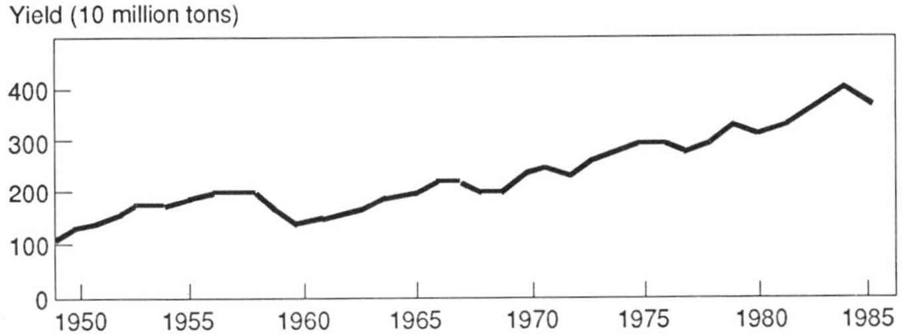

1. Time series of food crop yields in China, 1950-85 (total output is in million tons).

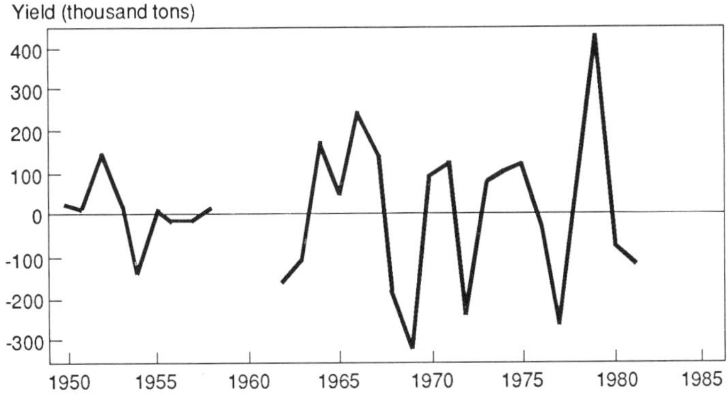

2. Fluctuation (thousand tons) of food crop yields in China, 1950-81.

Actual yield minus the yield trend gave the yield fluctuation. A positive value meant increased yield due to favorable weather and climate, a negative value signified decreased yields due to adverse weather and climate.

We added the negative values of each province, city, and autonomous region during a year to obtain the yield fluctuation for that year. In some years, yields in most provinces, cities, and autonomous regions increased, so that the national negative fluctuation was small, and was covered by the positive fluctuation (1951-53, 1955, 1965-67, 1970-71, 1973-75, 1978-79). In other years, yield decreases were widespread and the negative fluctuation was large. Total yield fluctuation was negative in periods when weather and climate were relatively abnormal and extreme meteorological events relatively frequent (1954, 1957, 1962-63, 1968-69, 1972, 1976-77, 1980-81). In about a third of the years between 1951 and 1985 (omitting 1959-61), negative yield fluctuations occurred.

Figure 2 shows yield fluctuations over the years. Meteorological conditions suitable or unsuitable to agricultural production appeared alternately, sometimes from one year to the next. Since the 1960s, the number of years with negative yield fluctuations increased, so that the total went up even as production was increasing. That indicates an increased threat to food production.

Regional distribution of yield fluctuations. Differences in yield fluctuations among different regions were large because of differences in production levels and climate.

To discount the effect of those differences, the coefficient of variation (CV) among yield fluctuations was calculated for each province, city, and autonomous region. The CVs of northeast, northwest, and north China were relatively high, and those of the Yangtze River Valley and south China were relatively low.

The primary agricultural regions are widely distributed from south to north over the eastern half of China. The relatively steady yields of major rice-producing provinces in the south are a beneficial factor for stable food production, but yield fluctuations are larger in the main food-producing regions in the north, especially in the northeast. Those fluctuations adversely affect economic growth.

Correlation of yield fluctuations among regions. The correlation coefficients of yield fluctuations in the whole nation and for 27 provinces, cities, and autonomous regions (except Ningxia, Xinjiang, and Taiwan, because data were not complete) were calculated. The relationship between the national yield fluctuation and that of each province, city, and autonomous region, as well as among most of the regions, was remarkable.

These correlations showed that the effects of weather and climate on yield are regional. China has a vast territory and climate varies greatly, weather and climate will be favorable in some regions while it may be adverse in others. Good harvests in one area help compensate for poor harvests in other areas. But the effectiveness of this trade-off is limited. Losses due to abnormal weather in large areas cannot be compensated for. We need to be able to anticipate these conditions and adopt remedial measures.

Classifying weather and climate over the years

The study of 5000 yr of climate in China by Prof. Zhu Kezhen (1964) showed larger fluctuations of weather in years with strong monsoon seasons (expressed as great differences in rainfall between years of abundant and little rain and large differences in temperature between cold and warm years). If abnormal climate were to occur in wide areas, it would be beyond the adaptation ability of food crops and the adjustment ability of agricultural techniques, and the consequences would be serious.

From our analyses of the practical influence of weather and climate on agricultural production, we identified three food production situations: good, average, and poor. *Good* means there was no serious meteorological event during the year, with unfavorable weather in only a few parts of an area. *Average* means there were a few serious meteorological events in some parts of an area or at certain times of the year, but they did not influence the overall situation. *Poor* means that serious meteorological events occurred across large areas and affected the overall situation.

Table 2 shows the classification of weather and climate from 1950 to 1982. On the whole, yields were always high in the good years and low in the poor years, with some exceptions. For example, in 1978—a poor weather year— drought occurred

Table 2. Classification of weather and climate in China from 1950 to 1982.

Year	Classification	Fluctuation	Year	Classification	Fluctuation
1950	Good	+	1967	Good	+
1951	Good	+	1968	Average	–
1952	Good	+	1969	Average	–
1953	Average	+	1970	Good	+
1954	Poor	–	1971	Average	+
1955	Average	+	1972	Poor	–
1956	Average	–	1973	Good	+
1957	Average	–	1974	Good	+
1958	Good	+	1975	Average	+
1959	Poor	–	1976	Average	–
1960	Poor	–	1977	Average	–
1961	Average	–	1978	Poor	+
1962	Average	–	1979	Good	+
1963	Poor	–	1980	Poor	–
1964	Good	+	1981	Average	–
1965	Average	+	1982	Good	+
1966	Average	+			

over a large area. But countermeasures were carried out effectively and a bumper harvest was obtained. A bumper harvest or a poor harvest may occur in an average weather year due to the influence of social factors.

Influence of meteorological events on yield fluctuations

The meteorological events of years with negative yield fluctuations were classified into three levels: serious, intermediate, and light (Table 3). In 10 of the 11 yr with negative yield fluctuations, at least four intermediate or serious meteorological events occurred. A correlation with significance at the 0.01 level was obtained in a regression analysis of the negative yield fluctuations and the areas affected or damaged by aberrant weather. The regression equation is

$$Y = -123.97 + 0.0058X_1$$

$$Y = -86.75 + 0.0125X_2$$

where Y = the fluctuation yield,
X_1 = the area influenced by bad weather, and
X_2 = the area damaged by bad weather.

The results showed that negative yield fluctuations coincided with bad weather. The data on bad weather in China show that drought and flood-waterlogging were the main factors influencing food crop yields.

From 1949 to 1981, about 30% of the national negative yield fluctuations occurred in the north China region and the middle-downstream reaches of the Yangtze River Valley, 13-14% occurred in the southwest and northeast regions, about 10% in the south China region, and less than 5% in the northwest region. By comparing the principal years of decreased yields in different regions with the meteorological events that occurred, we arrived at the following analyses:

Table 3. The circumstances of weather-caused calamities in years of low yields.

Year	Drought	Flood-waterlogging	Low temperature injury in Northeast China	Low temperature injury in Yangtze River Valley	Low temperature injury in South China	Freezing injury of wheat
1954	Light	Serious	Serious	Intermediate	Intermediate	Light
1957	Intermediate	Intermediate	Serious	Serious	Serious	Intermediate
1962	Serious	Serious	Light	Light	Intermediate	Intermediate
1963	Serious	Serious	Light	Intermediate	Intermediate	Light
1968	Intermediate	Serious	Intermediate	Intermediate	Intermediate	Intermediate
1969	Light	Serious	Serious	Intermediate	Light	Serious
1972	Serious	Light	Serious	Serious	Light	Serious
1976	Light	Light	Serious	Light	Intermediate	Light
1977	Intermediate	Intermediate	Light	Intermediate	Light	Serious
1980	Intermediate	Serious	Intermediate	Serious	Light	Serious
1981	Intermediate	Intermediate	Intermediate	Serious	Light	Light

- The north China region. The production of wheat, maize, soybean, and sweet potato in north China is important to the rest of the country. The region produces more than 50% of the nation's winter wheat and more than 35% of the maize. It has a very long history of high agricultural production, natural conditions are good, and the potential for production increases is great. But yearly yields are by no means stable.

  Drought and flood-waterlogging are the principal constraints to production. Before 1969, flood and waterlogging occurred in 5 of 6 yr of decreased yield. Since 1970, there have been serious droughts in 3 of 4 yr.
- The middle-downstream reaches of the Yangtze River Valley. A major region for food production, the Yangtze River Valley has an important influence on national economic development. Its food crop yield makes up one-third of the total national output. Rice is the main food crop, with wheat, maize, and potato grown in certain areas. This region has superior growing conditions and a suitable growing season.

  The main causes of negative yield fluctuations were drought, flood-waterlogging, and low temperature. Of the 12 yr that had negative yield fluctuations, 6 were years of drought and 6 of flood-waterlogging, and there was low temperature damage.
- The southwest region. Rice and maize are the main food crops in the southwest region. The Sichuanese basin is the main production base of commodity grain.

  Drought was the cause of low yields in 8 of 9 yr of decreased yields. In the remaining year, losses were the result of flooding and cloudy or rainy weather.
- The northeast region. Maize, soybean, and spring wheat are the main food crops of the northeast region. Grain production per capita is the highest in the country, and the region provides a great deal of commodity grain.

  Flood-waterlogging, low temperature, and drought were the main adverse meteorological events. Before 1976, there were 5 yr with cold injury, 4 with flood-waterlogging, and 2 with drought in 6 yr of decreased harvests. Since 1977, drought and flood-waterlogging have been the principal causes of damage.

- South China. Rice is the main food crop of south China, and more than 90% of the area is double-cropped.

  Rainy weather and low temperature damage to rice were the main causes of negative yield fluctuations.
- The northwest. Wheat is the main food crop in the northwest, maize is second. Yields are low, and depend on irrigation and the improvement of dryland agriculture.

  Almost every year with low yields correlated with drought.

Conclusions

Negative yield fluctuations occurred at the same time that total food crop production was increasing. For stable agriculture, negative fluctuations should be below 5% and yield decreases caused by weather and climate should be below 3%. During the 33 yr from 1949 to 1981, 15 yr were beyond this range: 7 yr were in the 5-10% range and 2 yr were higher than 20% (of course, not due entirely to weather and climate). Based on current production, there is a 33% possibility that in any given year, a reduction in yield of 15 million tons could be caused by weather and climate. That would be a negative fluctuation of 20-30 million tons. The possibility of a yield reduction of 20-25 million tons, leading to a fluctuation of 30-40 million tons, is about 20%. And the possibility of a yield fluctuation above 50 million tons cannot be excluded.

The yield fluctuation caused by climatic fluctuation is usually over wide areas. For example, there was flood-waterlogging damage in 14 provinces, cities, and autonomous regions in 1954 and drought damage in 19 provinces, cities, and autonomous regions in 1972. Before 1969, drought and flood-waterlogging were the main threats to production. Since 1969, several other natural calamities have occurred, and damage to wheat from low temperature in northeast China and the Yangtze River Valley has increased.

The areas affected by natural calamities increased from the 1950s to the 1970s. This increase was correlated with some inappropriate agricultural measures, such as raising of the multiple crop index, using high-yielding varieties that had too long growth durations, and moving the growth limit of crops north. The interaction of climatic factors and artificial factors has increased the possibility of yield losses.

Without a major breakthrough in agricultural technology, food production will be more restricted by weather once yields reach a certain level. In the high-yielding suburban districts of Shanghai, the trend for an increase of average food crop yields stabilized 1972-81, and correlations between yield and production technologies, such as the quantity of chemical fertilizer, the area irrigated effectively, or total agricultural machinery power, were not distinct. Food crop yields changed year by year with fluctuations of such meteorological conditions as days above 0 °C, annual precipitation, and average daily radiation. In the future, the same situation can be expected in other high-yielding regions.

Research on future climatic changes is being emphasized now. Climatic changes and their effect on grain production need to be studied in order to develop precautionary measures. We should examine the actual conditions of abnormal climatic events and the effect of calamities on crops, inquire into the relationship between the calamity and weather and climatic factors, establish mathematical models to express the relationship between food crop yield fluctuations and climatic change, and explore the possible influence of coming climate changes on food production. We should formulate such countermeasures as the distribution of crops on the basis of climate; the use of biotechnological methods in breeding of drought-resistant, flood-resistant, and low temperature-resistant varieties; the further development of farmland capital construction, to increase farmers' ability to combat calamity; and the development of a series of effective measures to prevent and counteract calamity.

Reference cited

Zhu Kezhen (1964) Some characteristic features of Chinese climate and their effects on crop production. Acta Geogr. Sin. 30(1).

Notes

Authors' addresses: Xin Naiquan, Department of Scientific Management, Chinese Academy of Agricultural Sciences (CAAS), Beijing, China; Chen Yannian, Agrometeorological Laboratory, CAAS, Beijing, China.
Citation information: International Rice Research Institute (1989) Climate and food security. P.O. Box 933, Manila, Philippines.

Effect of weather and climate on production and vulnerability of rice

D. V. Seshu, T. Woodhead, D. P. Garrity, and L. R. Oldeman

Rice is grown under the diverse environmental conditions of a wide range of latitude and altitudes. The major climatic factors affecting growth and yield include solar radiation, temperature, and rainfall (the latter is particularly important in rainfed rice culture). Seasonal and spatial variations in climate affect rice productivity. However, rice has the genetic diversity to tolerate adverse climates: an appropriate cultivar may enable a farmer to avoid or lessen a major climatic stress. From tests of a common set of cultivars at sites located at latitudes between 6° S and 37° N, functions have been determined that relate cultivar yield to weather variables. These can be used to assess the vulnerability of annual rice production to year-to-year weather variability. Recent variability in global rice production is considered in relation to trends in production vulnerability expected with climate change. The coefficient of variability of global production has decreased since the 1960s, despite an enormous increase in the quantity produced. Future changes in global climate would have major implications for rice production, but those changes would be less than the year-to-year weather differences with which farmers already contend.

Rice has been cultivated for more than 5,000 yr. About 3 billion people eat rice daily and about 300 million farmers grow rice in a wide range of ecosystems. Rice originated in the hot humid tropics with a strong monsoonal rainfall pattern and with relatively low direct solar radiation during the wet season. The crop subsequently adapted to a broad latitude range from 40° S in central Argentina to 53° N in northeastern China. Thus, rice is grown in more diverse environmental conditions than any other major food crop. It is cultivated in the cool climate of the high-altitude areas of Nepal and India and in the hot, arid climate that characterizes southern Pakistan and Iran. In Africa, Latin America, and parts of Asia, rice is commonly grown as an upland crop subject to frequent droughts. At the other extreme, floating rices thrive in annual flood waters exceeding 3 m depth in parts of Thailand, Bangladesh, Burma, and Vietnam.

On the basis of water regime, rice culture can be divided into five major categories: irrigated, upland, rainfed lowland, deepwater, and coastal wetlands. In most of these ecologies, rice is highly likely to encounter water shortage at some growth stage. In the rainfed lowlands, bunded fields retain a maximum sustained water depth of up to 50 cm. This regime is characterized by favorable moisture conditions or by the prevalence of drought, submergence, or stagnant water, or a combination of these factors. Deepwater rices are grown in water depths exceeding 50 cm. Where water depth exceeds 1 m, rices are referred to as floating types.

In 1985, total world production was 463 million tons of rough rice from 147 million ha of land, with an approximate yield of 3.1 t/ha. More than 75% of global production is in favorable areas. About 50% of the world's rice area is irrigated. In tropical countries, irrigated wetland occupies 30% of the rice area, accounting for about 50% of production. Rice in Africa and Latin America is cultivated predominantly under upland conditions.

A global perspective study, commissioned by the Food and Agriculture Organization of the United Nations, predicts that the world demand for rice will increase at a rate of 2.5%/yr from 1980 to 2000. This increase will match the increase in population, and will allow a small rise in per capita consumption. The increase must come largely from gains in rice production per unit area. In addition, production may expand into areas where rice is grown infrequently or not at all; the fact that such areas are not now used for intensive rice production indicates they are deficient in some aspect of soil or climate.

This paper examines the production vulnerability of rice to weather fluctuations and deals briefly with the prospective effects of longer-term climatic change. Emphasis will be given the role of modern rice technology in increasing rice production stability. First, we discuss recent work to quantify the relationship between rice performance and major weather variables. Second, genetic variability in the stability of response to weather fluctuations is examined for each of the major rice cultural systems. Third, we discuss evidence that global rice production variability has decreased in recent years, and that modern rice technology has played a role in this decrease. Finally, we draw upon experience in understanding rice-weather interactions to examine the implications of terrestrial climate change on future rice production.

Weather and climate determinants

Temperature, solar radiation, and rainfall are important weather factors that influence rice yields directly, by affecting the physiological processes involved in grain production, and indirectly, through their effects on disease and insect pressure. These factors are often difficult to differentiate in the field. In temperate regions, the rice calendar is circumscribed by the temperature regime. In the tropics, it is determined primarily by onset and withdrawal of the monsoon. Dry season rice in the tropics is limited to irrigated areas.

Based on the seasonal fluctuations in mean monthly minimum temperature (MINT), Oldeman et al (1986) recognized four groups of rice-growing regions.

- Highly seasonal [monthly MINT fluctuates more than 10 K (°C) during the year and/or the lowest MINT is less than 10 °C],
- Moderately seasonal [MINT fluctuates 8-15 K (°C)],
- Weakly seasonal [MINT fluctuates 3-8 K (°C)], and
- Nonseasonal [MINT fluctuates less than 3 K (°C)].

Extreme temperatures are destructive to plant growth. The critical low and high temperatures, normally below 20 °C and above 30 °C, vary among growth stages (Yoshida 1981). Within the critical limits, temperature affects grain yield by affecting tillering, spikelet formation, and ripening. Higher temperatures increase tiller number when light is adequate; low temperatures may produce more tillers under low light conditions (Yoshida 1973). During the reproductive phase, spikelet number per plant increases as temperature decreases. The temperature during ripening appears to affect weight per grain, which decreases with an increase in temperature (Murata 1976). The length of ripening is inversely correlated with daily mean temperature and positively correlated with grain yield.

Low temperature injury occurs not only in the temperate regions but also at high altitudes and in dry season crops in the tropics. High temperature stress occurs in the regular crops of tropical African countries and Iran; in the first rice crop in Pakistan; and in the dry season crops in Cambodia, Thailand, and India. Depending on growth stage, low temperature causes stunting, leaf discoloration, delayed flowering, incomplete panicle exsertion, and spikelet sterility; high temperature causes reduced height and tillering, white leaf tips, reduced spikelet number, reduced grain filling, and sterility (IRRI 1975). The indica varieties are better adapted to high temperatures, japonicas require low temperature for optimum ripening.

Rice yield potential is primarily determined by solar radiation in both tropical and the temperate regions. In the tropics, dry season rice yields are usually higher than in the wet season because of higher solar radiation. Solar radiation has the greatest effect on grain yield at the reproductive and ripening phases; the overall effect is very small at the vegetative phase (Yoshida and Parao 1976).

Too little or too much rainfall at any growth stage can cause partial or total crop failure. Water deficits may result in leaf senescence, impaired tillering, reduced height and leaf area, delayed flowering, and spikelet sterility. The rice plant is most sensitive to water deficits during reduction division to heading, resulting in high sterility. Excess water (partial submergence of the crop) may result in impaired tillering and decreased area of photosynthetic leaf surface (Yoshida 1981).

Rice, like other crops, is affected adversely if its soil-water balance manifests water shortage or extreme excess. The water requirement of rice depends on land topography, soil characteristics, and length of growing period and, for irrigated and rainfed lowland rice, on seepage-plus-percolation. In the rice-water balance, the requirement for water percolation and seepage may range from 1 to 10 mm/d; additionally, there is a pretransplanting water requirement for land preparation (puddling), typically 400 mm and exceptionally, for very dry soils, 1,100 mm (Wickham and Singh 1978).

For irrigated rice and for rainfed rice grown with standing water in bunded fields, the occurrence and duration without desired fieldwater (resulting perhaps from irrigation failure or drought) can be useful pragmatic predictors for the magnitude of loss of rice production (Wickham and Sen 1978). For dryland, unbunded rice, uptake of groundwater during dry topsoil conditions can, in some circumstances, give an increase in grain production of about 1 t/ha (IRRI 1986).

Weather variables that need to be monitored to support statistical and deterministic models for rice growth and yield include those used to determine the water balance for any crop: rainfall, air temperature and humidity, net solar irradiance, and wind speed. Additionally for rice, measurements of soil or fieldwater temperature, seepage-plus-percolation rate, and fieldwater and groundwater depth and duration (since groundwater affects both soil-water uptake and fieldwater percolation) are needed. For recharge of soil-water after soil drying, determination of by-pass flow (Bouma and de Laat 1981) also may be needed.

In irrigated rice, the weather factors that largely determine growth and yield are temperature and solar radiation. In rainfed culture, rainfall is the most limiting factor when temperatures are within critical limits.

A more detailed analysis of the dominant weather factors as they relate to rice cultural system follows. The data are drawn primarily from two sources, the International Rice Testing Program (IRTP)—an ongoing activity with nursery reports published annually—and the International Rice Research Institute/World Meteorological Organization International Rice-Weather Yield Nursery (IRWYN) conducted from 1983 to 1985 (Oldeman et al 1986).

Climate and irrigated rice

Temperature effects

Prolonged growth duration is one of the effects of low temperature. A comparison of duration from transplanting to maturity of variety IR36 at 4 sites, 1 each in Korea, India, Philippines, and Bangladesh, shows that with an increase in mean air temperature (transplanting to maturity) from 23.8 to 28.4 °C, duration is reduced by 37 d, from 124 to 87 d (Fig. 1). Within critical limits, lower temperatures favor higher yields because of higher net photosynthesis. This can be inferred from the above example, where IR36 produced 10.3 t/ha at Suweon and 2.7 t/ha at Joydebpur. Seshu and Cady (1984) observed a significant negative correlation between mean temperature during the 30-d period after flowering and grain yield in IRTP irrigated trials, with an indication of curvilinearity in the relationship. They estimated a yield decrease of 0.71 t/ha with an increase of minimum temperature from 18 to 19 °C, a decrease of 0.41 t/ha from 22 to 23 °C, and a decrease of 0.04 t/ha from 27 to 28 °C.

A comparison of grain yields at appropriate locations in temperate and tropical regions reveals the effects of temperature on growth duration and yield. Mean

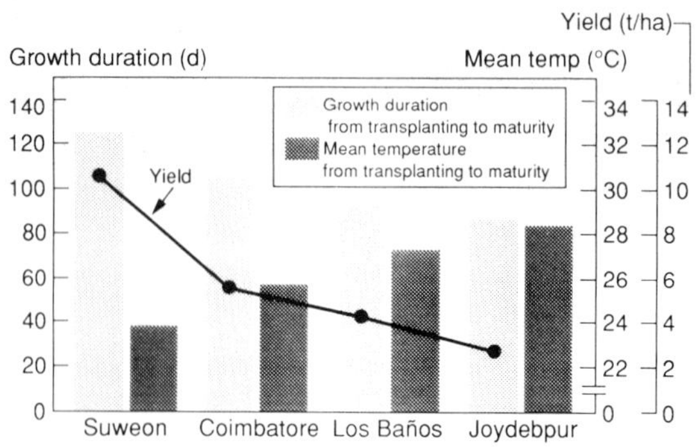

1. Effect of mean temperature on growth duration and yield of IR36 at four IRWYN sites.

temperatures from transplanting to maturity in the IRWYN trials at 2 sites, Suweon (37° N) and Cuttack (20° N), were 23.8 and 29.4 °C, respectively, while daily average radiation during that period at those sites were similar (Table 1). Site means for grain yield at Suweon were nearly three times those at Cuttack, whereas mean growth duration from transplanting to maturity was shorter at Cuttack by 3 wk. Low temperature reduces photorespiration, increases spikelet number when it occurs during the reproductive phase, and increases growth duration, including that of the ripening period. In this example, while daily radiation levels were somewhat similar at the two sites, because of the longer ripening period resulting from low temperature, the accumulated radiation during that important period was higher at Suweon, resulting in higher yields. This illustrates the direct and indirect effect of temperature on growth and yield.

Temperature summation varies with variety, latitude, and planting season. Early-maturing varieties have smaller temperature summations than late-maturing ones. Temperature summations tend to be higher at higher latitudes, particularly when the varieties grown have a certain degree of photoperiod sensitivity. IR36 showed 19.5% increase in temperature summation (transplanting to flowering) at Suweon (37° N) over that in the wet season trial at Los Baños (14° N); the corresponding increase for IR13429-196-1 was 5.75% (Table 2). Comparison of the

Table 1. Grain yields at tropical and temperate locations as affected by temperature in International Rice-Weather Yield Nursery (IRWYN) trials.[a]

Location	Mean temp (°C)			Radiation (mWh/cm^2 per d)			Growth duration (d)			Grain yield (t/ha)
	T-F	F-M	T-M	T-F	F-M	T-M	T-F	F-M	T-M	
Suweon (37° N)	23.7	24.0	23.8	461	399	439	76	41	117	10.8
Cuttack (20° N)	29.8	28.5	29.4	429	427	428	67	28	95	3.7

[a]T = transplanting, F = flowering, M = maturity.

Table 2. Temperature sum for 2 rice varieties at 2 different latitudes in IRWYN trials.[a]

Location	IR36			IR13429-196-1		
	Mean temp (°C)	Days T-F	Temp sum	Mean temp (°C)	Days T-F	Temp sum
Suweon (37° N)	23.9	80	1912	23.5	72	1692
Los Baños (14° N)	27.6	58	1600	27.6	58	1600

[a]T = transplanting, F = flowering.

% increase in temp sum at Suweon over Los Baños } = 19.50 for IR36, 5.75 for IR13429-196-1.

wet and cool dry season IRWYN trials at Joydebpur shows that the temperature summation for IR36 was 1,872 degree-days in the wet season and 2,224 degree-days in the dry season. Temperatures below certain thresholds are considered ineffective. Thus, increased temperature summation in the above examples may be due partly to addition of values in the ineffective ranges occurring on certain days.

The dry season rice crop in certain regions (e.g. Bangladesh and Eastern India) experiences cooler preflowering and higher postflowering temperatures. Under those conditions, vegetative growth is prolonged but the ripening period remains short (Fig. 2).

When temperatures are critically low and growth duration extremely prolonged (the degree of prolongation depending on the tolerance of the variety) yields are affected. A comparison of days to flowering of entries in an international upland nursery (IRTP 1985) at two tropical sites at two different altitudes illustrates the point (Table 3). Temperatures during the crop season ranged from 20 to 29 °C maximum and 9 to 13 °C minimum at Woretta, and 29 to 31 °C maximum and 22 to 23 °C minimum at Sitiung. The increase in flowering duration from Sitiung to Woretta for the selected varieties ranged from 27 to 93 d, suggesting varietal differences in tolerance for low temperature stress.

The diurnal difference in temperature is positively correlated with yield. A high diurnal difference leads to more efficient conversion and use of solar energy, resulting in higher net photosynthesis. The relationship between the diurnal difference in temperature and grain yield of variety MRC603-303 in the IRWYN trial is shown in Figure 3.

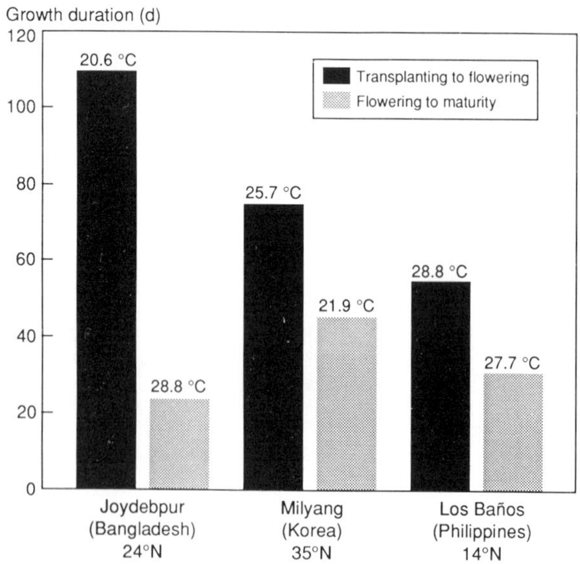

2. Growth duration of IR36 at three IRWYN sites. (Temperatures at the top of columns represent averages of mean daily temperatures for the period.)

Appropriate utilization of tolerant lines in breeding programs would be an effective approach to minimize adverse climatic effects. Through the IRTP nurseries, several promising breeding lines have been identified for tolerance for low temperature at different growth stages. Some have shown tolerance at all stages (e.g. Stejaree 45, Ching-shi 15, Tatsumi-mochi, Barkat, K335, K39-96-1, and China 1039). The performance of these lines is based on the expression of traits such as seedling vigor, panicle exsertion, spikelet fertility, and resistance to leaf discoloration and stunting. A comparison of flowering duration at selected sites in the tropical, subtropical, and temperate regions shows that K335 has a remarkable degree of stability in days to flowering (Fig. 4).

When temperatures exceed 35 °C, the rice crop is injured. N22, an upland rice from India, has higher than 80% spikelet fertility at 35 °C, BKN6624-46-2, a lowland selection from Thailand, only has about 10% spikelet fertility (Satake and Yoshida 1978). IRTP screening trials indicated distinct varietal differences in heat tolerance. Some promising lines include IR2006-P12-12-2-2, UPR9616-1-1-1, CR156-5021-207, and Pusa 2-21.

Table 3. Comparison of flowering durations of 8 selected entries from 1984 IRTP Upland Observational Nursery at 2 tropical sites differing in altitude (Woretta, 1,815 m; Sitiung, 180 m).

Entry	Days to flowering	
	Woretta, Ethiopia	Sitiung, Indonesia
BR201-51-1-J1	143	91
B3619c-Tb-8-1-4	106	79
Dourado Precoce	136	72
IRAT109	106	74
IRAT112	102	71
IRAT134	102	75
IRAT141	102	74
IRAT146	166	73

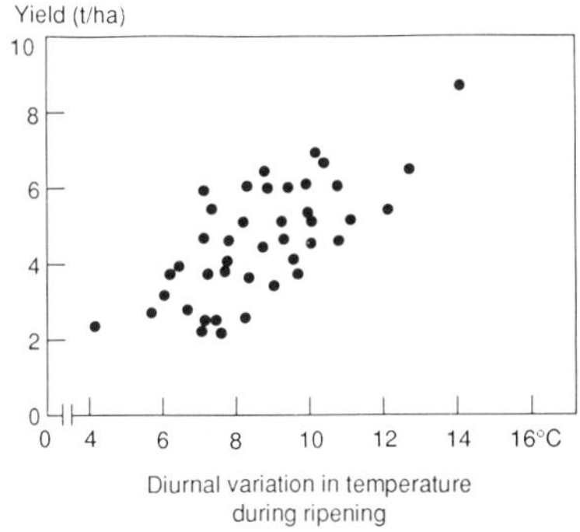

3. Scattergram showing effect of diurnal variation in temperature during ripening on yield (variety MRC603-303, IRWYN trials at 42 sites).

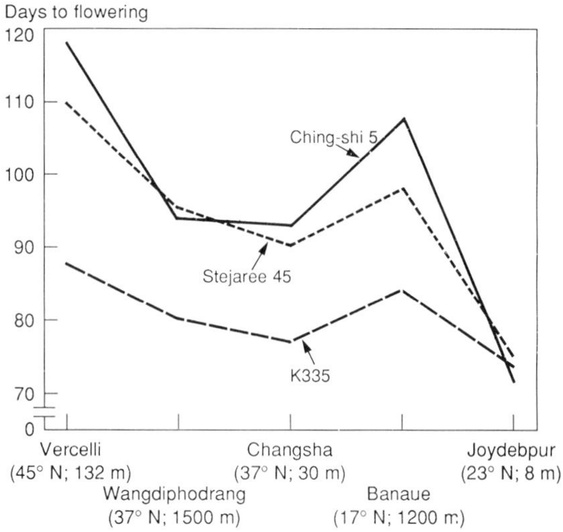

4. Varietal differences in stability of days to flowering, 1984 IRTP Cold Tolerance Rice Nursery. (Figures in parentheses below location name on x-axis indicate latitude and elevation.)

Table 4. Solar radiation and temperature during ripening phase and grain yields during wet and dry seasons at 4 IRWYN sites.

Site	Latitude (°N)	Solar radiation (mWh/cm²)		Mean temp (°C) at ripening		Grain yield (t/ha)	
		Wet	Dry	Wet	Dry	Wet	Dry
Masapang, Philippines	14	445	672	26.8	27.6	3.4	5.8
Sanpatong, Thailand	18	469	593	27.1	28.1	4.2	5.9
Hyderabad, India	17	385	544	24.5	26.4	3.0	5.3
Joydebpur, Bangladesh	24	363	561	27.4	28.8	2.5	4.9

Radiation effects

A comparison of wet and dry season grain yields at locations with moderate to low seasonal fluctuations in temperature reveals the influence of solar radiation. Dry season IRWYN trials produced higher yields than wet season trials at locations where dry season solar radiation during ripening was higher (Table 4). Temperatures during that growth phase did not differ appreciably between the two seasons in these trials. Radiation sum during ripening period bears a positive relationship with grain yield (Fig. 5). Number of filled grains per panicle is one of the yield components highly affected by level of radiation (Fig. 6).

Radiation and temperature interactive effects

Seshu and Cady (1984) analyzed yield-weather relationships in rice from results of a common set of early-maturing entries in IRTP irrigated trials conducted in 40

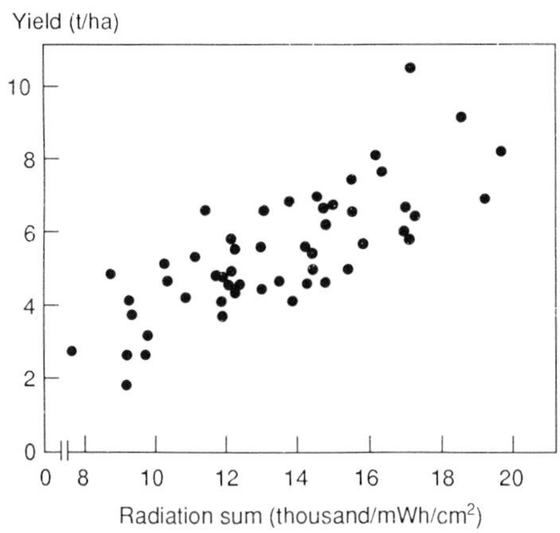

5. Scattergram showing effect of radiation sum during ripening on yield (variety IR36, IRWYN trials at 52 sites).

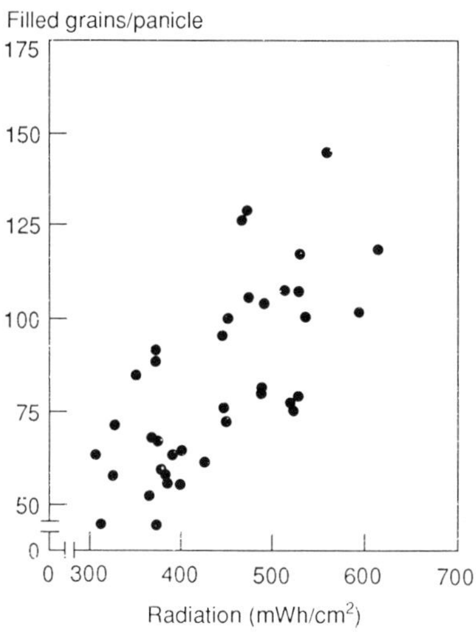

6. Scattergram showing effect of radiation during ripening on number of filled grains/panicle (variety MRC603-303, IRWYN trials at 38 sites).

environments. They formulated a prediction model (Fig. 7) that uses average daily solar radiation (RAD) in mWh/cm^2 and minimum temperature (MINT) in °C during ripening (30-d period after flowering).

$$\hat{Y} = \underset{}{30.2} + \underset{(.0009)}{0.0041}\ \text{RAD} - \underset{(0.18)}{2.10}\ \text{MINT} + \underset{(.004)}{0.038}\ (\text{MINT})^2 \qquad (1)$$

where $\hat{Y}$ is predicted yield and values in parentheses are standard errors of the estimated coefficients.

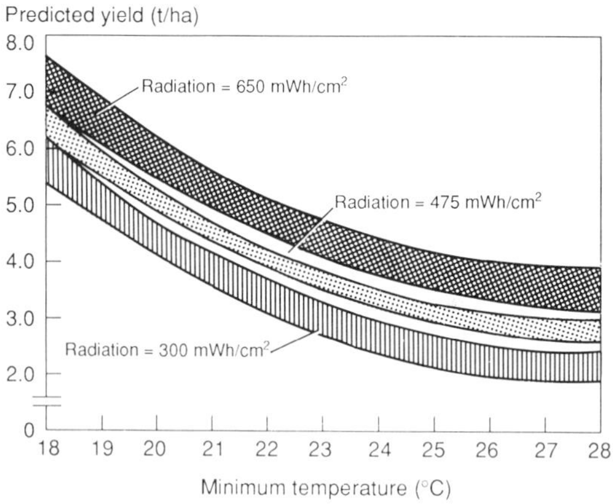

7. Yield response curve, calculated from prediction equation 1, which includes minimum temperature (°C) and solar radiation (mWh/cm^2) during ripening (30-d period from flowering). Shaded area for each radiation value is 90% confidence interval on the mean response (t/ha).

Oldeman et al (1986) considered that radiation sum representing the combined effects of temperature and radiation is an important variable. They used three preflowering and two postflowering weather variables to explain the yield-weather relationship based on results of the IRWYN trials and postulated the following prediction equation:

$$\hat{Y} = \underset{(0.05)}{4.80} + \underset{(0.07)}{0.57}\,(DNB - 3.1) - \underset{(0.03)}{0.15}\,(TDB - 28.1) + \underset{(0.009)}{0.17}\,(RSC - 14.0) + \underset{(0.02)}{0.17}\,(RSC - 14.0) - \underset{(0.02)}{0.13}\,(TNC - 24.) \quad (2)$$

where $\hat{Y}$ = predicted yield;

DNB = day-night temperature difference before flowering, with an average of 3.1;

TDB = day temperature before flowering (°C) with an average of 28.1;

RSB = radiation sum $\times 10^{-3}$ (in mWh/cm^2) before flowering, with an average of 30.4;

RSC = radiation sum $\times 10^{-3}$ (in mWh/cm^2) after flowering, with an average of 14.0;

TNC = night temperature after flowering (°C) with an average of 24.0.

Both IRTP and IRWYN trials showed a yield difference of more than 6.0 t/ha between the more favorable and the less favorable combination of radiation and temperature (Table 5).

Climate and rainfed rice

Rainfall is the most important weather factor in rainfed rice culture, which is limited to areas where annual rainfall exceeds 1,000 mm. Water losses in lowland ricefields largely result from transpiration, evaporation, percolation, and seepage; in upland

Table 5. Effect of radiation and temperature on yield of IR36 under optimum management at 2 IRWYN sites.

Weather factor	Growth stage	Suweon, Korea	Joydebpur, Bangladesh
Radiation sum (mWh/cm^2)	Transplanting to flowering	37,348	30,716
Radiation sum (mWh/cm^2)	Flowering to ripening	17,055	9,973
Minimum temperature (°C)	Flowering to ripening	18.7	24.8
Day-night temperature difference [K, (°C)]	Flowering to ripening	8.8	7.2
Mean temperature (°C)	Seedbed	16.0	28.2
Yield (t/ha)		10.3	2.6

Table 6. Grain yields at 4 1985 IRTP Upland Rice Yield Nursery sites representing different levels of drought stress—fully irrigated (Dokri, Pakistan), moderately severe stress (Vyara, India), and severe stress (Santo Tomas, Philippines, and Campeche, Mexico).

Selected variety	Grain yield (t/ha) at different drought stress levels			
	No stress	Moderately severe stress	Severe stress[a]	
	Dokri	Vyara	Santo Tomas	Campeche
IR9729-67-3	8.2	3.2	0.2	0.6
IRAT109	3.8	1.7	1.0	2.4
Site mean[b]	5.2	1.7	0.6	0.8
LSD (5%)	1.2	0.8	0.4	0.6

[a]Drought stress was severe during vegetative phase at Campeche and reproductive phase at Santo Tomas. [b]Mean of 22 entries.

fields, the major forms of water loss are transpiration and surface runoff. Transpiration and evaporation are related to meteorological factors; percolation, seepage, and surface runoff are controlled by soil characteristics and topography.

Upland rice

Drought is a dominant stress in upland rice culture. The duration and degree of stress and the occurrence of stress relative to crop growth stage determine the impact on crop performance.

Research is under way to combine drought tolerance with high yielding ability and other desirable traits. Table 6 compares grain yields at four sites of the International Upland Rice Yield Nursery (IURYN) (IRTP 1985) representing three levels of drought stress. Mean yields for the 23 entries were 5.2 t/ha at Dokri (Pakistan), where the trial was fully irrigated; 1.7 t/ha at Vyara (India), with

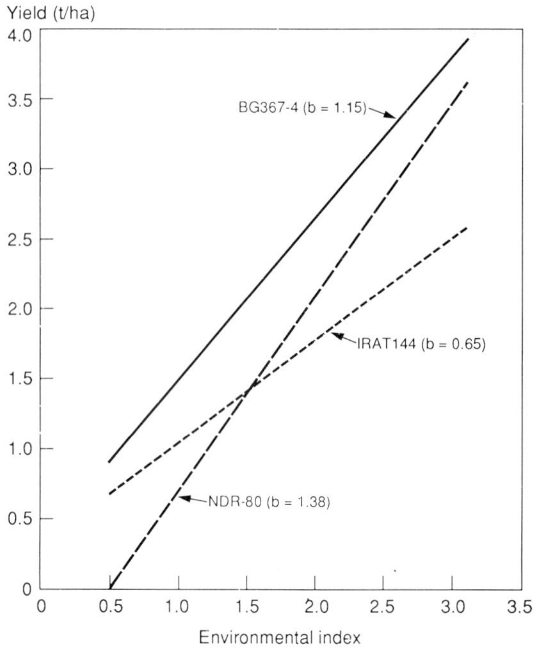

8. Regression of yield of 3 selected varieties on site means (23 varieties, 16 sites). IRTP upland yield nursery, 1985.

moderately severe drought for 10 d during postflowering; 0.8 t/ha at Campeche (Mexico), with severe drought during the vegetative stage; and 0.6 t/ha at Santo Tomas (Philippines), with severe drought for 22 d during flowering. Drought-tolerant IRAT109 produced significantly higher yields than nontolerant IR9729-67-3 at Campeche and Santo Tomas, where drought was severe. Yields of IR9729-67-3 were significantly higher under full irrigation at Dokri and under moderately severe drought at Vyara.

While some cultivars perform well under specific weather conditions, superior performance across a range of conditions would be more desirable. The performance of BG367-4 from Sri Lanka suggests the progress being made in that direction. Regression analysis of varietal yields on site means in the IURYN (IRTP 1986) at 16 sites representing different levels of drought stress showed stable performance by drought-tolerant IRAT144 (stability index = 0.65). However, BG367-4 outyielded IRAT144 at all levels of the environmental index (Fig. 8).

The importance of varietal tolerance to drought under upland conditions is illustrated by the results of a yield nursery (IRTP 1986) at Santo Tomas, Philippines. Entries were scored for drought tolerance on a 0-9 scale (0 = most tolerant, 9 = most susceptible). Yields were negatively correlated with drought tolerance scores (i.e. positively associated with level of tolerance to drought) (Fig. 9).

When upland rice is grown at higher altitudes, the crop may encounter both drought and low temperature, as happens in parts of India, Indonesia, and Ethiopia. Varieties such as B3619c-Tb-8-1-4 were tolerant of both stresses.

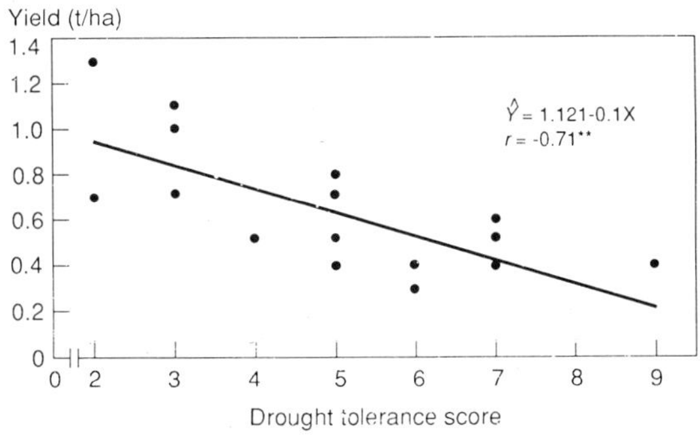

9. Regression of grain yield on drought tolerance score (0 = most tolerant, 9 = most susceptible), 1985 IRTP upland yield nursery, Santo Tomas, Philippines.

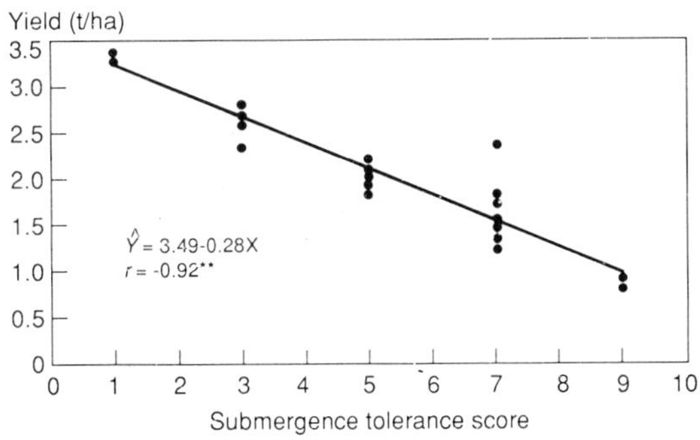

10. Regression of grain yield on submergence tolerance score (0 = most tolerant, 9 = most susceptible), 1985 IRTP rainfed lowland yield nursery, Phitsanulok, Thailand.

Rainfed lowland rice

Both drought and submergence are encountered in the rainfed lowlands, and may occur at any growth stage.

The Rainfed Lowland Yield Nursery at Phitsanulok, Thailand (IRTP 1986), illustrates the relationship between yield performance and level of varietal tolerance for submergence (Fig. 10). The trial was completely submerged (more than 1 m deep, maximum 2 m) for 10 d during booting and flowering. The mean yield was 2.0 t/ha; tolerant varieties RP1057-184-5-3-2 and Mat Candu produced more than 3.0 t/ha and susceptible lines IR19431-72-2 and IR14753-86-2 less than 1.0 t/ha.

In the multilocation rainfed lowland yield trials, the breeding line IR13146-45-2-3 showed a good yield combined with high stability (Seshu 1986). It compared to the popular variety Mahsuri in stability, but yielded higher under a wide range of environments (Fig. 11). IR19431-72-2 showed less yield stability than Mahsuri but yielded better at most locations.

Water stagnation for an extended period is another rainfed lowland stress. In the 1983 Rainfed Lowland Yield Nursery at Chinsurah, India, water level was more than 30 cm deep for most of the growing season, resulting in very low yields (Seshu 1986). The site mean for 24 entries was only 1.4 t/ha. Only NC492 and Mahsuri

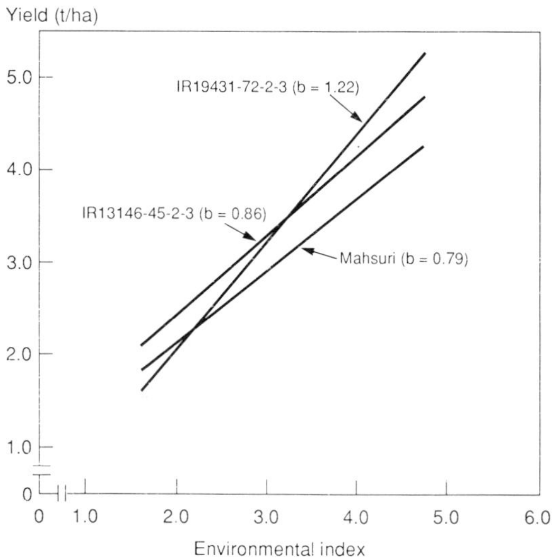

11. Regression of yield of 3 selected varieties on site means (24 varieties, 13 sites). IRTP rainfed lowland yield nursery, 1983.

produced yields higher than 3.0 t/ ha; the rest yielded less than 2.0 t/ ha. These results draw attention to the deficiency of stress tolerance in currently available modern varieties.

Deepwater rice

Deepwater rices are grown in water depths of 50-100 cm. Floating rices are cultivated where flooding exceeds 100 cm. These rice types have the ability to elongate (internode elongation) rapidly with rising floods and to survive complete submergence, an essential trait when the rise in water level exceeds the rate of internode elongation. Elongation rate is affected by water temperature. Kondo and Okamura (1932) reported that elongation is greatest at 25-30 °C and is retarded at 35-40 °C. The light penetration required for photosynthesis may also affect elongation. Survival under complete submergence is greater at lower temperatures (Palada and Vergara 1972), probably because the oxygen content in the floodwater is reduced as water temperature increases. Kondo and Okamura (1934) reported damage to rice plants under low oxygen content. If the rice plant is flooded in warm water for a long period during grain formation, the grains may germinate on the panicle.

Climate and incidence of diseases and insects

Climatic factors indirectly affect rice yield through their effect on the population dynamics of plant pathogens and insects. Shahjahan et al (1986) related the appearance of various diseases in Bangladesh to seasonal climatic changes, citing examples for tungro disease, bacterial blight, bacterial leaf streak, blast, sheath blight, and damping off. The bacterial and viral diseases occur primarily during the hot humid conditions of the wet season, the fungal diseases occur during all rice-growing seasons.

Temperature, relative humidity, rainfall, and mass air movements affect the distribution, development, survival, behavior, migration, reproduction, population dynamics, and outbreaks of insect pests of rice. For example, high temperature and low rainfall can cause severe stem borer infestation. Rainfall is important for the population increase of rice gall midges and green leafhoppers. Brown planthoppers annually spread from tropical to subtropical zones, where they multiply, and then migrate to temperature zones (Kisimoto and Dyck 1976).

Weather-induced production variability

Global rice production has increased enormously during the last 3 decades. Modern technology has been a major contributor to these productivity gains. Increases in yield per hectare now greatly exceed increases in area planted. Now the serious concern is whether modern technology has increased production variability (Hazell 1982, Mahra 1981).

This section briefly examines whether new technology may tend to exacerbate the instability of rice production in relation to seasonal weather variability. Of course, variation in production data does not reflect weather variability alone. Production stability is affected by a number of other important factors, such as market prices and input supply regularity. The discussion draws on a recent analysis by Flinn and Garrity (1985).

Global rice production and yields have increased more rapidly since the late 1960s than they did during before the introduction of modern varieties (Fig. 12). However, despite the dramatic increase in production, the coefficient of variability (CV) of global rice production has declined slightly (Hazell 1985) from 4.0 to 3.8% between the 2 periods 1960-61 to 1970-71 and 1971-72 to 1982-83. The probability of a production shortfall (5% below trend) was also lower in the second period. This trend toward lower production variability at the global level appears to be unique to rice among the major cereals (Hazell 1985). However, yield variance increased significantly in India and North Asia.

In India, the CV of rice production and yield increased, particularly in southern India. In northern India, where most riceland is irrigated and yields rose most dramatically, production variability and yield variability remained nearly constant. There was also no change in production and yield variability in eastern India, where rice is predominantly rainfed with much lower yields. The similarity in production and yield variability of the irrigated area in north India and the mostly rainfed area in eastern India is striking (Fig. 13). Irrigation may not necessarily imply more stable production: irrigation water supply may not be dependable in drought years. However, Carlson (1985) examined yield variability in 13 Asian countries. He found that, in general, the CV of both yield and production significantly decreased with irrigation development and the adoption of modern varieties. Fukui (1982) also examined yield variability of rice production in some major rainfed regions.

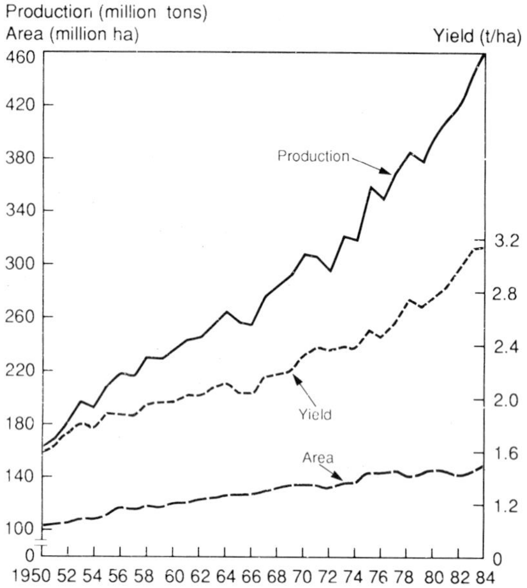

12. Global trends in production, area, and yield of rice, 1950-84 (Sources: FAO production yearbooks, various issues).

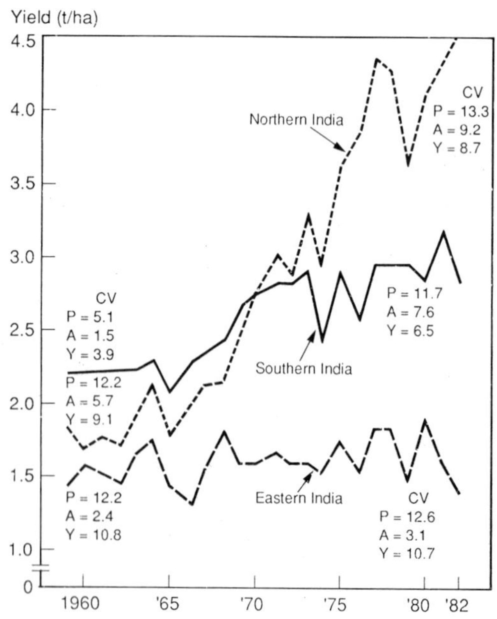

13. Trends in rice yields and coefficients of variation (CV) of production (P), area (A), and yield (Y) in India (Flinn and Garrity 1985).

Improving the stability of modern rice technology

The relentless increase in demand for rice in the tropics will necessitate further increases in production. Those increases must come through higher yields and increased cropping intensity. This can only be achieved by technological advances in water and nutrient management, other agronomic practices, and continued selection of new cultivars that respond efficiently to higher management levels.

Varietal development

Rice improvement programs are shifting to greater emphasis on the development of improved cultivars adapted to unfavorable physical environments, particularly drought-prone, flood-prone, and low temperature conditions. Cultivars with improved yields and yield stability are being generated for each of the major adverse hydrological and temperature environments.

The characteristics of modern rice cultivars allow them to respond to better nutrition and more uniform water supply by producing higher grain yield per day and per crop. But the structure and function of the modern rice plant may predispose it to be more severely affected by water deficit or excess than older varieties. In some drought-prone environments, the shorter stature, shallower root system, higher tillering, and photoperiod insensitivity of modern cultivars may result in severe damage or crop failure.

Within existing rainfed rice areas, yields during favorable weather should improve. However, yield instability may increase, since yields will be low when severe floods or droughts occur, irrespective of yield potential. The availability of cultivars better adapted to unfavorable environments also may extend rice cultivation to more marginal areas, increasing area, yield, and production instability.

The increasing genetic uniformity of rice cultivars planted over wide areas will continue to be a concern, with the additional implication of yield instability due to weather factors, as well as to pest and disease attack. A key factor in alleviating production instability will be the development of the research infrastructure in each country, so that national programs can develop varieties and management practices specifically adapted to local environments.

Crop and soil management

Increased stability of rice-based farming systems will be to a large extent the result of technology adapted to meet the needs of specific agroclimatic environments. Farmers in the more variable production environments must be offered a range of technical options, to enable dynamic adjustments in their farm enterprises to short-term weather and other perturbations.

Rice yield variability increases as nitrogen fertilizer rates increase (Evans and De Datta 1979). The interaction between the amount of N and random factors such as solar radiation, water regime, and pest incidence is strong (De Datta 1981). Profits increase as N rate increases, but profit variability also rises (Flinn and Garrity 1985).

For soil-constrained areas not yet supporting rice production, temperature and water-balance data are needed to anticipate riceland development. It is possible to estimate component variables of the water balance using proxy measurements, such as cloudiness and sunshine duration, to predict incoming shortwave and outgoing longwave irradiance (Woodhead 1966, 1967). The precision of such proxy methods has been determined for East Africa conditions (Woodhead 1970). He showed that 10 yr records of monthly mean cloudiness (as would often be available from a local civil airfield) can be used to predict monthly mean shortwave irradiance, ±10%.

Implications of climate change

For centuries, rice farmers have endured and exploited extreme variations in growing-season weather. These year-to-year variations dwarf any projected climatic change in mean weather variables. Nonetheless, some changes of climate that are projected do have implications for rice production.

The variables of a changed climate that would most affect rice production (in order of decreasing confidence of their probable changes) are atmospheric CO_2 concentration, air temperature, solar irradiance, and precipitation. The increase in partial pressure of CO_2 has potential to increase photosynthesis in crops. But for poor rice farmers, the potential gains may be limited by their inability to afford the increased nitrogen, phosphorus, and potassium needed. Moreover, higher production may be mainly through increased carbon-nitrogen ratios; the protein nutrition of rice-eating families may not be improved.

In rice breeding, 40-50 selection cycles could be completed before atmospheric CO_2 reached 600 ppm. That number of cycles would give breeders an opportunity to develop varieties with guard-cell morphology that takes advantage of increased water-use efficiency. Such varieties could rapidly translocate assimilates away from leaves, avoiding accumulation of starch. Rice crop management may become easier, in that the enhanced competitiveness of C3 species may result in decreased weed pressure. Evidence that rice, in contrast to other crops, may not decrease its shoot-root ratio at higher partial pressures of CO_2 is encouraging (Strain and Cure 1985).

The expected increase in water-use efficiency of rice might make it possible to slightly expand dryland rice areas and could decrease the water deficit vulnerability of some existing areas of dryland rice. Increased production in saline areas may also be possible (Acock and Allen 1985).

A negative implication of an elevation of mean daily temperatures due to global warming is the possible lowering yield potential in many areas. Rice yields are sensitive to small increases in minimum temperature during postflowering, particularly in situations where minimum temperatures are relatively low (i.e. less than 22 °C) (Seshu and Cady 1984).

Yield estimates for a range of sites varying in latitude from 31° N to 6° N were reduced 10-20% with a rise in temperature of 1-2 °C during the November-December harvest period (Table 7). Yield estimates were most severely affected at higher latitudes. In areas where the harvest period could be shifted as temperature rises, this effect would be offset. But where the harvest period cannot be shifted due

Table 7. The effect of increasing minimum temperature on estimated rice yields. Model of Seshu and Cady (1984) (equation 1).

	Amritsar, India	Patna, India	Bhubaneswar, India	Madras, India	Colombo, Sri Lanka
Latitude (°N)	31° 38′	25° 30′	20° 14′	13° 00′	06° 42′
Elevation (m)	234	52	45	16	50
November harvest					
MINT[a] (°C)	15.6	22.1	23.3	25.1	23.8
Yield at current MINT (t/ha)	8.7	4.3	3.9	3.1	3.6
Yield at +1 °C temp rise	7.8 (10)	3.9 (9)	3.5 (10)	2.9 (6)	3.3 (8)
December harvest					
MINT[a] (°C)	3.3	16.0	19.6	23.5	22.7
Yield at current MINT (t/ha)	–[b]	8.2	5.6	3.3	4.0
Yield at +1 °C temp rise	–[b]	7.3 (11)	5.0 (11)	3.0 (9)	3.6 (10)
+2 °C	–[b]	6.5 (21)	4.5 (20)	2.8 (15)	3.3 (18)
+3 °C	–[b]	5.8 (29)	4.0 (29)	2.7 (18)	3.1 (23)

[a]Values in parentheses are yield reduction percentages. Mean minimum temperature (MINT) during the month previous to the month of harvest. [b]Temperature below the range of model validation.

to direct or indirect dependence on the monsoonal rainy season, higher minimum temperatures may reduce grain yields. Harvest periods in many rainfed and partially irrigated areas of Asia cannot be shifted significantly because of exposure to severe terminal drought stress.

However, global warming will in principle allow a northward expansion of rice-growing areas and a lengthening of rice-growing seasons now constrained by low temperature. In some areas, these effects would allow more productive rice-based cropping systems and new opportunities for two rice crops where now only one is possible.

Regional changes in mean solar irradiance and mean precipitation may actually mitigate the effects of higher temperatures and increased CO_2 concentration. Unfortunately, predictions of regional changes in photosynthetic irradiance and precipitation are as yet uncertain. Given the vulnerability of rice, the implications of changes in precipitation and irradiance can be stated only in the most general terms: there may be some justification to construct new irrigation facilities in some regions, to revise management of existing facilities in other regions, and to diversify cropping patterns.

Regression models to predict the effects of temperature change or an altered precipitation regime have limitations in anticipating the overall impact of climatic change on production. Rosenberg (1983) has proposed studying the migration of major agricultural crops across strong climatic gradients as a way to gain insight into the dynamic nature by which crop production technology is progressively adapted to more extreme conditions. He examined the northward advance of the production belt of hard red winter wheat in North America. Production has expanded northward and westward by hundreds of kilometers during the last few decades, as cultivars and agronomic practices were developed to ensure more stable yields in progressively cooler climates.

The potential for agricultural research to contribute to adapting crop production systems to a gradually changing climate should not be underestimated. Over the long term, the research infrastructure is itself a dominant factor in stabilizing production. However, countries with weak agricultural research systems may be less able to make the adjustments that will be required.

Fortunately, the effects of a changing climate will occur slowly enough that farmers, breeders, agronomists, and agrometeorologists will have time to adapt. The long-term changes will be less than those that now occur with year-to-year weather variability. There will be time for adaptation in crop production systems and in international trade in cereals, not only for rice, but for the world's other major staples.

References cited

Acock B, Allen L H Jr. (1985) Crop responses to elevated carbon dioxide concentrations. Pages 53-97 *in* Direct effects of increasing carbon dioxide on vegetation. B. R. Strain and J. D. Cure, eds. U.S. Department of Energy, Washington, D.C.

Bouma J, de Laat P J M (1981) Estimation of the moisture supply capacity of some swelling clay soils in the Netherlands. J. Hydrol. 49:247-259.

Carlson G A (1985) Rice production variability: the role of pest resistant varieties and other inputs. Paper presented at the IFPRI/DSE Workshop on Sources of Increased Variability in Cereal Yields, 26-29 Nov 1985, Feldafing, Germany.

De Datta S K (1981) Principles and practices of rice production. John Wiley and Sons, New York.

Evans E T, De Datta S K (1979) The relation between irradiance and grain yield of irrigated rice in the tropics as influenced by cultivar, nitrogen fertilizer application and month of planting. Field Crops Res. 2(1):1-17.

FAO—Food and Agriculture Organization (1980-84) FAO production yearbook. Rome. (various issues)

Flinn J C, Garrity D P (1985) Yield stability and modern rice technology. Paper presented at the IFPRI/DSE Workshop on Sources of Increased Variability in Cereal Yields, 26-29 Nov 1985, Feldafing, Germany.

Fukui H (1982) Variability of rice production in tropical Asia. Pages 17-37 *in* Drought resistance in crops with emphasis on rice. International Rice Research Institute, P.O. Box 933, Manila, Philippines.

Hazell P B R (1982) Instability of Indian foodgrain production. IFPRI Research Report 30. International Food Policy Research Institute, Washington, D.C.

Hazell P B R (1985) Changing patterns of variability in world cereal production and their implications for price stability. International Food Policy Research Institute, Washington, D.C. (mimeo.)

IRRI—International Rice Research Institute (1975) Annual report for 1974. P.O. Box 933, Manila, Philippines.

IRRI—International Rice Research Institute (1986) Annual report for 1985. P.O. Box 933, Manila, Philippines.

IRTP—International Rice Testing Program (1985) Final report of 1984 International Rice Testing Program nurseries. International Rice Research Institute, P.O. Box 933, Manila, Philippines.

IRTP—International Rice Testing Program (1986) Final report of 1985 International Rice Testing Program nurseries. International Rice Research Institute, P.O. Box 933, Manila, Philippines.

Kisimoto R, Dyck V A (1976) Climate and rice insects. Pages 367-391 *in* Climate and rice. International Rice Research Institute, P.O. Box 933, Manila, Philippines.

Kondo M, Okamura T (1932) Relationship between water temperature and the growth of rice plants. V. The gradual difference in the damage to which rice plants are exposed by flooding with clear or muddy water [translation]. Ber. Ohara Inst. Landwirtsch. Forsch. 6:205-209.

Kondo M, Okamura T (1934) Relationship between water temperature and the growth of rice plants. VI. Cause of the death of the rice plant which has stood for some time under water [translation]. Ber. Ohara Inst. Landwirtsch. Forsch. 6:211-218.

Mahra S (1981) Instability in Indian agriculture in the context of the new technology. IFPRI Research Report 25. International Food Policy Research Institute, Washington, D.C.

Murata Y (1976) Productivity of rice in different climatic regions of Japan. Pages 449-470 *in* Climate and rice. International Rice Research Institute, P.O. Box 933, Manila, Philippines.

Oldeman L R, Seshu D V, Cady F B (1986) Response of rice to weather variables. Report of an International Rice Research Institute/World Meteorological Organization Special Project. International Rice Research Institute, P.O. Box 933, Manila, Philippines.

Palada M C, Vergara B S (1972) Environmental effects on the resistance of rice seedlings to complete submergence. Crop Sci. 12:209-212.

Rosenberg N J (1983) A synthesis and some questions on the possible effects of increased atmospheric CO_2 concentration on agriculture. Page 16 *in* Agronomy abstracts. American Society of Agronomy, Madison, Wisconsin.

Satake T, Yoshida S (1978) High temperature-induced sterility in indica rices at flowering. Jpn. J. Crop Sci. 47:6-17.

Seshu D V (1986) Rice varietal testing in rainfed lowlands of South and Southeast Asia. Pages 77-101 *in* Progress in rainfed lowland rice. International Rice Research Institute, P.O. Box 933, Manila, Philippines.

Seshu D V, Cady F B (1984) Response of rice to solar radiation and temperature estimated from international yield trials. Crop Sci. 24:649-654.

Shahjahan A K M, Duve T, Bonman J M (1986) Climate and rice diseases. Pages 125-137 *in* Impact of weather parameters on growth and yield of rice. International Rice Research Institute, P.O. Box 933, Manila, Philippines.

Strain B R, Cure J D (1985) Status of knowledge and recommendations for future work. Pages 205-213 *in* Direct effects of increasing carbon dioxide on vegetation. B. R. Strain and J. D. Cure, eds. U.S. Department of Energy, Washington, D.C.

Wickham T H, Sen C N (1978) Water management for lowland rice: water requirements and yield response. International Rice Research Institute, P.O. Box 933, Manila, Philippines.

Wickham T H, Singh V P (1978) Water management through wet soils. Pages 337-358 *in* Soils and rice. International Rice Research Institute, P.O. Box 933, Manila, Philippines.

Woodhead T (1966) Empirical relations between cloud amount, insolation and sunshine duration in East Africa. I. E. Afr. Agric. For. J. 32:211-213.

Woodhead T (1967) Empirical relations between cloud amount, insolation and sunshine duration in East Africa. II. E. Afr. Agric. For. J. 32:474-477.

Woodhead T (1970) Mapping potential evaporation for tropical East Africa: the accuracy of Penman estimates derived from indirect assessments of radiation and wind speed. Int. Assoc. Sci. Hydrol. 92:232-241.

Yoshida S (1973) Effects of temperature on growth of the rice plant (*Oryza sativa* L.) in a controlled environment. Soil Sci. Plant Nutr. 19:299-310.

Yoshida S (1981) Fundamentals of rice crop science. International Rice Research Institute, P.O. Box 933, Manila, Philippines. 269 p.

Yoshida S, Parao F T (1976) Climatic influence on yield and yield components of lowland rice in the tropics. Pages 471-494 *in* Climate and rice. International Rice Research Institute, P.O. Box 933, Manila, Philippines.

Notes

Authors' addresses: D. V. Seshu, T. Woodhead, and D. P. Garrity, International Rice Research Institute, P.O. Box 933, Manila, Philippines; L. R. Oldeman, International Soil Reference and Information Centre, Wageningen, The Netherlands.

Citation information: International Rice Research Institute (1989) Climate and food security. P.O. Box 933, Manila, Philippines.

Vulnerability of rice to climate

B. Venkateswarlu

Rice is vulnerable to climate and its cultivation continues to be a risky enterprise, despite advances in modern technology. Modern rice is growing commercially primarily in nonproblem irrigated conditions, and even there increases in production are not commensurate with scientific breakthroughs. Climatic factors like low light stress, high and low temperatures, rainfall, high humidity, dew and mist, and abundant and deficit water conditions reduce rice cultivation to subsistence farming. Crop establishment, growth and development, and productivity are affected by the intensity of these climatic factors. Tiller mortality, poor spikelet filling, sprouting, broken kernels, and grain spoilage are the major adverse effects. Salvaging measures, adaptive cultivars, and new skills are needed to increase rice productivity throughout Asia, Africa, and Latin America.

Climate influences plant life in many different ways, and can inhibit, stimulate, alter, or modify crop performance. Its components—temperature, solar radiation, rainfall, relative humidity (RH), and wind velocity—independently or in combination influence rice crop growth and productivity.

Many researchers (De Datta and Zarate 1970, Evans and De Datta 1979, Moomaw and Vergara 1964, Murata 1964, Murthy et al 1976, Sridharan 1975, Stansel et al 1965, Tanaka et al 1964, Venkateswarlu 1977, Yoshida and Parao 1976) have reported on the vulnerability of rice to climatic factors. Apparently rice is more sensitive to climatic aberrations than other cereals because it is a monsoon crop. Cyclonic weather and untimely rainfall result in severe losses in Asia, Africa, and Latin America. Scarcities, market gluts, and surplus-deficit situations (and the economic disorders that follow these events) are the effects of adverse climatic conditions.

Climatic factors

Several approaches have been taken to ascertain the critical levels, ranges, summations, and optima/maxima for the climatic factors that affect rice production. Choudhary and Ghildyal (1969), Matsushima et al (1958), Moomaw and Vergara (1964), Satake (1976), and Toriyama (1962) have determined the temperature optima, critical levels, differences, and ranges for different phenological stages at the tissue, organ, and plant level. For solar radiation, the critical levels and vulnerability of growth phases were reported by Stansel et al (1965), De Datta and Zarate (1970), Murthy et al (1976), Venkateswarlu et al (1977), and Yoshida (1978). However, the nature of productive and nonproductive climates has not been clarified.

Climate types

Nageswara Rao (1981) classified climates as 1) rising, 2) declining, and 3) constant (Fig. 1). In rising climates, the temperature (22-33 °C) and solar radiation (350-540 cal/cm^2 per d) gradually increase from the time of sowing to harvest; RH decreases at late growth stages. This leads to gradually increasing evapotranspirational demand. Productivity is higher and the influence of biotic factors is lower. Areas where crops are sown from December to February generally experience rising climates.

In declining climates, the situation is the opposite: temperature (31-24 °C) and radiation (500-250 cal/cm^2 per d) gradually decrease from time of sowing to harvest; RH increases from 40 to 70% during the reproductive phase. Early wet season crops confront this situation, and the grain yield usually declines from around 20% to as much as 60% of that harvested in a rising climate. Both temperature and solar radiation decrease during ripening. Those conditions are not favorable for efficient and effective grain filling. Thus, the declining climate works against crop efficiency and productivity. Increased RH and lower radiation encourage biotic factors. Declining climates are experienced in Italy, Portugal, Spain, Russia, USA, Philippines, India, Japan, and Korea.

In constant climates, temperatures are nearly constant (24-26 °C), solar radiation plateaus around 350-450 cal/cm^2 per d, and RH ranges from 70 to 80%. In some areas, crops sown from June to July experience constant and stable temperatures, prolonged lower solar radiation, and higher RH, which do not favor higher productivity. In fact, yields are 20 to 40% less than in rising climates. This suggests that tropical and subtropical conditions are not highly favorable for rice production. Thailand, Indonesia, Philippines, Sri Lanka, and some parts of China experience constant climates.

Relatively higher temperatures at early stages of rice, with a decline around primordial initiation and subsequent higher solar radiation (Yoshida and Parao 1976) do not happen in most situations in the tropics. In effect, in most rice-growing countries, farmers confront unfavorable and risky climates. In such situations, the yield potentials of modern rice are not realized. With intensive input management, the number of spikelets can be increased, but number of grains decreases (Fig. 2), resulting in 30-60% gaps.

This means that 2-3 t/ha is lost in a declining climate and 1.5-2.5 t/ha is lost in a constant climate, in a crop that normally should yield 5 t/ha. The gap increases more with high management levels, where the potential is 10 t/ha, but the loss due to climate runs to 5 t/ha. This is a serious gap. The loss also is silent. We often are more concerned about yield losses due to biotic factors, which are sporadic, irregular, and unpredictable. Losses due to climatic factors, which are regular and extensive, go unnoticed. Modern cultivars could be yielding only 30-50% of their potential.

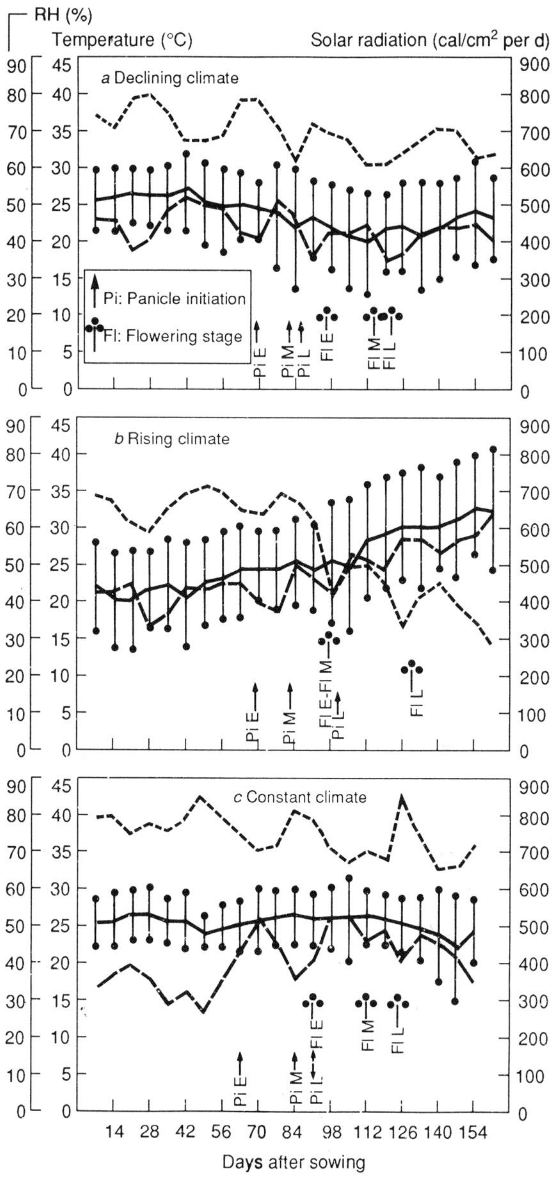

1. Climatic factors showing a) declining trend during crop season—"declining climate"; b) gradually rising trend during crop season—"rising climate"; and c) stable or constant trend during crop season—"constant climate."

Climate factor combinations

Combinations of climatic factors also should be considered. Intensity and stability vary from factor to factor and from situation to situation, and influencing crop growth and productivity differently. In general, low solar radiation results in lower

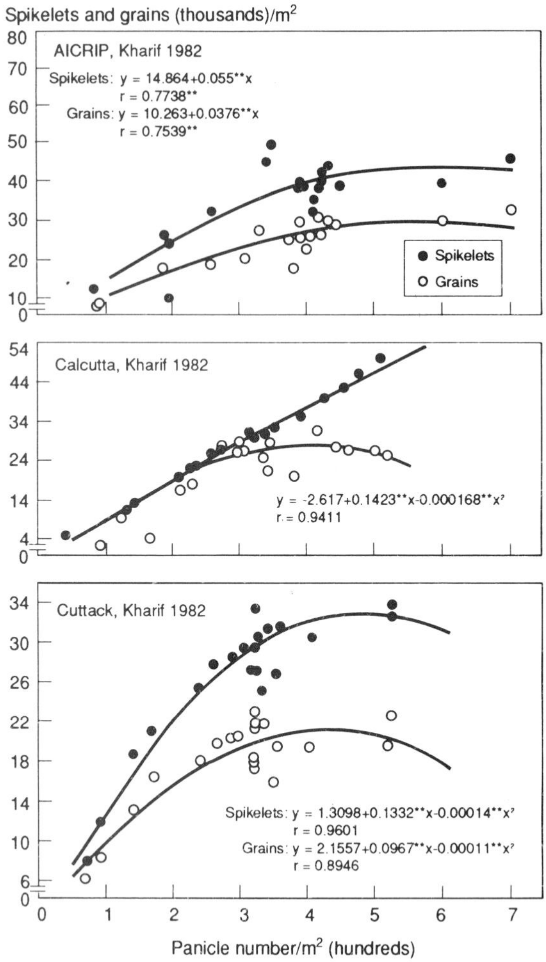

2. Nature of the gap between spikelets and grains at different panicle numbers in rice.

productivity. When it is associated with lower temperatures (about 18-22 °C), yields are still lower. If, in addition, RH decreases and wind velocity increases, the effect is compounded.

Climatic factors can compensate for one another. The adverse situation often experienced during the reproductive phase of summer rice in India, Pakistan, Kampuchea, Iran, Iraq, Laos, and Thailand is a result of the interaction of high temperatures + higher solar radiation + lower RH (Venkateswarlu et al 1977). In that situation, either flowering ceases, or grains do not fill, or panicles turn white, with ears resembling stem borer damage. If solar radiation is high and temperatures are moderate, the effect decreases. If RH increases are added to the decrease in temperature, the negative effect further abates, sometimes keeping losses to a minimum.

Adjusting to climate

Regular monitoring of crops and climatic factors would help in predicting problems, allowing measures to be taken to minimize losses.

Low light conditions

Cloudiness, with resulting lower radiation, is a problem in the coastal areas of Asia, Africa, and Latin America during the monsoon season (Murata 1964, Nayak and Murthy 1980, Stansel et al 1965, Venkateswarlu et al 1977, Yoshida 1978). Murthy et al (1976) found that 30-60% of yield losses under cloudy conditions occur primarily in medium- and long-duration rice cultivars. Tillering, grain filling, assimilation of CO_2, and grain yield are severely affected (Fig. 3). Low light intensity is a major stress during the rainy season, in the coastal and adjoining forest areas in Africa. It also encourages biotic factors, particularly planthoppers (Lin 1970, Nishida 1975, Pathak 1968, Venkateswarlu and Kalode 1981). Recently, it was observed that genotypes differ in adaptation to low light stress, indicating the possibility of developing cultivars with tissue tolerance and plant competence for this stress. Such an approach might boost yields in 60-70% of the rice-growing areas, helping to achieve another plateau in rice yield potential.

Rainfall and wet weather

Accumulation and stagnation of water in ricefields affect crop stand and yield. A major problem is when rainfall wets the mature grain of a standing crop, which then sprouts in the field. Hardly any modern rices are dormant, and this problem is a major threat in coastal areas. Spraying acetic acid (0.2%) at 20-25 d after flowering has been found effective in checking preharvest sprouting.

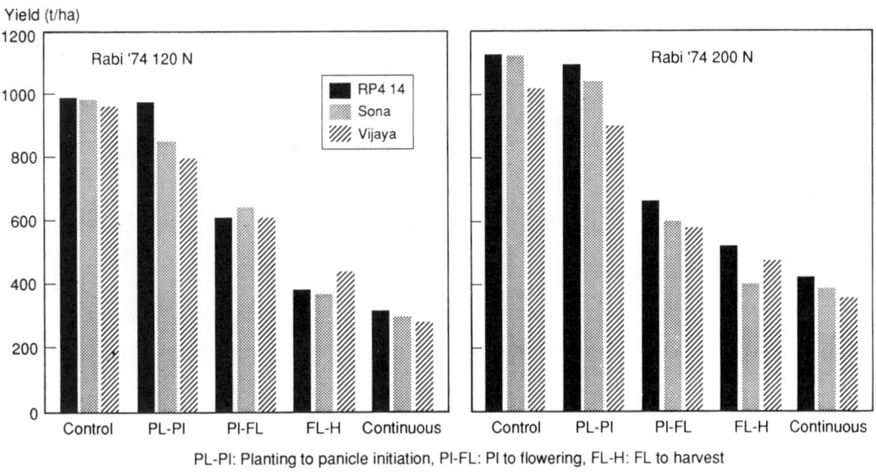

3. Effect of low-light intensity (40-50% of natural light) at different growth stages on grain yield (Venkateswarlu et al 1977).

The excess water problem is further compounded when harvested rice is spread in the field to dry. Rains at this stage soak the panicles, resulting in massive sprouting. In India, this problem occurs every 2-3 yr, particularly in the coastal areas. A 5% common salt solution can be used to inhibit sprouting (Venkateswarlu and Somasundara Rao 1981). Soaked panicles are immersed in 5% solution, then stacked in the field. The salt solution arrests germination and, when the weather is clear, the grain can be threshed and dried. Dipping in salt solution also helps prevent grain discoloration, mold development, and off-odor.

Wet-threshed or salt-treated wet grains can be kept in coarse fiber bags 10-13 d. This helps to arrest sprouting, discoloration, mold development, and off-odor. When the weather is clear, the rough rice can be dried and used as normal grain.

Drying rice sheaves in the field is a problem in both wet and dry seasons. In the wet season, harvest is in October and November when unpredictable showers and drizzle, as well as dew, mist, and fog, occur. Panicles are exposed to alternately wet and dry conditions. The exposed grain is vulnerable to fungi, bacteria, and algae, resulting in discoloration and spoilage. During the dry season, radiation and temperature increases lead to overdrying. That favors grain fissuring, leading to a large proportion of broken grains in milling. A simple and practical method of drying was devised by Padmaja Rao and Venkateswarlu (1986): sheaves are arranged continuously, with panicles of one bundle covered with the straw of the following bundle. This protects the grains from dew, mist, and light to moderate showers, and from high solar radiation. That helps prevent sprouting and enhances milling and head rice recovery.

Waterlogging is a major constraint in Southeast Asia. Crop stand is affected, the functional ability of the plant is paralyzed, and root activity decreases, leading to lower productivity. Semideep and deep water conditions are the problems. Planning to create drainage systems should be the long-range strategy, developing suitable cultivars and new management skills should be the immediate concern. Submergence tolerance, tiller survival, carbon storage, root activity, photosynthesis, and spikelet filling ability are characters that need improvement.

Deep marshy soils and areas susceptible to tidal waves pose salinity and alkalinity problems. Cultivar improvement has received some attention, but crop management practices, such as increased seedling age, number of seedlings per hill, and drainage still need refining.

Rainfed fields with 750-1,100 mm rainfall suffer in several ways: inadequate availability of nutrients such as N, P, Si, and Fe, and Al toxicity result in poor performance. A deep, thick, and actively penetrating root system is needed to tap moisture from deeper layers of soil. Thick cuticle, sensitive stomata, desiccation-tolerant tissue, and deep and active root systems are traits that suit these areas.

References cited

Choudhary T N, Ghildyal B P (1969) Germination response of rice seeds to constant and alternating temperatures. Agron. J. 61:328-330.

De Datta S K, Zarate P M (1970) Environmental conditions affecting the growth characteristics, nitrogen response and grain yield of tropical rice. Biometeorology 4:71-89.

Evans L T, De Datta S K (1979) The relation between irradiance and grain yield of irrigated rice in the tropics, as influenced by cultivars, nitrogen fertilizer application and months of planting. Field Crops Res. 2:1-17.

Lin K S (1970) Studies on the microclimatic factors in relation to the occurrence of the rice planthopper [in Chinese, English summary]. Plant Prot. Bull. (Taiwan) 12(4):184-189.

Matsushima S, Tsunoda K, Manaka T (1958) Effects of air temperatures, light intensity, difference between day and night temperature at different stages of growth on ripening in rice plants [in Japanese]. Agric. Hortic. 33:877-883.

Moomaw J C, Vergara B S (1964) The environment of tropical rice production. Pages 3-13 *in* Mineral nutrition of the rice plant. Proceedings of a symposium at the International Rice Research Institute. Johns Hopkins Press, Baltimore.

Murata Y (1964) On the influence of solar radiation and air temperature upon the local differences in the productivity of paddy rice in Japan [in Japanese, English summary]. Proc. Crop Sci. Soc. Jpn. 33:59-63.

Murthy K S, Nayak S K, Sahu G (1976) Effect of low light stress on rice crop. Pages 74-84 *in* Proceedings of a symposium on crop plant response to environmental stresses held at Vivekana Laboratory for Hill Agriculture, 4 July 1975, Almora, UP, India.

Nageswara Rao E (1981) Studies on growth, nutrient uptake and productivity in rice in relation to climatic rhythms. Ph D thesis, Osmania University, Hyderabad, India.

Nayak S K, Murthy K S (1980) Effect of varying light intensities on yield and growth parameters in rice. Indian J. Plant Physiol. 23(3):309-316.

Nishida T (1975) Causes of brown planthopper outbreaks. Rice Entomol. Newsl. 2:38.

Padmaja Rao S, Venkateswarlu B (1986) More rice if you dry your paddy this way. Intensive Agric. 24(3):19-20.

Pathak M D (1968) Ecology of common insect pests of rice. Annu. Rev. Entomol. 13:257-294.

Satake T (1976) Determination of the most sensitive stage to sterile-type cool injury in rice plants. Bull. 113. Hokkaido National Agricultural Experiment Station, Japan. 35 p.

Sridharan C (1975) Studies on the influence of climatological factors on rice under different water management practices. Ph D thesis, Orissa University of Agriculture and Technology, Bhubaneswar, India.

Stansel J W, Bollich C N, Thysel J R, Hall V L (1965) The influence of light intensity and nitrogen fertility on rice yield and components of yield. Rice J. 68:34-35.

Tanaka A, Navasero S A, Garcia C V, Parao F T, Ramirez E (1964) Growth habit of the rice plant in the tropics and its effect on nitrogen response. IRRI Tech. Bull. 3. 80 p.

Toriyama K (1962) Studies on cool tolerance of rice varieties with special reference to testing methods and its inheritance [in Japanese, English summary]. Bull. Aomori Agric. Exp. Stn. 7:109-153.

Venkateswarlu B (1977) Influence of low light intensity on growth and productivity of rice, *Oryza sativa* L. Plant Soil 47:713-719.

Venkateswarlu B, Kalode M B (1981) Influence of narrow and broad leaf canopies on population dynamics of brown planthopper (*N. lugens*) in rice. Indian J. Plant Physiol. 24(3):183-192.

Venkateswarlu B, Maddulety T Y, Subbaiah S V (1977) Influence of hot dry climate on sterility and ripening in irrigated rice (*Oryza sativa* L.). Indian J. Plant Physiol. 22:173-180.

Venkateswarlu B, Prasad G V S S, Rao A V (1977) Effects of low light intensity on different growth phases in rice (*Oryza sativa* L.). Plant Soil 47:37-47.

Venkateswarlu B, Somasundara Rao V (1981) Prevention of germination and spoilage of submerged paddy. J. Food. Sci. Tech. 18:161-163.

Yoshida S (1978) Tropical climate and its influence on rice. IRRI Res. Pap. Ser. 20. 25 p.

Yoshida S, Parao F T (1976) Climatic influence on yield and yield components of lowland rice in tropics. Pages 471-494 *in* Climate and rice. International Rice Research Institute, P.O. Box 933, Manila, Philippines.

Notes

Author's address: B. Venkateswarlu, Directorate of Rice Research, Rajendranagar, Hyderabad, India.

Citation information: International Rice Research Institute (1989) Climate and food security. P.O. Box 933, Manila, Philippines.

The climatic vulnerability of wheat

R. B. Austin

The interannual coefficient of variation in yield is suggested as a good measure of food security for wheat. Among regions with different production levels, low yield is associated with a high interannual coefficient of variation in yield. In part, this is because the more marginal climates for wheat production are also the more variable with respect to environmental factors that limit yield, particularly rainfall and temperature. Results of selection experiments show it is possible to breed varieties with higher yield potential and reduced susceptibility to climatic hazards, and so more stable yields. However, the best fitness in a particular range of environments will be achieved by breeding for that range. General adaptation to a wider range of environments will entail some sacrifice of fitness to a more particular subset of those environments. If wheat cultivation is to be extended to the lowland tropics, new sources of heat tolerance will need to be found, perhaps in related species or by molecular biological techniques.

In 1985, world wheat production was estimated to be some 510 million tons. Wheat can be grown under temperature extremes ranging from –40 °C in winter in parts of Canada and the USSR to +40 °C during grain filling in some low-latitude regions. It also grows in areas with wide differences in photoperiod and water availability. By these measures, it is a very adaptable species.

This adaptability arises partly because breeders have exploited the considerable genetic variation within the species to produce varieties suited to particular environments. Wheat is also phenotypically adaptable because, like other Gramineae, it can tailor its growth to make the best use of available resources, particularly nutrients and water. Of course, all plants possess this characteristic to some extent. But cereals, especially wheat, exhibit it to a high degree.

Wheat also is a very good staple dietary item, especially when the whole grain is eaten. It may be cooked in a variety of ways to suit diverse tastes and lifestyles.

In addition to being a function of climate, soil conditions, and husbandry, wheat yields depend on the extent to which the varieties grown are adapted to those conditions. In areas where lack of rainfall or extremes of temperature impose severe limits on yield, or in areas where husbandry is poor, yields are commonly only 1-2 t/ha. In favorable climates, on moisture-retentive soils, with good husbandry, such as are found in northwestern Europe, yields regularly exceed 10 t/ha.

This large geographical variation in yield is not of primary concern for food security, because the variation is well known. Population distribution and trading patterns take it into account. The gradual changes in yield and production that occur in response to economic and political pressures and climatic drift can easily be accommodated without endangering food security.

Of greater concern are year-to-year, or interannual, variations in yield. Some interannual variation is an unavoidable consequence of variations in weather, although both on a farm scale and nationally, variation can occur because of such factors as fertilizer availability and cost. Although improved agronomic practices and higher levels of inputs primarily tend to increase yields, they can also decrease the variability of yield.

Breeders can contribute to the decrease in variability of yield in a number of ways. If resistant varieties are provided, yield losses to pests and diseases, which typically vary in severity from year to year, can be reduced or eliminated. Varieties also need to be adapted to the climate, including variable climatic hazards, in the areas for which they are bred. It is possible that the degree of insurance against variable environmental hazards that varieties need to have involves some yield loss, but this is very difficult to quantify.

Variability of yield

The total variation of yield over a large number of fields and years can be partitioned into two main components: site-to-site and year-to-year. Church and Austin (1983) partitioned the site-to-site variance into two components: yield between fields on the same farm and yield between farms, and the year-to-year variance into three components: individual farm average yield, average district yield, and national average yield. Part of the variance in each component will be caused by environmental factors. Part of the interannual variation in national yield will be caused by the effects of changes in agronomic practices (including the varieties grown) over the years. This will be manifest as a slow trend and is often removed when the effects on yield of variation in year-to-year weather are considered.

The main concern here is the extent of the interannual variation in national yield and any evidence that it can be decreased directly, by agronomic practices, or indirectly, by breeding varieties with more stable yields. These issues were among those considered at a 1985 conference organized by the International Food Policy Research Institute and the German Foundation for International Development. At the conference, it was agreed that no method of expressing variation in yield or production was appropriate to all situations.

For countries that are large producers and consumers of wheat and where the area planted does not fluctuate much from year to year, appropriate measures of variability of supply are the coefficient of variation (CV) of national yield or the standard deviation of yield. Table 1 gives wheat yields and their interannual standard deviations for different areas of the world. In general, the larger the yield, the smaller the CV of yield (Fig. 1). To the extent that mean yields are a measure of the suitability of the physical environment (mainly temperature and rainfall) for wheat production, the less suitable environments also give the more variable yields.

Table 1. Mean yields and their standard deviations (SD) for wheat-producing areas of the world.

Country/area	Yield[a] (t/ha)	SD of yield[b] (t/ha)
USSR	1.78	0.27
USA	2.02	0.11
Western Europe (UK, France, FRG, Italy)	3.87	0.28
India, Pakistan	1.21	0.09
Eastern Europe (Poland, Romania, Hungary, DDR, Czechoslovakia, Bulgaria)	3.04	0.30
Canada	1.62	0.22
Australia	1.26	0.23
Brazil	0.85	0.18
Argentina	1.52	0.21
Egypt, Turkey, Spain	2.07	0.20
South Africa	0.98	0.14

[a]Averages for the 12 yr 1960, 1961, 1962, 1967, 1968, 1969, 1974, 1975, 1976, 1982, 1983, and 1984. Sources: FAO Production yearbook vol. 17, 29, and 38, FAO Rome, 1985. [b]1960-61 to 1982-83, linear time trend removed where significant. Source: Hazell 1985.

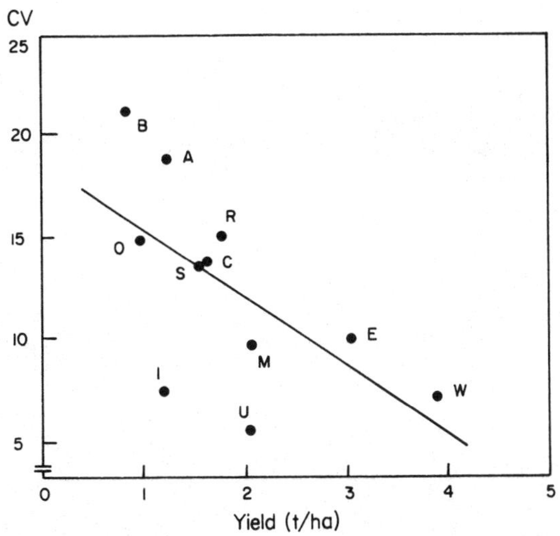

1. Relationship between interannual coefficient of variation (CV) of yield (linear time trend removed) and mean yield for different wheat-producing areas: B = Brazil; A = Australia; R = USSR; O = South Africa; S = Argentina; C = Canada; M = Egypt, Turkey, and Spain; E = Eastern European countries (Poland, Romania, Yugoslavia, Hungary, Bulgaria, Czechoslovakia, and DDR); I = India and Pakistan; W = Western Europe (UK, France, FRG, and Italy); U = USA. Data from FAO production yearbook, vol. 17, 29, and 38 and from Hazell (1985). Fitted regression shown, $r = -0.613$ ($p = 0.05$).

Clearly, socioeconomic factors also can contribute to variation in yield, because they determine whether farmers can afford inputs such as fertilizers. An example of this is given in Table 2. When plots were fertilized with nitrogen in the spring, but not in the fall, yields were the least variable. The primary reason probably was variable

Table 2. Yield (t/ha) and coefficient of variation (CV) of yield for some plots in the Broadbalk Experiment, Rothamsted, UK.

	No fertilizer	FYM[a] only	Inorganic fertilizer only[b]
1852-1918			
Grain yield (t/ha)	0.86	2.43	2.51
CV (%)	24.9	16.7	20.4
1970-78			
Grain yield (t/ha)	1.70	5.87	5.09
CV (%)	16.8	17.0	9.8

[a]Farmyard manure. [b]P, K, and Mg in autumn; 144 kg N/ha in fall 1852-1918 and in spring 1970-78.

leaching of nutrients, particularly nitrogen, in the winter following fertilizer application.

In general, the more control farmers have over yield-reducing factors, the higher their yields will be. Although they may be more variable in absolute terms, they will be relatively less variable in the particular situation. For example, in very simple systems, such as 13th century wheat farming in England, yields were about 0.5 t/ha. The interannual CV of yield was about 30%. From 1950 to 1985, the mean yield in England was 4.2 t/ha and the interannual CV about 8% (Austin and Arnold 1989). From these figures, we can see that security of wheat production in England is much better now than it was 7 centuries ago.

In England, present levels of interannual variation in yield for individual fields and between-field variation within a year are about 15-20% of mean yield (Church and Austin 1983). Experience tells us this variation is likely to be the result of the influence of many factors, each having a continuous distribution of effects and interactions.

Simulations show that the observed variation could rise if yield was affected by 10 or more factors with independent, additive effects, each having a 10% effect on yield (Fig. 2). These simulations are instructive because they show that removing the variation caused by 1 factor (with a 10% effect on mean yield, such as might occur after the introduction of a disease-resistant variety) would have a barely perceptible effect on the total variability of yield. However, in the simulation, mean yield would be increased a very significant 5%.

These broad conclusions are likely to be affected by interactions, as can be illustrated by a hypothetical example. Suppose that the effect of one factor, drought, is such as to prevent positive and additive responses to three other factors, nitrogen, variety, and cool weather. If the independent effects of each of these factors were 10%, then the effects of the interactions with drought could be simulated by replacing the 4 factors by 1 factor having a 40% effect. Mean yield would decrease by 20%, but the CV of yield would increase from 15% (10 factors each having a 10% effect) to 20% (6 factors each having a 10% effect and 1 having a 40% effect). In favorable areas for wheat, such as Western Europe, or in irrigated wheat production, interactions would not be as dramatic as this example.

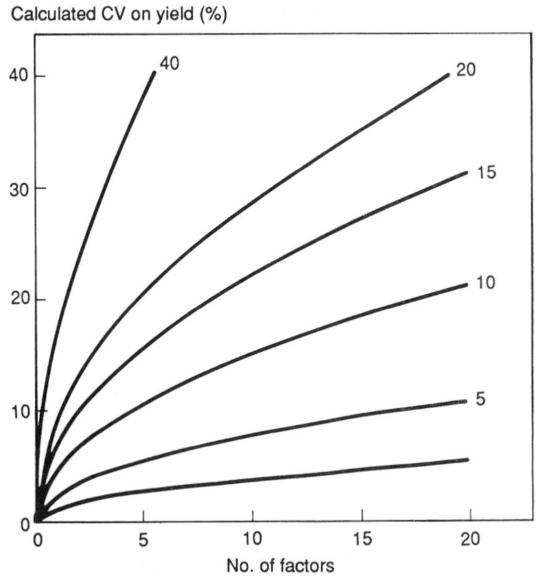

2. Calculated coefficients of variation (CVs) in yield as a function of the number of factors causing variation. Lines are for factors increasing yield by 2.5, 5, 10, 15, 20, and 40%.

In a less favorable area for wheat, it might be supposed that 1 factor, such as drought, might be dominant and reduce yield by 40%, with the interactions with other factors preventing positive responses to 3 of them, as in the previous example. But suppose that 6 additional factors also caused variation in yield, each having only a 5% effect. This combination of effects and interactions would give a 30% CV (32% of the variance would be accounted for by the major factor and its interactions). This is similar to the highest values of the interannual CV observed for national yield.

Interannual CVs of yield also will depend on the homogeneity of climate and cultural practices in the area for which they are calculated. The larger the area, the more the effects on yield of spatial variations in rainfall, temperature, etc., will average out. This is probably one reason the interannual CVs of yields in the USA and India plus Pakistan are less than would be expected for the yields obtained (Fig. 1). In practical terms, the larger the area and the more diverse the climate for wheat growing, the less will be the interannual variation in yield, and the lower will be the probability of a given shortfall. Whether the population will benefit depends on other factors, including the ability to transport grain within the area.

Environmental factors as causes of yield variation

Evidence from statistical studies

Experience tells us that, apart from episodic events (killing frosts, hail, etc.), wheat yields are a function of weather. The Romans recognized that wheat yields were greater after dry than after wet winters. This observation was confirmed 2,000 yr later by a formal statistical analysis of data from the Broadbalk experiment at

Rothamsted in the United Kingdom (Buck 1961). Importantly, this effect of winter rainfall was not evident in plots that had received 100 kg N/ha as inorganic fertilizer, indicating that the adverse effect of winter rainfall was a consequence of the leaching of nitrate from the soil. Buck also concluded that at Rothamsted, variations in sunshine had little effect and variations in temperature no effect on yields. Stated in other terms, in the normally equable climate of eastern England, the effects of variation in individual elements of the climate are small in relation to the effects caused by variation in all other factors. Broadly similar results have been found in other regions with equable climates (Pitter 1977).

Not surprisingly, the picture is different in environments more limiting for wheat production. For example, Cornish (1950) found in South Australia that 70-80% of the variance of yield over 296 locations for the years 1896-1941 was accounted for by variation in seasonal rainfall. More recently, Seif and Pederson (1978) found that rainfall during early vegetative growth strongly correlated ($r=0.87$) with grain yield in 37 trials at different locations in New South Wales, Australia, 1968-72.

Nix and Fitzpatrick (1969) analyzed yields for the 15 yr 1950-65 in Queensland, Australia, and found that grain yield correlated positively ($r=0.70$) with the sum of available water at planting plus rainfall from sowing to heading. In further analyses, these authors found that the ratio of available water in the root zone at the beginning of a 2-wk period around heading to potential evapotranspiration during the period was strongly correlated with yield ($r = 0.83$), indicating that wheat is especially sensitive to drought stress during this period. A great deal of experimental evidence confirming this has now accumulated (e.g. Fischer 1973). It shows that the component of yield most affected is number of grains per spikelet. In the Great Plains of the USA, Feyerherm and Paulsen (1981) found that moisture availability was more than three times more important than temperature in causing variation in winter wheat yields.

It can be concluded that variations in water availability are the dominant causes of spatial and temporal variation in wheat yields in many areas of the world. Water availability is determined by rainfall, solar radiation, temperature, and the other meteorological elements that affect evapotranspiration. In much of the wheat-growing areas of the world, the direct effects of variation in the nonrainfall elements of the environment are much less than their indirect effects through water availability. Much of the evidence suggests that agronomic practices can greatly modify all these relationships and reverse the sign of some of them.

Evidence from experiments

The use of regression analysis to relate yield to weather does not reveal the underlying causes of any significant relationships detected. Causal relationships need to be sought through experimentation. Innumerable experiments of this kind have been carried out. In field and glasshouse experiments, associations are commonly found between the factor under study and other factors, so that the results are difficult to interpret.

Experiments in controlled environments can be free of such associations but may suffer from other disadvantages. Plants grown in pots have different root development and water supply patterns from those grown in the field. Light intensity is commonly much lower than that experienced under many field conditions, and the spectral composition is often different. Wind, certainly gustiness, is virtually absent and spatial and temporal variations in temperature and vapor pressure are generally abnormal.

Despite these disadvantages, much has been learned about the effects of temperature, daylength, light intensity, water stress, vapor pressure, and atmospheric CO_2 concentration on photosynthesis, leaf initiation and expansion, floral initiation and spike development, and many other aspects of growth and ontogeny. However, it is generally impossible to apply this knowledge to any quantitative assessment of the effect of a given difference in environment (one location versus another, or one year versus another) on yield.

Simulation models should make it possible to integrate component processes and their dependence on environmental factors in such a way as to allow predictions to be made. Advocates of this approach claim a reasonable degree of success in predicting responses to climatic factors, although not absolute levels of yield (Godwin and Vlek 1985, Otter and Ritchie 1985). At present, however, physiologically based statistical models appear to give more reliable measures of the importance of particular factors as causes of variation in yield than do simulation models.

The adaptability of wheat

Hexaploid bread wheat is presumed to have originated in the fertile crescent region of the Middle East, where its presumed wild diploid progenitors must also have grown. As far as is known, all the present-day wild diploids that are related to those progenitors, and the wild tetraploid *Triticum dicoccoides*, grow in climates that have either cool or cold winters, warm springs, and hot summers (Harlan and Zohary 1966). The physiology of all the species is attuned to this type of climate. In general, the seeds have no pronounced dormancy. After ripening in May or June, they remain at rest until the onset of the winter rains, to germinate when wetted. The seedlings of some species and/or ecotypes may experience subzero temperatures for several weeks, and therefore need to be winter hardy. A vernalization requirement is often associated with this. In nature and in agriculture, vernalization prevents ear initiation before the onset of severe weather, which would otherwise damage the ear primordia. However, winter hardiness is not entirely due to the effects of vernalization genes. In low-altitude areas with mild winters, ecotypes lack both a vernalization requirement and winter hardiness. These environments are usually the ones with the earliest onset of hot dry weather, where it is important that the plants grow rapidly and reach the reproductive phase before the onset of hot weather (Kushnir and Halloran 1982). Genotypes from these areas grow rapidly in cool temperatures and short days, characteristics needed in wheat for many low-latitude regions, including India, Mexico, and Australia.

In their natural environments, plants of the wild species usually reach ear emergence and anthesis just before temperatures start to increase rapidly and reserves of soil moisture approach exhaustion. Wild diploid species have small grains and a short grain filling period to cope with these conditions.

Hexaploid wheat, which man has selected for its larger grains, benefits from a longer grain filling period. If wheat is prematurely killed by drought and high temperature, grains will not reach their maximum size. However, wheat is partially buffered from the effects of late drought via several mechanisms. First, drought in the environments where wheat (as well as the wild species) grows becomes progressively more severe as the season advances. Drought after the formation of tillers is complete will cause late-formed tillers to die; the more severe the drought, the more late tillers die. The effect of this tiller death is to reduce water loss and to trim the future demand of the crop for water to a level likely to be satisfied from what will be available. Second, drought advances development (times of ear emergence and anthesis), helping those shoots that survive to escape the worst effects of late drought. Third, drought reduces extension growth and the demand of the vegetative organs for assimilates at a time when crop assimilation rates are close to their fastest. As a consequence, a greater proportion of assimilates are stored as remobilizable carbohydrate (mono-, di-, and oligosaccharides), mainly in the stems. These carbohydrates are mobilized during grain filling to partly meet the demand of the grain for carbon for starch synthesis (Austin et al 1977, Bidinger et al 1977). There may be genetic variation in capacity to accumulate a buffer store of carbohydrates to provide protection against terminal drought, but it is difficult to select for this character.

We do not yet have the ability to make a complete inventory of the allelic variation in wheat and its relatives that has given the species the genetic flexibility needed for the evolution of ecotypes capable of exploiting the range of environments encountered. Many genes in wheat are triplicated (one each from the A, B, and D genomes). This gives a great deal of scope for continuous variation in attributes determined by common loci, even where there are only two allelic variants. Judging from documented cases (e.g. the high molecular weight glutenin subunits, Payne and Lawrence 1983), there may be multiple alleles at many loci, giving even greater opportunity for variation. Added to this is the certainty of interacting effects with genes at other loci and, for some important proteins (e.g. LHC II, the light harvesting chlorophyll a/b binding protein), the existence of many different loci throughout the genome that code for generically the same protein. For example, there are probably 40-50 genes that code for LHC II in wheat.

For most of the agronomically significant attributes that vary genetically in wheat, the genes involved are not known. Table 3 lists some of these attributes. Most of the attributes have pleiotropic interrelationships with one or more others as well as with yield. Usually, breeders will only be able to select explicitly for easily identifiable attributes, but breeding and trial procedures will give varieties that have the optimum expression of all attributes important for high and stable yield. Physiologists argue that progress in breeding would be faster and more certain with better knowledge of the significance for yield of particular attributes and how that depends on environmental factors. For some attributes, this knowledge is beginning to accumulate.

Table 3. Some physiologically significant characters in wheat.

Genetic variation and its consequences well established
Duration of life cycle
Sensitivity to photoperiod
Sensitivity to vernalizing temperatures
Earliness which is independent of (a) and (b)
Cold hardiness
Tiller production and survival
Mature plant height
Components of grain yield per ear
Leaf posture
Genetic variation and its consequences less well established, uncertain, or unknown
Photosynthetic characteristics, particularly maximum rates at light saturation
Rooting behavior
Leaf longevity
Biomass production
Responsiveness to major nutrients N, P, and K
Sensitivity to minor element deficiencies
Tolerance for toxic levels of elements, e.g. aluminum, and for salinity
Drought tolerance and water-use efficiency
Heat tolerance

Exploiting the adaptability of wheat

Life cycle duration

In low-latitude regions of the world, the growth and yield of wheat are limited either by long, hot, dry summers or by continuously superoptimal temperatures. The exceptions are where wheat is grown at high altitudes (Kenya highlands, Ecuador) with conditions that approximate those of temperate regions. In areas with hot dry summers, the life cycle of the crop must be tailored so that grain filling is not prematurely terminated by high temperature and drought (Hoogendoorn 1985). The lower the latitude, the shorter the growing season and the shorter the period from sowing to flowering when potential yield (numbers of fertile florets/m^2) is determined. Clearly, the shorter the season, the lower the potential and realized yield. Varieties adapted to those climates lack a vernalization requirement and are photoperiod insensitive. In addition, they possess precocity genes whose effects are independent of those of photoperiod and vernalization genes.

Photoperiod and vernalization insensitivity genes confer on wheat the ability to flower sooner in the short days of low-latitude regions (10- to 12-h days) than plants possessing the contrasting alleles. This insensitivity needs to be augmented by precocity genes in varieties grown in regions with very short seasons. Although precocity genes appear to be located on several chromosomes, the physiological basis of their action is not known. It is possible that their primary effects are on the temperature response of leaf and spikelet primordium initiation.

In general, there is much genetic variation in life cycle timing, and breeders have relatively little difficulty producing the genotypes with appropriate developmental patterns that are a prerequisite for breeding programs. There is likely to be considerable scope for fine tuning, however, especially when breeding programs are targeted to a relatively small geographical area over which the climate does not vary greatly. For example, in an analysis of a chronological succession of English varieties, it has been found that modern varieties reach the stage of ear emergence 2-3 wk earlier than traditional varieties (Austin et al 1989).

In part, this may be pleiotropically associated with reduced height and/or reduced number of leaves per shoot. However, in selection experiments, early flowering was found to have a beneficial effect on yield that was independent of variation in plant height caused by the *Rht*$_2$ gene (Innes et al 1985). Flowering time is especially critical in the Queensland wheat belt, where yields decline by an average of 1.2%/d when the plants flower either before or after the midwinter optimum (Woodruff and Tonks 1983).

In cool conditions, rapid leaf appearance is associated with rapid production of tiller primordium sites, and so with high tillering capacity. In environments where water and nitrogen are not too limiting, high tillering will benefit yield (Innes et al 1981).

Drought tolerance and water-use efficiency

Some physiologists have held out the promise that the drought tolerance of crops can be substantially improved, enabling crops to be extended into drier areas than heretofore. Often, the key to success has been advocated as screening for a particular morphological or biochemical trait. The literature abounds with descriptions of such screening tests and correlations with drought resistance. Wisely, many breeders have dismissed these claims as being unlikely to help them select for yield in moisture-limiting environments, either because they do not believe the claims or because they consider the screening tests impractical to use on a large enough scale to stand a reasonable chance of success.

Good drought tolerance may be equated to good yield in a particular moisture-limited environment. Thus, like yield in any other environment, it is likely to be determined by a great many genes and their interactions. In reasonably well-adapted breeding material, no substantial variation in performance (i.e. more than a 10% effect on yield) is likely to be found that can be attributed to a single gene or attribute. In addition, an attribute that is beneficial in one environment (or season) with a particular level and pattern of water availability may be detrimental in another one.

These principles are illustrated by the results of a series of selection experiments at the Plant Breeding Institute, Cambridge. Selections were evaluated at 3 levels of water availability: full irrigation, drought for about 4 wk before anthesis, and drought during grain filling. The conclusions are summarized nonquantitatively in Table 4. They show that selection in appropriate directions for the attributes tested would benefit yield in one or two of the environments, but usually not in all three. The existence of this G × E interaction means that it would only be worth selecting for particular traits as part of the process of breeding for high yield if the interannual variations in water availability were less than the regional variations in water availability. Where this is so, it may be worthwhile developing subprograms to produce varieties for particular soil types or for regions within a general area.

Table 4. Effect on yield of selection for different plant attributes as assessed in field experiments at the Plant Breeding Institute, Cambridge, UK.

Attribute	Reference	Full irrigation	Drought for 4 wk before anthesis	Drought during grain filling
High number of ears/m^2	Innes et al 1981	+	–	+
High number of grains per ear	Innes and Blackwell 1981	0	–	+
Early ear emergence	Innes et al 1985	0	0	+
Rht_1 and Rht_2 dwarfing genes	Innes et al, unpublished, cited in Austin 1987	+	+	+
Erect leaves	Innes and Blackwell 1983	+	+	+
High capacity to accumulate abscisic acid	Innes et al 1984	+	0	+

[a] + indicates a beneficial effect, – a detrimental effect, and 0 no effect.

It must be emphasized that in extremely water-limited environments, the scope for yield improvement is likely to be small, particularly if there are severe colimitations imposed by soil or climatic factors.

Heat tolerance

In the existing gene pool of bread wheat, researchers have not found sufficient variation in plant development to produce varieties for the lowland tropics. Yet there is a growing demand for wheat and its products in countries with such climates and the lack of foreign exchange in some of them causes their governments to be interested in promoting wheat production.

In the lowland tropics, wheat is severely affected by leaf and ear diseases, notably *Helminthosporium sativum*. Additionally, plant development is abnormal, with few tillers, and hence few ears, formed. The ears are small and set few grains. It is possible that these responses could be mitigated or avoided if genotypes with slower rates of development at high (20-30 °C) temperatures than existing varieties could be found. Limited evidence from studies on rates of initiation of leaf primordia at different temperatures is not inconsistent with this prediction (Fig. 3).

In the known geographical range of *Triticum* and *Aegilops*, all species and ecotypes appear to experience cool temperatures during the early vegetative stage. Unless ecotypes that grow in areas where the temperature is substantially higher can be found, breeders will not be able to turn to the wild species for sources of increased heat tolerance. However, a possible source of heat tolerance may be found in *Secale africanum*, a wild species of rye that appears to grow in the wet, warm season in parts of South Africa. This species is being studied and, if it is found to be heat tolerant, attempts will be made to transfer its genes to bread wheat via a triticale route.

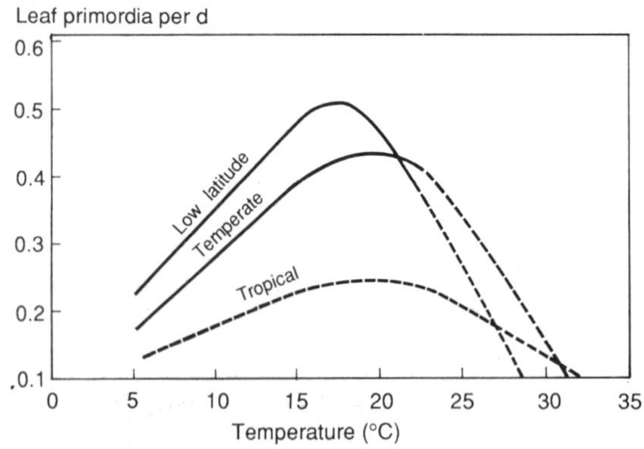

3. Response in rate of leaf initiation to mean temperature. Temperate area spring wheat; low-latitude spring wheat; hypothesized response of heat-tolerant genotype. Continuous lines are based on data, dotted lines are extrapolations.

Prospects

Because the modern diploid *Triticum* and *Aegilops* species, and presumably the diploid ancestors of wheat, have evolved in a region of diverse climate, and because wheat is allo-polyploid, it has wide variations that have been exploited by breeders to produce varieties that can be grown over a wide range of climates (Zohary et al 1969). The sustained rate of genetic improvement in yield suggests that progress will continue to be made by conventional breeding. However, higher yields imply greater nitrogen use. Thus, sustained benefit from genetically improved yield potential will only be realized if more nitrogen fertilizer is applied.

In the next few decades, introducing genes from unrelated species into wheat will likely be possible. These techniques also will enable reintroduction of modified wheat genes and modifications of the developmental and environmental control of gene expression (Austin et al 1986). By these means, it may be possible to engineer genetic variation that does not exist in nature and that can be exploited by breeders to produce varieties with decreased vulnerability to climatic hazards, which will enable the species to adapt to environments outside its present range. By attaching promoters responsive to exogenously applied chemicals to new or existing genes, the expression of the genes could be controlled and, in this way, some control over the response of the whole plant to environmental factors could be achieved. It is conceivable that, in the long term, a practical scheme based on a system that switches on genes will be devised, enabling farmers to modify the phenotype of wheat plants to suit particular weather conditions.

References cited

Austin R B, Arnold M H (1989) Variabilities in wheat yields in the UK: analysis and future prospects. *In* Variabilities in cereal yields and implications for agricultural research and policy. J. R. Anderson and P. B. R. Hazell, eds. Johns Hopkins University Press.

Austin R B, Edrich J A, Ford M A, Blackwell R D (1977) The fate of the dry matter, carbohydrates and ^{14}C lost from the leaves and stems of wheat during grain filling. Ann. Bot. 41:1309-1321.

Austin R B, Flavell R B, Henson I E, Lowe H J B (1986) Molecular biology and crop improvement: a case study of wheat, oilseed rape and faba beans. Cambridge University Press, London. 114 p.

Austin R B, Ford M A, Morgan C L (1989) Genetic improvement in the yield of winter-wheat: a further evaluation. J. Agric. Sci. Camb. 112:295-301.

Bidinger F R, Musgrave R B, Fischer R A (1977) Contribution of stored preanthesis assimilate to grain yield in wheat and barley. Nature 270:431-433.

Buck S F (1961) The use of rainfall, temperature and actual transpiration in some crop-weather investigations. J. Agric. Sci. Camb. 57:355-365.

Church B M, Austin R B (1983) Variability of wheat yields in England and Wales. J. Agric. Sci. Camb. 100:201-204.

Cornish E A (1950) The influence of rainfall on the yield of wheat in south Australia. Aust. J. Sci. Res. Ser. B 3:178-218.

Feyerherm A M, Paulsen G M (1981) An analysis of temporal and regional variation in wheat yields. Agron. J. 73:863-867.

Fischer R A (1973) The effect of water stress at various stages of development on yield processes in wheat. Pages 233-241 *in* Plant responses to climatic factors. United Nations Educational, Scientific and Cultural Organization, Paris.

Godwin D C, Vlek P L G (1985) Simulation of nitrogen dynamics in wheat cropping systems. Pages 311-332 *in* Wheat growth and modelling. W. Day and R. T. Atkin, eds. Plenum Press, New York.

Harlan J R, Zohary D (1966) Distribution of wild wheats and barley. Science 153:1074-1080.

Hazell P B R (1985) Sources of increased variability in world cereal production since the 1960s. J. Agric. Econ. 36:145-159.

Hoogendoorn J (1985) The physiology of variation in the time of ear emergence among wheat varieties from different regions of the world. Euphytica 34:559-571.

Innes P, Blackwell R D (1981) The effect of drought on the water use and yield of two spring wheat genotypes. J. Agric. Sci. Camb. 96:603-610.

Innes P, Blackwell R D (1983) Some effects of leaf posture on the yield and water economy of winter wheat. J. Agric. Sci. Camb. 101:367-376.

Innes P, Blackwell R D, Austin R B, Ford M A (1981) The effects of selection for number of ears on the yield and water economy of winter wheat. J. Agric. Sci. Camb. 97:523-532.

Innes P, Blackwell R D, Quarrie S A (1984) Some effects of genetic variation in drought induced abscisic acid accumulation on the yield and water use of spring wheat. J. Agric. Sci. Camb. 102:341-351.

Innes P, Hoogendoorn J, Blackwell R D (1985) Effects of differences in date of ear emergence and height on yield of winter wheat. J. Agric. Sci. Camb. 105:543-549.

Kushnir U, Halloran G M (1982) Variation in vernalization and photoperiod response in tetraploid wheat (*Triticum turgidum dicoccoides*) ecotypes. J. Appl. Ecol. 19:545-554.

Nix H A, Fitzpatrick E A (1969) An index of crop water stress related to wheat and grain sorghum yields. Agric. Meteorol. 6:321-337.

Otter S, Ritchie J T (1985) Validation of the ceres-wheat model in diverse environments. Pages 307-310 *in* Wheat growth and modelling. W. Day and R. T. Atkin, eds. Plenum Press, New York.

Payne P I, Lawrence G J (1983) Catalogue of alleles for the complex gene loci, Glu-A1, Glu-B1 and Glu-D1 which code for the high-molecular-weight subunits of glutenin in hexaploid wheat. Cereal Res. Comm. 11:29-35.

Pitter R L (1977) The effect of weather and technology on wheat yields in Oregon. Agric. Meteorol. 18:115-131.

Seif E, Pederson D G (1978) Effect of rainfall on the grain yield of spring wheat, with an application to the analysis of adaptation. Aust. J. Agric. Res. 29:1107-1115.

Woodruff D R, Tonks J (1983) Relationship between time of anthesis and grain yield of wheat genotypes with differing developmental patterns. Aust. J. Agric. Res. 34:1-11.

Zohary D, Harlan J R, Vardi A (1969) The wild progenitors of wheat and their breeding value. Euphytica 18:58-65.

Notes

Author's address: R. B. Austin, Institute of Plant Science Research (Cambridge Laboratory), Trumpington, Cambridge CB2 2JB, United Kingdom.

Citation information: International Rice Research Institute (1989) Climate and food security. P.O. Box 933, Manila, Philippines.

Effect of temperature on wheat in India

V. K. Dadhwal

The phenology, yield, and yield components of four wheat cultivars were studied for three yr (1979-80 to 1981-82) at New Delhi in a date-of-sowing experiment. Crop stages used were sowing (SOW), first spikelet initiation (FSI), terminal spikelet initiation (TSI), anthesis (ANT), and end of grain filling (MAT). Multiple regression analysis was used to relate the duration of four growth phases—SOW-FSI, FSI-TSI, TSI-ANT, ANT-MAT—to means of temperature, photoperiod, and solar radiation in each phase. Photothermal and thermal unit requirements of preanthesis and grain growth, respectively, were estimated. The latter could be used to predict duration of grain filling in different parts of India. Temperature effects on kernel weight and kernel number were quantified. A procedure based on factor analysis was used to estimate the effect of increments in temperature during each phase on final grain yield.

Wheat grown in the winter season in India is sown when temperatures cool down enough to allow establishment of a good stand and harvested after rising temperatures have terminated grain filling. Crop duration of wheat at any location is thus controlled by length of the winter season and by photoperiod. Photoperiod is determined by latitude and time of year. Temperature and available soil moisture also are dominant environmental variables that affect wheat yield.

A simple approach to quantifying the effect of temperature on wheat in the field is to sow the crop on different dates, thereby testing across the large seasonal changes in temperature for the different thermal regimes. Results of such a study carried out at New Delhi (India) are summarized here.

Materials and method

The experiment was conducted for three seasons (1979-80 to 1981-82) at the farms of the Indian Agricultural Research Institute, New Delhi. Four wheat cultivars (HD4502, Hindi 62, Kalyansona, Sonalika) were sown on 7-10 dates each season in 3- × 4-m plots, in a split-plot design with 3 replications. Date of sowing was the main plot. Cultural conditions and plant observations have been described by Saini and Dadhwal (1986a,b) and Saini et al (1986a,b).

The life cycle of wheat was divided into four phases: 1) sowing (SOW) to first spikelet initiation (FSI), 2) FSI to terminal spikelet initiation (TSI), 3) TSI to anthesis (ANT), and 4) ANT to end of grain filling (MAT). The phases were delimited from observations on mother shoots (MS). Average daily temperature (T), photoperiod (P), change in photoperiod (dP), and incident photosynthetically active radiation (IPAR) were averaged over each phase. Seasonal changes in T and P during 1980-81 are shown in Figure 1. (Hereafter, subscripts 1, 2, 3, and 4 used with weather data denote the 4 phases.)

Statistical analysis was carried out using the standard statistical software packages SPSS (Nie et al 1975) and SAS (Barr et al 1976).

Results and discussion

Major conclusions from the study (Dadhwal 1983) summarized here are only for cultivar Kalyansona. Detailed results, including varietal differences in response to temperature, have been described by Saini and Dadhwal (1986a,b) and Saini et al (1986a,b; 1987).

Preanthesis phenology

Sowing date. The durations of three preanthesis phases (SOW-FSI, FSI-TSI, TSI-ANT) obtained with different dates of sowing over three seasons are shown in Figure 2. Early or late sowings caused twofold to threefold reductions in phase durations. Phasic development is slowest from mid-December to January, the period of coolest temperatures and shortest photoperiod (Fig. 1). Thus, the longest SOW-FSI durations occur with the mid-December sowings, while the longest TSI-ANT phase is associated with the mid-November sowing. Beyond these optimum dates, SOW-FSI, SOW-TSI, and SOW-ANT were reduced by 1 d by 5.6, 2.9, and 2.0 d delay in sowing, respectively.

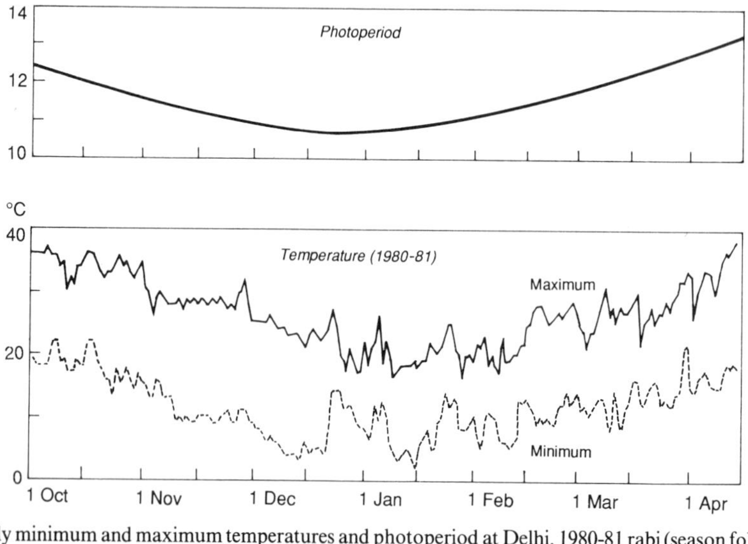

1. Daily minimum and maximum temperatures and photoperiod at Delhi, 1980-81 rabi (season following the wet season).

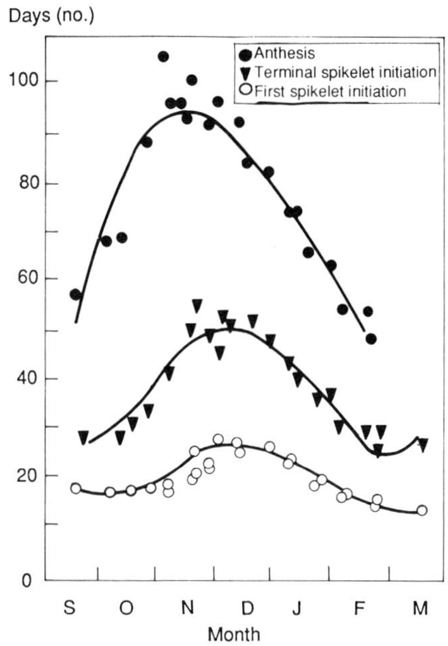

2. Duration of three preanthesis phases in cultivar Kalyansona sown on different dates (data for three seasons).

Regression analysis with weather data. Two multiple regression approaches were investigated for quantifying the effect of weather on duration of growth phases (D). In the fixed model approach, the regression model used (Hodgson 1978) was

$$D = a + b_1T + b_2T^2 + b_3P \qquad (1)$$

The estimates of coefficients and parameters of fit for Kalyansona are given in Table 1. Negative coefficients for b_1 and b_3 indicate shorter D under higher T and longer P. In the best subset approach, the 2-3 variables giving the highest R were chosen from the following equation:

$$D = f(^{+}\sqrt{T}, T, T^2, T^3, P, T{\times}P, dP, P{\times}dP, IPAR) \qquad (2)$$

The response of phase durations to temperature obtained from regression coefficients is plotted in Figure 3. The curves differ not only among phases, but also among cultivars at each phase (Saini et al 1986b). The response curves also highlight

Table 1. Prediction of phase duration (D) from temperature (T) and photoperiod (P) during that phase (cultivar Kalyansona). $D = a + b_1T + b_2T^2 + b_3P$.

Phase[a]	R^2	SEE	a	b_1	b_2	b_3
SOW-FSI	0.929	1.0	130.07	–3.86	0.095	–6.43
FSI-TSI	0.909	2.3	106.47	–4.33	0.080	–3.24
TSI-ANT	0.933	2.3	235.85	–13.83	0.303	–4.25
ANT-MAT	0.880	4.2	140.55	7.79	0.120	–

[a]SOW = sowing, FSI = first spikelet initiation, TSI = terminal spikelet initiation, ANT = anthesis, MAT = end of grain filling.

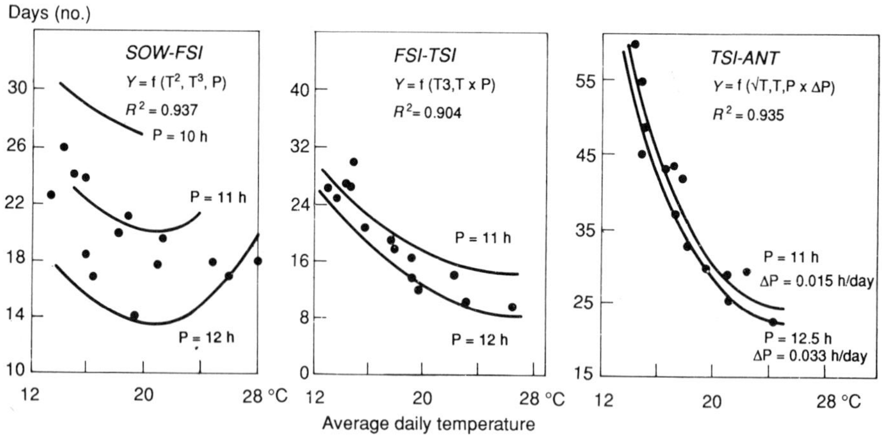

3. Effect of temperature on duration of three preanthesis phases estimated from multiple regression analysis.

the importance of the interaction between T and P and the role played by daily dP in controlling preanthesis phase durations. The dP has been shown to influence the rate of leaf appearance in wheat (Saini et al 1987).

Photothermal units. A simple approach for quantifying the effect of T and P on phase durations is through the use of photothermal units (PTU), defined as degree daylength hours (DDLH) required to complete a phase (Nuttonson 1955). A requirement of 10,000 DDLH above 4.4 °C for SOW-ANT was estimated for cultivar Sonora 64 (Pande et al 1975). In this study, PTU requirements of 7,652 DDLH above 0 °C for SOW-TSI, 5,763 DDLH above 4.5 °C for TSI-ANT, and 12,792 DDLH above 2.8 °C for SOW-ANT phases were obtained for Kalyansona.

Spikelet number

Final spikelet number is the product of rate and duration of spikelet initiation (SPI). The interrelationships between these variables for different sowing dates in Kalyansona are shown in Figure 4. Large variations in rate and duration of SPI resulted in only small effects on spikelet number, as rate and duration showed strong negative correlations. However, when data from four cultivars were pooled, high rates of SPI resulted in fewer spikelets, irrespective of environmental conditions (Dadhwal 1983).

Postanthesis phase

Dry matter accumulation. Grain filling was monitored by following changes in weight of central spikelets on mother shoot ears using short interval samplings. Results for Kalyansona for the 1980-81 season are shown in Figure 5. Because date of anthesis was delayed from December to April, the rate of accumulation increased, but a shorter duration of grain filling resulted in lower final weights. Sofield et al (1977) observed a similar pattern under controlled conditions.

Temperature and duration of grain filling. In this study, a T range of 13.6 to 30.1 °C caused grain filling durations to vary from 14.9 to 62.1 d. All cultivars had

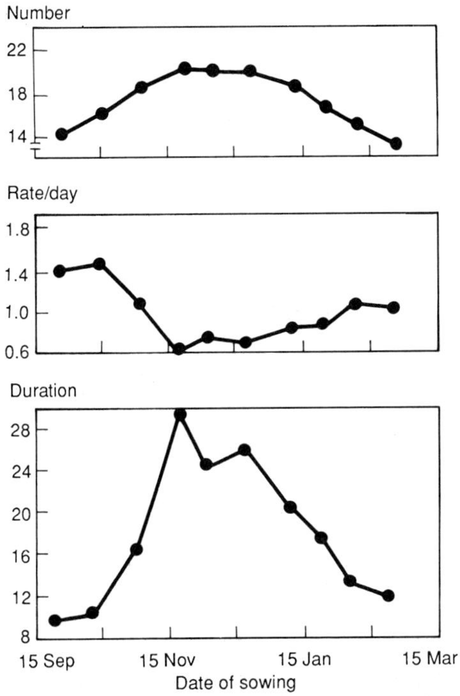

4. Interrelationships between rate and duration of spikelet initiation and spikelet number in wheat cultivar Kalyansona sown on different rates during 1980-81.

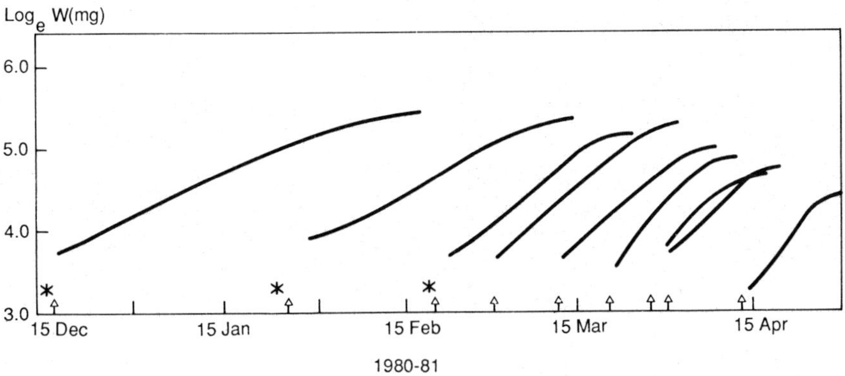

5. Dry matter accumulation in three central spikelets of mother shoot ears of wheat cultivar Kalyansona (1980-81). Arrows indicate date of anthesis. Curves are Richard's growth function or polynomial exponentials(*).

similar response curves. Each 1 °C rise in temperature reduced growth duration by 2.6 d, which is close to the 2.8 d obtained in other studies (Wiegand and Cuellar 1981). However, we found the response to be nonlinear (Saini and Dadhwal 1986a).

Temperature and grain weight. Increased rate of grain filling was not sufficient to compensate for reduced duration, and grain weight decreased by 1.58 mg for each 1 °C rise in temperature during ANT-MAT.

Heat unit for grain filling. Since the pooled ANT-MAT duration did not show any relation with the photoperiod during ANT-MAT, only thermal units need to be estimated to predict ANT-MAT duration. A requirement of 393 degree days above 7.5 °C for completion of grain filling was estimated (Saini and Dadhwal 1986b). This information can be used for predicting the duration of grain filling in different parts of the country (Fig. 6). Results show that below 20° N, temperature-induced short grain filling duration may be very significant. In the high altitude regions Jammu and Kashmir, Himachal Pradesh, and hills of western Uttar Pradesh, yields would not be limited by this factor.

Grain yield

The effect of temperature on pooled grain yield (g/m^2) of three cultivars (HD4502, Kalyansona, Sonalika) over two seasons (1979-80, 1980-81) was quantified using procedures based on multiple regression and factor analysis.

Multiple regression analysis. The estimates of regression coefficients for predicting grain yield from mean temperatures during four phases were

$$GY\ (g/m^2) = 1364.4 + 14.4T_1 - 15.3T_2 - 35.0T_3 - 13.4T_4 \quad (N = 33,\ R^2 = 0.925,\ SEE = 59.28). \tag{3}$$

SEE was 12.7% of the mean grain yield in this data set. The largest coefficient for T_3 showed that TSI-ANT was the most susceptible to increases in temperature.

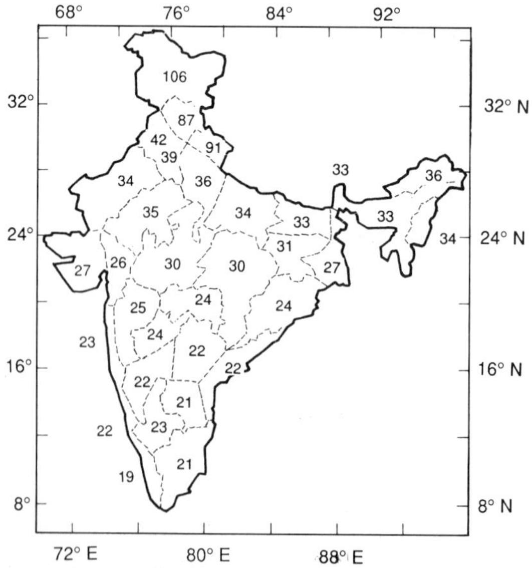

6. Predicted duration of grain filling of wheat in 33 meteorological divisions of India using heat units. (Assumed date of anthesis = 1 Feb)

Table 2. Impact coefficients of daily temperatures on grain yield of wheat for 4 growth phases.

Phase[a]	Mean (°C)	SD (°C)	Impact coefficients	
			g/m²	%
SOW-FSI	18.4	3.70	5.6	1.48
FSI-TSI	17.2	3.06	– 7.2	–1.77
TSI-ANT	17.5	3.11	–35.1	–7.69
ANT-MAT	21.3	4.65	–17.9	–3.51

[a]For explanation of abbreviations, see Table 1.

Factor analysis. Impact coefficients (i.e. the effect of unitary increments in independent variants on a dependent variable) of T were estimated using a procedure suggested by Jones (1982) consisting of 1) factor analysis to convert correlated independent variables to orthogonal factor scores, 2) regression analysis between factor scores and the dependent variable, and 3) estimation of impact coefficients from regression coefficients. The results (Table 2) show that, while a 1 °C rise above mean T in TSI-ANT reduced grain yields by 35.1 g/m^2, the corresponding figure for ANT-MAT was only 17.9 g/m^2. The large sensitivity of TSI-ANT to T is due to the reduction in grain number by lower phase duration and higher T, which for cv Kalyansona was estimated to be 1173 grains/m^2 per 1 °C rise in temperature.

References cited

Barr A J, Goodnight J H, Sall J P (1976) A user's guide to SAS76. SAS Institute, Raleigh, North Carolina, USA.

Dadhwal V K (1983) A study of phenology, growth and yield in wheat in relation to environmental factors, especially photoperiod and temperature. Ph D thesis, Indian Agricultural Research Institute, New Delhi. 298 p.

Hodgson A S (1978) Rapeseed adaptation in northern New South Wales. I. Phenological responses to vernalization, temperature and photoperiod by annual and biennial cultivars of *Brassica campestris* L., *Brassica napus* L., and wheat cv Timgalen. Aust. J. Agric. Res. 29:693-710.

Jones D R (1982) A statistical inquiry into crop-weather dependence. Agric. Meteorol. 25:91-104.

Nie N H, Hull C H, Jenkins J G, Steinbrenner K, Bent D H (1975) Statistical package for social sciences. McGraw Hill, New York.

Nuttonson M Y (1955) Wheat climate relationships and the use of phenology in ascertaining the thermal and photothermal requirements of wheat. American Institute of Crop Ecology, Washington, D.C. 388 p.

Pande H K, Lahiri S, Ravindranath E (1975) Thermal and photothermal units to predict wheat crop maturity. Indian J. Agric. Sci. 45:210-212.

Saini A D, Dadhwal V K (1986a) Influence of sowing date on grain growth duration and kernel size in wheat. Indian J. Agric. Sci. 56(6):439-447.

Saini A D, Dadhwal V K (1986b) Heat unit requirement during the period of grain growth in wheat and its application for adjusting sowing dates in different regions. Indian J. Agric. Sci. 56(6):447-452.

Saini A D, Dadhwal V K, Nanda R (1987) Thermal and day-length changes affecting leaf number and their appearance in wheat. Indian J. Agric. Sci. 57:815-820.

Saini A D, Dadhwal V K, Phadnawis B N, Nanda R (1986a) Influence of sowing date on pre-anthesis phenology of wheat. Indian J. Agric. Sci. 56:503-511.

Saini A D, Dadhwal V K, Phadnawis B N, Nanda R (1986b) Thermal and photoperiod effects on phase durations of four wheat varieties grown on different sowing dates. Indian J. Agric. Sci. 56:646-656.

Sofield I, Evans L T, Cook M G, Wardlaw I F (1977) Factors influencing the rate and duration of grain filling in wheat. Aust. J. Plant Physiol. 4:785-797.
Wiegand C L, Cuellar J A (1981) Duration of grain filling and kernel weight of wheat as affected by temperature. Crop Sci. 21:95-101.

Notes

Acknowledgment: Award of Senior Research Fellowship during the course of this study by the Director, Indian Agricultural Research Institute, is gratefully acknowledged.
Author's address: V. K. Dadhwal, Space Applications Centre (ISRO), Ahmedabad 380 053, India.
Citation information: International Rice Research Institute (1989) Climate and food security. P.O. Box 933, Manila, Philippines.

Climatic vulnerability of maize in China

Li Jingxiong

Five maize regions in China were characterized by climatic factors. Annual growth rates of yield and production averaged 2.9% 1949-84, while total production fluctuated by 60 million t. Sixty percent of the production losses were due to adverse weather in 8 years. The coefficient of variability for yield fluctuation was lower in the southernmost region. Maize in northeast China is vulnerable to cold weather; the leading production provinces there lost 13.1-30% (0.5-1.5 million t) in each of 4 severe years. Correlations between crop yield and mean temperatures for May-September are significant. Cold injury often retards plant growth, delays silking, and prolongs grain filling. To alleviate the effects of adverse weather, covering seedlings and transplanting seedlings have been used, primarily in high altitude and cold dry areas.

The tremendous increase in maize productivity around the world during the last half century has been attributed to the application of modern agrotechnology. However, the overall impact of climatic factors on yield was often neglected, until a crop failure followed bad weather. In a country like China, where a large proportion of the crop is grown at high latitudes and high altitudes, the climatic vulnerability of maize is apparent.

Characterization of maize regions

The areas of maize cultivation in China lie between 50° and 22° N latitude, stretching from the northern border (a cold temperate zone) to the vicinity of Hainan Island (a subtropical climate). Maize is cultivated from along the seacoast in the east to the arid highlands in the west. We traditionally expect good maize crops in the plains regions of northeast and north China. Nearly 25% of the maize hectarage is distributed in the hilly mountains of the southwest, where average yields are relatively low. Many factors—climate, soil types, hybrids, cultural methods, and even social customs—interact in determining productivity.

Five maize regions, characterized primarily by climate, but designated by their geographical location, have been delineated (Fig. 1) (Li 1986).

Region I, in the northeast, lies north of 40° N. Its crop season of 120-150 d is suitable for a single crop of spring maize. Mean temperature during crop growth is 19-22 °C, with the mean temperature of the warmest month never exceeding 24 °C.

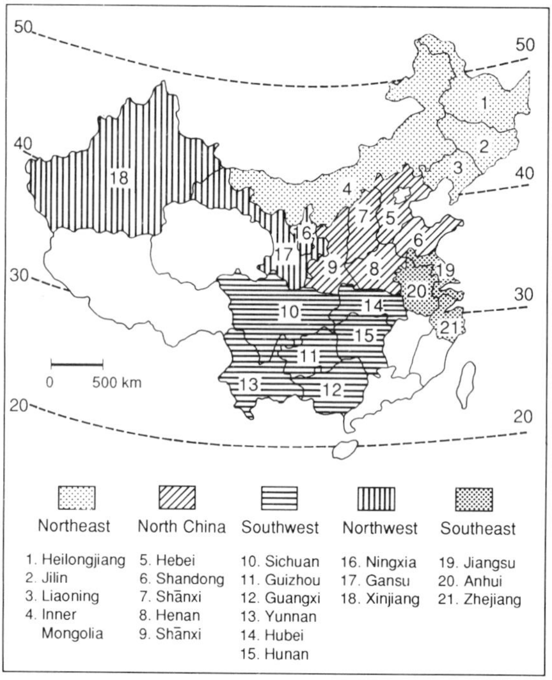

1. Leading maize regions in China.

Annual precipitation ranges from 300 to 500 mm, but may reach 800 mm in some of the eastern parts. Cumulative solar radiation is 800-1,000 h. During grain filling, the daily maximum-minimum temperature difference may reach 13 °C. Temperatures low enough to cause injury occasionally occur in the north.

Region II, in the north, is referred to as the valley region of the Yellow, Huai, and Hai Rivers. It lies between 33° and 40° N and contains slightly more than 40% of the total maize hectarage. With 170-240 frost-free days, the area is favorable for double cropping. Some spring maize is grown in the surrounding mountains, and summer maize is widely cultivated with irrigation in the plains. Maize is closely associated with winter wheat in a system of intercropping or sequential double cropping. Average temperature during the growing months of summer maize is 24-26 °C; mean temperature of the warmest months is 26-28 °C. At grain filling, the daily maximum-minimum temperature difference is 7-10 °C. Annual precipitation varies between 400 and 600 mm, 60-70% of it falls in July-August. Both drought and waterlogging are frequent problems.

Region III, in the southwest, has a warm and humid climate with a mean temperature of 21 °C (higher than in the basin areas) during the growing period. Spring maize is the important crop on hilly mountains. Annual precipitation often reaches 600-1,000 mm. Foggy and misty days almost year-round cut solar radiation accumulation to 500-700 h.

Region IV, in the northwest highlands, is known for its cold, arid climate. The growing period for maize is only 120-170 d, with a mean temperature of 20-25 °C

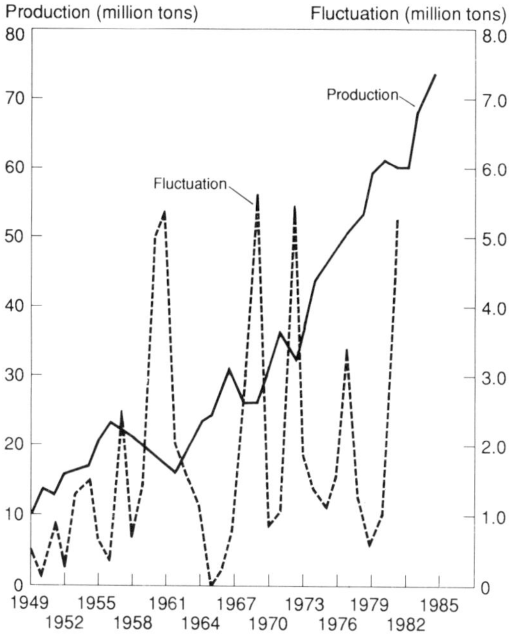

2. Maize production during 1949-84 and its negative fluctuation during 1949-81.

and a daily maximum-minimum difference of 13-17 °C. Solar radiation of 900-1,000 h is higher than in other regions. Annual rainfall is 150-300 mm, too low for growing maize, and almost all maize crops are irrigated.

Region V, in the southeast, has only 3.4% of the nation's maize hectarage. With more than 250 frost-free days in the southern part of the region and a daily mean temperature of 25-28 °C, its main crop is irrigated rice.

Production and fluctuation

Trends in maize production

Growth in maize production from 1949 to 1984 is shown in Figure 2. For technical reasons, the growth curves of hectarage and yield are not plotted, but they parallel the production curve closely, except for a shrinkage in hectarage since 1981. This may be because farmers recently have been encouraged to grow cash crops that are more economical than cereals. During the same period 1981-84, favorable weather led to a number of bumper harvests, with average maize yields reaching nearly 4 t/ha. Total maize production leveled off at 73.3 million t in 1984, after annual growth rates of 2.9% in yield and production and 2.8% in hectarage (an annual 75.1 kg/ha yield increase and 1.8 million t production increase). Maize hectarage increased 170,000 ha/yr to 1981.

The impact of weather on maize production was strikingly illustrated in the fall of 1985, when a typhoon hit the two leading provinces in the northeast region. The big loss in that year's crop was compensated for by favorable weather the following year.

The significance of yield fluctuation

Chen (1986) collected production data for 1949-81 from 23 provinces, representing 98% of the total hectarage and 99% of the total output. Production during 1949-58 and 1962-81 and overall were determined for each province by means of orthogonal polynomial regression, plotted to show the general trend. The fluctuation term was obtained by subtracting the time trend term for a year from actual yield. A positive value signifies the increment attributable to favorable weather, a negative value is the increment attributable to adverse weather. The sum of provincial negative values within a year represents the negative yield fluctuation for the whole country. Production fluctuation over time is plotted in Figure 2.

Maize yields since 1949 showed a total of 60 million t of negative fluctuations attributable to adverse weather. Maize yields fluctuated down an average 1.8 million t/yr. There were 5 yr of 5-million-t and 3 yr of 2.5- to 5.0-million-t fluctuation. If a negative fluctuation of 2.5 million t is set to indicate a poor crop, then the sum of the 8 yr (35.7 million t) would account for 60% of the total fluctuation. We found that maize production suffered from all kinds of weather disasters in 1957, 1960, 1961, 1968, 1969, 1972, 1977, and 1981.

Fluctuations among regions. To eliminate the effect of differential yield potential among provinces, coefficients of variability (CV) were calculated (Table 1). The lowest CV was for Yunnan, followed by Guangxi. As a group, southwest region provinces had the lowest variability. Perhaps the warmer climate gave some stability

Table 1. Coefficients of variability (CV) for production fluctuation of maize in different provinces of China.

Province	CV
Region I (northeast)	
Inner Mongolia	20.6
Liaoning	19.3
Heilongjiang	18.9
Jilin	16.9
Region II (north)	
Shanxi	17.2
Beijing	16.9
Hebei	15.9
Shandong	15.7
Henan	14.8
Shanxi	14.0
Region III (southwest)	
Hubei	15.1
Sichuan	13.1
Guizhou	12.7
Guangxi	10.7
Yunnan	8.2
Region IV (northwest highlands)	
Gansu	20.3
Xinjiang	13.9
Region V (southeast)	
Jiangsu	14.0

to maize production. All northeast region provinces had much higher variability, indicating the hazardous climate there. The data also suggest an effect of temperature gradient on production. This seems reasonable, because the intermediate CV in Region II corresponds to its intermediate temperatures.

Relation of fluctuation to other factors. Fluctuations in maize production correlated significantly with areas that suffered from extreme weather and areas where crop failed due to adverse weather (Chen 1986).

In the leading maize regions, production fluctuations were fairly consistent. We compared provinces with similar correlations and found them to be neighbors with similar climates.

We also found significant correlations between production fluctuations of maize and other cereals (Table 2). Maize production coincided with rice, wheat, and other cereals production. Maize and wheat are the primary crops in northern China; maize and rice are important in the southwest.

Climatic vulnerability

Effect of temperature

In the higher latitudes, cold waves that occur once every three or four years cause heavy damage to maize, rice, soybean, and sorghum. In 1969, 1972, and 1976, Jilin lost an average 13.1%, 425,000 t, of maize each of those years, following normal production the previous years. Heilongjiang Province, lying farther north (43°25′-53°33′ N) suffered a 30% loss, 1.0-1.5 million t, each of four cold years (1969, 1972, 1976, 1977).

Critical stage of cold injury. Pan et al (1983) used linear regression equations and found a significant correlation between crop production in Jilin and average temperatures during May-September at three localities and between the mean temperatures of August and June-August at different locations (Table 3) (Pan et al 1983).

In warmer Huiteh, in the mean temperature during May-September dropped 1 °C below normal (equivalent to a loss of 153 °C available temperature), yield would be reduced by 289 kg/ha. In cooler Dunhua, 514 kg/ha yield loss would occur. In the last 10-15 yr, the distinction between a bumper harvest year and a poor harvest year was found to be a difference of 0.6-0.8 °C in mean temperature during June-August and May-September (Table 4).

Table 2. Correlation coefficients of maize production fluctuation with production fluctuations of other cereals.

Period	Rice	Wheat	All cereals averaged
1949-58	0.50	0.06	0.62
1962-81	0.54	0.40	0.86**
1949-81	0.68**	0.44*	0.81**

Significant levels: * 5%, ** 1%.

Table 3. Correlation coefficients between maize production and mean monthly temperature (Jilin, 1978-1980).

Site	May	Jun	Jul	Aug	Sep	Jun-Aug	May-Sep	Period	No. of years
Jilin Province	0.388	0.379	–0.073	0.479*	0.369	0.466*	0.582**	1950-75	26
Huiteh	0.247	0.333	0.152	0.498**	0.403*	0.541**	0.521**	1950-75	26
Dunhua	0.337	0.467*	0.331	0.309	0.370	0.599**	0.735***	1953-75	23
Jilin Province	14.9	19.9	23.0	21.5	14.9	21.5	18.9	1950-75	26
Huiteh	15.5	20.3	23.4	22.0	15.4	21.9	19.3	1950-75	26
Dunhua	11.8	16.1	19.7	19.0	12.3	18.3	15.8	1953-75	23

Significant levels: * 5%,'** 1%, *** 0.1%.
Linear regression for May-Sep in the province: y = –592.5 + 31.4x;
in Huiteh: y = – 742.51 + 38.53x; in Dunhua: y = – 1085.3 = 68.54x.

Table 4. Average temperature and production differences by season across time.

Site	Harvest	Average temp (°C)		No. of years investigated
		Jun-Aug	May-Sep	
Huiteh	Bumper	22.0	19.7	11
	Poor	21.6	19.0	15
	Difference	0.6	0.7	
Dunhua	Bumper	18.8	16.2	10
	Poor	18.0	15.4	13
	Difference	0.8	0.8	

The critical time for maize to be injured by cold is in August, when plants are in the tasseling-silking and early kernel filling stages. Cold waves in early September affect only kernel plumpness. Meteorological data for the mid-northeast show that air temperature there begins to drop in August.

Type of cold injury. The Jilin experiment also showed the effect of low temperature on growth duration. When the soil temperature at 5 cm was 20 °C, seedling emergence took only 6 d. At 15 °C, it was 10 d and at 10 °C, 21 d. For every 1 °C lower than normal, duration from emergence to silking was delayed 6 d, and from silking to full maturity, 4 d. The most common symptoms of cold injury were delayed flowering and prolonged maturity. When growth is delayed, maize plants cannot mature before early frost. cannot mature before early frost.

Another experiment in Heilongjiang with different planting dates also illustrated the effect of low temperature on maize growth and performance (Table 5) (ICC 1980).

The same maize hybrid planted in a cooler environment (Hailun) took 6-9 more days to reach tasseling than in the warmer environment (Harbin). By extrapolation, when daily temperature is lowered 1 °C, duration from emergence to tasseling is prolonged 5-7 d. But when planting was postponed, inflorescence development was

Table 5. Effect of planting date on duration and yield of maize. Heilongjiang, 1980.

Planting date (site)	Daily temp (°C)	Emergence to tasseling (d)	Grain wt (g)/ear	100-kernel wt (g)	Yield (t/ha)
5 May					
(I)	20.1	64	152.3	26.4	7.9
(II)	18.9	73	100.0	16.7	4.5
20 May					
(I)	21.1	56	130.4	21.8	6.1
(II)	19.5	65	105.0	17.0	4.5
4 June					
(I)	22.0	53	111.3	19.5	5.9
(II)	20.7	59	60.0	10.6	1.8

speeded up under higher temperatures, and tasseling came earlier. Lower temperatures during the development period also had an adverse impact on rate of dry matter accumulation and yield components.

An early frost does not necessarily mean a crop failure if it occurs independently, but it is a threat to maize if it occurs with drought in the spring and/or with cold weather in the summer. Other factors in combination with cold injuries also harm crops. We found that cold temperature with heavy rainfall in 1957 led to delayed ripening of maize; cold temperature with drought in 1972 and cold temperature with early frost in 1969 severely damaged rice and sorghum; and low temperature with reduced sunlight caused heavy injury to rice in the humid region of East Heilongjiang.

Precipitation and yield

Drought, flood, and waterlogging are more threatening to maize production than cold weather. Since 1949, there have been six years of severe weather across the country. An early August 1963 flood in central Hebei caused 500,000 t losses in maize alone. (Daily precipitation was recorded at 865 mm and total rainfall in 7 d reached 2,050 mm.) In 1972, a serious drought covered a large part of 11 provinces in the north, northeast, and northwest regions, and three leading maize provinces had negative yield fluctuations of 300,000 t each. Hebei experienced another flood in 1977. In 1981, the western part of the country experienced droughts; the east part had waterlogging.

The occurrence of adverse weather differs with latitude. In Yunnan, the southernmost province with more than 1 million ha of maize, no serious drought has ever been reported. In the extreme north, 26.6 million ha of cereal crops in Heilongjiang suffered from drought and flooding during 1949-77, causing a loss of 20.1 million t.

The demand and supply of water during the developmental stages of maize grown near Beijing are summarized in Table 6. These figures are for normal years, when the annual precipitation was 644.2 mm. In north China, the most critical

Table 6. Distribution of rainfall and water demands of maize plant in the vicinity of Beijing, 1983-84.

Growth stage	Rainfall		Evapotranspiration		Shortage variability	
	mm	%	mm	%	mm	%
Planting – elongating	61.1	12.4	70-80	20.0	47.3	1.0
Elongating – silking	134.8	27.5	105-120	30.0	11.7	1.4
Silking – ripening	295.6	60.0	175-200	50.0	4.4	3.2
Planting – ripening	491.5	100.0	350-400	100.0	63.4	1.0

requirement of maize for water is around tasseling (from 10 d before tasseling to 20 d after). Under the semiarid conditions of north China, there is little rainfall in April and May. This makes planting spring maize difficult, especially in the mountainous areas, unless soil moisture is available from the previous fall. If the rainy season comes late, after the end of June, canopy leaves of spring maize show wilting, and summer maize cannot be planted without irrigation. This summer drought adversely affects grain filling and yield.

If total precipitation in July-August exceeds 400 mm, or if more than 100 mm falls in June, waterlogging (4-5 d of continuous rain followed by strong evapotranspiration on the next bright day) is inevitable. Waterlogged spring maize is characterized by premature wilting, which interrupts grain filling. Soil fungi such as *Fusarium* in association with waterlogging may contribute to premature wilting and stalk rot. Maize is vulnerable to waterlogging only at the late milk or dough stage.

Alleviating approaches

Little work has been done on tolerance for drought. Castleberry and Lerette (1979) found a new type of drought tolerance in maize (called latente) which originated from the Mexican germplasm Michocan 21. Inbred lines and hybrids bearing this genetic character have been developed. However, the higher performance of the latente hybrid under drought stress was not consistent under normal conditions.

To deal with cold injury, agronomists in the northeast region often recommend using short-duration varieties which can fully mature 10 d before early frost. But some farmers risk planting longer duration hybrids to take advantage of their higher yields.

The traditional practice of fall plowing can cause the loss of 45-75 t soil moisture/ha by the time of planting. That is 2-3 times the water required for seed germination and seedling growth. A recent experiment in south Heilongjiang indicated a 10.3% yield increase in plots that were plowed every 4 yr over those plowed every fall.

Another experiment carried out at 45°26′N showed that maize could be planted when the air temperature was at 7 °C for several days. Maize planted 15 d earlier than usual utilized the extra cumulative temperature of 150-200 °C for emergence, and the crop matured 3 d earlier. Applying a high rate of P fertilizer in combination with N was recommended; silking date was 5-6 d earlier than the check plot.

Covering the maize field with polyvinyl chloride (PVC) membrane, a recent development, has attracted many farmers in the cold, dry regions and mountainous areas. PVC is effective in preserving soil moisture, raising soil temperatures, and suppressing weeds. This practice could alleviate the risks of cold injury and drought. So far, it has been used on 100,000 ha of maize in six provinces. Yields as high as 11.3 t/ha have been reported. The yield increase has ranged from 30 to 60%. In Lichuan county, Hubei Province, about 870 ha of maize scattered in cold, dry mountains at altitudes of 1,400-1,900 m were grown under cover. Average yields were 6.2 t/ha, 4.3 t/ha higher than the check. Net return was 535.5 yuan/ha after deducting the cost of the PVC.

In the northern part of the country, another method of avoiding cold damage—transplanting maize seedling in pots—has been practiced on 480,000 ha. It achieved the same effect as using PVC. With pot culture, long-duration maize hybrids with higher yield potential can be grown to full maturity.

Summary

China's total maize production in 1984 was 73.3 million t. The annual growth rate during the last 35 yr has been 2.9%. Maize production fluctuated 60 million t during eight of those years, 60% of it due to adverse weather. The northeast region had a larger coefficient of variability than the southernmost provinces.

Maize in the northeast was vulnerable to low temperature. Each of the leading provinces lost 13-30%, 0.5-1.5 million t, of grain in each of four severe years. Significant positive correlations were found between maize production and mean temperatures May-September, August, and June-August. For each degree below normal of mean temperature in May-September, the yield loss was 289-514 kg/ha. Low temperatures retarded growth and delayed silking and kernel filling.

Drought, flood, and waterlogging are more important to maize production than low temperature. Since 1949, these adverse weather events have occurred in 6 yr.

Polyvinyl chloride covering and transplanting have helped protect crops from cold injury and drought stress on 580,000 ha of maize in the cold dry mountainous regions.

References cited

Castleberry R M, Lerette R J (1979) Latente, a new type of drought tolerance? Pages 46-56 *in* Proceedings of the 34th annual corn and sorghum industry research conference. American Seed Trade Association, Washington, D.C.

Chen Y N (1986) Production fluctuation of maize in relation to agrometeorological factors [in Chinese]. Meteorology Laboratory, China Agricultural Academy of Sciences, Beijing.

ICC—Institute of Crop Cultivation, Heilongjiang Academy of Agricultural Sciences (1980) Types of cold injury in maize and their preventive measures in Heilongjiang [in Chinese]. Paper presented at the Symposium on Preventive Measures against Cold Injuries in the Northeast Region. Harbin, Heilongjiang.
Li Jingxiong (1986) Maize production and improvement in China. Pages 23-34 *in* Proceedings of the second Asian regional maize workshop. International Maize and Wheat Improvement Center, Mexico.
Pan T F, Chao H K, Fong T S, Lu C P (1983) Cold injuries and their prevention in crop plants [in Chinese]. Agricultural Press, Beijing, China.

Notes

Author's address: Li Jingxiong, Chinese Academy of Agricultural Sciences, Beijing, China.
Citation information: International Rice Research Institute (1989) Climate and food security. P.O. Box 933, Manila, Philippines.

Climatic vulnerability of major food crops: soybeans

H. E. Kauffman and J. E. Gleason

Soybean is the world's leading source of high-protein meal and edible oil. Unlike staple food crops such as the cereals, more than 90% of soybean production is in exclusively rainfed growing areas in only 3 major growing regions: the midwestern United States, the Southern Cone countries of South America, and northeastern China. Concentration of global production in these regions has created an inherently risky situation for soybean-importing countries: world soybean prices are easily affected by changes in supply conditions of these areas. World prices increase significantly with weather-driven production shortfalls. Less developed countries that depend on large volumes of imported edible oil suffer most severely from wide swings in prices.

During the last century, the major world production areas for soybean (*Glycine max* (L.) Merr.) have moved from their center of origin in the Orient to midwestern United States and the Southern Cone countries of South America (Argentina, Brazil, Paraguay, and Uruguay). For many centuries, China was the leading soybean producer. Today, 75% of world production is concentrated in North and South America.

Just as the important production centers have shifted, the primary uses of soybean have changed, from a simply processed human food in the Orient to highly refined products derived from solvent extraction processing in the western hemisphere. Soybean meal now accounts for more than 50% of the world's high-protein meal. Soy oil has the largest share of the edible oil market (about 30%), despite the gains made by palm and rapeseed oils during the last decade (Table 1). About half the world production is traded in international markets, making soybean a dominant force in determining prices for high-protein meal and edible oil (USDA 1986).

The risk of concentrated production

The U.S. produces more than half the world's soybean and exports about 50% of its production. Brazil, Argentina, and Paraguay produce nearly 25% of the world total and export more than 75% of the soybeans they produce (USDA 1986).

Europe is the largest importer of soybean and soybean products (about 60% of the world total), followed by the East Asian countries of Japan, Korea, and Taiwan

Table 1. World production of edible meals and oils, 1985 (USDA, 1985).

Source	Production (% of total)	
	Edible meal	Edible oils
Soybean	50.2	30.3
Cotton	15.5	7.3
Peanut	11.2	6.9
Sunflower	9.4	12.8
Rapeseed	8.6	11.5
Coconut	2.3	5.5
Palm	–	14.7
Linseed	–	1.7
Olive	–	3.9
Fish	–	3.5
Palm kernel	1.2	1.9

(about 20%). An important percentage of the European soybean imports are processed and exported as oil and meal. Less developed countries import most of the remaining soybean products. Soybean consumption during the next decade is expected to increase most rapidly in the less developed countries (ASA 1983). Table 2 shows imports and exports of soybean, soybean meal, and soy oil for different world regions.

The concentration of global soybean production in only three world regions, North and South America and China has created an inherently risky situation for soybean-importing and -exporting countries. Sudden and large changes in annual world production due to weather variation would have an important bearing on food security and the economic stability of consumers and producers in many countries.

Effects of climate on soybean production

Drought and temperature stresses are the weather-related stresses that affect the growth of soybean, and therefore production, most significantly (Howell 1956, Runge and Odell 1960). In addition, several general climate conditions have limited soybean production primarily to the temperate regions of the world.

Specific weather-caused stresses

In general, drought stress causes more soybean yield reduction than any other weather-related stress. Soybean is nearly always grown without irrigation, making it vulnerable during periods of inadequate rainfall. Although soybean can survive short periods of drought stress because of its long flowering period and extensive root system, the water deficiency limit is 100 mm for growing soybean without irrigation (Da Mota 1978). The pod-filling stage is most sensitive to drought (Shaw and Laing 1966). Moisture deficits for 2-3 wk immediately after flower bud differentiation reduce growth and cause heavy flower and pod dropping.

Soybean can tolerate short periods in waterlogged soils better than maize and many other food crops. However, heavy rains after seeding cause significant yield

Table 2. Imports and exports of soybean, soy meal, and soy oil by region, 1984 (FAO 1984).

Region	Soybean (thousand t)		Soy meal (thousand t)		Soil oil (thousand t)	
	Imports	Exports	Imports	Exports	Imports	Exports
North America	1,711.7	19,640.8	1,147.8	4,471.3	200.4	1,041.3
South America	396.8	5,166.7	646.3	10,189.0	458.7	1,412.2
Europe	14,194.7	98.3	16,899.3	4,811.2	790.0	1,460.4
Africa	65.0	–	531.6	11.0	472.9	–
Asia	7,780.8	858.4	2,151.8	1,195.7	1,964.3	105.1
Oceania	36.4	.1	12.1	–	46.1	.2
USSR	615.2	–	600.0[a]	–	119.7	–

[a]Estimated.

losses. Plant stand is adversely affected by excess moisture, and weeds are very difficult to control. Serious damage to seed quality and yield results when excessive moisture occurs after pod differentiation.

Production also is affected by extreme temperatures. In temperate regions, soybean production has been limited to areas with temperatures higher than 10 °C during germination and about 30 °C for the vegetative stage. Before the crop matures, frost can cause serious yield loss. Temperatures higher than 38 °C during flowering can also cause significant yield losses.

Climate-related constraints unique to the tropics

Historically, soybean yields in the tropics have been low. In tropical China and Indonesia, where soybean has been grown for many centuries, yields average less than 1 t/ha. Yields in northeast China and the temperate regions of North and South America average more than 2 t/ha. Only during the past decade has it been shown that high soybean yields are possible in the tropics (Kiihl et al 1984, INTSOY 1986).

Soybean has not yielded well in the tropics because germplasm adapted to the tropics has not been widely bred. Given that the center of origin is in North Central China, evolution of most soybean germplasm has taken place in temperate regions. Furthermore, although many active breeding programs have developed cultivars for temperate regions over the past 50 yr, breeding programs for adapting soybean for the tropics began only within the last 15 yr, with comparatively little effort toward improving tropical soybean cultivars during the last decade.

Soybean is regarded as a short-day plant, flowering only when daylength is below a critical value. This photoperiod response limits soybean to very narrow belts of latitude. That has allowed development of a diverse range of soybean cultivars in temperate climates but has severely restricted soybean production in the tropics.

Seed viability also is much more serious problem in the tropics than in cooler climates. Rapid degradation of seed prevents storage for planting in the next season. Many tropical countries that have not been able to establish a domestic seed industry have been forced to depend on annual imports of seed stock (IITA 1982).

Soybean grown in tropical environments also tends to have more disease and insect problems than that grown in temperate regions. Control measures require expensive inputs, so that production cost is higher in most tropical countries than in temperate countries. Brazil, which has had the most success in growing soybean in tropical environments, has established an outstanding integrated pest management program to keep pest incidence low.

Relation of prices to production

Weather-related factors cause production uncertainty in soybean-producing countries. For countries that import soybeans, those weather conditions generate market or pricing risks because of the close relationship between prices and production levels.

Agricultural prices depend on a number of factors. Income, prices of competing or complementary goods, tastes and preferences, and technological change in production contribute to the establishment of prices over time. Yearly price variation is heavily dependent on changes in annual supply conditions; supply conditions in turn are based on current production, imports or exports, and carry-over from the previous year. We are concerned here with isolating the relationship between annual soybean production in the world's major growing regions and average annual prices. Because production is greatly influenced by weather, the relationship between weather and prices is apparent.

With soybean production in the U.S., Brazil, and China accounting for approximately 90% of total world production, soybean-importing countries face a cost risk due to the production concentration. Changes in supply, caused in part by weather variation, in any of these three countries contribute heavily to changes in world soybean prices.

Deflated world soybean prices for 1965-84 are shown in Figure 1. During those 20 yr, the mean price was US$238.02/t, with the highest price US$418.90 in 1973 and the lowest US$145.50 in 1982. The straight line (slope = -6.78, r^2 = .43) indicates the price trend for the period. Prices exhibited a pronounced downward trend with considerable variation (coefficient of variation [CV] 25.6%).

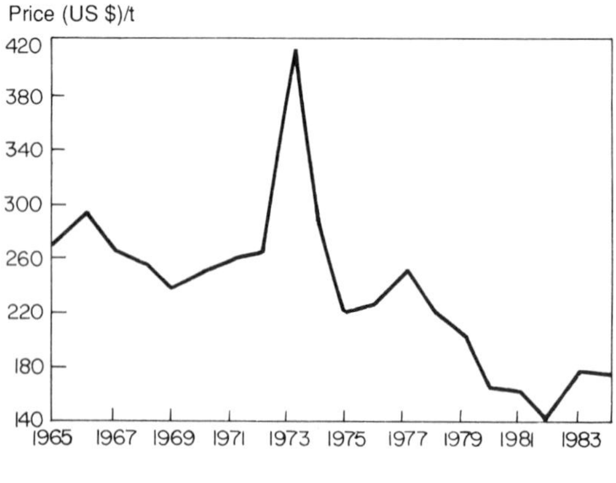

1. World soybean prices, 1965-84.

Although prices followed a downward trend, production increased steadily. Production in the U.S. rose from 19 million tons in 1965 to more than 60 million tons in 1980; the increase was due to gains in both productivity and hectarage. Brazil's soybean production increased at an even faster rate. Brazil produced 0.5 million tons in 1965; in 1980 and 1981, production had risen to 15 million tons. Hectarage increases accounted for those gains, yield increases were relatively unimportant. China's production showed only a slight upward trend, remaining in the 7-8 million ton range (Fig. 2, 3, 4).

Despite the upward production trend, sizable variation occurred during the 20 yr. The CV for U.S. soybean production was 31.8%, for Brazil 76.4%. The CV for China, which is less important as an exporting country, was 11.8%. The combined production of the U.S., Brazil, and China over time had a CV of 32.8%. (Prices and production figures are summarized in Table 3.)

To evaluate the relation between prices and production in the three countries, simple linear regression was used. Highly correlated production figures (Table 4) for

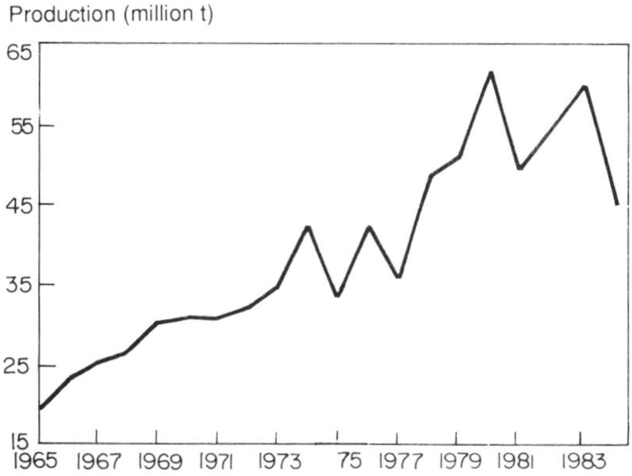

2. Soybean production, United States, 1965-84.

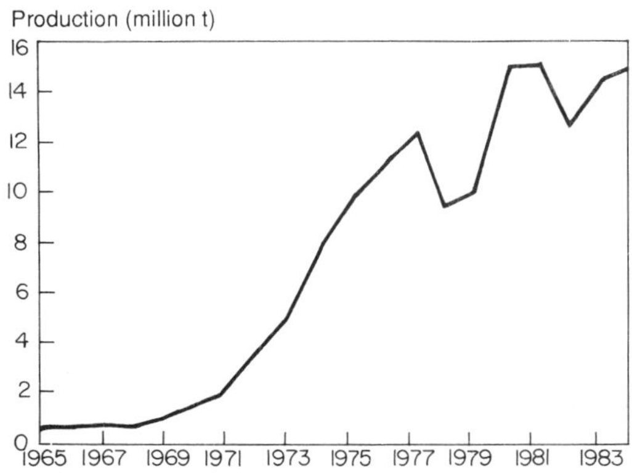

3. Soybean production, Brazil, 1965-84.

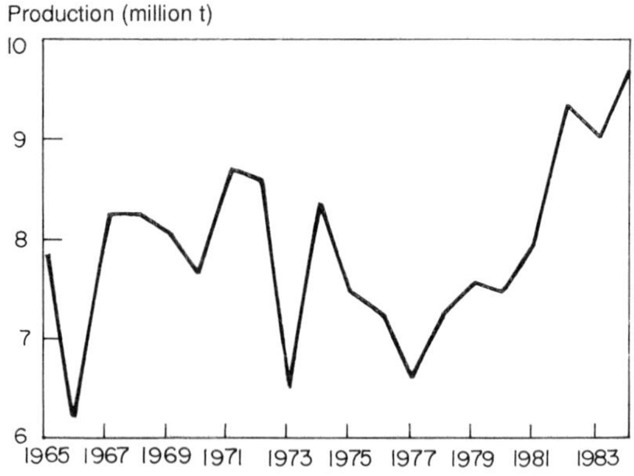

4. Soybean production, China, 1965-84.

Table 3. Soybean production in the U.S., Brazil, and China, and world prices.[a]

Year	Production (million t)			Price (US$/t)
	U.S.	Brazil	China	
1965	19.08	.52	7.87	272.2
1966	23.01	.60	6.14	293.0
1967	25.27	.72	8.27	267.0
1968	26.58	.65	8.27	257.0
1969	30.13	1.06	8.04	238.2
1970	30.84	1.51	7.63	250.1
1971	30.68	2.08	8.71	259.0
1972	32.01	3.67	8.61	265.1
1973	34.58	5.01	6.45	419.0
1974	42.12	7.88	8.37	286.9
1975	33.10	9.89	7.47	220.0
1976	42.14	11.23	7.24	226.3
1977	35.07	12.51	6.64	252.0
1978	48.10	9.54	7.26	222.0
1979	50.86	10.24	7.57	205.1
1980	61.53	15.16	7.46	166.0
1981	48.92	15.20	7.94	164.0
1982	54.14	12.84	9.33	146.0
1983	59.61	14.75	9.03	178.0
1984	44.52	15.40	9.76	175.1

[a]Production: United States Department of Agriculture; Price: U.S. #2, Yellow Bulk, NFS c.i.f. Rotterdam (UK up to December 1972).

the U.S. and Brazil prevented the analysis of price and production variables for each country from yielding meaningful results. Therefore, the relationship between summation of production in the three countries and prices was evaluated. That analysis indicates that the combined production in the U.S., Brazil, and China for 1965-84 has a highly statistically significant relationship with price variation during those same years. The regression coefficient is –2.3509, showing that a one million ton increase in overall production is associated with a price decline of US$2.35/t.

Table 4. Correlation matrix.

	U.S. production	Brazil production	China production	World production	Price
U.S. production	1.00				
Brazil production	.873*	1.00			
China production	.263	.179	1.00		
World production	.987*	.936*	.292	1.00	
Price	– .646*	– .649*	– .512*	– .683*	1.00

*Statistically significant at .05 level. n = 20.

The effect of weather on soybean production in the U.S. in 1984 provides an example of how weather fits into the price-production equation. Hot and dry weather in the American Midwest during the summer of 1984 significantly decreased total U.S. production. The average price of soybean remained relatively high into the early 1980s. Although other factors, such as prices of competing oils or carry-over from previous years, can also affect the magnitude of price swings, weather is a dominant factor.

The volatile nature of prices has an impact on a nation's food security through foreign exchange expenditures. Foreign exchange reserves are especially important for low- and middle-income countries that use foreign currencies not only to purchase imported products but also as a risk management tool.

Low-income countries often have deficits of edible oils and may rely heavily on imports of soybean or soy oil. For example, India, a low-income country with insufficient production of edible oils, imports large quantities of soy oil. In 1984, according to FAO statistics, Indian imports of soy oil were valued at more than US$500 million (current dollars). This dependence on imports of soy oil to satisfy consumer needs is an obvious drain on foreign exchange. The risk of volatile prices due to concentrated production can have an enormous impact on year-to-year foreign exchange expenditures.

Another example is Indonesia, which in 1984 imported more than 400,000 t of raw soybean valued at US$129.5 million. According to our analysis, a one million ton change in combined soybean production in the U.S., China, and Brazil will affect Indonesia's expenditures on soybeans by ± US$940,000.

When discussing changes in foreign exchange expenditures, it is important to refer to the average amount of production variation that has been exhibited. The average magnitude of production variation from 1965 to 1984 was 32.8% for the U.S. and Brazil, which leads to the likelihood that large swings in foreign exchange expenditures on soybeans will occur. Overall, low-income countries in 1984 imported 3.4 million tons of raw beans and 2.4 million tons of soy oil. Table 5 shows the quantities and values of soybean and soy oil imported by low-income countries in 1982-84.

Exporting countries such as the U.S. and Brazil also are affected by weather-caused risks. In midwestern U.S. in particular, heavy dependence on soybean exports has resulted in specialized cropping patterns that do not allow flexible

Table 5. Soybean and soy oil imports of less developed countries, 1982-84 (FAO 1984).

Year	Near East		Far East		Africa		Latin America	
	Quantity (thousand t)	Value (thousand US$)	Quantity (thousand t)	Value (thousand US$)	Quantity (thousand t)	Value (thousand US$)	Quantity (thousand t)	Value (thousand US$)
				Soybean				
1982	107.8	31,841	1,219.2	332,965	35.1	9,502	2,026.2	532,174
1983	87.4	28,509	1,137.4	316,121	15.6	3,632	1,298.3	329,157
1984	213.2	76,686	1,338.2	445,080	22.0	6,000	1,823.1	564,665
				Soy oil				
1982	519.0	279,450	886.6	448,770	443.5	236,839	587.2	334,317
1983	770.7	391,989	917.1	505,105	404.4	214,551	540.7	295,116
1984	608.8	464,517	1,361.6	976,201	396.9	291,763	669.6	504,157

responses to changes in prices or other economic conditions. Decreasing prices and unfavorable exchange rates have caused many U.S. farmers considerable hardship; this has brought about the need for expensive government policy. Stabilization of prices in the soybean market is as important to U.S. and Brazilian farmers and governments as it is to less developed countries.

Response of Taiwan

A steady supply of soybean at reasonable prices is important for newly established livestock industries in middle-income countries and areas, such as in Korea and Taiwan, China. Economic and political conditions for agricultural policymakers in Taiwan are such that exposure to the instability of international agricultural trade is viewed as high-risk behavior. Rather than bear that risk, policies have been instituted that commit large amounts of resources to insulate Taiwan from possible shocks associated with agricultural trade. Soybean, which Taiwan requires as a raw material for many food products as well as for animal feed, is covered by a price and trade policy that has food and national security as explicit goals.

Taiwan imports about 1.5 million tons of soybean a year from the U.S. Purchases in the future are expected to remain high. Purchases of soy oil and soy meal, at one time significant, are now very low, in part because Taiwan's crushing industry is capable of producing oil and meal from imported beans.

The soybean policy in Taiwan uses prices and trade barriers as mechanisms to increase domestic soybean production capacity. In 1984-85 when the policy was first instituted, soybean support prices were US$625/t, about double the world price. The policy is financed with funds acquired through a US$243/t surcharge on soybean imports. There is also a 7% import tax.

In addition to high prices, Taiwan provides farmers with convenient collection facilities for harvested soybean, then pays the cost of transportation to processing plants. All farm families in Taiwan are free to participate in this program. A similar policy exists for maize and sorghum, which Taiwan also imports in large quantities.

Increasing production for food security

Taiwan pays a high price for food security. Most low-income countries are unable to commit such large amounts of resources to protection policies. Other measures are needed to protect those countries from volatile prices and an unstable supply of commodities. Cooperation or support from exporting countries is essential.

Multilateral or bilateral commodity agreements protect both importers and exporters from price changes and large, sudden swings in supply or demand. Unfortunately, the history of commodity agreements has been uneven at best; most such agreements have been between less developed exporting countries and more developed importing countries. With regard to soybean or grains, the roles are reversed.

A better, more long-lasting response to unstable international trade would be to diversify world agriculture so that soybean production is not confined to the U.S.,

Brazil, and China. Soybean can be produced economically in many countries. New processing technologies are being introduced that will eliminate the highly capital-intensive processing currently used in high-income and large soybean-producing countries. New technologies are available to process soybean into human food, an aspect more important in less developed countries than in North America and Europe.

It is in the mutual self-interest of less developed countries and soybean-exporting countries to minimize risks associated with soybean trade. Increases in soybean production in less developed countries would lessen the severity of price swings, making production less risky for farmers in the U.S. and Brazil.

References cited

ASA—American Soybean Association (1983) 2002: A blueprint for soybeans. St. Louis, Missouri. 4 p.
Da Mota F S (1978) Soya bean and weather. Tech. Note No. 160. World Meteorological Organization, Geneva, Switzerland.
FAO—Food and Agriculture Organization (1984) FAO trade yearbook. Vol. 38. Rome.
Howell R W (1956) Heat, drought and soybean. Soybean Dig. 16(10):14-16.
IITA—International Institute of Tropical Agriculture (1982) Annual report. Ibadan, Nigeria. 217 p.
INTSOY—International Soybean Program (1986) The soybean solution: meeting world food needs. Urbana, Illinois. 32 p.
Kiihl R A S, de Almeida L A, Dall'Agnol A (1984) Strategies for cultivar development in the tropics. Pages 301-304 *in* World soybean research conference III. R. Shiblee, ed. Westview Press, Boulder, Colorado.
Runge E C A, Odell R T (1960) The relation between precipitation, temperature, and the yield of soybeans on Agronomy South Farm, Urbana, Illinois. Agron. J. 52:245-247.
Shaw R H, Laing D R (1966) Moisture stress and plant response. Chapter 5. Plant environment and efficient water use. Am. Soc. Agron. Spec. Publ., Madison, Wisconsin.
USDA—United States Department of Agriculture (1985) Oilseeds and products. Foreign Agriculture Circular. Washington, D.C.
USDA—United States Department of Agriculture (1986) Oilseeds and products. Foreign Agriculture Circular. Washington, D.C.
USDA—United States Department of Agriculture (1987) Oilseeds and products. Foreign Agricultural Circular. Washington, D.C.

Notes

Authors' address: H. E. Kauffman and J. E. Gleason, The International Soybean Program, University of Illinois at Urbana-Champaign, USA.
Citation information: International Rice Research Institute (1989) Climate and food security. P.O. Box 933, Manila, Philippines.

Yield stability of sorghum and millet across climates

N. G. P. Rao, G. R. K. Rao, and H. S. Acharya

Seventy percent of the world's area under sorghum and millet, about 70 million ha concentrated in the less developed economies of Asia and sub-Saharan Africa, is climate dependent. In India, technological innovations involving genetic alterations coupled with input management practices and a better understanding of the process of adaptation have resulted in moderate advances in the productivity and stability of sorghum and pearl millet cultivated during the rainy season. Such a change has yet to be accomplished in sub-Saharan Africa or in postrainy season sorghum in India. This paper uses the rainy season sorghum of India as a case study to analyze the climatic limits and limitations to production and demonstrates how manipulations of genotype and environment have resulted in accelerated growth rates and stability in production despite rainfall fluctuations. A similar analysis of the sub-Saharan situation analyzes the technological potential for a change in productivity and stability there. An analysis of the stagnant postrainy sorghum situation in India is attempted. Brief reference is made to such factors as resource limitations, demand and supply, pricing, and alternative uses for sorghum and millet, which have a bearing on promoting production.

Treatises on the green revolution seldom make a positive reference to dryland agriculture. The lack of progress in dryland production has been analyzed in terms of the nature and significance of risk imposed by environmental factors, the design of appropriate technologies, socioeconomic constraints, and policy implications. When references are made to small and marginal farms and resource-poor farmers, sometimes the so-called livelihood technologies also are mentioned. It appears we are still largely at an analytical and experimental stage in this sector of agriculture.

Stagnation or slow growth in the production of coarse cereals, pulses, and oilseeds in predominantly rainfed agriculture has become a serious concern for planners and policymakers. The poor performance of these commodities has been attributed to their low value, adaptation to poor habitats, and production and consumption by the poorer sections of the society (Jodha and Singh 1982). Are these factors responsible for low production, despite the availability of technology, political will, and effort? Is viable technology available to accomplish better growth rates in the same habitats? Is the market demand and price situation for coarse grains a constraint? Currently, there is talk of spreading the green revolution to rainfed agriculture in India and of an Indian-type revolution in Africa's mostly rainfed agriculture.

A significant and sustained productivity change has taken place in a portion of the rainy season (kharif) sorghum belt in India. This transformation of rainfed kharif sorghum is relevant to the climate-dependent agriculture of semiarid regions.

As much as 70% of the world's sorghum and millet area, about 70 million ha, is situated in the less developed economies of the semiarid tropics of Asia and sub-Saharan Africa. The coarse grains are cultivated primarily as rainfed crops. Grain yields are low and climatically vulnerable. Sub-Saharan Africa, West Asia, and the millet and postrainy season sorghum areas of India are the probable regions of future coarse grain deficits (Ryan and Von Oppen 1984). Can yield levels of sorghum and millet in those areas be stabilized despite climatic anomalies? The following analysis is based on results achieved in India and on some limited experience in Africa.

The India case study

State and country level

Of the more than 16 million ha currently under sorghum in India, 60% is grown during the rainy season and the rest during the postrainy season (rabi). Black soil dominates the sorghum belt, with a limited extent of Alfisols. The black soils may be shallow, medium, or deep; medium deep soils form the largest segment. A large portion of the sorghum belt is situated between 700 and 1,000 mm isohyets.

Maharashtra is the most important sorghum state, with 6.3 million ha; about 3 million ha are planted to sorghum during the monsoon and the rest during the postrainy season. Sorghum hybrids became available after 1965-66 and their adoption was spreading at a slow pace until their performance during the 1972-73 drought gave impetus to hybrid sorghum cultivation during kharif, and brought about a productivity change.

Using the method of monotonic approximations (Acharya and Kulkarni 1986), trends in productivity of rainy season sorghum have been analyzed (Fig. 1). After 1973-74, a quantum jump in productivity continued to 1978-79, then leveled off. While the trends for sorghum and pearl millet are similar, the productivity increases in sorghum are more marked. The mean yields and coefficients of variation for the two periods are presented in Table 1. After 1973-74, sorghum yields almost doubled and variability was reduced. The trend was similar for pearl millet, but the magnitude of change was lower.

A comparative analysis of the growth and stability of kharif and rabi sorghums in Maharashtra for the periods 1961-62 to 1972-73 and 1973-74 to 1983-84 shows advances in productivity during the rainy season but nearly stagnant production in the postrainy season. The advances during the rainy season were the result of hybrid cultivation.

The overall impact of the sorghum hybrids, whose coverage in India varies from 28 to 38% of the area in different states, compared with the progress made with other cereals like irrigated wheat and rice (Narain et al 1984), is presented in Table 2. The growth rates of sorghum production after the development of high-yielding dryland varieties are comparable to those of irrigated wheat and rice, although the yield levels are still low.

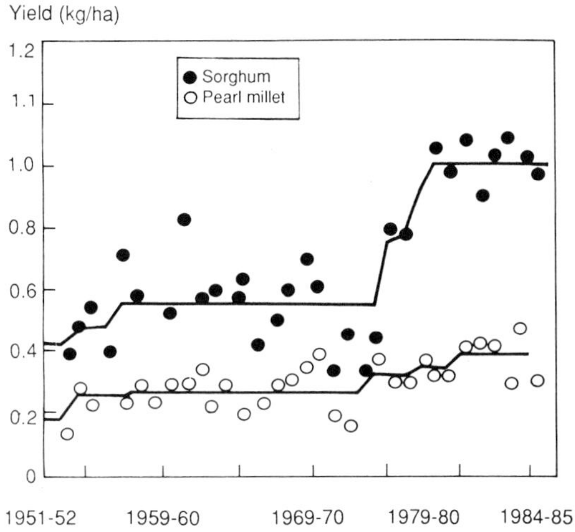

1. Trends in productivity of rainy season sorghum and pearl millet in Maharashtra.

Table 1. Area, production, and productivity of kharif sorghum and pearl millet in Maharashtra during the periods 1956-73 and 1974-84.

	1956-73		1974-84	
	Mean	CV (%)	Mean	CV (%)
	Sorghum			
Area (thousand ha)	2531	3.86	2958	3.33
Production (thousand t)	1496	24.17	3048	11.50
Productivity (kg/ha)	589	23.50	1029	9.64
	Pearl millet			
Area (thousand ha)	1823	11.41	1663	8.98
Production (thousand t)	501	30.61	644	21.09
Productivity (kg/ha)	268	23.17	386	16.89

Table 2. Compound growth rates of different crops – All India (1968-69 to 1978-79).

Crop	Area	Productivity	Production
Sorghum	-1.458*	4.491**	2.958*
Pearl millet	-1.291	1.670	0.353
Maize	0.132	0.504	0.637
Rice	0.777	1.790	2.584*
Wheat	2.853	2.355*	5.281**
Pulses	0.991	-0.053	0.939
Total food grains	0.481	2.259*	2.750**

*Significant at 5% level, **significant at 2% level.

After the advent of hybrids, an increase in productivity together with some stability (reflected by reduced coefficients of variation) has been accomplished on an area basis. During these years, both normal and aberrant years of rainfall occurred.

Experimental level

Since 1963, experimental hybrids and varieties have been tested in annual replicated field trials in 35-50 sites spread all over the kharif area. All India average yields of released hybrids CSH-1, CSH-5, CSH-6, and CSH-9 varied between 2.5 and 4.0 t/ha; higher order yields ranged between 5.0 and 6.0 t/ha.

Several analyses (Rao 1970; Rao and Harinariyana 1968; Rao et al 1982, 1986) clearly showed the yield superiority and stable performance of hybrids over local varieties. Hybrids also were superior to improved varieties, particularly under drought stress. The performance of CSH-1 compared to traditional local varieties across about 300 experiments over several years is presented in Figure 2.

Because risk aversion is an important criterion in rainfed agriculture, farmers' risk aversion (which considers both yield and stability) was used as a criterion to

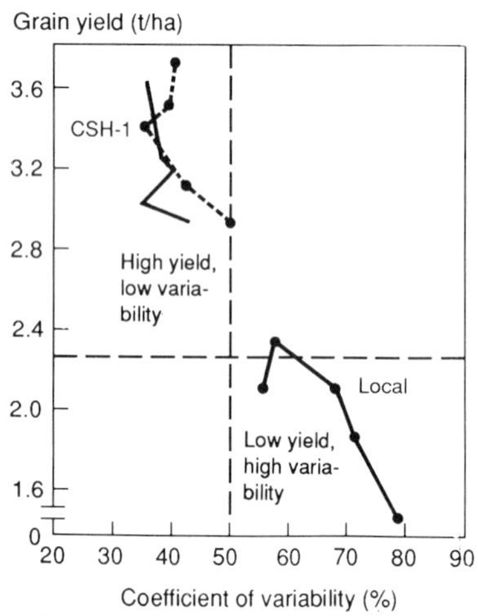

2. Stability of sorghum hybrids and traditional local varieties.

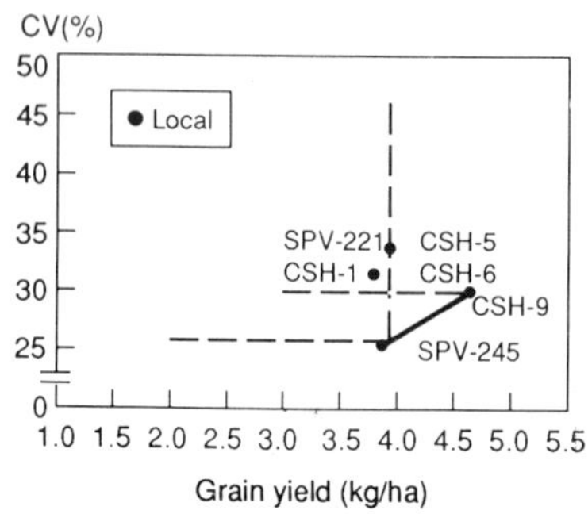

3. Adaptability efficiency of some selected hybrids and local varieties.

rank genotypes (Fig. 3). Preference-based rankings did not differ markedly from yield-based rankings (Barah et al 1981). Further, there was no rank reversal of top yielders grown under low and high management conditions (Vidyasagar Rao et al 1981). The transformation from risk-prone traditional sorghums to high-yielding stable performers has been analyzed in detail by Rao (1982).

The yield gap

Average yields of kharif sorghum in Maharashtra State are still less than 1.0 t/ha, while experimental averages in all years are more than 2.5 t/ha and higher order yields more than 5.0 t/ha. Similar results have been obtained in several other states. This gap is certainly not a technological gap. At the same site and under similar climatic conditions, one farmer may harvest 2.5-4.0 t/ha, while his neighbor obtains only marginal yields. The factors responsible are such things as lack of resources, timing, prevailing prices, near absence of interstate or external marketing, and negligence.

Immediate efforts to enhance production will have to be in bridging the farmer-to-farmer gap in a given region, rather than in reducing the gap between experimental yields and farmer yields. Further technological research in areas where constraints have been identified is needed.

However, if the production and productivity of sorghum in India are to go up, demand will have to increase. At present, we seem to be self-sufficient as far as the food requirement is concerned. Exploitation of alternative uses of sorghum for animal feed, starch, malting, and energy might generate additional demand.

Climatic limitations and opportunities

Of the climatic elements, rainfall is the most potent influence on crop growth in the semiarid tropics. High temperatures at sowing in the Sahel of West Africa and low temperatures in the rabi sorghum belt of India are also significant, together with soil type, crop duration in relation to rainy season duration, and management practices. Rainfall aberrations include 1) delayed onset of monsoon, 2) premature cessation of rains, 3) alternate dry and wet spells during crop growth, and 4) years of lower or higher precipitation than normal.

A single location example

Parbhani in Maharashtra receives an average seasonal rainfall of 824 mm between 15 Jun and 15 Oct. Soils are deep to medium Vertisols. The area is considered to be in the assured rainfall zone. Sorghum is grown during both the rainy season and the postrainy season. A polynomial curve of the normal rainfall, with deviations on both sides, for the years 1944-86 is depicted in Figure 4.

Distribution of rainfall during some aberrant years is depicted in Figure 5. The years 1975 and 1983 were high rainfall years and 1972, 1984, 1985, and 1986 were low rainfall years at Parbhani and in the country as a whole (Kulkarni 1986). The range of rainfall was from 443 mm to 1,436 mm. Spread over time, this pattern could represent the rainfall situations likely to be encountered in sorghum-growing areas.

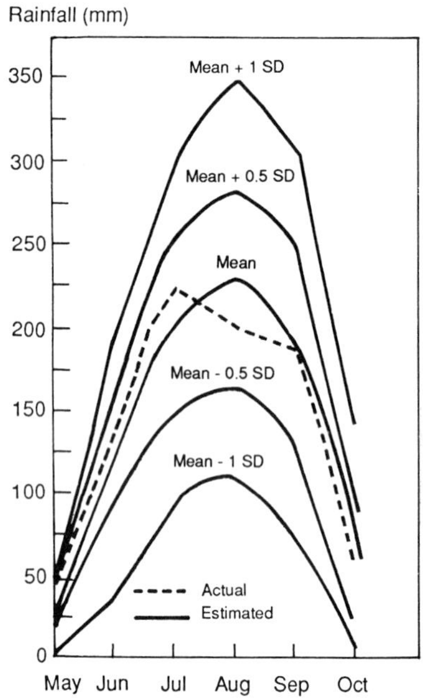

$$\text{Mean} = -233 + 136.2t - 1.59t^3 - 6.53 \times 10^{-3}t^5 \qquad (R^2 = 0.970)$$

4. Monthly rainfall distribution with deviations (Parbhani, Maharashtra, 1944-86).

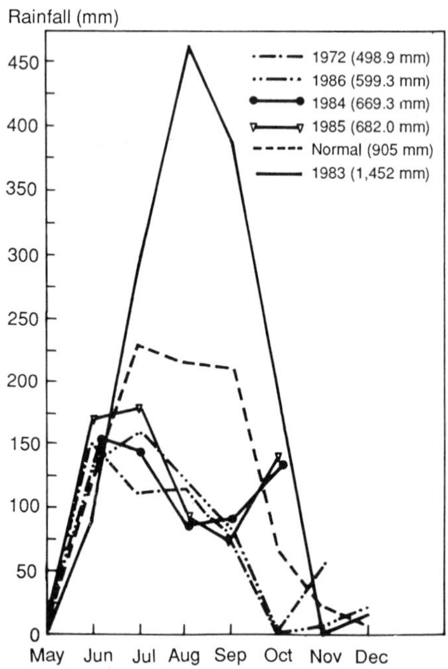

5. Monthly rainfall distribution during some aberrant years (Parbhani, Maharashtra).

Table 3. Water budgeting (mm) through field measurements during rainy season, meteorological weeks 22-44 at Parbhani.

Parameter	1972	1983	1985	1986
Rainfall	442.7	1399.0	508.3	578.4
Potential evapotranspiration	852.9	697.6	788.0	738.2
Evapotranspiration	398.7	545.0	465.6	454.2
Runoff	0	295.0	10.3	26.0
Percolation beyond root zone	0	300.0	0	2.0
Soil water storage at end of period	44.0	259.0	32.4	96.2
Soil water recharge	255.0	577.6	210.6	295.9

Table 4. Rainfall limits in sorghum- and pearl millet-growing regions of India.

Crop	Rainfall (mm)			Potential evapotranspiration (mm)		
	Mean	Range	0.5 SD	Mean	Range	0.5 SD
Kharif sorghum	915	805-1024	109.6	1682	1500-1864	182.0
Rabi sorghum	764	674-855	90.8	1771	1582-1880	108.7
Pearl millet	521	433-610	88.6	1687	1402-1892	204.9

Data on weekly water budgeting during the growing season of kharif sorghum for the years 1983 (high rainfall), 1986 (moderate drought), and 1972 and 1985 (severe drought) were obtained from studies in runoff plots, percolation tanks, and lysimeters (Table 3). Water consumption for 110-d sorghum in the rainy season is 386 mm and in the postrainy season, 295 mm (G.R.K. Rao 1981). During the rainy season, this requirement is met even during years of severe stress when 150-d traditional local varieties would fail.

Multisite testing in India

A similar analysis was carried out for 34 sites in the kharif sorghum area, 16 in the rabi area, and 19 in the pearl millet-growing area (Table 4).

Rainfall over the kharif area (Fig. 6) more or less resembles the single-location analysis. Rainfall deviations are within 0.5 SD. This means the area as a whole is reasonably safe for cultivars of 100- to 110-d duration. Rainfall during the postrainy season is less, but has two peaks during July and September, and relatively more rainfall is recorded during October. Rainfall in the pearl millet belt is less than in the rabi sorghum belt, with a single peak during July.

The potential evapotranspiration (PET) of rabi and pearl millet areas is higher than kharif sorghum (Fig. 7). Both pearl millet and rabi sorghum are cultivated in

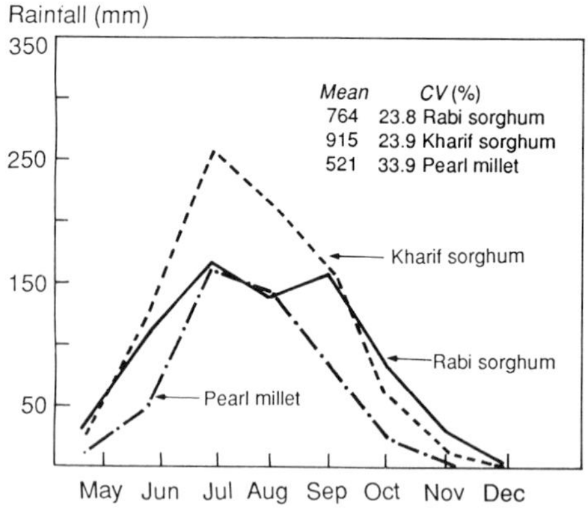

6. Rainfall distribution in rabi sorghum, kharif sorghum, and pearl millet areas in India.

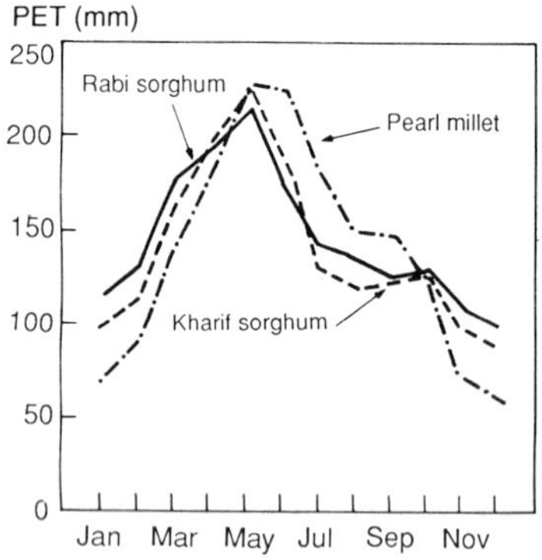

7. Potential evapotranspiration (PET) in rabi sorghum, kharif sorghum, and pearl millet areas in India.

the shallow black soils of Maharashtra, and both tend to become risky, with rabi sorghum more vulnerable. Short-season hybrids of pearl millet tend to be reasonably safe over most of the area.

The distribution zones of sorghum and pearl millet are reasonably distinct. Kharif sorghum is confined to better rainfall black soil areas, rabi sorghum to low rainfall medium and deep black soils, and pearl millet to low rainfall light soils.

During the postrainy season, minimum temperatures drop rapidly after 1 Nov (Fig. 8); this is not conducive to crop growth. Between 15 Nov and 21 Dec, the lowest minimum temperature ranges from 5 to 8 °C. Unless the crop is sown by mid-September in single-crop rabi areas and by 10 Oct in deep Vertisols where both kharif and rabi cropping are practiced, there will not be adequate time for initial rapid growth. Delayed plantings also will result in delayed maturity and greater drought stress. Depending on the progress of monsoon rains, the soil moisture profile, and soil workability, sowing dates need to be advanced.

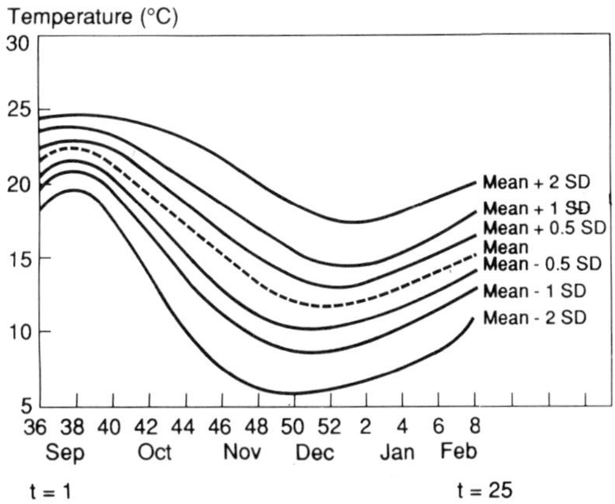

$$Temp = 20 + 1.9\,t - 0.49\,t^2 + 3.44 \times 10^{-2} t^3 - 1.01 \times 10^{-3} t^4 + 1.13 \times 10^{-5} t^5$$

$$(r^2 = 0.99)$$

8. Weekly minimum temperature distribution with deviations (Parbhani, Maharashtra).

West African multisite analysis

Rainfall distribution at 75 West African sites representing rainfall regimes ranging from 300 to 1,800 mm is shown in Figure 9. The curves conform to the single and multisite analyses of India except for the spread of the rainy season. West Africa has a single peak in August. PET rates are higher than in India (Fig. 10). Another difference is that West African soils are lighter and are described as tropical ferruginous. Sudanese soils are similar to the black soils of Deccan. There is a typical south-north orientation of rainfall, with a time lag between the start of rains in the south and the north of as much as 3 mo. The end of the rainy season is sharp, and occurs approximately 1.5 mo earlier in the north than in the south.

Millet is concentrated in the 400-700 mm rainfall zone, with sorghum where rainfall is 700-1,300 mm and higher (up to 1,600 mm). Both sorghum and millet also coexist in a relay intercropping system. By and large, the duration of the sorghum crop is 30-40 d longer than the duration of the rainy season. If the rains cease prematurely, yields in all zones are affected.

The climatic trends for tropical Africa have been analyzed by Farmer and Wigley (1985). They believe that for West Africa, Sudan, and Ethiopia, continuation of the low levels of rainfall of the 1970s and 1980s is more likely than a return to earlier wet spells. Because most rainfall in tropical Africa comes from convective clouds, rainfall tends to be highly variable in both time and space.

Relationship of rain to crop performance

Start of rains and sowing time

In India, after the onset of the southwest monsoon, the date by which an accumulated rainfall of 75 mm is received is considered the date of sowing. At Parbhani, this amount is normally received by 20 Jun, the earliest by 13 Jun, and the latest by 4 Jul (Fig. 11).

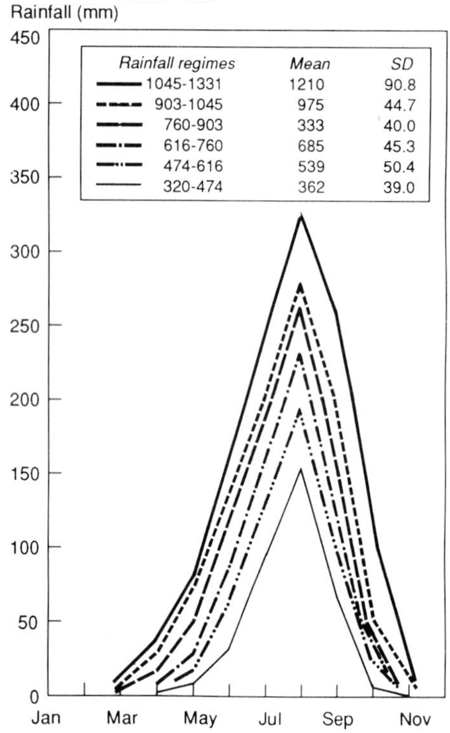

9. Rainfall distribution for different regions of West Africa.

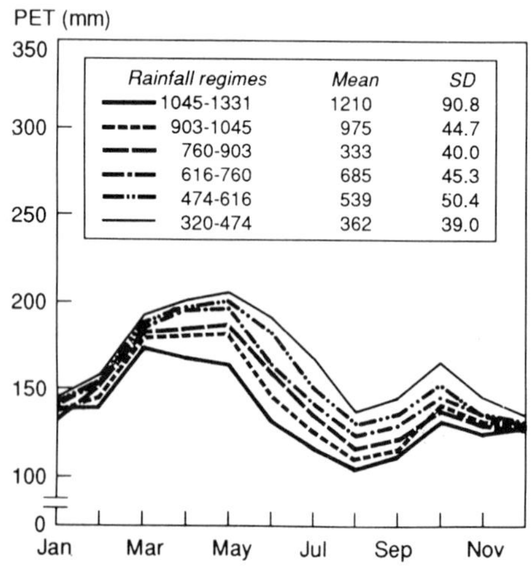

10. Potential evapotranspiration distribution for different regions of West Africa.

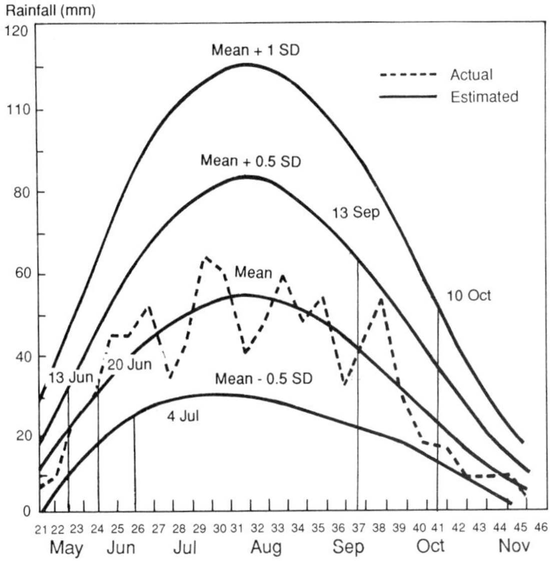

Mean $= 4.34 + 6.37t - 1.79 \times 10^{-2}t^3 + 1.24 \times 10^{-5}t^5$ $R^2 = 0.812$

11. Weekly rainfall distribution with deviations in commencement, and cessation (Parbhani, Maharashtra).

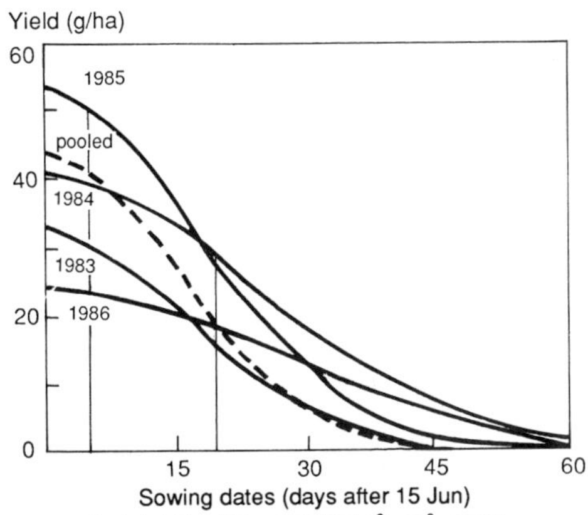

Pooled yield $= 44.6 \exp(-0.0021\ x^2)$ ($R^2 = 0.644$)

12. Relationship between sowing time and yield of kharif sorghum.

If sowing rains are received but sowing is delayed, there is a progressive yield decline (Fig. 12). The reasons are both genotypic and due to the buildup of shootfly (*Atherigona soccata* Rond), a serious seedling pest of sorghum. In a given area, if some farmers plant on time and others late, the latecomers have to face an aggravated pest situation. In high-rainfall years, incidence is high and the rate of

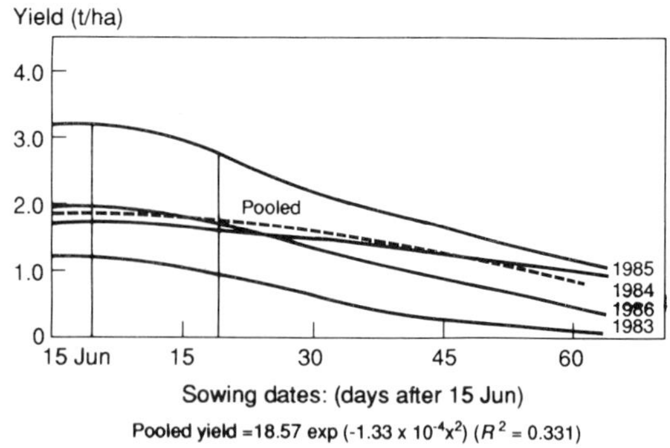

13. Relationship between sowing time and yield of pearl millet.

buildup fast. If the start of the rain is delayed, the pest buildup is also delayed. Depending on the arrival of rains, sorghum can be sown up to 7 Jul. Beyond that, it is not advisable and, if done, should be under plant protection. Genetic resistance to shootfly would help in all situations, but currently is available only at moderate levels. Indian farmers largely avoid shootfly by early sowing.

In pearl millet, yields also decline with progressive delays in sowing, but the rate of decline is slower than in sorghum (Fig. 13). Unlike sorghum, seedling pest problems are not serious, and pearl millet can stand relatively later sowings. If sorghum sowings are delayed, pearl millet would be a good alternate crop.

Premature cessation of rains

At Parbhani, the rains normally cease by the end of September, although small amounts of rain may fall during the first half of October. In aberrant years, the rains may be over by mid-September (Fig. 11).

When rains stop early, long-duration sorghum fails to make grain, but 90- to 110-d sorghum will yield satisfactorily. For rabi crops, it has been estimated that 150 mm of rainfall received after sowing and stored moisture will return satisfactory yields. Depending on progress of the rainy season, the optimal time of sowing for rabi sorghums is from mid-September to 10 Oct. In the single-cropped rabi sorghum belt with medium black soils, mid-September sowings are preferable because pest problems are not serious. In deep Vertisols and where kharif and rabi sorghums are grown in the same region, a certain delay is needed to permit soil cultivation and to avoid heavy shootfly periods. Under all circumstances, delay beyond 10 Oct results in reduced yields.

Continuous dry and wet spells during crop growth

At Parbhani, hybrid sorghums normally are grown from 11 Jun to 21 Oct. During drought years, continuous rainless periods ranging from 12 to 32 d have been encountered during different growth stages (Table 5).

Despite several weeks of drought stress during the severe dry years, 90- to 110-d hybrids did not fail. However, crops of local cultivars of 150-d duration and longer failed. Alfisols are relatively more vulnerable, but in general, total crop failures have been avoided. The situation may vary in African climates.

Table 5. Severe dry spells at Parbhani, India.

Year	Date	Duration (d)
1972	17 Jul-4 Aug	19
	25 Aug-4 Sep	12
	16 Sep-15 Oct	31
1984	17-29 Jun	13
	10 Aug-10 Sep	32
	27 Sep-8 Oct	13
	16 Aug-11 Sep	27
1986	26 Jun-13 Jul	19
	16-31 Aug	16
	29 Sep-16 Oct	18

Table 6. Rainfall regimes at Parbhani, 1944-86. (Normal rainfall between June and October: 824 mm)

		Below normal	Above normal
Normal	Range	686-824 mm	824-963 mm
mean ± 0.5 SD	Years	1947, 1950, 1952, 1960, 1962, 1964, 1965, 1976, 1981	1944, 1948, 1953, 1956, 1958, 1959, 1969
Moderately	Range	547 mm - 686 mm	963 - 1,102 mm
low/high mean ± (0.5-1 SD)	Years	1945, 1952, 1954, 1966, 1967, 1968, 1971, 1974, 1977, 1978, 1982, 1984, 1985	1949, 1957, 1970, 1973, 1979, 1980
Severe - low/high	Range	<547 mm	>1,102 mm
mean ± >1 SD	Years	1946, 1972, 1986	1955, 1961, 1963, 1975, 1983

Heavy and continuous rains in Vertisol areas have done more damage to sorghum than dry spells. Yield levels during 1975 and 1983, which were heavy-rainfall years, were lower than during 1984 and 1985, which were characterized by moderate drought. This is due to impeded drainage, leading to nutrient loss and associated factors.

Low- and high-rainfall years

In Parbhani, rainfall during crop growth may be classified as normal (Mean $\pm 1/2$ SD), moderately low or moderately high (mean $\pm$ SD), and very low or very high (when rainfall is below or above 1 SD) (Table 6). All such years have occurred during the 1980s. Classification of years by rainfall indices in India (Kulkarni 1986) is close to this.

Apart from the effects of drought or impeded drainage, dry and wet years are generally characterized by the occurrence or absence of certain insect pests and diseases, although the trends can vary. Shootfly incidence and buildup are generally higher in high-rainfall years. The peak period of shootfly infestation in all years is

August-September. Earhead pests tend to be aggravated during wet years. Midge incidence was high during high-rainfall years 1975 and 1983. Because the occurrence of midge is usually severe after 15 Sep, late-flowering genotypes are affected. When the flowering in an area is prolonged because of early hybrids and late locals planted side by side, the problem is aggravated. Midge has been avoided by planting hybrids on an area-wide basis, rather than by mixing them in with local late-maturing cultivars.

Charcoal rot disease has appeared in both dry and wet years, more severely during dry years. Charcoal rot aggravates lodging. Good resistance to charcoal rot is available in varieties such as E 36-1 and SPV 511. Reasonable levels of resistance are being incorporated into commercial hybrids and varieties.

Grain molds and grain deterioration also are important problems in wet years. These conditions become acute with cloudy and wet spells during harvest. While CSH-1 is highly susceptible to grain molds, hybrids like CSH-5 and CSH-9 have good tolerance. If harvests are not delayed, the mold problem can be contained. Good levels of resistance to downy mildew and leaf spots have now been incorporated into cultivars. In normal years, wet years, or years of late rainfall, a second crop of safflower or chickpea is possible after sorghum in Vertisols.

In pearl millet, downy mildew in particular and ergot became setbacks in India and early advances in productivity were slowed. Currently available pearl millet hybrids have good resistance to downy mildew.

Net effects of climate-imposed limitations

Grain yield data for Parbhani and Maharashtra during 1951-85 were related to six rainfall regimes. Using the method of monotonic trends (Acharya and Kulkarni 1986), time series data were split into a technological trend and deviation from the trend, largely attributable to climatic factors.

In the deviations from the trend, mean squares between groups were significantly larger than mean squares within groups. The means are diagrammatically presented in Figure 14. The deviations are large in both very low- and very high-rainfall years. There is indication that the advent of hybrids is helping bridge this gap.

Genotype-environment-input manipulations

For practical purposes, climate and weather are uncontrollable while the soil and the plant can be manipulated. Our understanding of climatic fluctuations, extreme weather events, and possible climate change is increasing. By manipulating and modifying the genotype-environment-input relations and interactions, we can minimize crop vulnerability across a range of climatic events.

Genotype alteration

Genotype alteration as the basis of agricultural transformation in tropical drylands has been analyzed in detail by N.G.P. Rao (1981). The attributes of traditional tropical cultivars have been tallness, long duration, photoperiod sensitivity, low harvest indices, and poor community performance. The timing of the main stages in

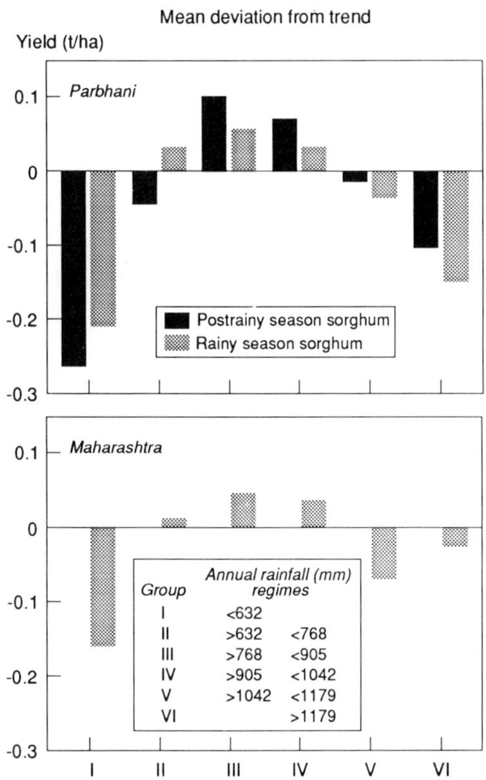

14. Deviations from the trend in sorghum yield in years of different rainfall patterns.

the growth cycles is optimized in relation to seasonal conditions through control mechanisms that are extremely sensitive to daylength and temperature. The design and development of productive and stable agriculture require changing the cycles of growth and reproduction in such a way that the emphasis is on the economic product rather than on total dry matter, with the more critical phases of growth coinciding with favorable periods of climate.

Duration, dry matter production, and distribution. Most traditional kharif sorghum of the Deccan and central Indian plateaus matures in 140 d or longer. Duration of a normal rainy season is from mid-June to late September, with July the peak rainfall month. Sorghum is usually in the vegetative stage until the 1st or 2d week of October. If the rains cease early, yield losses are heavy. Total dry matter produced may be as high as 450 g/plant, nearly 70% of it accumulated in the stalk before flowering. The new hybrids and high-yielding varieties produce less dry matter per plant with greater accumulation in the earhead, resulting in better harvest indices, as high as a ratio of 50:50.

Superior sorghum hybrids and varieties (CSH-1, CSH-5, CSH-6, and CSH-9) with shorter growth durations (100-110 d) consistently yield well. Average yields

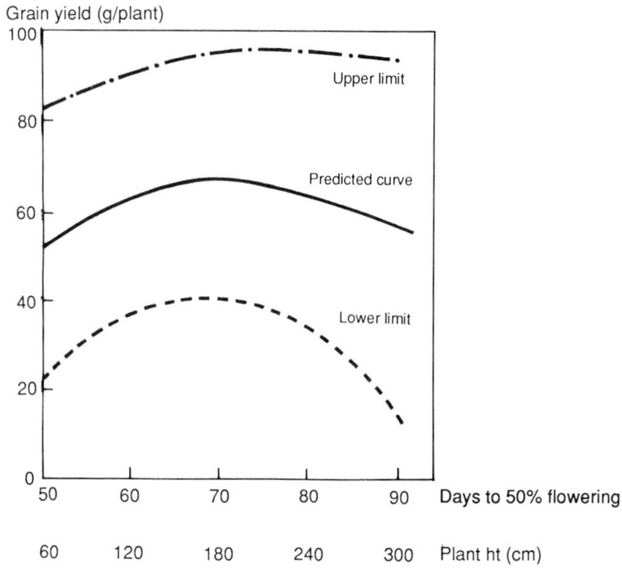

15. Optimum phenotype, related to plant height and days to flowering.

under rainfed conditions ranged from 2 to 4 t/h, with more than 6 t/ha under favorable rainfall. The critical growth stages—seedling, flower primordia, and grain filling—coincide with periods of assured rainfall or satisfactory soil profile moisture. In breeding for efficient water use in grain production, corrections for duration, dry matter production, and differentiation at optimal times of the season should constitute the first steps in modifying traditional tropical sorghums (N.G.P. Rao 1981; Rao 1972, 1982; and Rana 1982).

Optimum plant type. On the basis of studies of a wide range of variation for plant height, maturity, and yield, Rana et al (1984) established that an optimum plant type for high yields was one with 68 d to flowering and 175 cm plant height (Fig. 15). The current high-yielding and widely adapted rainy season hybrids in India are intermediate in height (170-200 mm) and days to flowering (67-75 d) and are described as intermediate optima. The cultivars found promising in other parts of the world also vary around this optimum. Tall, late types have no yield advantage.

In the postrainy season, modified cultivars with resistance to shootfly that can tolerate low temperatures under delayed planting are yet to become available. Currently available hybrids such as CSH-8R exhibit superiority under controlled early planting but lose their advantage if planting is delayed. Rabi planting cannot take place before mid-September. Hybrids and varieties that exhibit general superiority over the prevalent cultivar M35-1 have not yet been developed.

Hybrid homeostasis and adaptability. Our studies also have established that hybrids have homeostatic advantages over improved varieties, particularly under moisture stress. The hybrids CSH-9, CSH-6, CSH-5, and CSH-1 yielded the highest in kharif areas of India and were the most widely adapted. Improved varieties were superior to local varieties in yield and adaptability but were not comparable to hybrids. Local varieties were characterized by low yields and high coefficients of variability.

Yield- and risk-preference-based rankings of the hybrids are closely related. Adaptability and stability also are correlated, lending support to breeding toward genotype alteration and multisite testing in pursuit of low risk and high yields (Barah et al 1981; Rao et al 1975, 1982, 1986).

Genotype-input management relations

Nutritional adaptation is widespread in nature, with distinct genotypic differences in responses to nutritional elements as well as to toxicities. The response of altered genotypes to fertilizer and population levels is spectacular and, coupled with their lower susceptibility to climatic variables, notably rainfall (Rao et al 1975), their adoption is on the increase, although fertilizer use on commercial sorghum fields is still low.

Considering all agronomic inputs, including use of fertilizer and pesticides, the performance of high-yielding hybrids and varieties remains satisfactory only under optimal inputs (including irrigation water) and management. Consequently, "high yield agriculture" is associated with "high input agriculture." Whether this technology is applicable to the small farmers in developing countries has been questioned. Vidyasagar Rao et al (1981) examined this aspect in multisite experiments over several years. The top ranking hybrids and varieties maintained their relative ranks under both high and low levels of inputs and management. Other studies indicate that agriculture based on altered genotypes is not incompatible with lower inputs. The use and level of inputs are related more to availability, supply, and credit than to limitations imposed by technology.

Sorghum-based cropping systems

The component cultivars of traditional intercropping systems are themselves products of climate-vulnerable subsistence agriculture. Except for the spread of risk over species, they are essentially replacement systems characterized by low yields. Unless the components themselves undergo radical alteration, the system will not change.

Rao and Rana (1980) demonstrated that single crop stability and productivity are prerequisites for productive intercropping systems. Several All India intercropping trials with pigeonpea, soybean, and peanut as intercrops showed that sorghum was the primary crop; it yielded 90-95% of its single crop yield. New and more profitable crops like onion and garlic are now being tested. Traditional intercropping systems which had been replaced by single crops of hybrid sorghum are now being oriented toward more profitable intercropping.

While such intercropping systems are advantageous in areas of relatively low rainfall, multiple cropping is more profitable in high-rainfall areas with moisture retentive soils. A vast portion of the black soil belt of the Deccan and central Indian plateaus where 800 mm annual rainfall sustained 5- to 6-mo crops of traditional sorghum can now produce an assured crop of a short-season hybrid in all years and a following crop of safflower or chickpea in normal and above-normal rainfall years. Kharif sorghum followed by safflower and mungbean followed by rabi sorghum have proven to be the most feasible and profitable cropping sequence in Vertisols.

Rao (1985) and Rao and Rana (1980) demonstrated that present shortages of grain legumes and edible oilseeds could be met by sorghum-based intercropping and sequence cropping in existing sorghum areas. That productivity will probably be achieved by maximizing the number of crops rather than by increasing the yield of each crop in areas in India and Africa where long season sorghums have traditionally been cultivated. Emphasis on manipulating the cropping system with modified cultivars will be more fruitful, as was demonstrated in India, than attempts to breed improved cultivars comparable to long-duration local cultivars.

Soil and water conservation measures

Resource conservation and effective utilization of all components of production are now being emphasized in watershed-based dryland farming in both black and red soils. These measures will provide a long-term base for reaping sustained benefits from modified cultivars, cropping systems, and production practices.

Analysis and conclusions

Vulnerability has been defined as the capacity to suffer from harm or to react to adversity (Timmerman 1981). In considering the vulnerability of crops to climatic elements, surface hydrology, soil processes, and systems of exchange between the atmosphere and biota also need to be considered. An approach that will obtain reasonably good yields under adversity and optimum yields if favorable conditions prevail is probably best. A production system using modified cultivars, cropping systems, and management practices has conferred considerable stability to the productivity of kharif sorghums across a range of geographic areas and climatic regimes in India.

Relevance of the case study to semiarid tropics

Components responsible for a change in the productivity of kharif sorghum in India include

- Development of hybrids with wide adaptation and stability, reflecting a quantum jump in yield level across a range of geographical areas and climatic regimes. Plants with intermediate height and growth duration met the demand for food and fodder and performed well in good and bad years.
- Production technology with emphasis on time of planting to avoid shootfly damage, moderate use of fertilizer, enhanced plant populations, and a block approach to hybrid coverage to avoid earhead pests like midge and head bugs.
- Improvements in grain quality, fodder yield, and tolerance for pests and diseases.
- Development of hybrid sorghum-based intercropping and sequence cropping systems that incorporate grain legumes and edible oilseeds.

The question has been raised, whether such a program is relevant and feasible in the rest of the semiarid tropics. One argument from West Africa has been that most technologies have been oriented toward improving single crops rather than complete farming systems, and that single crop technologies have not been compatible with the socioeconomic rationale of practices developed by the farmers (Abalu and D'Silva 1980).

Common features of several African agricultural systems are planting in wide rows; using long-duration, tall cultivars; cultivating low plant populations, not using fertilizer, cultivating predominantly with hand tools, and mixing several crops in space and time. The traditional component crops and practices themselves have been vulnerable to climatic fluctuations and have not provided the necessary yield increase or risk avoidance. Crop components must be altered in such a way that they have the capacity to yield well across varying conditions. Once single crop performance and stability are accomplished, the new components can be integrated into the cropping systems. Cultivar alteration must occur before farming system alteration (Rao 1980).

Another argument is that sorghum work in India may not have relevance to Africa. Rao (1977) studied the sorghum situation in Sudan, Somalia, and South and North Yemen. The Sudan situation is similar to the Deccan Vertisol conditions of India. He found no serious barriers to adaptation per se. Hybrid materials developed at the International Crops Research Institute for the Semi-Arid Tropics (ICRISAT) were found promising, and a hybrid sorghum was released. Yield performance of a number of introduced varieties did not indicate any special barriers. In Somalia and the Yemen, introduced materials did well, although no serious efforts were made for large-scale adaptation.

A large number of advanced breeding material developed in India, at ICRISAT and various programs, were screened in West Africa for yield, pest complex, and grain quality over 3 yr. Selections S-34 and S-35 were released for general cultivation in Cameroon. S-35 performed well in Nigeria, Niger, and other countries. The trials also demonstrated that the long growing seasons of the Sudan and the northern and southern Guinean zones are amenable to more efficient cropping sequences over a single 6- to 7-mo sorghum crop.

In the southern African countries Zimbabwe and Zambia, breeding material developed at ICRISAT in Hyderabad, India, show promise and could be made available to farmers after production technology is developed. In East African countries with long and short rains, short-duration sorghum could be adapted to fit into the cropping systems.

In the dryland cereal areas of Africa, maize has made inroads into sorghum areas. It is vulnerable to climatic fluctuations. To stabilize cereal production, marginal areas of maize should be returned to sorghum or millet.

It is necessary to aim at quantum increases in yield levels rather than at marginal increases. Components of the cropping system need to be altered, then fitted into properly designed inter- and multiple-cropping systems with moderate fertilizer use. Only revolutionary changes, rather than evolutionary changes, can bring about the needed productivity and stability.

The resource-poor farmer

By and large, farmers who grow coarse grains, pulses, and oilseeds in mixed or sole cropping systems are in general resource poor. The coarse grains often are referred to as "poor man's cereals." Holdings of these farmers are marginal (less than 1 ha). In India, the net cultivated area of an average holding was 1.8 ha in 1976-77. Marginal holdings account for 55% of all holdings. Food crops predominate, with 80-90% coarse grains and mixed cropping in smaller holdings (Sarao 1983). In West Africa, most farms are small. In Mali, the number of hectares cultivated per resident ranges from 0.5 to 1.1; in Niger, it is 0.6 to 1.7. In Sudan, there are some large mechanized farms. In some southern African countries, both commercial large-size farms and small communal farms exist. Argentina has some large farms.

Analysis of technology option patterns in irrigated and rainfed areas of India shows that the operators of small- and medium-size farms have led in adopting new technology. Studies of kharif sorghum and dryland crops at Indore (Chaudhari 1980) and other areas showed that labor-intensive technology favored small farmers than large farmers.

The demand-price situation

Food crops, which occupy 80-90% of the small and marginal farms, are first for home consumption, then for the market. Prices of coarse grains in general are lower than those of grain legumes, edible oilseeds, or other commercial crops. The analysis of Binswanger et al (1980) indicated that variability in production is the major source of income risk, rather than price variability. The first priority has to be improved and stabilized production. While prices of legumes and oilseeds increased several fold, food grain prices, particularly those of coarse grains, were almost constant. Hence, there is a tendency to shift to other crops. In India, when sorghum production dropped below 8 million tons, imports were needed. But now, production of 11-12 million tons of sorghum per year meets the national demand. If that yield is exceeded, prices fall and farmers have no motivation to increase production.

Our approach in India has been to manipulate both intercropping and sequence cropping in such a way that net returns are maximized while meeting the need for coarse grains and fodder. We are trying to introduce high-value crops into the system. We also are attempting to diversify the use of sorghum into animal feeds and other areas, such as malt, starch, and energy production, to move in the direction of whole plant utilization. Our experience with the low-value crops has been that first, production has to increase and stabilize to meet national requirements. Then attention can be paid to income enhancement through manipulation of cropping systems and diversification of commodity uses.

Per capita consumption of coarse grain in general has decreased. In West Africa, there is evidence that pricing policies have contributed to shifts in consumption, from domestically produced coarse grains to imported staples.

A generalized approach

The vulnerability of most tropical sorghums stems from their growth duration relative to the rainy season, their pattern of dry matter production and distribution, and their highly localized adaptation. To develop better approaches to crop improvement, several agroclimatic regions have been defined and the duration of crop growing seasons computed. According to Krishnan (1974), the crop growing season in the sorghum belt of India varies between 130 and 206 d. In similar estimates in Africa, a season of up to 240-260 d has been computed.

One question is, when 90- to 110-d sorghum can yield better than traditional sorghum of 150-210 d, why should we grow excessively long-duration sorghum. Today, the adaptability barriers have been broken. In India, in place of a multitude of agroclimatic regions for the kharif season, we treat it as 1 zone and grow 90- to 110-d duration sorghum. Modified sorghum cultivars developed in India and screened for adaptation to the soil and climatic conditions of Africa have proved promising. The heterozygote advantage established with hybrids is being gradually built into inbred varieties. Coordinated international/national programs for screening, selection, and testing can identify cultivars that are more stable and resistant to prevalent and potential insect pests and diseases, with adaptability for a range of situations.

Despite experience with sorghum, wheat, and rice, some still tend to discourage a generalized approach and to create restrictions. The concept of site-specific research is being extended too far. In Africa, we still seem to quarrel with our goals, some pleading largely for preservation of existing systems, with only minor changes, and others for radical genotype and system changes (Rao 1985). Our choice is for the latter.

To us, the rainy season sorghum belt of the tropics is a coherent region. When 90-to 110-d sorghum yields as much as sorghum with double that duration, it is wasting the energies of the plant and ourselves to breed sorghum comparable in maturity to the traditional cultivars of different regions. The climatic resources in long-season areas could be better exploited by developing more efficient cropping sequences based on modified cultivars. Globally coordinated programs in cooperation with regional and national networks could bring about near elimination of the climatic vulnerability of sorghum.

The same approach may hold for millet. Breeding broad-based cultivars could provide the basis for refinements, with agronomic adjustment at the regional level.

References cited

Abalu G O I, D'Silva B (1980) Socio-economic analysis of existing farming systems and practices in northern Nigeria. Pages 3-10 *in* Proceedings of the international workshop on socio-economic constraints to development of semi-arid tropical agriculture. International Crops Research Institute for the Semi-Arid Tropics, Patancheru, India.

Acharya H S, Kulkarni B S (1986) Performance of wheat and sorghum in Parbhani district: an application of monotonic approximation. J. Maharashtra Agric. Univ. 11:60-62.

Barah B C, Binswanger H P, Rana B S, Rao N G P (1981) The use of risk aversion in plant breeding: concept and application. Euphytica 30:451-458.

Binswanger H P, Jodha N S, Barah B C (1980) The nature and significance of risk in semi-arid tropics. Pages 303-306 *in* Proceedings of the international workshop on socio-economic constraints to development of semi-arid tropical agriculture. International Crops Research Institute for the Semi-Arid Tropics, Patancheru, India.

Chaudhari A K (1980) Socio-economic field assessment of farming technology in the Indo-UK project. Pages 97-106 *in* Proceedings of the international workshop on socio-economic constraints to development of semi-arid tropical agriculture. International Crops Research Institute for the Semi-Arid Tropics, Patancheru, India.

Farmer G, Wigley T M L (1985) Climatic trends for tropical Africa. Pages 1-136 *in* A research report for the Overseas Development Administration.

Jodha N S, Singh R P (1982) Factors constraining growth of coarse grain crops in semi-arid tropical India. Indian J. Agric. Econ. 37:346-354.

Krishnan A (1974) Some climatological features of the semi-arid tropical regions of the world. Pages 53-91 *in* Proceedings of the international workshop on farming systems. International Crops Research Institute for the Semi-Arid Tropics, Patancheru, India.

Kulkarni A P (1986) Rainfall indices: systematic tool for forecasting. Econ. Times, Jan. 25.

Narain P, Pandey R K, Santisarup (1984) Perspective for foodgrains production in 2000 A.D. Pages 879-883 *in* Commerce, June 30.

Rana B S, Barah B C, Binswanger H P, Rao N G P (1984) Breeding optimum plant types in sorghum. Indian J. Genet. 44:385-398.

Rao G R K (1981) Agrometeorological studies for better agronomic practices. Ph D thesis, Andhra University, Waltair.

Rao N G P (1972) Genotype × environment interactions in grain sorghum hybrids. Indian J. Genet. 30:25-30.

Rao N G P (1972) Sorghum breeding in India: recent developments. Pages 101-142 *in* Sorghum in the seventies. N. G. P. Rao and L. R. House, eds. Oxford and IBH Publishing Co., New Delhi.

Rao N G P (1977) Sorghum research and production in Sudan, Somalia, Yemen Arab Republic and People's Democratic Republic of Yemen. FAO Report. Near East Regional Office, Cairo.

Rao N G P (1980) West African agricultural systems and sorghum improvement. International Crops Research Institute for the Semi-Arid Tropics, Patancheru, India. 32 p. (mimeo.)

Rao N G P (1981) Genotype alteration: the basis of agricultural transformation in tropical drylands. Indian J. Genet. 41:437-449.

Rao N G P (1982) Transforming traditional sorghums in India. Pages 39-59 *in* Sorghum in the eighties. International Crops Research Institute for the Semi-Arid Tropics, Patancheru, India.

Rao N G P (1985) Sorghum production in relation to cropping systems. Pages 3-11 *in* Proceedings of the international sorghum entomology workshop. International Crops Research Institute for the Semi-Arid Tropics, Patancheru, India.

Rao N G P, Harinariyana G (1968) Phenotypic stability of hybrids and varieties in grain sorghum. Curr. Sci. 38:92-98.

Rao N G P, Rana B S (1980) Sorghum based cropping systems to meet shortages of pulses and oilseeds in India. Curr. Sci. 49:622-626.

Rao N G P, Rana B S (1982) Selection in temperate-tropical crosses of sorghum. Pages 403-419 *in* Sorghum in the eighties. International Crops Research Institute for the Semi-Arid Tropics, Patancheru, India.

Rao N G P, Rana B S, Rao V J M, Reddy B B (1982) Development of new hybrids and varieties and their performance. Indian J. Genet. 42:213-219.

Rao N G P, Rao V J M, Reddy B B (1986) Progress in genetic improvement of kharif sorghums in India. Indian J. Genet. Plant Breed. 46(2):348-354.

Rao N G P, Subba Rao S, Vidyasagar Rao K (1975) Rainfall fluctuations and crop yields. Curr. Sci. 44:694-697.

Ryan J G, Von Oppen M (1984) Global production and demand for sorghum and millet to the year 2000. Pages 41-61 *in* Agrometeorology for sorghum and millet in the semi-arid tropics. International Crops Research Institute for the Semi-Arid Tropics, Patancheru, India.

Sarao B S (1983) All India report on agricultural census 1976-77. Pages 1-119 *in* Government of India report. New Delhi.

Timmerman P (1981) Vulnerability, resilience and the collapse of the society. Environmental Monograph 1. Institute of Environmental Studies, University of Toronto, Canada.

Vidyasagar Rao K, Subba Rao S, Rao N G P (1981) Genotype × input-management interactions in sorghum. Indian J. Genet. 41:54-58.

Notes

Authors' address: N. G. P. Rao, G. R. K. Rao, and H. S. Acharya, Marathwada Agricultural University, Parbhani 431 402, Maharashtra State, India.

Citation information: International Rice Research Institute (1989) Climate and food security. P.O. Box 933, Manila, Philippines.

Climate vulnerability of sorghum and millet

M. V. K. Sivakumar

Sorghum and millet are the most important cereals for resource-poor farmers of West Africa. They are the only staple crops that can withstand the ravages of weather in that area. Temporal and spatial variations in rainfall and the persistence and patterns of rainfall shortages, especially since 1969, have contributed to low productivity over the last 2 decades. Increased yields can only be achieved through effective management of available resources, both physical and biological. Ongoing research in West Africa offers hope, but significant changes in present farming methods will be needed to reduce the climatic vulnerability of sorghum and millet.

Given the excellent analysis of the influence of climate on sorghum and millet production in India (Rao et al, this volume), I would like to focus on another region of the world—West Africa—where these crops are the most important cereal food for millions of resource-poor farmers. In northern Nigeria, sorghum contributes 73% of total calorie intake and 52% of per capita protein (Simmons 1976). Sorghum and millet play an important role in rural economies and are put to various uses: the grain is used to prepare a variety of local foods and drinks, the hay is used as animal feed, and the stalks are used to construct fences and thatched houses.

West Africa is the poorest region in the world, with the lowest gross national product per capita. About 90% of the population in this region live in villages and depend on subsistence agriculture for their survival. The population growth rate in the 1970s averaged 2.7% and is projected to remain about 3% across 1980-2000 (FAO 1981). This is the only region in the world where capita food production declined over the last two decades (World Bank 1984) and the ratio of food imports to total food increased.

Several factors are responsible for low agricultural productivity in West Africa. Some are climatic, principally low and highly variable rainfall and high demand for water imposed by the consistently high temperatures and radiation. In a large belt across West Africa, there were serious crop failures during 1968-73.

Rainfall variability

Rainfall in West Africa is low and variable. The scale of variability determines the magnitude of crop vulnerability and the extent of regional crop failures. Temporal or time-dependent variations in rainfall are common, and can be represented by three time scales: annual, monthly, and daily. The coefficient of variation of annual rainfall ranges between 15 and 30%. Variability in monthly rainfall is larger, since the rainfall is usually limited to 3-5 mo May-October. Rainfall variability reaches its maximum at the level of daily rainfall.

Rainfall in the semiarid regions is characterized by high spatial variability. Over the 500-ha research farm at the International Crops Research Institute for the Semi-Arid Tropics (ICRISAT) Sahelian Center (ISC), Sadore, Niger, we consistently observe 30-40% deviations in individual rain gauges from the central observatory; in isolated cases, deviations up to 80% have been recorded.

One unusual feature of the rainfall in West Africa is rainfall shortages persisting over one to two decades (Nicholson 1982). Rainfall fluctuations also are associated with a geographic pattern. For example, reduction in mean annual rainfall in both Niger and Burkina Faso after 1969 was characteristic of the entire region. After 1969, rainfall isohyets were displaced farther south showing that rainfall changes affect large areas.

Crop vulnerability

Rainfall variability leads to instability in traditional methods of crop production. In Burkina Faso, sorghum and millet production was stagnant (Fig. 1) between 1963 and 1984. Total sorghum production decreased while millet production stayed the same. However, average productivity/ha of both crops declined (Fig. 2). India achieved a more than 40% increase in sorghum and millet productivity between 1963 and 1984, Niger and Burkina Faso showed a negative productivity (Table 1).

When average rainfall in the Lower-Casamance region of Senegal decreased between 1969 and 1984 to levels consistently below normal, the area planted to sorghum and millet in the Bignona department decreased from 21,000 to 7,700 ha

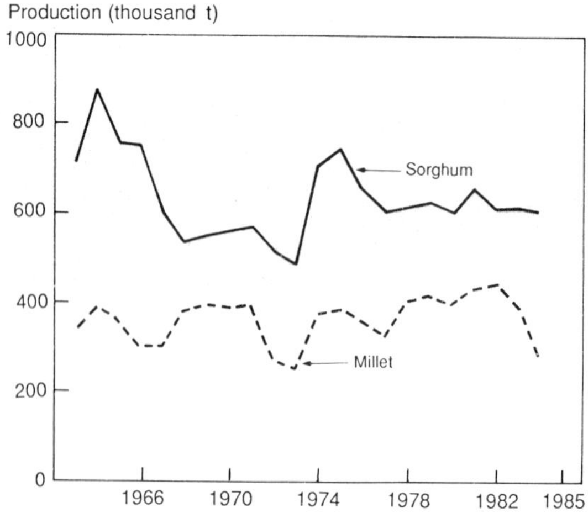

1. Total sorghum and millet production in Burkina Faso, 1963-84.

and production declined from 16,000 to 5,000 t (Posner et al 1985). As a result, instead of the food surpluses of 100 kg per capita recorded 1962-63, Bignona now faces food shortages of 170 kg per capita.

Based on these experiences, one may wonder why the farmers grow these crops. But sorghum and millet are the only staple crops that can withstand the climatic constraints of West Africa. For the short growing season of the Sahelian region, the crop that will give the highest relative yield is millet; in the regions with longer growing seasons, sorghum is ideal. Under drought stress, millet, with its shorter developmental phases, rapid regrowth, and greater plasticity conferred by asynchronous tillering, can make better use of the short periods of water availability. Water-use efficiency of sorghum also is higher than that of most agronomic crops (Kanemasu et al 1984). Reported optimum and maximum temperatures for the different phenological stages of sorghum and millet are higher than for other cereals.

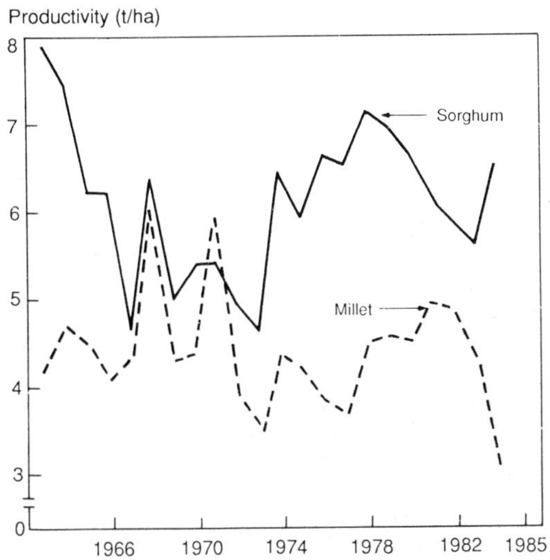

2. Average productivity of sorghum and millet in Burkina Faso, 1963-84.

Table 1. Expected and actual yields of sorghum and millet in 1984 based on 1963 yield levels.

Country	Crop	Yield (thousand t) Expected	Yield (thousand t) Actual	Difference (%)
Niger	Millet	1,494	900	-40
	Sorghum	619	200	-68
Burkina Faso	Millet	389	280	-28
	Sorghum	729	600	-18
India	Millet	8,048	11,800	+47
	Sorghum	8,250	11,800	+43
	Wheat	17,417	31,564	+81

That makes them ideal choices in the high-temperature environments of West Africa.

From a socioeconomic standpoint, an important consideration for growing these crops is that production in West Africa is primarily subsistence-oriented. Farmers who grow sorghum and millet market only 10-20% of their production; they consume up to 700 g per capita per d (Ryan and Von Oppen 1984).

Current situation, potential, and research needs

Despite the important role of sorghum and millet in the rural economies of West Africa and the need for increased production to feed the growing population, little progress has been made. During 1975-79, cereal self-sufficiency in Africa was only 75%. By 2000, it is expected to decrease to 56% (FAO 1981). The vulnerable countries in West Africa could have food deficits more than three times the deficit experienced in recent years. With little hard cash to pay for costly imports, the situation appears bleak.

The production increases needed to sustain food needs, according to FAO (1981) projections, will be achieved through higher yields (51%), increasing arable land (27%), and improving cropping intensity (22%). Ryan and Von Oppen (1984) question that projection for higher yields. Between 1961 and 1980, yield and intensity increases together contributed only 50% to production growth in West and eastern Africa.

The yield jumps projected by FAO as needed for 2000 can be achieved only through effective management of available physical and biological resources. Considerable potential for raising the yields of sorghum and millet above the current averages exists. At the ISC, Niger, during a 5-yr period (1982-86), yields of CIVT, an improved variety of millet, were 100-470% above average millet yields (0.5 t/ha) and 16-78% above yields of the local cultivar (Sivakumar, this volume). In 1984, when the lowest annual rainfall of this century was recorded and when 2.5 million people were directly affected (Timberlake 1985), we harvested 1.1 t/ha with a seasonal rainfall of only 213 mm. Significantly, the maximum yield advantage (78%) of the improved cultivar was obtained in this severe drought year. With moderate application of nitrogen and phosphorus fertilizer, we achieved a threefold to fourfold increase in millet yields at ISC (ICRISAT 1985).

High yields on research stations, such as those at ISC, give rise to considerable optimism and provide evidence that sorghum and millet are not that vulnerable to climate after all. However, farmers in West Africa who have gone through the worst droughts during 1970-84 do not share this optimism. A critical examination shows that their farming methods are hardly suited to harvesting stable and high yields.

Farming systems in much of West Africa are land-extensive, diversified, and fragmented. Low man-land ratios have encouraged long bush-fallow systems with little or no use of nonlabor inputs. Matlon (1985) gave the average NPK applied in 1978-82 as 3 kg/ha among the 8 Sahelian countries and less than 5 kg/ha for West Africa as a whole. Less than 5% of the area planted to sorghum and millet is plowed before planting and fewer than 5% of the farmers use improved varieties. Their poor resource base (particularly capital, labor, and management) makes adoption of improved varieties difficult, as was shown by Banta and Bbuyemusoke (1985) in the "Guided Change Project" in northern Nigeria. There, only 47% of 153 farmers adopted an improved sorghum variety.

The World Development Report (World Bank 1982) stated: "Yield increases still depend on the subtle interaction between soil, water, seeds, and sunlight, but the process is not as well understood under rainfed conditions as it is with irrigated land." Agrometeorologists play an important role in mapping adaptation zones for sorghum and millet varieties of different growth cycle lengths and in identifying areas with maximum potential.

Much research remains to be done in developing stable and high-yielding varieties for the marginal areas. The more marginal the ecological conditions, the more important the need for adaptation of a variety to specific conditions. The agenda for action for sub-Saharan Africa (World Bank 1981) states that agricultural research has not succeeded in producing varieties adapted to these special conditions. Research carried out by the sorghum and millet improvement teams of ICRISAT in West Africa is aimed at breeding varieties adapted to these marginal conditions. The resource management team is addressing the development of appropriate technologies for optimal use of the human, biological, and physical resources of the region. Multidisciplinary interinstitutional research should be able to reduce the climatic vulnerability of sorghum and millet.

References cited

Banta A, Bbuyemusoke S (1985) Improved sorghum production technology in northern Nigeria. Pages 204-217 *in* Appropriate technologies for farmers in semi-arid West Africa. H. W. Ohm and J. G. Nagy, eds. Purdue University, Indiana.

FAO—Food and Agriculture Organization (1981) Agriculture: toward 2000. Rome, Italy.

ICRISAT—International Crops Research Institute for the Semi-Arid Tropics (1985) Annual report 1984. Patancheru, Andhra Pradesh, India.

Kanemasu E T, Piara Singh, Chaudhuri U N (1984) Water use and water use efficiency of pearl millet and sorghum. Pages 175-182 *in* Agrometeorology of sorghum and millet in the semi-arid tropics. Proceedings of the international symposium, 15-20 Nov 1982, ICRISAT Center, India. International Crops Research Institute for the Semi-Arid Tropics, Patancheru, Andhra Pradesh, India.

Matlon P J (1985) A critical review of objectives, methods and progress to date in sorghum and millet improvement: a case study of ICRISAT/Burkina Faso. Pages 154-179 *in* Appropriate technologies for farmers in semi-arid West Africa. H. W. Ohm and J. G. Nagy, eds. Purdue University, Indiana.

Nicholson S E (1982) The Sahel: a climatic perspective. Club du Sahel, Paris, France.

Posner J, Kamuanga M, Sall S (1985) Production systems in the lower Casamance and strategies adopted by farmers for the water shortage. Pages 179-203 *in* Appropriate technologies for farmers in semi-arid West Africa. H. W. Ohm and J. G. Nagy, eds. Purdue University, Indiana.

Rao N G P, Rao G R K, Acharya H S (1989) Yield stability of sorghum and millet across climates. Pages 165-186 *in* Climate and food security. International Rice Research Institute, P.O. Box 933, Manila, Philippines.

Ryan J G, Von Oppen M (1984) Global production and demand for sorghum and millet to the year 2000. Pages 41-62 *in* Agrometeorology of sorghum and millet in the semi-arid tropics. Proceedings of the International Symposium, 15-20 Nov 1982, ICRISAT Center, India. International Crops Research Institute for the Semi-Arid Tropics, Patancheru, Andhra Pradesh, India.

Simmons E B (1976) Rural household expenditure in three villages of Zaria province, May 1970-July 1971. Samaru Miscellaneous Paper No. 56. Institute for Agricultural Research, Ahmadu Bello University, Zaria.

Sivakumar M V K (1989) Some strategies to cope with drought in the Sahelian zone. Pages 541-555 *in* Climate and food security. International Rice Research Institute, P.O. Box 933, Manila, Philippines.

Timberlake L (1985) Africa in crisis. The causes, cures of environmental bankruptcy. Earthscan, International Institute for Environment and Development, London.

World Bank (1981) Accelerated development in sub-Saharan Africa: an agenda for action. Washington, D.C.

World Bank (1982) World development report 1982. Oxford University Press, New York.

World Bank (1984) Toward sustained development in sub-Saharan Africa. Washington, D.C.

Notes

Author's address: M. V. K. Sivakumar, International Crops Research Institute for the Semi-Arid Tropics, B.P. 12404, Niamey, Niger.

Citation information: International Rice Research Institute (1989) Climate and food security. P.O. Box 933, Manila, Philippines.

Operational agroclimatic assistance to agriculture in sub-Saharan Africa: a case study

K. Traore and M. Konate

The severe drought that affected the Sahelian countries in the 1970s made the scientific community and political authorities aware of the need to consider climatic information in designing agricultural projects. A pilot experiment has been underway in Mali since 1982, so far in a limited area. A multidisciplinary group uses climatic, agronomic, and field information to advise a test group of farmers working on experimental plots. The same farmers are also asked to use only their own traditional knowledge in work on control plots. The experimental plots produced a yield increase of about 25%. The first stage of the project consisted of determining the best methodologies to use. Assistance now will be extended to progressively cover the entire farming area of the country with selected methodologies.

The severe drought in the Sahel in the 1970s prompted several affected countries to reinforce their meteorological and hydrological services, to assist agriculture in the fight against drought.

In just a few years, our agrometeorological service in Mali had developed the capability to provide information for assessing the impact of weather on food production and for early weather warnings. However, illiterate or low-educated farmers obviously could not by themselves use that information in their field operations. So we still were not assisting agriculture.

In 1982, we initiated a pilot project to determine an effective methodology for operational agrometeorological assistance to agriculture. We set up an experiment in which a multidisciplinary team used agronomical and climatic information to advise a selected group of farmers, to determine the impact of this type of information on yield.

The pilot project was chosen for the following reasons:

- After several years of devastating drought, the country needed a strategy to secure food production. The experiment, if successful, could contribute to that goal.
- A great deal of research had been done on the relations between soil, water, plants, and the atmosphere, and some of the published results seemed suitable for farming advisories.
- Such a project would be low cost and could be carried out by existing staff.

The pilot project was located southwest of the capital city, Bamako, in a sector of Bancoumana that is part of the agricultural extension organization Operation Haute Vallée (OHV). This sector has an area of 4,290 km^2, 5,980 ha of it planted to millet and sorghum, Mali's main food crops. Mean annual rainfall is about 800 mm.

Protocol

The objective was to provide agrometeorological information to farmers that would enable them to make more efficient decisions in their agricultural activities.

Four villages were selected: Bancoumana, Kenieroba, Kongola, and Makandiana (Fig. 1). The experiment was based on the following steps:

1. Basic meteorological parameters (rainfall, temperature, wind, humidity, sunshine) were measured at a reference agrometeorological station. Because of its high spatial variability, rainfall also was measured in rain gauges installed close to selected fields. Soil data (water-holding capacity, runoff coefficient) and plant data (phenological stages, crop coefficients, pests and diseases, weeds) were collected.

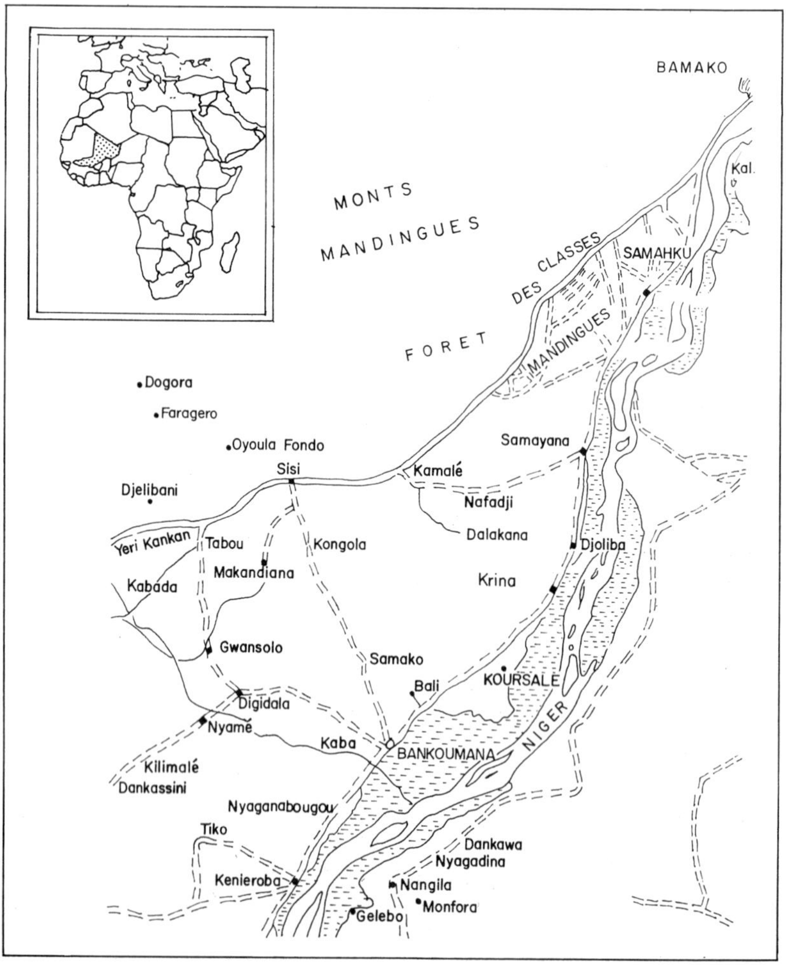

1. The agrometeorological pilot zone.

2. Agroclimatological functions (potential evapotranspiration, water balance models) were computed.
3. Advisories based on soil-water-crop relationships and agricultural practices (tilling, planting, weeding, application of fertilizer) were issued.
4. Directives were sent out through extension workers in a form understandable to farmers.

An improved variety of sorghum (Tiemarifing, 120 d) was grown in Bancoumana and Kenieroba while a local variety of millet (Toutoucoun, 130 d) was grown in Kongola and Makandiana. Both cultivars are widely grown in the region.

In each village, farmers were classified by OHV according to the level of equipment they possessed. Two farmers were selected who had one pair of oxen, one plow, and a multi-use plow and two who had two pairs of oxen, plows, multi-use plows, seeders, and even one or more carts.

The selected farmers each cultivated a 0.5-ha field divided into two equal plots.

- On the "traditional" plot, each farmer used his own crop calendars and traditional knowledge.
- On the "test plot," the farmer followed the directives of a multidisciplinary team of agronomists, meteorologists, agrometeorologists, and crop protection and functional literacy specialists. Directives were given the farmers by OHV extension officers, who received them by radio.

Both plots were tilled with a plow. At tilling, ammonium phosphate was applied on the whole field at 100 kg/ha. Urea (50 kg/ha) was applied at stem elongation.

Test plots were sown in rows 80 cm apart. Seed holes were 45 cm apart. Plants were thinned to two to three per seed hole.

For the traditional plots, the farmers determined sowing and thinning procedures.

Crop calendars for the traditional plots were fixed by the farmers' own criteria. At present, we have only after-the-fact knowledge of these calendars, but we are surveying the farmers for more details.

For the test plots, activities were carried out according to instructions from the multidisciplinary team, which based its recommendations on agrometeorological conditions and field observations.

Methodologies

The water balance principle

Decisions for plowing, sowing, weeding, and fertilizer application were based on 10-d water balance calculations. The model used was the one suggested by Forest (1974):

$$P_i + SM_{i-1} - RO_i = AE_i + SM_i + D_i$$

where P_i is the amount of rain during day i,

SM_{i-1} is the soil moisture content estimated at the start of day i,

SM_i is the soil moisture content for the end of day i,

RO_i is the runoff for day i,

AE_i is the actual evapotranspiration for day i, and

D_i is the draining for day i.

Soil moisture (SM) refers to the root depth Z_i, which is estimated from a linear model.

Draining (D) is set at zero except for cases where the total water supply $P_i + SM_{i-1} - RO_i$ is greater than the maximum storage capacity MS_i of the root zone. In that case, we have

$$D_i = P_i + SM_{i-1} - RO_i - MS_i$$

and the total supply is reduced to MS_i.

The actual evapotranspiration AE_i is estimated following Eagleman (1971):

$$AE_i = 0.732 - 0.05(ME_i) + (4.93(ME_i) - 0.661(ME_i)^2)\ MR_i - (8.57(ME_i) - 1.56(ME_i)^2)\ MR_i^2 + (4.35\ ME_i - 0.88(ME_i)^2)\ MR_i^3$$

ME_i is the maximum evapotranspiration $ME_i = K_i\ PE_i$ where K_i is the crop coefficient and PE_i the potential evapotranspiration estimated using Penman's formula.

MR_i is the moisture ratio

$$MR_i = (\text{Min}\ (P_i + SM_{i-1} - RO_i,\ MS_i) - WP_i)/(FC_i - WP_i)$$

where WP_i is the wilting point and FC_i is the field capacity of the root zone.

Water balance was calculated daily. For each 10-d period, the model gave actual evapotranspiration AE, soil moisture at the end of the decade, and the crop water requirement satisfaction rate AE/ME. Using statistics from the rainfall file, the team determined the probability of receiving the required rainfall to meet crop needs during the next 10-d period.

Decision to plow

Water balance was calculated from the time of the first rainfall. If at the end of a 10-d period "i," an AE_i/PE_i of 0.2 was obtained, and if there was a probability of at least 80% that the AE_{i+1}/PE_{i+1} would be 0.2 for the next 10-d period, soil moisture conditions are favorable for plowing (Franquin 1973). The instruction went out to start plowing during the 10 d to come.

Decision to sow

A study of sowing dates based on a frequential analysis of climatic balance (cf. Meteorologie et IER 1984) shows that at Bancoumana, sowing on 20 May is successful in 1 of 2 yr (probability 0.5); after 5 Jul, the soil becomes too moist for sowing in 1 of 2 yr (agroclimatic event B1, defined by Franquin 1973). Sowing was therefore planned between 20 May and 5 Jul.

Further, taking the water balance calculated from the plowing period, if AE = 0.3 PE at the end of a 10-d period between the two dates above, and if there was at least 80% probability that this would also be so for the next 10-d period, then sowing could take place. That is, the decision to sow was made when it was likely that soil moisture conditions would be good for sprouting.

Of course, when these conditions for sowing occurred late, or if for some reason sowing could not take place in time, insofar as possible seed was chosen accordingly.

The 80% probability thresholds used for the plowing and sowing decisions corresponded to minimum failure risks of 20% (2 of 10 yr). Nonetheless, it would have been more effective to couple these forecasts based on statistical rainfall analysis, with dynamic meteorological forecasts covering several days. The Directorate of Meteorology is still seeking to resolve this technical problem.

Weeding

The first weeding was done after the third leaf appeared and the second weeding at stem elongation. Additional weeding was done as needed. Weeding was done on days without rain and with plenty of sunshine. Here again, it would have been preferable to have had access to 2- or 3-d forecasts. A day may be sunny in the morning, but the weather may change during the afternoon. Or weeding may be done one day, and it may rain the following day, rendering the effort useless.

Spreading urea

Urea was applied at stem elongation on a day when the soil was considered sufficiently moist. Here, too, it would have been ideal to carry out this operation when a long dry spell was not forecast, to avoid burning the crops with unabsorbed urea.

Crop protection

The primary enemies of crops in the pilot zone are parasitic plants and diseases and insects.

The principal parasitic plant is striga, which we uprooted, preferably before it flowered. This prevented seeds from forming and germinating late.

We were limited to recording diseases and insects. The main diseases were mildew and smut, but effective treatments have yet to be developed. Nevertheless, the correlations between the appearance of these diseases and the various agrometeorological parameters will be studied, with a view to forecasting disease.

Dissemination of information

Because the objective was to advise farmers on the optimal dates for agricultural operations, time was precious and the exchange of information between multidisciplinary team and the field people (farmers and extension workers) was of great importance. Figure 2 gives a schematic representation of the way information was collected and managed in the pilot project. It consisted of three main stages: data collection, data processing, and transmission of recommendations to the field.

Each 10 d, two observers collected agrometeorological and crop protection data in the pilot zone. Any pest and disease samples were identified by the Integrated Pest Management section of the Agronomic Research Office. Agrometeorological data were processed by the Meteorology Service. The multidisciplinary team met 2 d later to analyze the data and issue instructions.

Directives were immediately transmitted by radio to the pilot zone by the OHV representative.

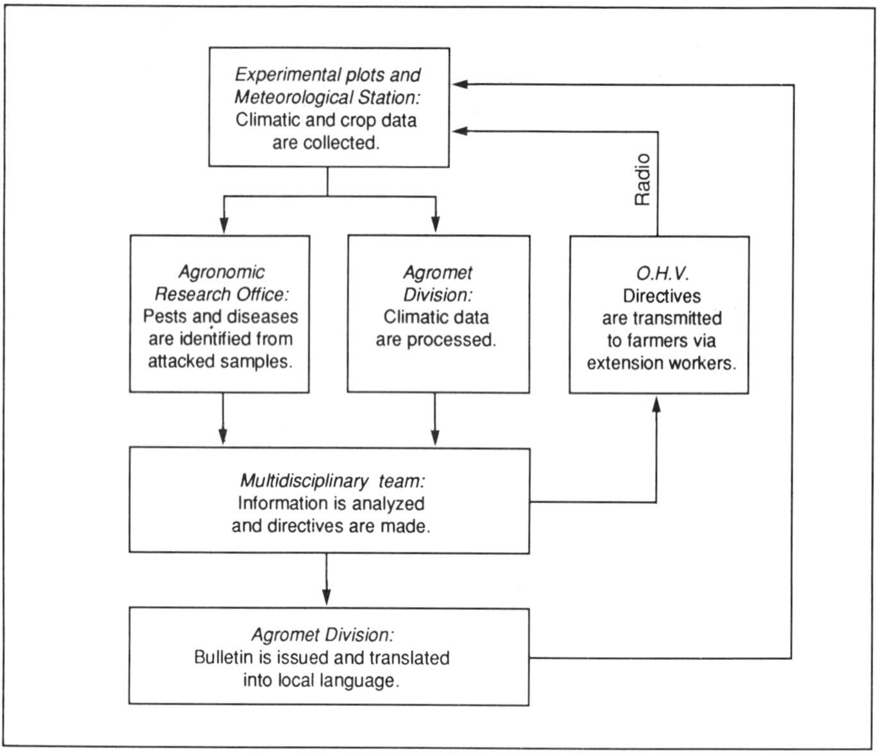

2. Schematic representation of the circulation of information in the pilot project.

Results

To carry out the experiment, project management and the pilot farmers agreed on the following arrangements: Project management provided inputs, but not manpower nor farming tools. At harvest, yields were estimated from sampled squares (called yield squares) and seeds were selected for the next farming season. The balance of the sample was returned to the farmer.

Although procedures for this experiment were relatively simple, some problems arose. After the first year, some farmers began applying the experimental team's advice on their control plots as well as on test plots. A survey showed that either those farmers did not understand the procedure, or they did it on purpose to maximize profits, knowing from their first-year experience that the test plots seemed more productive than the control ones.

Tables 1 and 2 show yields in both plots for three consecutive years. Yields from test plots were generally higher than yields from control plots for both sorghum and millet. The increase in yield had a mean value of 17% for sorghum and 26% for millet.

Because the farmers did not always follow directions for the test plots, we also compared yields from the test plots with yields from other fields in the village where fertilizer was used. Tables 1 and 2 also show the results of this comparison (we removed all plots that were severely damaged by animals or birds). The test plots achieved a 41% yield increase in sorghum and 26% in millet.

A survey is under way to better understand what determines farmer decisions.

Feedback

Several field visits were organized to assess the impact of the project. During those visits, it appeared that the experiment had had a positive effect on farmers. Statements like "we are ready to use part of our production to pay for the multidisciplinary team's advice if necessary" were often heard. Some said they got higher production from the experimental plots than from their family farms, which were sometimes two to four times larger, and concluded that they could increase their production while farming a smaller area by using agroclimatic advice. Some farmers said they had tested the directives on parts of their farms and achieved satisfactory results.

Other village farmers told us they had paid special attention to the way our pilot farmers worked, and had used the same methods.

At any rate, during different assessments of the project by international experts and in the training seminars we organized, more and more farmers in the area asked us to give them the benefit of the farming advice we were giving our pilot farmers.

Prospects for the future

The pilot project was designed for four stages: experimental, assessment, extension, and generalization. The first stage, which has been completed, consisted of data collection and testing of agroclimatic methodologies for farming advisories. During the assessment stage, the data will be analyzed and methodologies proved relevant for operational agroclimatic assistance to agriculture will be selected for future use. At the extension stage, training courses and seminars will involve farmers and extension workers in data collection and use of agroclimatic information. Some literate farmers have been trained in their local language to measure rainfall. In the fourth stage, assistance will be progressively extended to all the farmlands in the country.

In the future, our efforts will focus on extending the process, first to cover all of the Bancoumana sector and then to Banamba, which has a typical Sahelian climate. Other sectors of the OHV will be covered progressively. After that, we can assess the impact of the design over an entire extension area.

Rural development organizations (or agricultural extension organizations) are responsible for transferring technologies developed by research to farmers. The agroclimatic information must be seen as part of the total agricultural technological package to be transferred to farmers.

Dissemination of information could be improved by establishing a network of radios in the zones to be covered. Farmers would collect rainfall data and extension workers would do observations on crops.

Table 1. Sorghum yields (kg/ha) in the pilot fields.

Year	Bancoumana								Kenieroba							
	Site 1		Site 2		Site 3		Site 4		Site 1		Site 2		Site 3		Site 4	
	Test	Control	Test	Control	Test	Control	Test	Control	Test	Control	Test	Control	Test	Control	Test	Control
1983	1635	1485	1675	1617	827	725	1818	1777	1345	863	763	872	405[a]	345[a]	1075	533
1984	1340	1230	1543	1220	1940	1910	1300	1403	1398	920	1050	795	1310	1408	1380	1203
1985	1310	810	773	1785	1588	1233	1203	1223	1195	320[b]	475[c]	440[c]	960[d]	820[d]	504[e]	483[e]
Mean	1428	1175	1664	1541	1452	1289	1440	1468	1313	701	763	702	892	858	986	740
Percent increase[f]																
$100 \times \frac{Y_T - Y_C}{Y_C}$	22		8		13		–2		87		9		4		33	
$100 \times \frac{Y_T - Y_A}{Y_A}$	43		66		45		44		46		1		46		36	

[a]Planting was very late. [b]Plot was abandoned for 1 mo. [c]Young plants died for undetermined reasons. [d]Damaged by cattle. [e]Water puddles persisted in the field. [f]Y_T = yield in test plot, in kg/ha; Y_C = yield in control plot, in kg/ha; Y_A = yield (kg/ha) in the village (1,000 for Bancoumana, 900 for Kenieroba).

Table 2. Millet yields (kg/ha) in the pilot fields.

Year	Kongola								Makandiana							
	Site 1		Site 2		Site 3		Site 4		Site 1		Site 2		Site 3		Site 4	
	Test	Control	Test	Control	Test	Control	Test	Control	Test	Control	Test	Control	Test	Control	Test	Control
1983[a]	525	383	755	738	595	483	700	538	803	750	1032	855	585	388	513	410
1984	1050	863	1025	1000	975	908	1028	828	1463	915	1645	733	925	763	995	800
1985	1083	701	1000	780	968	1120	868	785	1300	1278	1325	1046	773[a]	505[a]	903	683
Mean	886	649	927	839	846	837	865	717	1189	981	1334	878	761	552	804	631
Percent increase[b]																
$100 \times \frac{Y_T - Y_C}{Y_C}$	37		10		1		21		21		52		38		27	
$100 \times \frac{Y_T - Y_A}{Y_A}$	19		13		8		5		63		75		9		12	

[a]Damaged by birds. The yields were not considered in computing average yield. [b]Y_T = yield in the test plot, in kg/ha; Y_C = yield in the control plot, in kg/ha; Y_A = av yield (kg/ha) in the village (900 for Kongola, 850 for Makandiana).

During a recent seminar, we noted that older farmers, who are the traditional agricultural decisionmakers in their families, were less receptive to the suggested technology. They could be reached by special radio broadcasts and, if necessary, by demonstrations on plots in their family farms. The directives could be disseminated through existing extension groups of farmers in the villages.

We also could improve the methods by introducing short- and medium-range forecasts, operational agrophysiological models, and warnings on pest and disease appearances.

Conclusion

The results obtained during the three yr of experimentation were convincing. Although we do not claim that the entire increase in yield was due solely to agroclimatic information, gathering specialists in several agricultural disciplines to advise the farmers was an effective way to increase crop production.

The fact that some farmers applied the directives on all plots somewhat impeded our scientific analysis, but it showed that the farmers accepted the technology. That, in itself, is something dramatic, considering how difficult it is to make our farmers accept new ideas.

We think we could have achieved even higher yields if weather forecasting and agrophysiological models had been available. Also, surveys among farmers are needed to improve our understanding of the realities of their environment.

We hope that with these improvements, the extension of operational agroclimatic assistance to agriculture will help Sahelian farmers gain control over the specter of drought-induced famine.

References cited

Eagleman J R (1971) An experimentally derived model for actual evapotranspiration. Agric. Meteorol. 8(4/5):385-394.

Forest F (1974) Bilan hydrique efficace et prospective décadaire des besoins en eau des cultures pluviales en zone soudanosahélienne. Ministère Francais de la Coopération,

Franquin P (1973) Analyse agroclimatique en regions tropicales: methode de intersections et periode frequentielle de vegetation. Agron. Trop. 28:6-7.

Meteorologie et IER (1984) Contribution a l'etude des dates de semis des principales cultures pluviales au Mali. Comite Nate de la Rech Agronomique annee 1984, Bamako.

Notes

Authors' address: K. Traore and M. Konate, National Meteorology Office, P.O. Box 237, Bamako, Mali.

Citation information: International Rice Research Institute (1989) Climate and food security. P.O. Box 933, Manila, Philippines.

Assessing climatic variability from long-term crop yield trends

D. E. McCloud

Crop yields depend primarily on cultivar, level of technology, incidence of plant pests, climate, and soil factors. We developed a method to separate the effects of technology and environment on crop yields, and to measure the relative adaptation of peanut and soybean compared with rice, wheat, and maize. A technology index was calculated using a linear regression of years to yield. Standard error of the estimate was used to measure fluctuations around the trend line, to establish an environmental index. A linear model of crop yield to year was used to assess climatic variability and yield trends. This method separated technology from environmentally induced crop yield variability. Peanut and soybean had environmental indices similar to rice, wheat, and maize.

Crop yields in a given area depend primarily on the level of technology, climatic and soil conditions, and incidence of plant diseases, weeds, and other pests. Plotting long-term crop yield series will show the variability due to level of technology as well as that due to the environment. We developed a method to separate the effects of technology and environment on crop yields, and to measure the relative adaptation of peanut and soybean compared with rice, wheat, and maize.

Methodology

Stanhill (1976) reported the trends and deviations in English wheat yields for more than 750 yr. He found the best fit was a parabolic equation relating the log of yield to year, $r = 0.94$. McCloud (1976), using similar techniques, reported yield trends of eight Florida crops over a 57-yr period using polynomial equations to fit years vs yields. Linear and log yield models were fitted, and r values from 0.25 to 0.96 were obtained. In five of eight crops, the linear model gave the highest r value. Russell (1973) fitted fifth degree polynomial equations to calculate the yield trends of several Australian and United States crops. His equations explained variances ranging from 17.8% for Australian barley to 98.5% for U.S. rice.

When a long-term series of crop yields for any given site is analyzed, the first step is to establish whether a technology trend exists. A microcomputer with a plotting program facilitates examination of the data for yield trends and year-to-year variability (Fig. 1). When crop yields overall show a continual increase with years, this increase is generally a result of superior technology (i.e. better cultivars, higher fertilizer levels, or other improved management practices). A high year-to-year variability indicates environmental fluctuations.

Various equations were compared to illustrate the methodology. Over the period 1919-85, a third order polynomial produced an excellent fit, $r = 0.97$ ($P \leq 0.001$) (Fig. 2). A semilog relationship between years and log yield (Fig. 3) model also had a

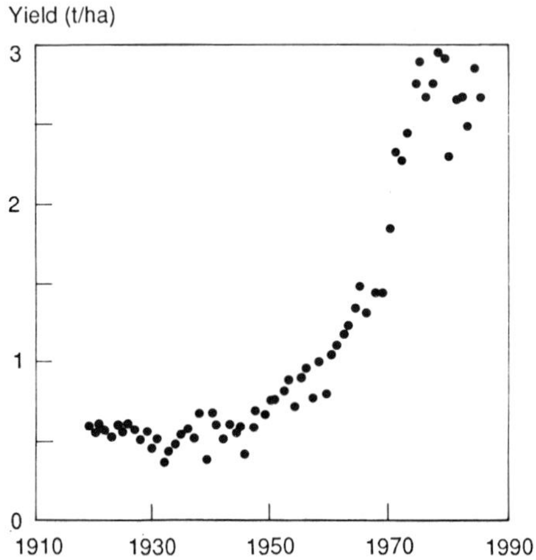

Fig. 1. Florida peanut yields, 1919-85.

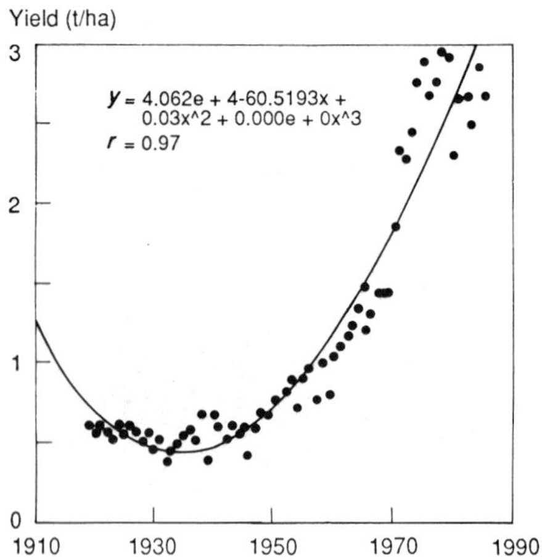

Fig. 2. Florida peanut yields, 1919-85, polynomial.

highly significant $r = 0.92$ ($P \leq 0.001$). The exponential model produced a very good fit, $r = 0.91$ ($P \leq 0.001$) (Fig. 4). A simple linear model of years vs yield gives a reasonable fit $r = 0.87$ ($P \leq 0.001$) (Fig. 5). However, inspection of yield data shows major changes in technology. Those changes should be separated into periods. Figure 5 shows two periods in Florida peanut yields: 1919-49 and 1950-85.

These two periods are reasonably linear, and should be treated separately. Within each period, the technology trend is calculated by a simple linear regression analysis of yield vs year. The regression slope is a measure of the actual yield increase per year in the units in which the yields are given, (e.g. t/ha). Since the mean yield is given in these same units, the technology trend is dimensionless. Multiplication by

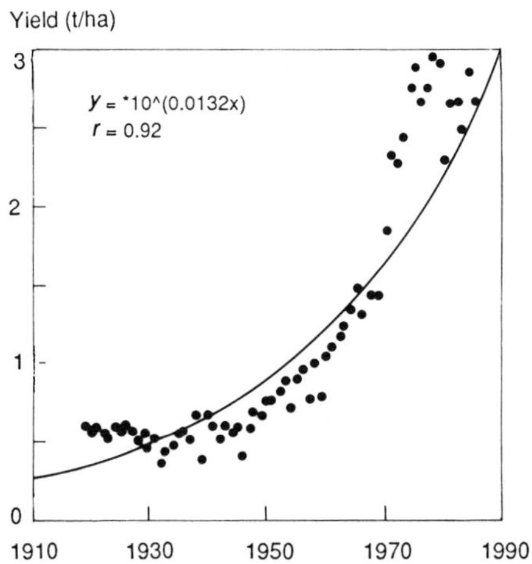

Fig. 3. Florida peanut yields, 1919-85, log y vs year.

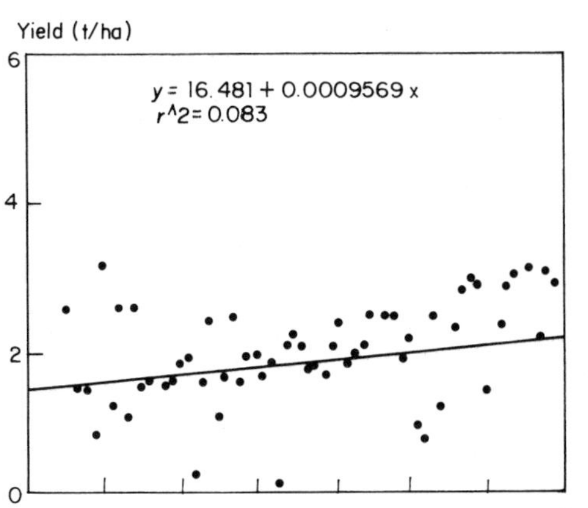

Fig. 4. Florida peanut yields, 1919-85, exponential.

100 results in a percentage index. The technology index is calculated using equation 1.

$$\text{Technology index} = \text{regression slope}/\text{mean yield} \times 100 \quad (1)$$

The second step is to determine the year-to-year fluctuations in yield around the trend line, which measures the environmental variability. The standard error of the estimate from the regression analysis can be used to measure the fluctuations around the trend line. These year-to-year variations may be due to climatic differences; weeds, diseases, insects, or other pest-induced yield effects; or variable inputs. The units for the standard error of the estimate are the same as those for mean yield: the environmental index also is dimensionless. Multiplication by 100 gives a percentage index. The environmental index is calculated using equation 2.

$$\text{Environmental index} = \text{standard error estimate}/\text{mean yield} \times 100 \quad (2)$$

In a study of weather variability and maize production, Thompson (1986) reported that the two major technology factors contributing to increases in U.S. maize yields are genetic improvements and larger fertilizer applications (particularly nitrogen). Other technology factors include higher plant densities; chemicals to control insects, diseases, and weeds; better mechanical practices; and improved management. Thompson found that the year-to-year yield fluctuations after the technology trend was removed were predominantly a result of precipitation and temperature variations. Consequently, the environmental index is mostly a result of year-to-year climate fluctuations.

Results

Peanut

Peanut yields were obtained from the Florida Crop and Livestock Board (1968, 1968-85). Florida peanut yields over the last three-quarters of a century have been

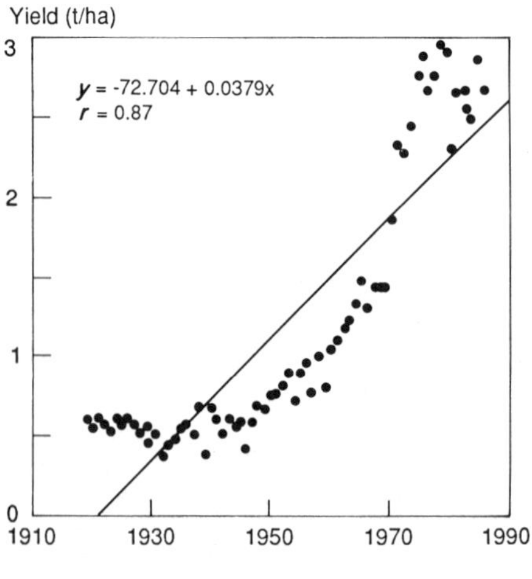

Fig. 5. Florida peanut yields, 1919-85, linear.

markedly affected by cultivar improvement, particularly partitioning of assimilates to the yield fraction (Duncan et al 1979). Scientific breeding of peanut in Florida started in 1928, with the first artificial crosses of peanut plants by Dr. Fred H. Hull. It took 15 yr for the first multiline cultivar, Dixie Runner, to be ready for seed increase (Fig. 6).

Rapid adoption of Dixie Runner by growers showed that it was a major improvement over the common Florida runner peanut planted previously, but there is no reliable way to estimate the yield improvement that resulted. The second important release, in 1952, was Early Runner, which produced 25% higher yield than Dixie Runner. Florunner, released in 1969, soon became the most widely grown peanut in the U.S. It had a 25% yield advantage over Early Runner. The fourth in the series, Early Bunch, released in 1977, gave nearly 10% higher yield than Florunner. Florida peanut breeders successively increased the partitioning factor for each new cultivar: Dixie Runner 40.5%, Early Runner 75.7%, Florunner 84.7%, and Early Bunch 97.8%.

Advances in peanut cultivar improvement now will have to come from a different yield increase strategy. The latest cultivar from the University of Florida breeding program is Sunrunner, released in 1983. This cultivar has improved market qualities and a slightly higher yield (3%) than Florunner.

An analysis of Florida peanut yields for 1919-49 disclosed almost static yields: slope is 0.001 t/yr, r was nonsignificant (Fig. 7). The technology index was 0.2% (Table 1). During this period, the environmental index was 14.6%. This index is rather high. After climate-induced yield fluctuations (largely drought) are separated out, most of the year-to-year variability in peanut yields in Florida results from leaf spot diseases (in some years, these can severely reduce yields). The years 1950-85 were marked by rapid increase in Florida peanut yields, the linear relation was good, $r = 0.94$ ($P \leq 0.001$) (Fig. 8). The technology index was 4.2%, the highest of any that I have calculated. During this period, the environmental index was 16.5%, again a reflection of climate fluctuations and the susceptibility of peanuts to leaf spot disease.

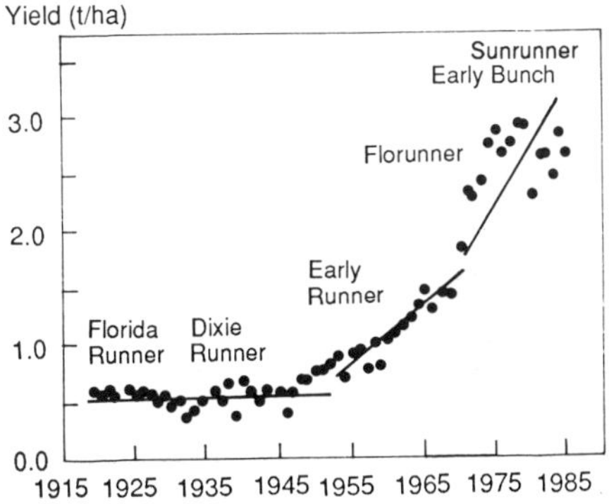

Fig. 6. Florida peanut cultivars and yields, 1919-85.

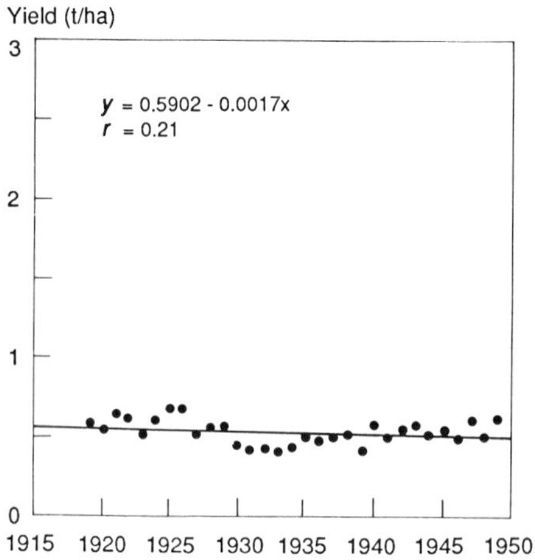

Fig. 7. Florida peanut yields, 1919-49.

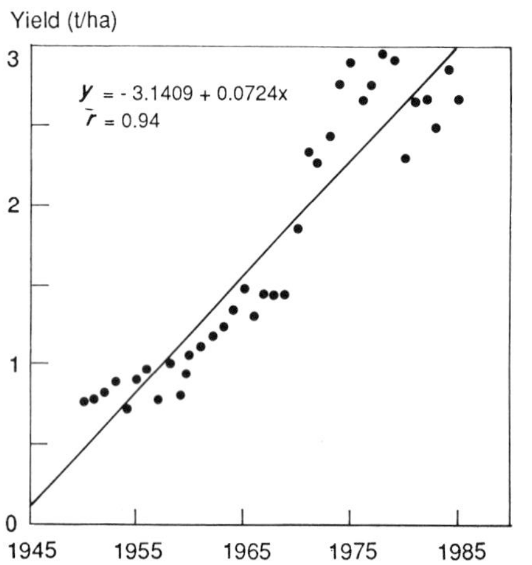

Fig. 8. Florida peanut yields, 1950-85.

Soybean

Florida soybean yields were obtained from the Florida Crop and Livestock Board (1968, 1968-85). Yields between 1949 and 1985 show a highly significant linear relation between year and yield, $r = 0.55$ ($P < 0.001$) (Fig. 9). Much less progress has been achieved in the improvement of Florida soybean yields than of peanuts. Florida peanut had a high technology index of 4.2%, Florida soybean had a very low technology index, 0.7% (Table 1). The environmental index for Florida soybean was 12.2%, somewhat lower than for Florida peanut (Table 1). Florida soybean is similar to peanut in drought resistance, but soybean is not susceptible to leaf spot and hence have a lower environmental index. Iowa is a leading state in U.S. soybean

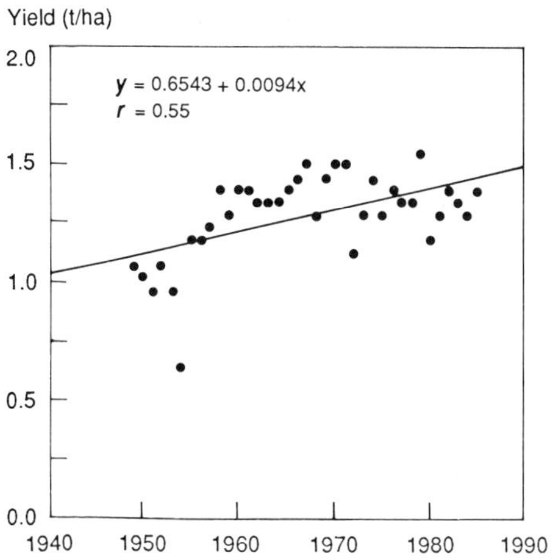

Fig. 9. Florida soybean yields, 1949-85.

Table 1. Mean yields, technology index, and environmental index for peanut, soybean, rice, wheat, and maize at eight sites.

Site	Crop	Period	Mean yield (t/ha)	Technology index (%)	Environmental index (%)
Florida	Peanut	1919-49	0.554	0.2	14.6
Florida	Peanut	1950-85	1.746	4.2	16.5
Florida	Soybean	1949-85	1.281	0.7	12.2
Iowa	Soybean	1948-84	1.467	1.6	9.0
Shizuoka	Rice	1885-1985	2.957	0.8	13.2
Saga	Rice	1885-1985	3.496	0.9	16.6
Hokkaido	Rice	1885-1949	1.864	0.5	32.6
Hokkaido	Rice	1950-85	3.824	2.0	19.8
India	Wheat	1898-1964	0.719	0.03	10.3
India	Wheat	1965-84	1.342	3.5	6.8
Canada	Wheat	1948-84	1.560	1.5	17.2
Florida	Maize	1919-50	0.537	−0.1	14.9
Florida	Maize	1951-85	1.989	3.5	15.9
Illinois	Maize	1947-84	4.138	2.3	13.5

production, and Iowa soybean yields for 1948-84 were compiled from USDA, Agricultural Statistics (1948-85). Iowa soybean yields show a highly significant linear relationship over the period 1948-84, $r = 0.89$ ($P<0.001$) (Fig. 10). The technology index is 1.6% (Table 1). Midwestern soybean breeders have been more successful in improving their soybean cultivars. For this period the environmental index is 9.0%, indicating that soybean is better adapted to Iowa than to Florida.

Rice

Rice grows primarily in wet tropical and subtropical regions. An excellent long-term rice yield series has been compiled by the Hokkaido Department of Agriculture

(1985). Additional data for two sites in Japan—Shizuoka, located in central Japan's coastal plains, facing the Pacific Ocean, and Saga in southern Japan (Kyushu), a wet plains area where Japan's highest yields are usually produced—were provided by A. Udoguchi (January 1987, pers. comm.).

Rice yields for Shizuoka from 1885 to 1985 show an excellent linear relationship between years and yields, $r=0.88$ ($P<0.001$) (Fig. 11). Over the entire 101-yr period, the technology index was 0.8% (Table 1), although there is visual evidence to substantiate a leveling off of yields during World War II, and an increased technology index after 1950. The environmental index for Shizuoka was 13.2%. The rice data for Saga, where Japan's highest yields are often produced, substantiated a

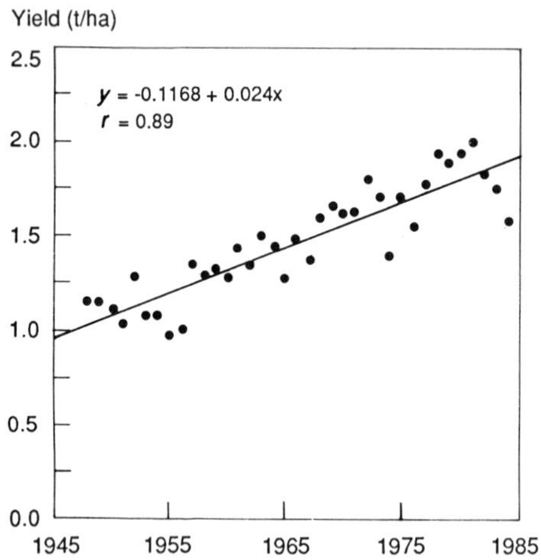

Fig. 10. Iowa soybean yields, 1948-84.

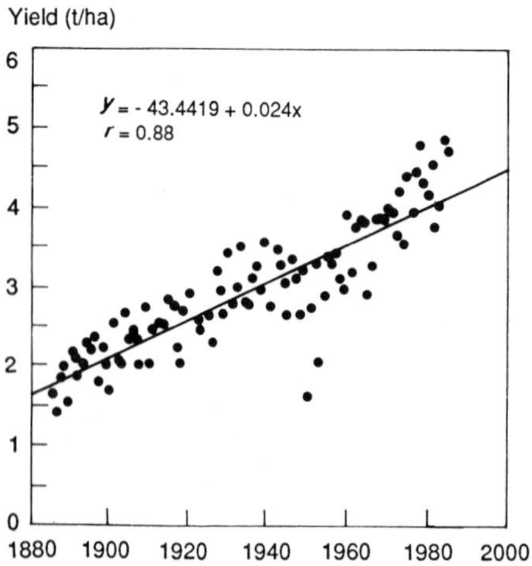

Fig. 11. Shizuoka rice yields, 1885-1985.

linear relation between years and yields (Fig. 12), $r = 0.85$ ($P<0.001$). The technology index was 0.9% (Table 1), the environmental index was 16.6%. Here, too, the effects of the World War II years are clearly evident, and the postwar increase in technology is even more pronounced than for Shizuoka (Fig. 11). Thus, three periods could be used to more precisely fit the Saga rice yields.

In the Hokkaido Prefecture, a considerable area of rice is grown farther north than in any other region of the world. In Hokkaido, rice is on the northern fringe of its adaptation. Because of the short growing season, rice is planted in farm greenhouses and transplanted to the field when conditions have become warm enough for plant growth. Rice yields for Hokkaido from 1885 to 1985 demonstrate a

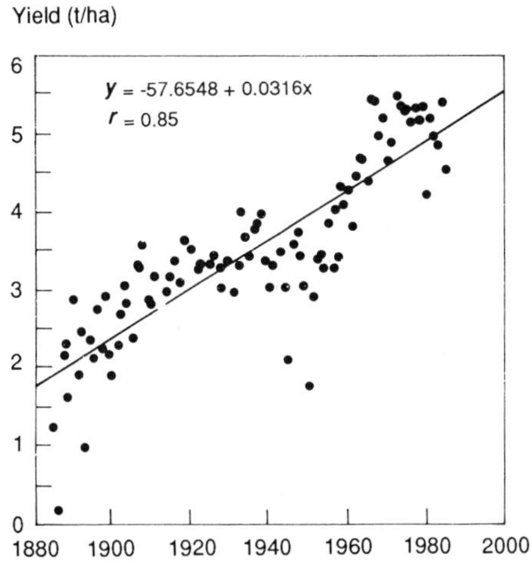

Fig. 12. Saga rice yields, 1885-1985.

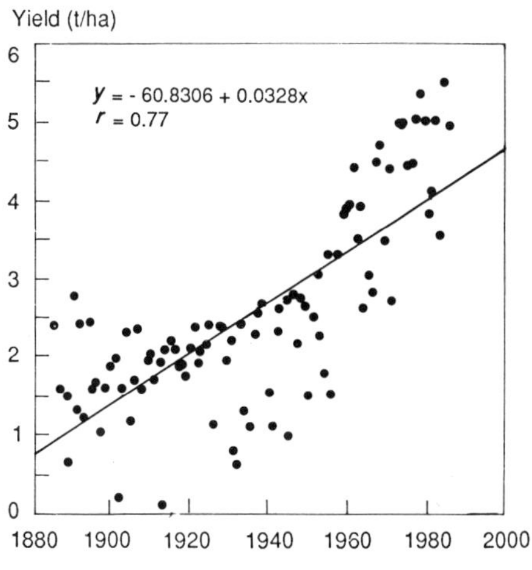

Fig. 13. Hokkaido rice yields, 1885-1985.

linear trend, $r = 0.77$ ($P < 0.001$) (Fig. 13). This 98-yr period had 27 cold years. An examination of the data reveals two distinct periods: 1885-1949 and 1950-85.

Rice yields from 1885 to 1949 did not increase appreciably (Fig. 14). The slope was 0.0095 t/ha, $r = 0.08$ (P was nonsignificant), and the technology index was 0.5% (Table 1). The environmental index for 1885 to 1949 was an extremely high 32.6%, indicating very large year-to-year variability in yields and showing that rice was not well adapted to Hokkaido. In two years, 1902 and 1913, temperatures were extremely cold during the rice growing season. In 1902, rice yielded 0.2 t/ha and in 1913, produced an average yield of only 0.1 t/ha.

Beginning about 1950, Hokkaido rice yields showed a linear yield take-off (Fig. 15), $r = 0.74$ ($P < 0.001$). The technology index for 1950-85 was 2.0% (Table 1). For

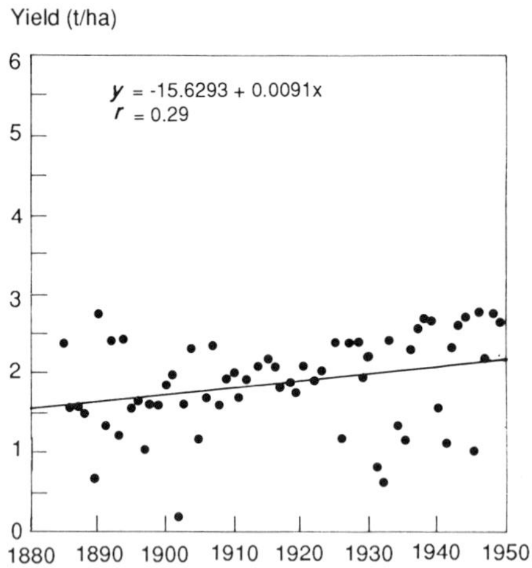

Fig. 14. Hokkaido rice yields, 1885-1949.

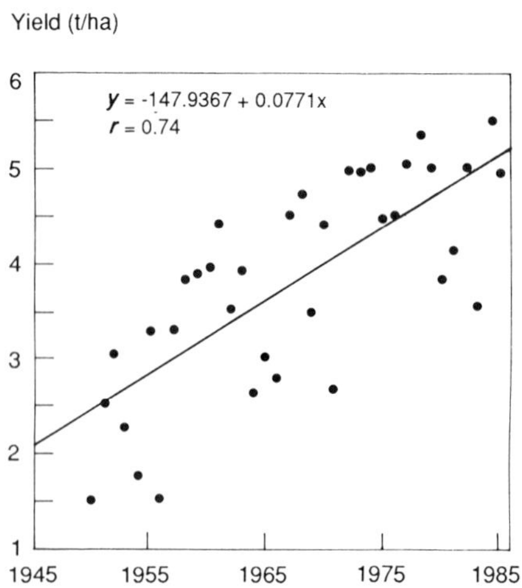

Fig. 15. Hokkaido rice yields, 1950-85.

1950-85 the environmental index was 19.8%, about half the 32.6% of the earlier period, indicating better adaptation of cultivars. Three years during 1950-85 were especially cold: 1956, 1971, and 1983. In those years, yields were 1.5, 2.7, and 3.6 t/ha, respectively. Rice is now reasonably well adapted to the Hokkaido Prefecture, and shows the best adaptation in Shizuoka.

Wheat

India wheat furnish another interesting long-term yield series. Wheat yields in India from 1898 to 1984 were compiled from Brown (1965) and FAO (1966, 1965-84). For almost 100 yr, yields remained nearly constant (Fig. 16), averaging 0.7 t/ha. The slope was 0.0002 t/ha, r was nonsignificant, and the technology index was 0.03%

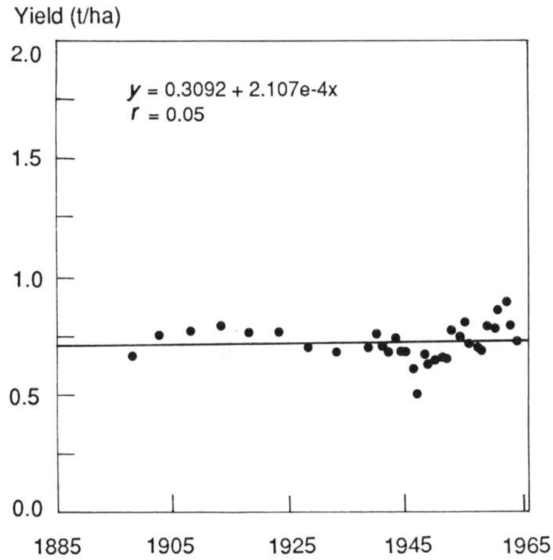

Fig. 16. India wheat yields, 1898-1964.

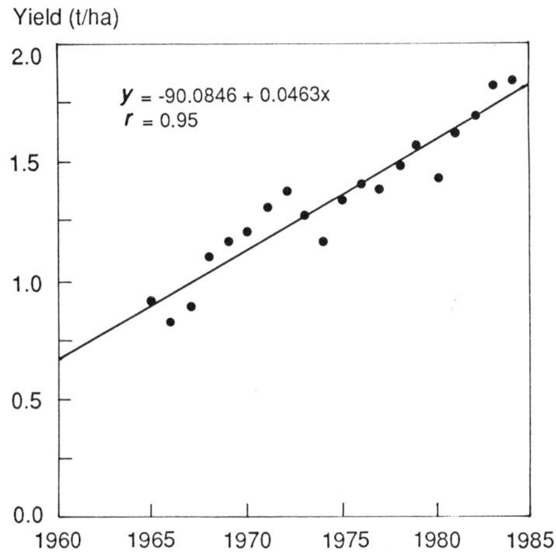

Fig. 17. India wheat yields, 1965-84.

(Table 1). The environmental index was 10.3%, indicating good adaptation of wheat to India.

In 1965, wheat yields began to increase linearly, as more and more farmers began to use the new fertilizer-responsive cultivars of the Green Revolution. Over the 1965-84 period, there was a highly significant linear relationship between year and yield, $r = 0.95$, ($P<0.001$) (Fig. 17). The technology index was 3.5% (Table 1), indicating a rapid adoption of the new technology package by Indian growers, and the environmental index was 6.8%, showing that the new cultivars have excellent adaptation to India.

Canada wheat yields were obtained from FAO (1966, 1965-84). The period 1948-84 exhibited a highly significant linear relationship between year and yield, $r=0.69$ ($P<0.001$) (Fig. 18). The technology index (Table 1) was 1.5%, indicating a gradual increase in yield over the years (0.0232 t/ha per yr). The environmental index was not as high as expected, 17.2%. Because of the high environmental risk of droughts, diseases, and cold temperatures in Canada, wheat yields would be expected to be more variable.

Maize

Florida maize yields were obtained from the Florida Crop and Livestock Board (1968, 1968-85). From 1919 to 1950, yields remained static, r was nonsignificant (Fig. 19). Average yield was 0.5 t/ha (Table 1). The technology index was low, –0.1% (Table 1). The environmental index was 14.9%. Indices for Florida peanuts for the same period were similar. For 1951-85, year and yield showed a linear relationship, $r=0.92$ ($P<0.001$) (Fig. 20). The average maize yield was 2.0 t/ha. The technology index was high, 3.5% (Table 1), largely the result of new nitrogen fertilizer-responsive high-yielding hybrids (in 1950, only 30% of Florida's maize area was planted to hybrids). The environmental index was 15.9%. Droughts were the principal cause of lower yields in Florida maize.

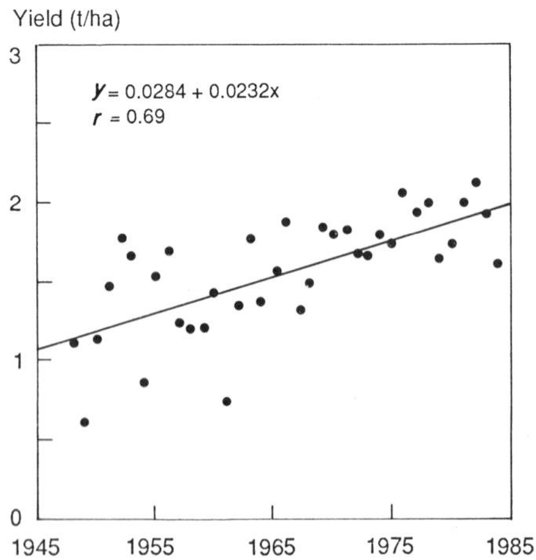

Fig. 18. Canada wheat yields, 1948-84.

Illinois, which is in the heart of the U.S. Corn Belt, usually leads in U.S. maize production. Illinois maize yields from 1947 to 1984 were obtained from USDA (1948-85). For 1947-84, there was a very good linear relationship between years and Illinois maize yields, $r = 0.89$ ($P<0.001$) (Fig. 21). The technology index was 2.3% (Table 1). By 1947, Illinois maize hectarage was 99.5% planted to hybrids. The principal factors causing yield increases after 1947 were additional nitrogen fertilizer

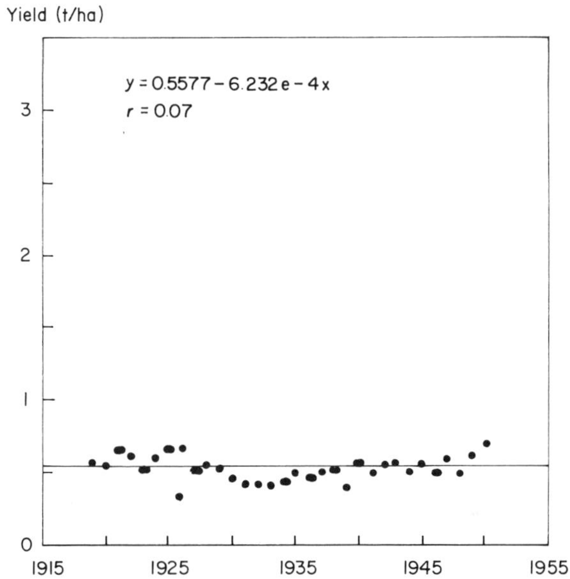

Fig. 19. Florida maize yields, 1919-50.

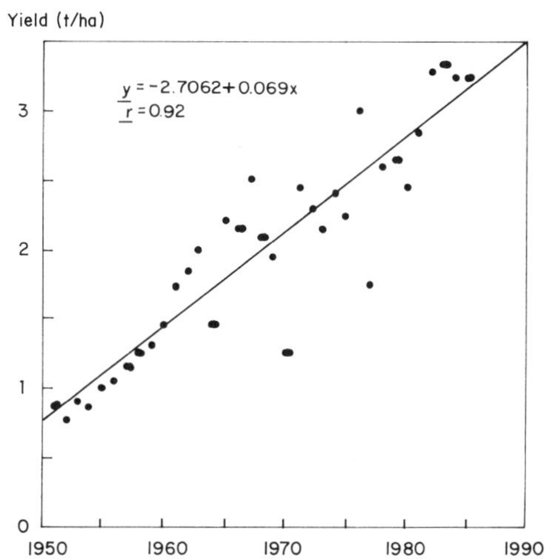

Fig. 20. Florida maize yields, 1951-85.

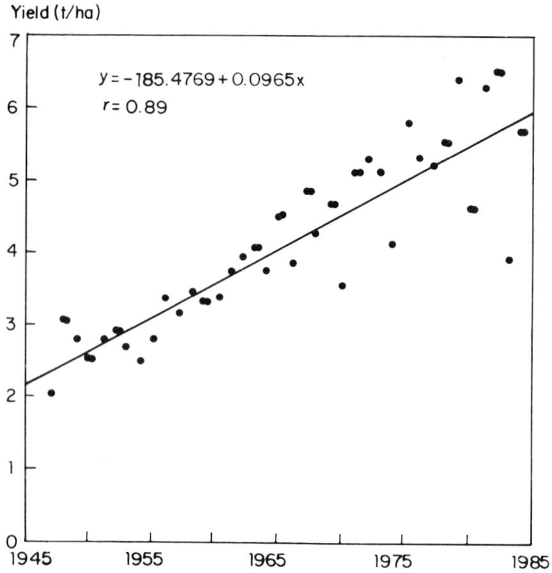

Fig. 21. Illinois maize yields, 1947-84.

and superior maize hybrids. The environmental index for Illinois maize was 13.5%, about the same as for Florida maize. However, average maize yield for Illinois was 4.1 t/ha compared with 1.9 t/ha for Florida (Table 1).

Summary

A simple linear model of crop yields vs years was used to assess climatic variability and yield trends from a long-term yield series. Technology index was calculated using linear regression slope divided by mean yield expressed as a percentage. Fluctuations around the trend line were a measure of environmental variability (mostly climate related). Environmental index was calculated using standard error of the estimate divided by mean yield expressed as a percentage. This method separated technology from environmentally induced crop yield variability. It also measured relative technology level and the comparative adaptation of crops. Peanut and soybean had similar environmental indices to rice, wheat, and maize. Soybean is well adapted to Iowa, and India wheat showed excellent environmental adaptation, especially since 1965.

References cited

Brown L R (1965) Increasing world food output. U.S. Department of Agriculture Economic Research Service, Foreign Agricultural Economic Report No. 25. Washington, D.C. 140 p.

Duncan W G, McCloud D E, McGraw R L, Boote K J (1979) Physiological aspects of peanut yield improvement. Crop Sci. 18:1015-1020.

Florida Crop and Livestock Board (1968) Florida agricultural statistics, field crops summary 1919-1967. Orlando, Florida. 22 p.
Florida Crop and Livestock Board (1968-85) Florida agricultural statistics, field crops summary. Orlando, Florida. 22 p.
FAO—Food and Agriculture Organization of the United Nations (1965-84) FAO production yearbook. Rome. (various issues)
FAO—Food and Agriculture Organization of the United Nations (1966) World crop statistics, area, production and yield 1948-64. Rome. 458 p.
Hokkaido Department of Agriculture (1985) Present Hokkaido agriculture. Sapporo, Japan. 221 p.
McCloud D E (1976) Florida field crop yield trends with a changing climate. Soil Crop Sci. Soc. Fla. 36:200-204.
Russell J S (1973) Yield trends of different crops in different areas and reflections on the sources of crop yield improvement in the Australian environment. J. Aust. Inst. Agric. Sci. 39(3):156-166.
Stanhill G (1976) Trends and deviations in the yield of the English wheat crop during the last 750 years. Agro-Ecosystems 3:1-10.
Thompson L M (1986) Climatic change, weather variability, and corn production. Agron. J. 78:649-653.
USDA—U.S. Department of Agriculture (1948-85) Agricultural statistics. Washington, D.C.

Notes

Author's address: D. E. McCloud, Department of Agronomy, Newell Hall 304, Institute of Food and Agricultural Sciences, University of Florida, Gainesville, Florida, USA.

Citation information: International Rice Research Institute (1989) Climate and food security. P.O. Box 933, Manila, Philippines.

Impact of climate variation on production of pulses

R. Khanna-Chopra and S. K. Sinha

Chickpea, pigeonpea, cowpea, mungbean, urd bean, and other leguminous crops are major sources of vegetable protein in subtemperate and tropical regions. Unstable production and relatively low productivity have led to a sharp decline in the pulses-to-cereal production ratio, causing an almost 50% drop in per capita availability of pulses in last two decades. Pulse crops in subtemperate and tropical regions have a relatively short history of improvement compared with wheat and rice. Cultivars respond to photoperiod, temperature, and water availability. Even good rainfall seasons are not necessarily associated with good yield. Water availability below a certain limit causes poor crop stand and affects biomass and pod set. Excess water results in excessive vegetative growth but poor yield, conditions that favor diseases and pest infestation. The situation is further complicated by temperature regime. Greater effort is needed to identify location-specific biological and environmental constraints.

Food security is a broad concept. The United Nations Food and Agriculture Organization (FAO 1983) defined the following objectives of food security:

1. The ultimate objective of world food security should be to ensure that all people at all times have both physical and economic access to the basic food they need.
2. Food security should have three specific aims:
   a. Ensuring production of adequate food supplies.
   b. Maximizing stability in the flow of supplies.
   c. Securing access to available supplies on the part of those who need them.
3. While cereals will continue to be the main focus of attention, action should cover all basic food stuff necessary for health.

These objectives recognize not only security for food per se but for all basic food stuff. Thus, food security has nutritional security as an important component. Conceptually, this is important to developing countries, particularly to India. Grain legumes, or pulses, constitute a significant component of the Indian diet. With the majority of the population vegetarian, they are an important source of protein. Pulses complement cereals to provide a nutritionally balanced diet.

It is imperative that the production of both cereals and pulses be improved and maintained simultaneously. However, the so-called "green revolution" in India has been mostly confined to cereals, particularly wheat. In 1951-52, production of pulses was 17.3% that of cereals; it is now 9.0% (Fig. 1). Annual availability of cereals and

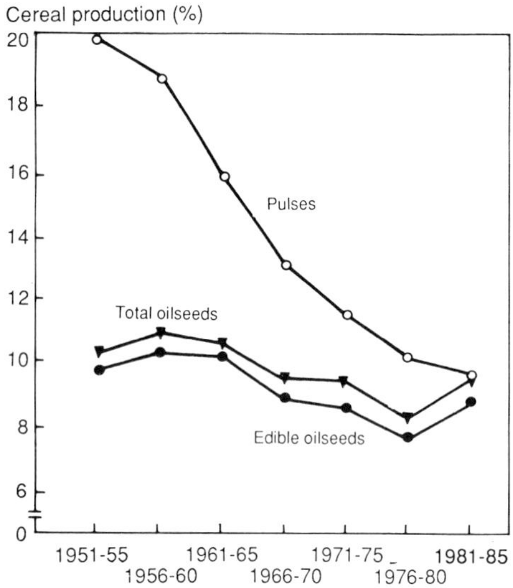

1. Pulse and oil seed production as ratio of total cereal production in India, 1951-85.

pulses per capita was 134 and 23 kg, respectively, in 1951-52. This has changed to 192 and 17.3 kg in 1984-85. The reduced availability of pulses has led to poor accessibility, because the prices of pulses have increased more than wheat. From 1975 to 1987, the price of wheat increased 40% while that of pulses increased 100-300%. The nutritional balance of Indian diets has been affected.

Production and productivity of pulses in India

More than 12.3 million tons of pulses are produced on 23.0 million ha in India (Table 1). Chickpea, lentil, and pea are grown in the winter season, the rest of the pulses are grown in the monsoon season. Mungbean and urd bean are planted in the summer as well as in the monsoon season.

Chickpea production is 4.75 million tons, 38.7% of total pulse production. Pigeonpea is the next, with 2.95 million tons, 20% of the total pulse production. Productivity of pulses is fairly low, ranging from 412 to 822 kg/ha.

Pulse production increased from 8.4 million tons in 1951-55 to 12.5 million tons in 1983-84, primarily due to an increase in area under pulse production (from 18.8 million to 23.4 million ha). Pulse yields have been stagnant in the last 3 decades, with an average of 486 kg/ha. Annual variation in pulse yields is fairly low, showing a high degree of stability in productivity.

A comparison of pulse yields in India, Asia, and the world from 1961 to 1983 shows similar productivity and stability trends (Table 2). Average pulse yields in Asia and the world are only 646 and 672 kg/ha with a coefficient of variation of 5.4 and 3.6%, respectively. India accounts for 25% of the world pulse production; its

Table 1. Production and yield of major pulse crops in India.

Scientific name	Common	Production (thousand t)	Yield (kg/ha)
Cicer arietinum	Chickpea	4755	651
Cajanus cajan	Pigeonpea	2440	768
Vigna radiata	Mungbean	1352	447
Phaseolus mungo	Urd bean	1192	412
Dolichos lablab	Horsegram	764	417
Lens esculenta	Lentil	531	544
Lathyrus sativus	Grass pea	481	409
Vigna aconitifolia	Moth bean	405	271
Pisum sativum	Pea	363	822

Table 2. Standard deviations, means and coefficient of variance of area, production, and yield of pulses, 1961-83 (FAO 1985).

	Area (million ha)	Production (million t)	Yield (kg/ha)
World			
SD	2.8	2.8	24.4
Mean	67.7	45.5	671.7
CV (%)	4.2	6.2	3.6
Asia			
SD	3.5	3.3	34.8
Mean	38.1	24.6	645.7
CV (%)	8.7	13.5	5.4
India			
SD	7.6	0.9	32.0
Mean	22.9	11.1	486.0
CV (%)	3.3	8.5	6.6

chickpea production contributes 77% of world production. Pigeonpea is primarily an Indian crop, and India the center of diversity for this crop.

Despite the importance of pulse crops in Indian diets, the production growth rate during 1949-50 to 1983-84 has been only 0.33%/yr, area has increased by 0.23%/yr, and yield (productivity) by 0.08%/yr. The growth rate of wheat during this period was 2.80%/yr for production, 6.02%/yr for area, and 3.12%/yr for yield. The growth rate in production, area, and yield for chickpea from 1967-68 to 1983-84 was negative (Fig. 2). Pigeonpea had a low growth rate during the same period. In the context of the need to improve production and productivity of pulses in the coming decades, this trend is alarming.

Productivity/environment correlations

Chickpea is grown in India between 15° N and 35° N. Most desi-type chickpea cultivars are grown between 20° N and 30° N, kabuli types are grown above 30° N. This latitudinal distribution causes significant differences in photoperiod, tempera-

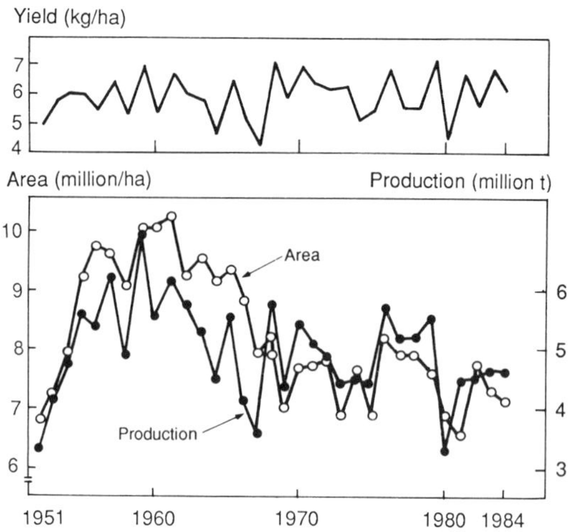

2. Yield, area, and production of chickpea in India, 1967-84.

ture, and precipitation, which individually and in combination, affect growth and differentiation. Both chickpea and pigeonpea are mostly rainfed crops. Pigeonpea is planted in the monsoon season, and depends on the rains for its water requirement. It is grown primarily as a mixed crop or intercrop, seldom as a monocrop. Early-duration varieties flower in mid- or late September and pod development occurs when the monsoon ends. Depending on the extent and distribution of rainfall, pigeonpea may experience drought stress during flowering and pod development. The probability of postflowering stress increases in medium- and long-duration cultivars. Because short-duration cultivars are important in intensive agriculture, water requirement is particularly important.

In the All India Coordinated Project on Improvement of Pulses, short-duration varieties Prabhat and T-21 were grown at 11 sites in 1975. Yields were not related to water availability (Table 3). Pigeonpea yields at Hissar, with 398 mm rainfall, were 1,389 kg/ha for variety Prabhat and 1,354 kg/ha for T-21. At Pantnagar, with 1,412 mm rainfall, yields were 1,236 and 944 kg/ha. Maximum yields of both varieties were obtained at Rahuri, where the rainfall was 618 mm. If grain yield is not related to total rainfall, then rainfall distribution or temperature could be influencing yield.

Chickpea is grown on the stored soil moisture received through rainfall in the monsoon season. Some rain also can be expected in the crop growth period. The relationship between chickpea yields and rainfall was examined for the Delhi and Kanpur districts in Uttar Pradesh, a major chickpea production state (Fig. 3). There was no relationship between yield and winter rainfall. Chickpea yields varied from 300 to 400 kg/ha irrespective of the amount of rainfall. Chickpea yields are strongly influenced by residual moisture from the preceding monsoon, weakening the relationship between yield and winter rainfall.

Wheat is also grown in the same season. In the Delhi and Kanpur districts, wheat yields showed a weak but positive relationship with rainfall, possibly because 80% of

Table 3. Grain yield of pigeonpea at different sites in India.

Site	Rainfall (mm)	Yield (kg/ha)	
		Prabhat	T-21
Ludhiana	682	1010	1028
Delhi	1035	833	1076
Hissar	398	1389	1354
Pantnagar	1412	1236	944
Junagarh	807	1007	1806
Baroda	668	936	1640
Bangalore	492	557	502
Badnapur	1029	1218	1377
Rahuri	618	1622	2323
Gulbarga	1442	999	–
Kudumiamalai	851	441	231

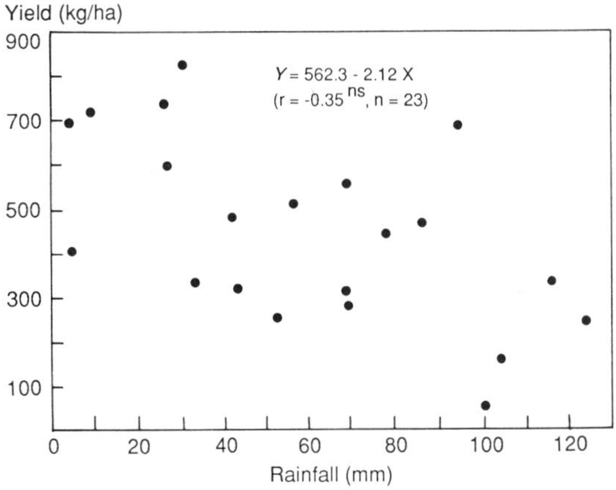

3. Relationship between rainfall and chickpea yield in Delhi and Kanpur.

the wheat crop is grown under irrigated conditions, and varietal selection has been for an irrigated environment.

Chickpea yields do not show any relationship with monthly means of maximum and minimum temperatures in the Kanpur district. However, both temperature and photoperiod affect crop phenology, with a consequential effect on yield, and those correlations should be examined. Water availability influences the crop canopy temperature. We observed in our laboratory that irrigated chickpea crop canopies take longer to flower than nonirrigated canopies, with a difference among varieties ranging between 13 and 20 d (Fig. 4). One effect of irrigation is the cooling of crop canopies by 3-7 °C, depending on ambient temperature and irrigation intensity. Nonirrigated crop canopies maintain temperatures higher than the ambient temperature. Phenological response, such as in flowering, is possibly due to the temperature of the plant itself, rather than to the ambient temperature per se. It

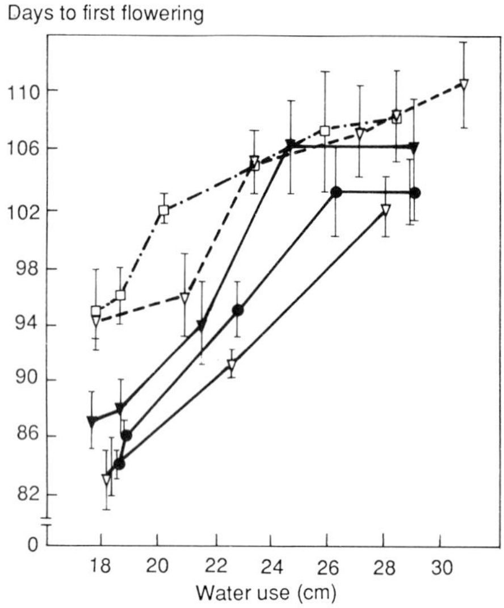

4. Effect of water availability on number of days required for flowering in chickpea cultivars.

would be difficult to find a relationship between ambient temperature, crop duration, and crop yield. However, definitions of the appropriate relationships between environmental factors and phenological responses would help in assessing crop responses to climatic variations.

Our analysis identifies these points:

- Productivity of pulses is low but stable.
- There is no significant correlation between yield and rainfall and temperature during the growing season.

An understanding of the physiological basis of yield in these crops would help us project the impact of changing climate on nutritional supplies.

Productivity stagnation

Average yields of chickpea are only 0.6-0.7 t/ha, but in experimental fields and agronomic trials in northern India, yields of 2.5-3.5 t/ha are common. In the Indian plateau, however, yields of only 1.5-2.0 t/ha are obtained. The potential productivity of cultivars is not realized in farmers' fields because of biotic and abiotic constraints. Legumes are susceptible to a large number of insects, pests, and diseases (the major biotic constraint). Abiotic factors include the environmental variables temperature, photoperiod, and water availability. Water availability shows large annual variation and has the maximum influence on seed yield.

Physiological basis of low yield

Yield is the culmination of plant processes involved in growth, development, and differentiation. Phenological stages also are influenced by environmental variables and have considerable influence on yield. Chickpea has four major phenological stages—germination, seedling growth, flowering, and pod development. (Legumes are indeterminate and do not have distinct vegetative and reproductive phases.) Leaf and stem growth continues during pod development and competes with the developing seeds for assimilates, leading to low productivity.

Germination and seedling establishment

Chickpea is planted after monsoon rains. Early planting cannot be adopted because relatively higher temperatures and longer daylength result in early differentiation and a limited vegetative phase. If planting is delayed, soil moisture recedes. Soil moisture at sowing could be a critical factor in obtaining optimal germination and plant stand.

Plant stand establishment in chickpea also depends on soil moisture content, quality of seed material, and other soil characteristics.

Saxena et al (1983) examined the effect of soil moisture on germination in desi and kabuli cultivars of chickpea. Germination percentage dropped from 85 to 60% as soil moisture varied from 26 to 21%. The study also identified considerable genetic variation in germination and plant stand under low soil moisture content.

Farmers often use their own seed, which may be of poor quality. In a village-level study, viability of farmers' seed ranged from 6 to 99%. Even when an adequate amount of seed was sown, the plant stand varied from 60 to 10% (Saxena et al 1980).

Salinity and sodicity are other problems in chickpea-growing areas. Salinity due to chloride and sulfate is often sulfate dominated. Chickpea is sensitive to both salinity and sodicity. Chickpea plants do not survive at more than 60 meq chloride salinity (Manchanda and Sharma 1980). Sodicity affects seedling mortality more than it does germination (Kumar et al 1983). No serious effort has been made to select salt-tolerant chickpea lines, as has been done in some leguminous species (Rains et al 1980).

Growth and leaf area development

Dry matter accumulation in chickpea is slow for a relatively long time after planting. Dry matter accumulation depends on photosynthetic surface and the rate of net assimilation. Chickpea is a C_3 plant and has a photosynthesis rate in the range of 200-400 μg CO_2/cm per h, comparable to other C_3 dicotyledonous crops (Khanna-Chopra and Vidya Lakshmi 1987, Van der Maesen 1972, Winter 1981). Leaf area development is very slow for the first 80-85 d after sowing (DAS). In Delhi, the leaf area index (LAI) was still <1 at 80 DAS, when flowering began (Aggarwal et al 1984). This feature is common to all chickpea-growing areas. At Hyderabad, maximum LAI was 0.7 without irrigation and 1.5 after flowering with 2 irrigations. At Hyderabad, maximum LAI occurred 40-80 DAS; total crop duration was about 110 d. Without irrigation, there is no chance of full interception of light. With irrigation, LAI may reach 2.5-4 for a brief period between 120 and 150 d at Delhi where full light interception may occur (Saxena and Sheldrake 1980).

LAI declines due to leaf senescence during pod/seed development. Leaf fall can account for 20-30% loss in plant dry weight at postflowering. At harvest, the chickpea plant is devoid of leaves. This monocarpic senescence also is observed in other legumes. The shed leaves are a loss of both assimilates and nitrogen (senescent leaves contain 2% nitrogen). In Delhi and Hissar, the period of fruit development and the rate of leaf senescence is related to the onset of drought stress and increasing ambient temperatures. If there are some rains, the pods mature over a longer duration and the leaves senesce at a slower pace. Lack of rains coupled with a sudden rise in temperature can result in premature senescence of the crop. In Hyderabad, no rain falls during fruit development, and the growth is terminated by increasing temperature.

Dry matter accumulation, nitrogen accumulation, and partitioning

Dry matter accumulation in chickpea shows a typical sigmoid curve, with a slow vegetative phase, a log phase after flowering, and a decline during fruit development (Khanna-Chopra and Sinha 1987). In Delhi, chickpea is planted with the onset of cooler temperatures in late October. Daylength and temperatures decline as the crop is established. Seedlings grow at a very slow rate for almost 100 d after planting. Crop growth rate (CGR) varies from 0.96 to 2.34 g/m^2 per d. The slow development of leaf area and the lack of seedling vigor also are observed in other legumes (Khanna-Chopra et al 1985, Williams et al 1981). In Hyderabad, chickpea is planted in October; early seedling growth is at higher temperatures than in Delhi and Hissar. Consequently, the CGR of a 50-d-old chickpea crop is 9 g/m^2 per d. The vegetative period is 50 d, compared to 100 d at Delhi (Table 4), and CGR increases exponentially to 150-160 DAS. This period coincides with flowering and early fruit set. At Hyderabad, the exponential phase is very short and CGR declines as the crop matures under increasingly high temperatures.

Crop duration is considerably longer in the north than at Hyderabad (Table 4). This results in higher dry matter accumulation at Delhi. The potential for dry matter accumulation in chickpea is considerably lower than in wheat and soybean. Soybean accumulates 10 t/ha in 126 d, wheat accumulates 14.3 t/ha in 156 d.

Nitrogen accumulation parallels dry matter accumulation during the pre-flowering phase, but declines subsequently. Loss in nitrogen fixation capacity during pod development is accelerated by the onset of drought stress. The nitrogen demand

Table 4. Variation in dry matter accumulation, yield, and other characters in chickpea grown in two contrasting environments in India.

Character	Hyderabad	Delhi
Total growth duration (d)	75-100	165-180
Vegetative period (d)	35-50	60-100
Ineffective flowering period (d)	NA	30-60
Duration of podding (d)	45-50	45-50
Total dry matter (t/ha)	2.0-2.6	5.5-8.8
Total dry matter (kg/ha per d)	20-30	30-40
Yield (t/ha)	0.9-1.4	2.2-3.0
Yield (kg/ha per d)	10-20	15-20
Harvest index (%)	40-60	35-40

of developing seeds is met by mobilization of stored N from the subtending leaves, leading to leaf senescence. Total nitrogen accumulated by a chickpea crop is higher at Delhi than at Hyderabad. At Delhi and Hissar, nitrogen accumulation continues until harvest, resulting in 118-142 kg N/ha accumulated under nonirrigated conditions (Sinha et al 1983). At Hyderabad, the crop accumulates only 57 kg N/ha, limiting yields.

Yield components and harvest index

The major yield components of chickpea are pod number, seed number, and seed weight per unit area. Each flowering node usually carries only one pod. These nodes are predominantly on the primary and secondary branches. Pod number per unit area is strongly correlated with yield (Bahl and Jain 1977). Variation in chickpea yields at Hyderabad and Hissar was analyzed in relation to yield components and total biological yield. Chickpea yields at Hissar are higher due to higher node number, pod number, and seed number/m^2 (Table 5). The longer vegetative growth duration at Hissar enables development of infrastructure to support the reproductive sink. Duration of effective podding varies from 40 to 50 d at both Hyderabad and Hissar. The increased sink potential at Hissar is supported by higher dry matter and nitrogen accumulation during the postflowering phase. This complementarity of source and sink ensures higher yields at Hissar.

The harvest index (HI) of chickpea also varies from place to place (Table 4), from 40 to 60% in different cultivars at Hyderabad and from 35 to 40% at Hissar and Delhi. The relatively low HI of chickpea at Hissar and Delhi is due to excessive vegetative growth, with investment of dry matter in the stem only partly utilized for seed development. Similarly, the nitrogen harvest index in chickpea is only 50% compared with 75% in wheat (Khanna-Chopra and Sinha 1987). The present-day plant type of chickpea and pigeonpea does not mobilize carbon and nitrogen efficiently from the vegetative plant parts to the developing seeds. The need is to breed a legume plant type that efficiently mobilizes nutrients toward seed development.

Table 5. Component analysis of chickpea yield variation at Hyderabad and Hissar in India.

Character	Hyderabad		Hissar	
	Desi CPS 1	Kabuli L 550	Desi JG 62	Kabuli L 550
Yield (t/ha)	1.4	0.8	3.0	3.6
Pods/m^2	683	584	2309	2465
Seeds/m^2	683	607	2541	2739
Seeds/pod	1.00	1.04	1.06	1.11
100-seed weight (g)	15.7	18.4	12.4	18.8
Harvest index (%)	55.7	48.3	51.0	53.0
Nodes/plant	121	186	262	292
Total dry matter (t/ha)	2.6	1.7	5.9	6.9

Effect of temperature and photoperiod

Chickpea is a quantitative long-day plant (Van der Maesen 1972). Long days promote flowering, short days delay flowering. Plants growing under short days produce more vegetative growth and ultimately more yield. Temperature also influences flowering. Chickpea grown in Delhi experiences short days coupled with low temperatures, which promotes vegetative growth and delays flowering. Chickpea grown in Hyderabad experiences higher temperatures and relatively longer days, which restricts vegetative growth and enhances flowering. These interactions of environment on phenology explain the longer crop duration of chickpea at Delhi.

Temperature and humidity influence fruit set in chickpea considerably. Flowers formed early in Delhi do not set fruits. Fruit set percentage of first-formed flowers was 80-90% at Indore where day/night temperatures were 25°/15 °C, 70-80% at Hyderabad where day/night temperatures were 31°/17 °C, and 0 at Delhi where day/night temperatures were 20°/6 °C. As ambient temperatures increased, fruit set at Delhi increased (Fig. 5). That failure of early-formed flowers to form fruit extends the vegetative phase by 35-50 d (Table 4).

The inability of early-formed flowers to form fruit could be related to a failure of fertilization or to a limitation in seed development. Studies on pollen germination and pollen tube elongation revealed marked temperature sensitivity. Pollen germination was not observed at 15 °C but was promoted at higher temperatures. The optimal temperature for pollen germination and pollen tube elongation was 20-25 °C.

Fruit set also is affected by relative humidity and light intensity. High relative humidity and low light intensity do not favor fruit set (personal observation).

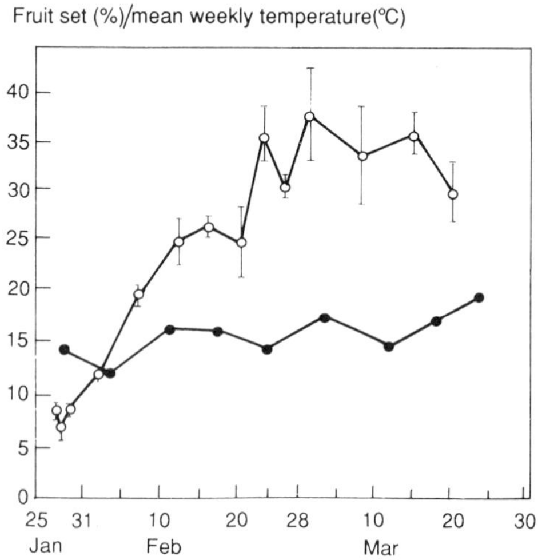

5. Relationship between fruit set percentage (o) in chickpea and mean weekly temperatures (•) during the flowering phase at Delhi.

Effect of water availability

Legume yields do not respond favorably at either low or high levels of available water. Most legume crops are rainfed. During the monsoon season, rains recede as flowering begins. Depending on the amount of rainfall, pod and seed development may occur during increasing drought stress. In one such year in Delhi, pigeonpea yields were increased considerably in early-, medium-, and late-duration cultivars by one irrigation at flowering (Table 6). The response to irrigation was greater in late-duration than in medium- and early-duration cultivars. Irrigation increased dry matter accumulation more than yield. Pigeonpea does not respond favorably to excessive rainfall or high irrigation levels (Sinha et al 1985). Excessive water availability leads to excessive vegetative growth, poor fruit set, and low yield.

Winter legumes grown on stored soil moisture exhaust the available water as flowering and fruit development begin. Without winter rainfall, water availability may be a limiting factor, leading to drought stress. Chickpea yielded best with two irrigations, one during vegetative growth and one during pod filling (Saxena and Yadav 1975, Sharma et al 1974), but yields declined at higher irrigation levels (Sinha et al 1985). The response of chickpea cultivars to a range of water availability showed a narrow optimal range for yield and HI (Fig. 6,7). Decline in yield at higher levels of

Table 6. Effect of single irrigation on biomass and grain yield in three cultivars of pigeonpea at two populations.

Cultivar	Duration (d)	Days to flower	Population density/m^2	Irrigation	Biomass (t/ha)	Yield (t/ha)	Harvest index (%)
Prabhat	132	67	16	No	6.68± 0.71	1.60± 0.20	23.95
				Yes	10.10± 1.12	3.33± 0.25	32.97
			32	No	6.90± 0.60	1.71± 0.17	24.78
				Yes	10.08± 0.90	3.10± 0.19	30.75
DL-74-1	150	77	16	No	9.07± 0.39	2.36± 0.20	26.01
				Yes	9.76± 0.18	2.93± 0.18	30.02
			32	No	9.99± 0.70	2.56± 0.20	25.62
				Yes	15.66± 0.60	4.25± 0.30	27.13
No. 148	180	94	16	No	10.58± 0.34	2.10± 0.13	19.84
				Yes	18.94± 1.78	4.29± 0.52	22.65
			32	No	8.95± 0.90	1.99± 0.06	22.23
				Yes	12.92± 1.61	3.12± 0.28	24.15

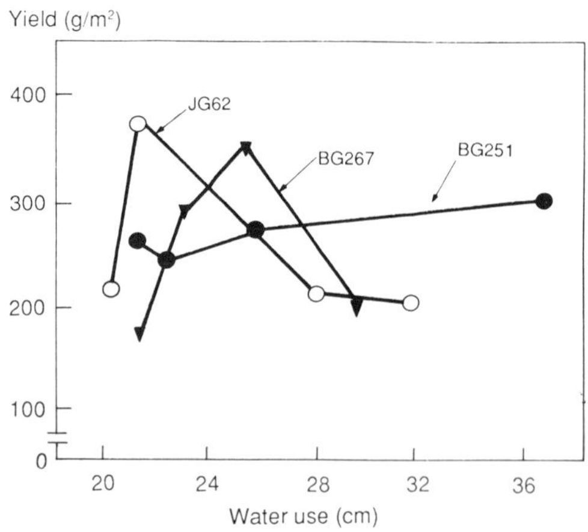

6. Yield of chickpea cultivars at different levels of water availability.

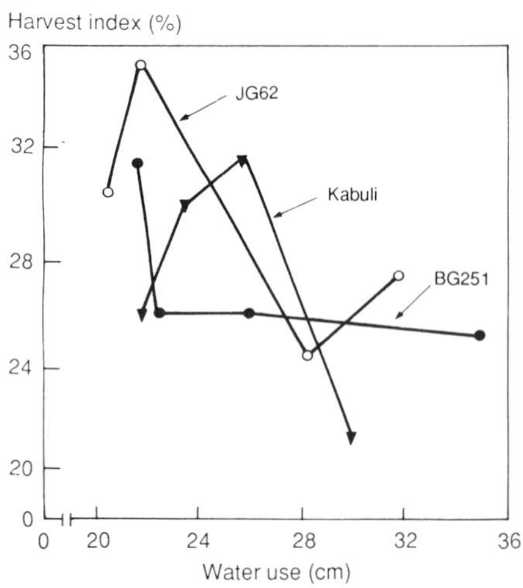

7. Harvest index of chickpea cultivars at different levels of water availability.

water availability was related to excessive vegetative growth and poor fruit set (Fig. 8).

In cereals, the amount of water available after anthesis has a linear relationship to yield (Passioura 1976). However, this was not observed in legumes, which are indeterminate crops. Excess water triggers vegetative growth, which acts as a vegetative sink for developing pods/seeds. Irrigation at flowering cools the crop canopy and increases the relative humidity, which reduces fruit set.

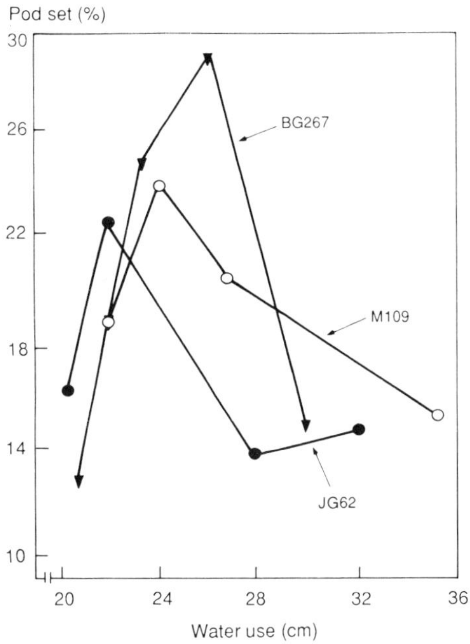

8. Pod set of chickpea cultivars at different levels of water availability.

Biotic stress

Effects of insect pests

Chickpea and pigeonpea are affected by many insect pests that appear at different growth and development stages. Singh and Singh (1978) observed 16-17 insect species during the crop season in chickpea and pigeonpea grown as monocrops or intercropped. Pod borer (*Heliothis armigera*) and pod fly (*Melanagromysa obtusa*) cause significant damage. In pigeonpea cultivars, damage varied with the duration of the variety. In short-duration cultivars, pod borer was more damaging; in the long-duration group, pod fly was more damaging. Damage to pods from all insect pests ranged from 25 to 91% in different pigeonpea cultivars (Davies and Lateef 1978).

In chickpea, *H. armigera* significantly damages both foliage and pods, causing 10-60% losses (Vaishampayan and Veda 1980). Desi cultivars appear more susceptible than kabuli and erect cultivars. Delayed sowing increased the incidence and damage of pod borer at Delhi. Basu and Pramanik (1969) reported similar observations in Bengal.

In nature, the distribution and abundance of insects are determined by the combined effect of different components of the environment; weather factors have the most influence. Few studies relate the population dynamics of the major insect pests of legumes with the weather. Vaishampayan and Veda (1980) analyzed the population dynamics of gram pod borer and the activity of its major larval parasite

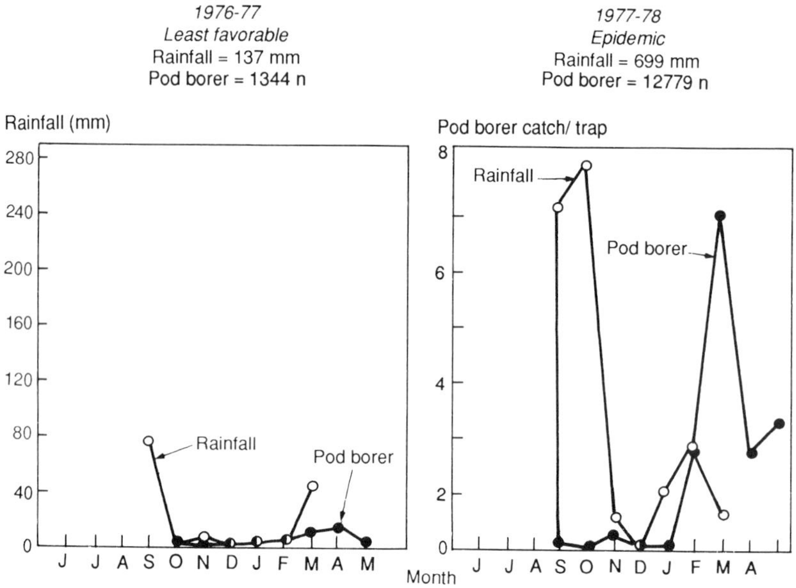

9. Relationship between rainfall pattern and pod borer population in chickpea fields at Jabalpur, India.

Ecphoropsis against weather parameters during epidemic and nonepidemic years at Jabalpur. During epidemic years, heavy rainfall early in the season stimulated luxuriant crop growth, which was conducive to heavy egg laying. Well-distributed rainfall in the successive months helped larval buildup, forming the basis for heavy insect population during pod formation (Fig. 9). Relative humidity below 75% and oscillating temperatures before the onset of flowering also helped larvae and pupae to survive. A count of 15-24 first-generation larvae/m^2 was considered a good indicator of a coming epidemic. Larval counts at regular intervals can help forecast outbreaks of gram pod borer, which can be averted by the use of insecticides.

In nature, larval populations of gram pod borer are suppressed by the parasite *Ecphoropsis predistinctus*. Activity of the parasite is directly dependent on density of the pest larvae. However, when density of the host larvae increased beyond a certain threshold level, the parasite was unable to control the pest (Vaishampayan and Veda 1980). It could be that different weather variables favor buildup of the insect pest and the parasite. Studies on insect pest populations and the biotic and the abiotic factors responsible for their buildup in legume crops are urgently needed. Epidemic outbreaks of pod borer reduced gram yield considerably at Jabalpur, and gram crops failed in other districts.

The susceptibility of chickpea to pod borer is influenced by water availability. As water availability increased, damage due to pod borer increased in desi and kabuli cultivars (Fig. 10); a modern tall cultivar was relatively resistant.

Pest management through mixed cropping

Intercropping chickpea and pigeonpea with nonlegume crops like linseed and pearl millet reduces and delays the incidence of major insect pests of legumes (Reed et al

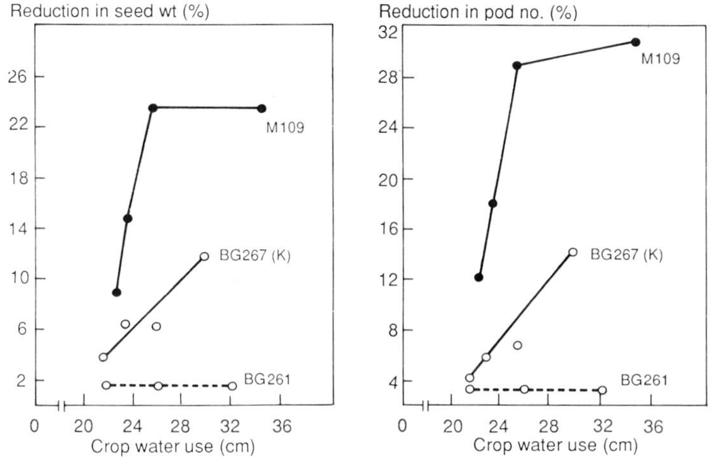

10. Relationship between pod borer damage and crop water use in chickpea cultivars at Delhi, India.

1987, Singh and Singh 1978). Pod borer and pod fly incidence was reduced 45-96% when pigeonpea was intercropped with pearl millet. Intercropping pigeonpea with black gram and cowpea also reduced the incidence of pod borer and other insects, but not as effectively as pigeonpea - pearl millet. Intercropping results in a change in crop canopy and thereby in microclimate, which ultimately affects the succession and pest buildup in a crop. Such intercropping combinations are practiced by farmers.

Implications of climatic change

CO_2 concentration in the atmosphere is rising at the rate of 1.5-2.0 ppm/yr. It has already reached 340 ppm, and is expected to double by the year 2045. Among the many uncertainties regarding resultant climatic changes, there is considerable agreement on the following effects of a doubling of CO_2 concentration:

1. Average earth temperature would rise by 2-3 °C.
2. Precipitation would increase in some regions, although the change in rainfall patterns is uncertain. Some scenarios suggest that regions between middle and high altitudes would be affected. These are the regions where many developing countries are located.
3. Evapotranspiration tendencies would increase in parts of the world, perhaps more than rainfall.

Regional and specific changes are not so certain, and several models depict contrasting scenarios. The effects of future climatic changes on agriculture at a given place can be ascertained only if the magnitude of climate change at that place is known. This is not possible with our present state of knowledge. However, on the basis of global generations, we can speculate on the effect of climatic changes on pulse productivity in India.

A number of greenhouse and field studies have shown that the productivity of C_3 plants increases when CO_2 concentration is increased (Sinha 1982). Pulse productivity may increase as the CO_2 concentration in the atmosphere increases. However, this increase in CO_2 will be coupled with an increase in temperature. Chickpea exhibits different temperature optima for vegetative growth, fruit set, and seed development. These optima influence the duration of the various phenological phases. Extension of the vegetative phase due to failure of fruit set at low temperatures in Delhi led to extended crop duration and higher yields than in Hyderabad. An increase in global temperatures may shorten the duration of pulse crops, thereby nullifying the beneficial effects of increased CO_2 concentration. Seed development at higher temperatures also will affect yields adversely.

Pulses depend on rainfall for their water requirement. Predictions on the effects of climatic change also envisage more uncertainty in rainfall patterns (Wittwer 1983). If precipitation decreases in pulse-growing areas, productivity would decline. Increases in precipitation in the Indian plateau or in northern India would be beneficial to pulse productivity.

Changes in temperature and precipitation will certainly influence the buildup of biotic stresses due to insects, pests, and diseases. Excess precipitation in association with warmer temperatures may favor development of insects and pests which reduce pulse yields. Excess rainfall would promote vegetative growth more than seed yield.

To ensure stability in pulse production, varieties that are resistant to insect pests and diseases and tolerant of environmental stresses are needed. However, the increase in CO_2 concentration could increase the capacity of pulses to tolerate environmental stresses (Wittwer 1983).

Conclusion

Food security includes nutritional security. Pulses in India are not only a major source of proteins, they also complement cereal proteins to provide the balance of essential amino acids. A vegetarian diet has the advantage of providing proteins at a low cost. Animal proteins are expensive, and are not easily accessible to most of the population.

The productivity of pulses in India has remained stagnant in the last 40 yr. Both abiotic and biotic factors are responsible for the low productivity. Among the biotic factors, insect damage of leaves in most of the *Vigna* species is common; in chickpea and pigeonpea, pod damage by pod borer is widespread. Farmers know that these pests can be controlled by pesticides, but for various reasons they do not practice pest control. Consequently, productivity of these crops in farmers' fields continues to be low.

International and national programs have had considerable success in improving the plant type of cereals, which are determinate crops, for greater productivity. This has not happened in grain legumes (with the possible exception of soybean and peas). Those essentially temperate crops are not exposed to uncertain weather conditions during their growth cycle. However, the grain legumes grown in India experience considerable intraannual and interannual variation in temperature, humidity, and precipitation. Apparently, they are more sensitive and vulnerable to these factors than are cereals. Because future climatic variations will include changes in temperature, humidity, and precipitation, it is likely that pulse crops will continue to be more vulnerable than cereals. This would create greater imbalances in nutrition in vegetarian Indian diet. Much greater efforts are needed to understand the influence of environmental factors on the mechanism of pulse productivity and to breed for improved plant types.

References cited

Aggarwal P K, Khanna-Chopra R, Sinha S K (1984) Changes in leaf water potential in relation to growth and dry matter production. Indian J. Exp. Biol. 22:98-101.

Bahl P N, Jain H K (1977) Association among agronomic characters and plant ideotypes in chickpea (*Cicer arietinum* L.). Pflanzenzichtg. 7:154-159.

Basu A C, Pramanik L M (1969) Effects of planting dates of gram on the infestation of *Heliothis armigera* (Hubner) and the gram yield. Indian J. Entomol. 31:181-183.

Davies J C, Lateef S S (1978) Recent trends in grain legume: pest research in India. Pages 25-31 *in* Pests of grain legumes: ecology and control. S. R. Singh, H. F. Vanemdeu, and T. Ajibolataylor, eds. Academic Press, London.

FAO—Food and Agriculture Organization (1983) World food report 1983. Rome, Italy.

FAO—Food and Agriculture Organization (1985) Production yearbook. Rome, Italy.

Khanna-Chopra R, Koundal K R, Kansal M (1985) Comparative behavior of seedlings of sorghum and some tropical legumes in relation to leaf expansion and growth. J. Agric. Sci. (Camb.) 104:625-630.

Khanna-Chopra R, Lakshmi V (1987) Photosynthesis rate does not relate with dry matter accumulation in chickpea. Proc. Indian Natl. Sci. Acad. B53:545-549.

Khanna-Chopra R, Sinha S K (1987) Chickpea: physiological aspects of growth and yield. Pages 163-190 *in* The chickpeas. M. C. Saxena and K. B. Singh, eds. CAB International, U.K.

Kumar D, Singh M P, Buttar B S (1983) Effect of sodicity on emergence, establishment and growth of chickpea. Int. Chickpea Newsl. 8:15-17.

Manchanda H R, Sharma S K (1980) Differential effects of chloride and sulphate soil salinity in chickpea. Int. Chickpea Newsl. 3:7-8.

Passioura J B (1976) Physiology of grain yield in wheat growing on stored water. Aust. J. Plant Physiol. 3:559-565.

Rains D W, Croughan T P, Stavarek S J (1980) Selection of salt tolerant plants using tissue culture. Pages 279-292 *in* Genetic engineering of osmoregulation: impact on plant productivity for food, chemicals and energy. D. W. Rains, R. C. Valentine, and A. Hollaender, eds. Plenum Press, New York.

Reed W, Cardona C, Sithanantham S, Lateef S S (1987) The chickpea insect pests and their control. Pages 283-318 *in* The chickpea. M. C. Saxena and K. B. Singh, eds. CAB International, U.K.

Saxena N P, Kapoor S N, Bisht D S (1983) Emergence of chickpea seedlings in sub-optimal moisture. Int. Chickpea Newsl. 9:12-14.

Saxena N P, Krishnamurthy L, Sheldrake A R (1980) Chickpea physiology. Page 112 *in* Pulse physiology. Annual report 1979-80. Part II. International Crops Research Institute for Semi-Arid Tropics, Hyderabad, India.

Saxena N P, Sheldrake A R (1980) Physiology of growth, development and yield of chickpeas in India. Pages 106-120 *in* Proceedings of the international workshop on chickpea improvement. International Crops Research Institute for Semi-Arid Tropics, Hyderabad, India.

Saxena M C, Yadav D S (1975) Some agronomic considerations of pigeon peas and chickpeas. Pages 31-62 *in* Proceedings of the international workshop on grain legumes. International Crops Research Institute for Semi-Arid Tropics, Hyderabad, India.

Sharma H C, Singh T, Ram Mohan D S (1974) Response of gram varieties to irrigation. H.A.U.S. Agric. Res. 4(4):255-260.

Singh R N, Singh K M (1978) Influence of intercropping on succession and population build up of insect pests in early variety of red gram *Cajanus cajan* (L.) millsp. Indian J. Entomol. 40(4):361-375.

Sinha S K (1982) The impact of carbon dioxide on agriculture in developing countries. Impact Sci. Soc. 32:311-323.

Sinha S K, Aggarwal P K, Khanna-Chopra R (1985) Phenological and physiological basis of irrigation in India. Advances in irrigation. Science 3:129-212.

Sinha S K, Lai B, Koundal K R, Khanna-Chopra R (1983) Nitrogen fixation does not limit yield in pulses. Curr. Sci. 52:916-918.

Vaishampayan S M, Veda O P (1980) Population dynamics of gram pod borer *Heliothis armigera* (Hubner) and its outbreak situation on gram, *Cicer arietinum* L. at Jabalpur. Indian J. Entomol. 42(3):453-459.

Van der Maesen L J G (1972) *Cicer* L. a monograph of the genus with special reference to the chickpea (*Cicer arietinum* L.), its ecology and cultivation. Medelelingen Landbouwhoge School, Wageningen. 342 p.

Williams L F, Dejong T M, Phillips D A (1981) Carbon and nitrogen limitation on soybean seedlings development. Plant Physiol. 68:1206-1209.

Winter K (1981) CO_2 and water vapor exchange malate content and ^{13}C value in *Cicer arietinum* L. grown under two water regimes. Z. Pflanzenphysiol. 101:421-430.

Wittwer S H (1983) Rising atmospheric CO_2 and crop productivity. Hortscience 18(5):667-673.

Notes

Acknowledgment: Vidya Lakshmi provided the data on chickpea.

Authors' address: R. Khanna-Chopra and S. K. Sinha, Water Technology Centre, Indian Agricultural Research Institute, New Delhi-110012.

Citation information: International Rice Research Institute (1989) Climate and food security. P.O. Box 933, Manila, Philippines.

Global climate variability: interannual and intraseasonal time scales

E. M. Rasmusson

In responding to internal and external forcing mechanisms, the atmosphere exhibits preferred regional and global modes of variability, which appear over and over again in generally similar forms. These teleconnections, a fundamental feature of atmospheric variability, dictate taking a global view of climate dynamics and prediction. The evolving teleconnection patterns that appear as part of two global-scale oscillations are particularly important elements of intraseasonal and interannual variability and prediction. They are the Madden-Julian (40-50 d) oscillation and the El Niño/Southern Oscillation (ENSO) cycle of coupled ocean atmosphere interactions (which evolves on the interannual time scale). Both oscillations have maximum amplitude in the Pacific trade wind-monsoon sector of the tropics and also affect other regions of the globe. Current prospects for predicting monthly and seasonal means by empirical-statistical techniques are discussed in light of improved understanding of short-term climate variability and its relation to ocean and land surface anomalies. Possible developments in short-term climate prediction are reviewed, with emphases on the possibility of the development and use of general circulation models of the coupled atmosphere-ocean-land surface climate system.

The atmosphere varies on time scales ranging from a span of minutes to climatic epochs spanning millions of years. The causes of these fluctuations change with the time scale of variability. Figure 1 is a highly schematic representation of sources of climate variability on time scales ranging from a few days to a few years. On the high frequency end of this range, day-to-day weather fluctuations are primarily controlled by internal atmospheric dynamics. They are manifestations of the growth, decay, and propagation of meteorological disturbances, which derive their energy from the internal structure of the atmosphere (i.e. horizontal or vertical gradients of wind, temperature, and moisture) (Shukla 1985).

Atmospheric dynamics also play a major role in determining the time-averaged, planetary-scale circulation features over longer periods. However, as the time scale of variability increases, the anomalous exchange of thermal and mechanical energy between the atmosphere and the upper ocean and land surface "memory components" of the climate system increases in importance. This exchange is largely controlled by the boundary conditions at the interface (i.e. sea surface temperature [SST], snow and sea ice, soil moisture, albedo [the percentage of incoming solar

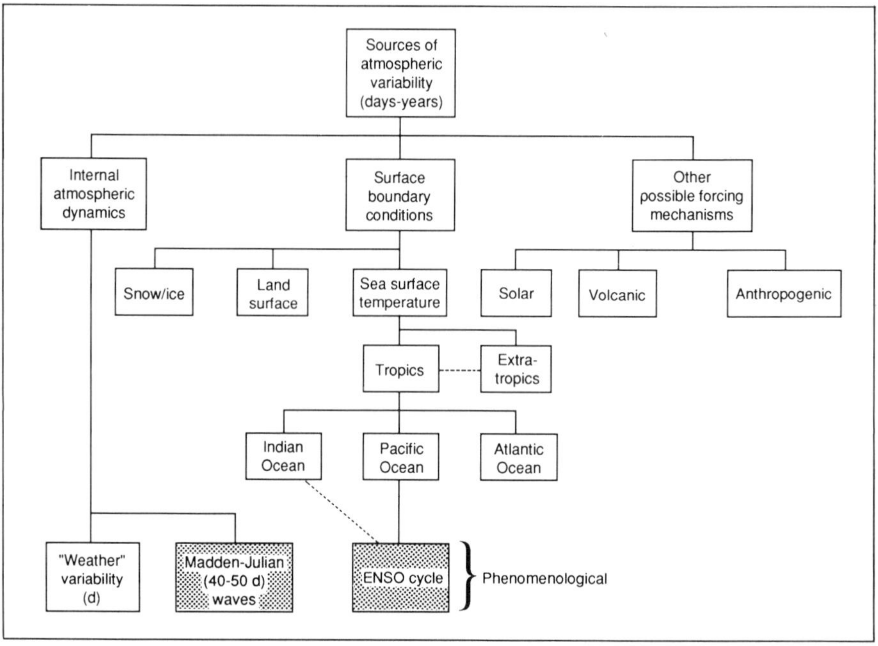

1. Schematic illustration of the causes of atmospheric variability on time scales ranging from days to years. The bottom tier of boxes represents selected phenomenological aspects of variability: two of them (shaded boxes) are discussed here.

radiation reflected from the earth's surface], and vegetative cover). Thus, the effect of both ocean and land surface boundary conditions, particularly in the tropics, is an important issue in short-term climate variability and prediction.

Modes of atmospheric variability

Although it seems highly unlikely that any two weather maps will ever be exactly alike, the atmosphere does exhibit preferred regional or even global patterns of response to the internal and external forcing mechanisms displayed in Figure 1. These atmospheric teleconnections appear over and over again, in broadly similar forms. They reflect a preferred pattern of coherent links between weather and climate variability over vast areas of the globe. This fundamental feature of atmospheric variability clearly dictates a global view of climate variability and prediction.

Teleconnection patterns appear to arise from a variety of causes. Diagnostic and theoretical studies (e.g. Simmons et al 1983), indicate that they are often initiated outside or on the periphery of the geographical region in which the train of anomalies appears. In the Northern Hemisphere extratropics, a small number of teleconnection patterns accounts for a significant portion of the variance in monthly and seasonal means (Wallace and Gutzler 1981). Some of these teleconnections appear to have links to tropical precipitation. Teleconnection patterns also have been identified in the Southern Hemisphere (e.g. Rodgers and van Loon 1982), and pronounced east-west oriented anomaly couplets, which span a major fraction of the earth's circumference, are a common feature of the equatorial belt (Yasunari 1985).

The month-to-month or year-to-year evolution of teleconnection patterns may, at first glance, appear to be entirely random. However, a careful examination of historical data identifies two modes of climate variability that are large in amplitude and global in scale and that exhibit a remarkably consistent pattern of evolution. They are most pronounced in the tropics, but affect parts of the extratropics as well. The first is the El Niño/Southern Oscillation (ENSO) phenomenon (Rasmusson 1985). ENSO's major swings, which span a period of about 2 yr, recur irregularly, at intervals of 2-7 yr. The second is the Madden-Julian (40-50 d) oscillation (Madden and Julian 1971, 1972).

The evolving teleconnection patterns associated with these two oscillations are of fundamental importance to short-term climate prediction. ENSO has important implications for seasonal and interannual prediction, especially in the lower latitudes. The Madden-Julian oscillations are most pronounced in the Indian Ocean-western Pacific monsoon sector, but also affect other areas in both the tropics and extratropics (Knutson et al 1986, Weickmann et al 1985).

El Niño/Southern Oscillation

As it is currently used, El Niño refers to an anomalous warming of the surface water of the eastern equatorial Pacific. The Southern Oscillation (Walker and Bliss 1932) is a global-scale seesaw in atmospheric surface pressure with centers of action around Indonesia-North Australia and in the southeast Pacific. The two phenomena are atmospheric and oceanic parts of an elegant and pervasive global system of climate fluctuations (for a review and bibliography, see Rasmusson 1985.)

ENSO is the most notable and pronounced example of global climate variability on the interannual time scale. However, its effects are by no means uniform; they vary both regionally and seasonally. In addition, both period and amplitude vary from one swing of the ENSO cycle to another. Nevertheless, the nearly simultaneous appearance of pronounced climate anomalies around the world, together with disruption of the marine ecosystems along the west coast of the Americas that occurs with the higher amplitude swings, can have serious adverse effects on regional and global food production.

The ENSO cycle owes its existence to large-scale, ocean-atmosphere interactions in the equatorial Pacific. The world's most extensive region of warm surface water (SST greater than 28 °C) is located in the western tropical Pacific-Indonesian region. Heavy rainfall and associated heating of the atmosphere over this huge warm pool represents a major source of the atmospheric heating that drives the large-scale circulation.

During the cold water phase of the cycle, the Pacific equatorial easterlies are strong, the warm pool is confined to the area west of the date line, and dry conditions prevail over the equatorial zone to the east. As a typical warm episode develops, SST in the central equatorial Pacific rises (i.e. the west Pacific warm pool extends eastward, accompanied by an eastward extension of the associated region of heavy rainfall). With the eastward shift of this atmospheric heat source, low-level winds in the western equatorial Pacific become more westerly. This warm phase of the cycle has received the most attention, the warm or ENSO episodes typically are more event-like in their evolution and the global pattern of regional anomalies that occurs (Fig. 2) leads to serious socioeconomic disruptions around the world.

The most pronounced ENSO anomalies appear in the monsoon-Pacific trade wind regions (i.e. the Indian Ocean-tropical Pacific sector). During a warm episode, parts of this region, as well as the northeast Pacific subtropics, are plagued by drought, while abnormally heavy rainfall occurs over the eastern equatorial Pacific. The global pattern of anomalies is illustrated by the 1982-83 warm episode—one of the strongest of the past century. In 1982, India experienced deficient monsoon rainfall, a pattern typical of these episodes. Drought conditions over Australia and Indonesia, which became serious in mid-1982, eased early in 1983, but continued over Melanesia and spread across the north Pacific subtropics, enveloping the southern Philippines and the Hawaiian Islands. Farther east, satellite data indicated far below normal precipitation from the Amazon Basin eastward across the Atlantic. Although the typical ENSO-related drought pattern which occurred over southeast Africa during the rainy season of 1982-83 was among the worst of the century, it was only one aspect of a more pervasive and longer period of drought that afflicted much of sub-Saharan Africa between 1982 and 1984 (Rasmusson 1986).

Few warm episodes exhibit anomalies approaching the magnitude of those observed during 1982-83. For example, both the episode of 1976-77 and the most recent warming of 1986-87 represent relatively mild occurrences. Within the last 50 yr of this century, only the episodes of 1940-41, 1957-58, and 1972-73 approached the intensity of the 1982-83 episode.

Madden-Julian (40-50 d) oscillations

The first evidence of systematic, intraseasonal oscillations was provided by Madden and Julian (1971, 1972). They identified a global-scale oscillation with an average period between 40 and 50 d that propagates eastward in the low latitudes. As with the ENSO cycle, this oscillation is particularly evident in large amplitude rainfall and wind fluctuations over the low latitude Indian Ocean-western Pacific monsoon sector. In fact, this oscillation in many ways mimics the atmospheric ENSO fluctuations, but on a much shorter time scale.

Global aspects of this oscillation have been described by a number of authors (Knutson et al 1986, Weickmann et al 1985). Although most pronounced in the tropical belt, the Madden-Julian oscillations also appear to be associated with wintertime weather regime flip-flops over parts of the Northern Hemisphere (Weickmann et al 1985). There is no evidence yet of a significant summertime signal in the Northern Hemisphere extratropics (Knutson et al 1986). The oscillation is strongest in the summer hemisphere of the tropics and may be intimately related to the timing of the onset, withdrawal, and active/break periods of the Indian monsoon (Yasunari 1980).

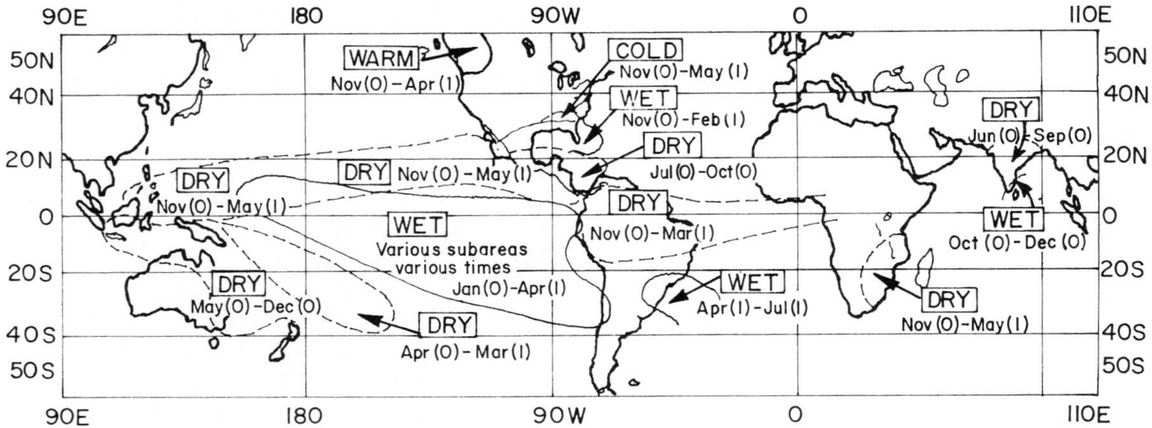

2. The pattern of anomalies during a typical warm episode in the ENSO Cycle of variability. Year 0 is the year of the initial warming of the ocean surface along the South American coast, year 1 is the following year. (For more detailed information on warm episode anomalies, see Ropelewški and Halpert 1986, 1987.)

It is difficult to evaluate the role of anomalous boundary conditions in the 40-50 d oscillation from observational data alone. An invaluable theoretical aid is atmospheric General Circulation Models (GCM). These simulation models are based on mathematical representations of the physical laws governing large-atmospheric motions. They are similar to those used in day-to-day numerical weather predictions, but are integrated for a much longer time, and require the use of large computers. State-of-the-art GCMs are capable of realistically simulating many aspects of the global climate. Models have been interactively coupled, in an admittedly crude manner, with land surface processes, including hydrology, and are beginning to be coupled with companion oceanic GCMs.

GCM model experiments (Lau and Lau 1986) point to internal atmospheric dynamics as the basic cause of the Madden-Julian oscillations. However, the longitudinal variation in the response arises from nonuniform, lower-boundary conditions, the strongest response being over the region of highest moisture content and surface temperatures (i.e. the monsoon sector). The north-south seasonal shift in activity follows monsoon activity and reflects the response of surface temperatures to the annual cycle of heating. Interannual variability is associated with the east-west shifts of the western Pacific warm pool; those shifts are associated with the ENSO cycle.

Decadal time-scale variations

Pronounced decadal-scale fluctuations occur in many regions of the world. The degree to which they exhibit systematic or predictable behavior is hard to establish from the relatively short time series of instrumental records. This is particularly true of multiyear wet and dry spells, which can have particularly pronounced socioeconomic impacts. Most attempts to relate multiyear variations to various forcing mechanisms have been less than convincing.

This paper does not address the problem of anthropogenic effects (e.g. desertification, deforestation, greenhouse warming); those effects are more closely related to questions of climatic change than of climatic variability. Nevertheless, the potential human effect on climate must be kept in mind when interpreting past and present climate data and when projecting longer time scale trends.

The role of solar variability and volcanic aerosols in short-term climatic variability remains a lively and controversial issue. Most of the many claims of cyclical variability related to solar or lunar cycles are based on short or inadequate precipitation or on river flow records. The study of Mitchell et al (1979) is a notable exception. They developed a proxy "Drought Area Index (DAI)" for the western United States by calibrating a long time series of tree ring data (1600-1962) with a 32-yr overlapping record of Palmer Drought Severity Index values (Palmer 1965). They found an apparently significant phase locking between the 22-yr Hale sunspot cycle and large amplitude variations in the DAI. Subsequently, Bell (1981) and Currie (1981) claimed the existence of an 18.6-yr period in the DAI data, phased with the lunar nodal regression cycle.

Questions have been raised regarding the reality or persistence of these rhythms. Following a recent reexamination of the data, Mitchell (1986, pers. comm.)

concluded that both signals exist in the DAI series and are of roughly equal strength. The 22-yr feature is regular in phase, but undergoes long-term changes in amplitude. The 18.6-yr feature also varies in amplitude, and, in addition, exhibited a disconcerting reversal of phase around 1800. Mitchell concluded that both components, as bona fide climatic signals, merit further study to clarify their physical origins and their potential predictive value. However, it should be kept in mind that "the connection between drought and solar behavior (also lunar tides?) is scarcely to be described as a reliable basis for operational climate prediction" (Mitchell et al 1979).

Unfortunately, many investigators are not as careful as Mitchell et al in evaluating the prediction potential of their results. Particularly notable in this regard are some claims of cyclic variability of Sahel rainfall and its predictability. The fact that fluctuations of a particular period have occurred 2 or 3 times during the length of record may provide an estimate of natural variability, but imply little else. The meteorological literature is replete with descriptions of "cycles" found *a posteriori* which failed miserably when used predictively.

Until the controlling processes are better understood, prediction of runs of predominantly dry years or multiyear climate trends will continue to be hazardous and highly controversial.

Empirical prediction

As early as 1922, L. F. Richardson proposed that, given an accurate description of its initial state, the future evolution of the atmosphere could be predicted by solving the physical laws which govern its behavior (Platzman 1967). For a number of reasons, this numerical/dynamical approach to prediction was not implemented until the 1950s. Since then, numerical prediction has revolutionized the operational forecasting of weather as far as a week in advance. Beyond this range, however, the use of dynamical methods is still experimental. Forecasts of short-term climate variability continue to be based on statistical relationships describing aspects of past behavior of the system.

Unfortunately, no two past climate states are identical, and the instrumental record available for identifying predictive relationships is relatively short, rarely approaching a century in length. The success of empirical prediction schemes depends on the existence and identification of strong, stable, statistical relationships that involve relatively few predictors.

Empirical techniques appear to be most applicable in the tropics. While day-to-day weather fluctuations are less predictable in the tropics, the opposite appears to be true for seasonally averaged conditions. This appears to result from a closer coupling between the atmosphere and the underlying surface conditions in the warmer regions of the tropics, and the related existence of the systematic low-frequency fluctuations discussed earlier. In retrospect, one of the great advances in our understanding of the climate system during the past few years has been the discovery, based on observational studies and ocean and atmospheric modeling experiments, that a significant share—and in the tropics, probably a major share—of the atmospheric variability on time scales of months to a few years is associated with variations in tropical SST. This is largely the basis for increased optimism regarding the possibilities of seasonal prediction in at least some parts of the tropics.

Statistical relationships, based at least in part on ENSO parameters, have been exploited in developing a number of prediction schemes for tropical and near-tropical regions (e.g. Nicholls 1985, Shukla and Mooley in press, Shukla and Paolino 1983). The best of these relationships typically show correlations with dependent data of 0.6 and 0.8 between seasonally averaged conditions over large regions (such as the Indian peninsula) and predictive parameters from the previous season. Nicholls (1985) related climate parameters one or more seasons in advance directly to Australian crop production and still found correlations of this magnitude. Beyond this, any real breakthroughs, if they occur, will likely require the use of coupled ocean-atmosphere dynamical models.

Predictive relationships based on slowly varying tropical parameters not directly related to the ENSO cycle have also been developed. A notable example is the relationship between Atlantic SST and rainfall over northeast Brazil (Hastenrath and Heller 1977, Moura and Shukla 1981).

Current empirical forecasts of monthly and seasonally averaged conditions in the extratropics exhibit modest skill at best, and it is unclear what further progress can be expected from empirical-statistical methodology alone. A few useful regional relationships with ENSO exist, but these are primarily limited to the Pacific sector and neighboring land areas and are, for the most part, weaker than the statistical relationships in the tropical belt.

Dynamical prediction

During the last 3 decades, day-to-day weather forecasting, particularly in the extratropics, has evolved from an art or, more positively, an empirical science, to a process based largely on the objective solution of mathematical equations which describe in a simplified manner the evolution of the atmosphere. This major scientific advance was made possible by the development of a satellite-based, global system of atmospheric observations, the use of high-speed computers capable of rapidly solving the governing equations, and the development of simplified finite difference forms of the governing equations suitable for use in numerical prediction.

Operational meteorological centers are now able to forecast temperate latitude fluctuations several days in advance with considerable skill. However, because of our imperfect ability to observe the initial state of the global atmosphere, the complexity and unstable nature of atmospheric flow, and our inability to write and solve the exact physical equations governing its evolution, there is an upper limit to the lead time of useful predictions of the instantaneous state-of-the-atmosphere. The theoretical limit of predictability of day-to-day variability does not exceed a few weeks in the extratropics and may be as short as a week or less in the tropics (Shukla 1985). Consequently, forecasts of day-to-day weather variations a month or more in advance are beyond the theoretical limit of predictability.

If, however, we relax our objective from the prediction of instantaneous states to predicting time averages over a specific period, the limits of predictability may increase substantially (Shukla 1985). The large-scale, more slowly varying components of the atmospheric circulation, which primarily determine its space- and time-averaged state for periods longer than a few days, are potentially more predictable than the more transient synoptic features that dominate the patterns on daily weather maps. Consequently, it may be possible to extend dynamical prediction methodology to climate time scales. Some encouraging research results have already been obtained (Miyakoda and Sirutis 1985, Shukla 1985).

As was previously pointed out, the space- and time-averaged state-of-the-atmosphere for these longer periods also reflects the influence of the lower boundary. Since the surface boundary conditions are likely to play a progressively greater role as the averaging period increases from monthly to seasonal and beyond, monthly and seasonal forecasting methodology will likely diverge during the next decade. The most important development in monthly forecasting will likely be a continuation of the movement from a predominantly empirical approach to a mixture of empirical and dynamical methodology that includes an increasing use of extended-range weather predictions from dynamical models. In this way, the monthly forecast will benefit from the prediction of at least a portion of the weather variability during the first part of the period. A broad range of prediction products covering the entire world can be produced, including confidence limits for each forecast.

Longer lead-time dynamical forecasts appear more feasible for seasonal averages, since the weather sampling problems become less important relative to the lower-frequency, potentially more predictable, climate signal. However, extension of dynamical methodology to the prediction of seasonal means requires improved, more complex GCMs, which will accurately simulate both the internal dynamics of the atmosphere and the effects of surface boundary conditions. In addition, predicting the evolution of the boundary conditions requires the development of coupled ocean-atmosphere models and better parameterization of atmosphere-land surface interactions. The development of coupled ocean-atmosphere models is a major thrust of the 10-yr Tropical Ocean Global Atmosphere Program of the World Climate Research Program (WMO 1985) which began last year.

Final remarks

The theoretical framework for addressing the question of weather predictability was developed during the 1960s, and a consensus on the ultimate limits of predictability has been reached. No such framework or consensus exists for short-term climate prediction, although considerable progress toward defining the prediction problem and identifying promising avenues of research has occurred during the last decade.

Consistently skillful monthly/seasonal forecasts over the entire world for all seasons are, in all probability, an unattainable goal. The potential predictability of monthly and seasonal means varies with region, season, parameter, and climatic regime. The statistics of atmospheric variability provide some guidance for evaluating, at least qualitatively, the potential limits of climate predictability.

Lack of adequate observations is among the most important factors that limit our understanding of climate dynamics. Establishing the limits of short-term climate prediction and exploiting the predictability that exists within the system require not only a better monitoring of the global atmosphere, but also a far better description of variations in the temperature and currents of the upper ocean, together with a more accurate description of land surface boundary conditions. These requirements can only be met by a more comprehensive and stable global observational network, in which meteorological and oceanographic satellites play a key role.

It cannot be stressed too strongly that the value of a forecast to a serious user depends not only on the quality of the forecast and the particular needs of that user, but also on the user's knowledge of that quality and of how the forecast should be applied (Gilman 1985). No matter what the methodology, only a portion of the monthly and seasonal variability is even potentially predictable. Therefore, short-term climate forecasts are inherently probabilistic in nature. In some cases, the probabilities attached to a forecast might be high enough to allow it to be used in making decisions, in others not. No forecast is complete without information on its level of confidence. If the forecast is to be used properly, it is essential that the user clearly understand the climatological background which the forecast modifies, the limitations of the forecast, and the economic or social impact of links with climate fluctuations. Improper decisions may be made if a low confidence forecast is used as a simple, categorical "yes/no" statement. Without the appropriate background and knowledge, there is a real danger that the user may apply a potentially useful forecast in a harmful way.

The degree to which climate variability is predictable has yet to be established. Nevertheless, with our increasing understanding of the behavior of the global climate system, we can make far better use of the data that now exist to minimize the adverse socioeconomic impacts of climate variability. This effort clearly requires close cooperation and interaction among scientists from a variety of disciplines.

References cited

Bell P R (1981) The combined solar tidal influence in climate. Predominant periods in the time series of drought area index for the western High Plains, A.D. 1700 to 1962. Pages 241-264 *in* Variations of the solar constant. S. Sofia, ed. NASA Conference Publication 2191. Washington, D.C.

Currie R G (1981) Evidence for 18.6-year M_N signal in temperature and drought conditions in North America since A.D. 1800. J. Geophys. Res. 86:11055-11064.

Gilman D L (1985) Long-range forecasting: the present and the future. Bull. Am. Meteorol. Soc. 66:159-164.

Hastenrath S, Heller L (1977) Dynamics of climatic hazards in northeast Brazil. Q. J. Roy. Meteorol. Soc. 106:447-462.

Knutson T R, Weickmann K M, Kutzbach J E (1986) Global scale intraseasonal oscillations of outgoing longwave radiation and 250 mb zonal wind during the Northern Hemisphere summer. Mon. Weather Rev. 114:605-623.

Lau N-C, Lau K-M (1986) The structure and propagation of intraseasonal oscillations appearing in a GFDL general circulation model. J. Atmos. Sci. 43:2023-2047.

Madden R A, Julian P R (1971) Detection of a 40-50 day oscillation in the zonal wind in the tropical Pacific. J. Atmos. Sci. 28:702-708.

Madden R A, Julian P R (1972) Description of global scale circulation cells in the tropics with a 40-50 day period. J. Atmos. Sci. 29:1109-1123.

Mitchell J M, Stockton W C, Meko D M (1979) Evidence of a 22-year rhythm of drought in the western United States related to the Hale solar cycle since the 17th century. Pages 125-143 *in* Solar-terrestrial influences on weather and climate. B. M. McCormac and T. A. Seliga, eds. D. Reidel, Dordrecht.

Miyakoda K, Sirutis J (1985) Extended range forecasting. Pages 55-86 *in* Issues in atmospheric and oceanic modeling. Part B: Weather dynamics. S. Manabe, ed. Academic Press, New York.

Moura D M, Shukla J (1981) On the dynamics of droughts in northeast Brazil: observations, theory and numerical experiments with a general circulation model. J. Atmos. Sci. 38:2653-2674.

Nicholls N (1985) Impact of the Southern Oscillation on Australian crops. J. Climatol. 5:553-560.

Palmer W C (1965) Meteorological drought. Research Paper 45. U.S. Weather Bureau, Washington, D.C.

Platzman G W (1967) A retrospective view of Richardson's book on weather prediction. Bull. Am. Meteorol. Soc. 48:514-550.

Rasmusson E M (1985) El Niño and variations in climate. Am. Sci. 73:168-178.

Rasmusson E M (1986) Global climate change and variability: effects on drought and desertification in Africa. Pages 3-22 *in* Drought and hunger in Africa: denying famine a future. M. H. Glantz, ed. Cambridge University Press, Cambridge, New York, N.Y.

Rodgers J C, van Loon H (1982) Spatial variability of sea level pressure and 500 mb height anomalies over the Southern Hemisphere. Mon. Weather Rev. 110:1376-1392.

Ropelewski C F, Halpert M S (1986) North American precipitation and temperature patterns associated with the El Niño/Southern Oscillation (ENSO). Mon. Weather Rev. 114:2352-2362.

Ropelewski C F, Halpert M S (1987) Global and regional scale precipitation patterns associated with the El Niño/Southern Oscillation. Mon. Weather Rev. 115:1606-1626.

Shukla J (1985) Predictability. Pages 87-122 *in* Issues in atmospheric and oceanic modeling. Part B: Weather dynamics. S. Manabe, ed. Academic Press, New York.

Shukla J, Mooley D A (1987) Empirical prediction of the summer monsoon over India. Mon. Weather Rev. 115:695-703.

Shukla J, Paolino D A (1983) The Southern Oscillation and long-range forecasting of the summer monsoon rainfall over India. Mon. Weather Rev. 111:1830-1837.

Simmons A J, Wallace J M, Branstator G W (1983) Barotropic wave propagation and instability and atmospheric teleconnection patterns. J. Atmos. Sci. 40:1363-1392.

Walker G T, Bliss E W (1932) World weather 5. Mem. Roy. Meteorol. Soc. 4:53-84.

Wallace J M, Gutzler D S (1981) Teleconnections in the geopotential height field during the Northern Hemisphere winter. Mon. Weather Rev. 109:784-811.

Weickmann K M, Lussky G R, Kutzbach J E (1985) Intraseasonal (30-60 day) fluctuations of outgoing longwave radiation and 250 mb streamfunction during northern winter. Mon. Weather Rev. 113:941-961.

WMO—World Meteorological Organization (1985) Scientific plan for the Tropical Ocean and Global Atmosphere Programme. WCRP-3. Geneva.

Yasunari T (1980) A quasistationary appearance of 30-40 day period in the cloudiness fluctuations during the summer monsoon over India. J. Meteorol. Soc. Jpn. 58:225-229.

Yasunari T (1985) Zonally propagating modes of the global east-west circulation associated with the Southern Oscillation. J. Meteorol. Soc. Jpn. 63:1013-1020.

Notes

Author's address: E. M. Rasmusson, University of Maryland, Cooperative Institute of Climate Studies, Department of Meteorology, College Park, MD 20742, USA.

Citation information: International Rice Research Institute (1989) Climate and food security. P.O. Box 933, Manila, Philippines.

Monsoon variability and its relationship to agricultural strategies

S. Gadgil

The agrometeorological tasks involved in incorporating an understanding of monsoon variability into the determination of optimal agricultural strategies are considered. Methodologies developed for identifying agricultural zones as clusters of cropping patterns and delineating coherent rainfall zones are illustrated by case studies of the state of Karnataka, India. Agrometeorological zones can be obtained by superimposing agricultural and rainfall zones. It is suggested that these agrometeorological zones are the logical units for derivation of the optimal cropping patterns, management practices, etc, and for exploring the possibility of generating agriculturally relevant predictions.

The dependence of agriculture in monsoonal regions on climate, and particularly on rainfall, is well known. In years of severe monsoon rainfall deficit (i.e. 1965, 1972), average yields in India decrease 15% or more. The importance of monsoon prediction stems from this strong link between agricultural productivity and the performance of the monsoon.

The Indian monsoon has been studied in depth for more than a century, and considerable insight has been accumulated into the nature of its space-time variability (Das 1986, Hasternath 1985, Ramage 1971, Rao 1976). Methods of generating long-range predictions of average summer monsoon rainfall for two large zones of the country and short- and medium-range predictions of average rainfall for each meteorological subdivision have been developed. However, the value of the predictions in terms of their impact on agricultural productivity has not been assessed.

In fact, it is not clear if the kind of predictions needed for choosing the appropriate crop variety or the optimal crop management strategy is being generated by meteorologists. This is not surprising: the nature of the meteorological predictions needed for managing the different crops has yet to be precisely spelled out by the agricultural scientists. That is a prerequisite to generating agriculturally relevant forecasts.

If we are to utilize the knowledge of monsoon variability to identify the optimal cropping strategy that would ensure stable maximum productivity and, when possible, to generate the predictions required for each crop involved, a genuinely interdisciplinary attack on the problem is essential, for which this is a particularly opportune time. During the last two decades, major advances have been made in the ability to manipulate varieties through plant genetics and many new insights into the intraseasonal and interannual variation of the monsoon have been generated. Deeper understanding of the planetary scale monsoon is a consequence of tremendous progress in tropical meteorology, due in part to the advent of satellites and computers.

Thus, while on the one hand it has become possible to manipulate the life-span and other life-history parameters of crop varieties to suit climatic conditions, on the other hand deeper understanding of climatic variability and a substantial increase in the potential for generating meteorological predictions on different time scales are imminent.

Agrometeorological strategies for monsoon regions

The ultimate aim of an agrometeorological strategy is to attain maximum and stable agricultural productivity in the face of large variability in rainfall, from wet to dry spells within the season and from good monsoon years to drought years. To achieve this, cropping patterns, specific varieties, and management practices appropriate for the climatic regime and for the predicted behavior of the monsoon during any specific year for each region have to be chosen.

It is important to note that climate has to be characterized not only by average behavior but also by intraseasonal and interannual variability. If the variability of the agriculturally relevant features of the monsoon, such as duration of the rainy season, total seasonal rainfall, and probability of occurrence of such events as dry spells, is adequately known, it should be possible to generate general prescriptions for the agricultural strategy of each region.

For example, if dependable rainfall toward the end of the rainy season is inadequate to meet the requirements of traditional varieties, short-duration varieties may be recommended. If the likelihood of a dry spell during the critical life stages of a crop variety is high, an alternative variety or management practices such as farm water improvement for supplementary irrigation could be prescribed. If the number of years in which rainfall patterns are most favorable for a specific crop or variety can be deduced from an analysis of interannual variation in amount and pattern of rainfall, an appropriate mix of crops or varieties can be chosen. We expect a significant enhancement in productivity when the agricultural strategy is appropriate for the known climatic variability.

The agricultural strategy thus derived would depend only on the known variability in climate and rainfall for each region, and would not change from year to year. The next step is to refine the strategy on the basis of meteorological predictions so that it is appropriate for a specific period or season. For example, if long-range forecasts of total rainfall during a season or length of the rainy season could be generated, crops or varieties could be adjusted appropriately. Recommendations on farming operations could be made on the basis of medium-range forecasts of monsoon performance. Predictions of dry spells associated with breaks in the monsoon and the resumption of rains could be useful in scheduling supplementary irrigation. If meteorological predictions could be used for deciding among a set of well-defined alternatives, it would be possible to estimate the value of the forecasts generated, as well as the losses incurred due to errors or inaccuracies in the forecasts.

But can such predictions be generated? It is important to note that there are intrinsic limits to meteorological predictability (Lorenz 1969, Shukla 1985). Space and time scales are inexorably linked: for each time scale, only the average over a certain spatial scale can be predicted. Long-range forecasts involving predictions about monsoon performance in a forthcoming year are only possible for the average situation over rather large regions.

Thus irrespective of computational power, it will not be possible to forecast rainfall over a particular agricultural field station on a given date a year ahead. It is necessary to investigate whether forecasts of the rainfall during a forthcoming season can be generated on a spatial scale small enough to be useful in choosing among alternative crops or varieties for a specific year. Medium-range forecasts of the rainfall during forthcoming weeks over smaller regions would be possible, and might be of use in planning farming operations.

Choosing the appropriate agricultural strategy on the basis of the rainfall variability over a region is a prerequisite to using predictions for a specific season. Here, we restrict our attention to the first task. That involves the following scientific problems:

- Identifying the crops that can be grown in different regions. This is equivalent to delineating the agricultural zones that are homogeneous in cropping patterns.
- Identifying rainfall zones that are coherent in rainfall variation in a season or a year, so that the year-to-year or season-to-season variation of average rainfall over the zone reflects the variation over any subregion. The variation of rainfall (mean and dependable rainfall at different probability levels during different seasons and for the critical periods for the different crops) have to be derived for each rainfall zone. Predictions based on conditional statistics also can be generated. Superimposing the agricultural and rainfall zones derived would yield agrometeorological zones that are homogeneous in cropping patterns and coherent in rainfall variation. Optimal cropping strategies then could be derived for each agrometeorological zone.
- Analyzing yield-rainfall relationships. Because yield depends on soil moisture (a complex function of current as well as preceding rainfall), the problem of identifying the optimum rainfall pattern for each crop or variety is not trivial. A two-pronged attack is needed to identify the relationship between yield and rainfall: 1) empirical studies using available data on yield under different rainfall regimes prevalent in different years, and 2) yield variability resulting from specified variations in rainfall patterns, assessed using computer models that simulate the growth of different crops.
- Deriving the optimal agricultural strategy on the basis of the information and insights gained, as either a minimax strategy or one that will result in yields within specified limits.

Identifying the agricultural zones

If the tolerance range of each crop for each factor is precisely known, it is theoretically possible to identify the crops that can be grown over a region by analyzing the spatial variations in soil and important climatic elements. However, the number of factors involved is very large and current knowledge about the sensitivity of different crops to each factor is limited. We adopted an alternative approach to identify regions that are sufficiently homogeneous in the relevant soil-climatic factors (the so-called agroclimatic zones) to be homogeneous in the types of crops grown.

We assumed the cropping pattern that has evolved historically from the crops or varieties available to be optimal for the soil and climate of a particular region, and identified agroclimatic zones as regions that are internally homogeneous in traditional cropping patterns. In the case reported here, the basic data set is the relative area under 12 major crops in 176 divisions of Karnataka.

Analyzing the spatial variation in cropping patterns among the different divisions is facilitated by a reduction in dimension, using principal component analysis. This involves working with new variables, each of which is a linear combination of the original variables—in this case, the fraction of the area under each crop. The weighting factors in this linear combination were chosen to rank the variables by the size of the variance (Gadgil and Joshi 1983). Figure 1 shows the

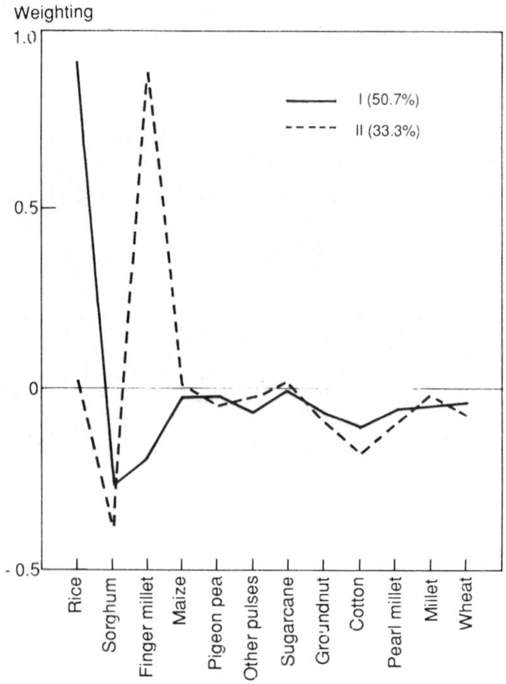

1. Crop weightings implied by the first 2 principal components.

weightings for the first two principal components. Note that the maximum weighting in the first component is for rice, in the second, for finger millet and sorghum. The spatial variations in cropping patterns can be represented as the distribution in amplitude of the first two principal components; together they explain 84% of the variance.

An analysis of this distribution shows distinct clusters of cropping patterns. The geographical locations of these clusters, i.e. the agricultural zones, and average cropping patterns are shown in Figures 2 and 3. It is interesting that cropping pattern diversity is high in the interior parts of the state, where the rainfall is low and has high variability. (The algorithm developed for identification of crosses and the relationship of these agricultural zones to such factors as soil, rainfall, and topography are discussed in Gadgil et al 1987.)

The agricultural zones are the basic unit for estimating the yields of different crops in a particular year. However, they are not necessarily the appropriate units for unraveling the interactions between rainfall variability and yield. Variation in rainfall may not be coherent within each zone, so that while part of the zone experiences drought, the rest of the zone may have normal or excessive rainfall. Because average rainfall over an agricultural zone may not reflect the situation in

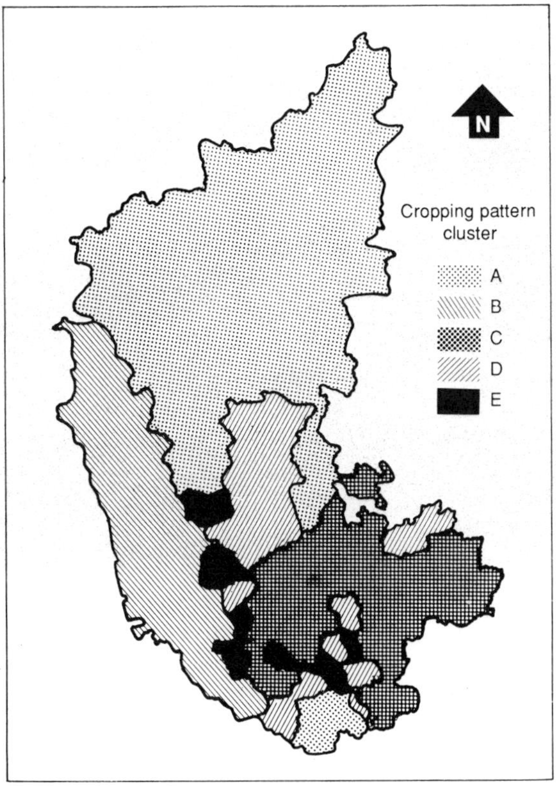

2. Agricultural zones of Karnataka.

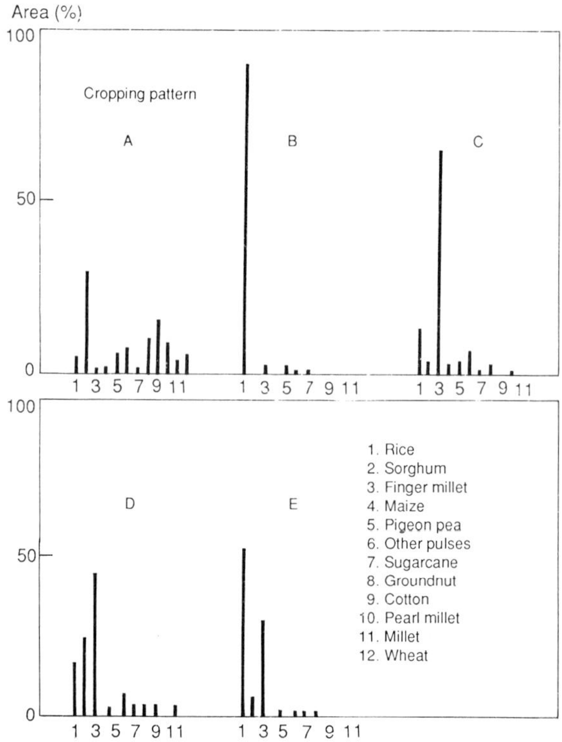

3. Average cropping patterns for the agricultural zones of Karnataka.

subregions, it cannot be used to estimate the impact of rainfall variability on productivity. The appropriate scale for analyzing rainfall variability is the spatial scale over which rainfall variations in a season or a year are coherent.

Rainfall variability and coherent zones

Monsoonal regions have large interannual variability of rainfall, from drought years to years with a good monsoon. Drought severity can be assessed in terms of the rainfall deficit in a season/year or in terms of its impact on critical resources. Usually drought is defined in terms of deficit, because it is not easy to quantitatively assess its impact on specific resources. Because below-normal seasonal or annual rainfall is seldom restricted to a single station, but occurs over larger regions, rainfall deficit averaged over regions (rather than stations) is used to assess drought.

The performance of the summer monsoon is usually evaluated by the India Meteorological Department from the average seasonal rainfall of each meteorological subdivision. However, we found that the three meteorological subdivisions of Karnataka (Fig. 4) are not coherent with respect to variation in annual rainfall (Gadgil and Yadumani 1987). Zones over which rainfall variations are coherent should form the basic unit for deriving the detailed statistics of the intraseasonal and interannual rainfall variation needed in our search for optimum cropping strategies.

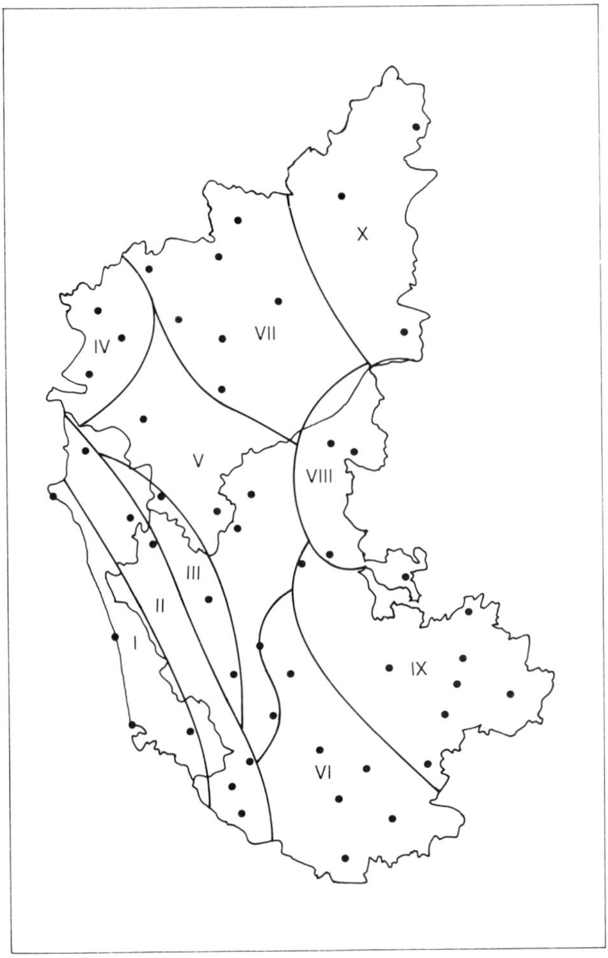

4. Station network, the three meteorological subdivisions (thin lines) and nine coherent rainfall zones (thick lines) of Karnataka.

Identifying coherent rainfall zones is also needed to define the occurrence of drought, to estimate drought impact, and to plan remedial measures.

Coherence between two stations, i.e. the degree to which rainfall variations at one station are related to those at another station, can be estimated by the cross-correlation. When the cross-correlation of annual rainfall is high, anomalies in annual rainfall (drought or excess rains) are likely to occur simultaneously over the two stations. Because the rainfall distribution pattern is also important, particularly for agriculture, coherence of the monthly rainfall pattern also has to be considered. In addition to the annual cross-correlation, we consider the disparity between the normalized monthly rainfall patterns at the two stations. (The methodology is described in Gadgil et al 1988.)

Analysis of monthly data for 1901-85 for 50 well-distributed stations in Karnataka shows that the state can be divided into 9 coherent zones (Fig. 4). With the coherent zones delineated, details on the observed variability (mean and standard deviation, as well as dependable rainfall at different levels of probability) of average rainfall during this century can be provided for each zone. In addition to generating information on variability, we found it possible to make some conditional predictions of average rainfall for some zones during some seasons. This suggests that coherent regions may also be the appropriate spatial scale for seasonal predictions.

A comparison of Figures 2 and 4 shows that agricultural zone B (corresponding to the rice belt) comprises two distinct coherent rainfall zones while zone A (the sorghum belt) comprises three coherent rainfall zones. In the southeastern and eastern parts of the state, the situation is more complex; within each coherent rainfall zone, there is a diversity of cropping patterns associated with large variations in topography and soil. The impact of rainfall variability on yield will have to be worked out separately for each agrometeorological zone.

The next task is to identify the type of rainfall variability information needed for each agrometeorological zone, taking into account the available crop varieties and management practices of the agricultural system. Considerable expertise is available, on developing models that simulate the growth of different crops (Huda 1984) and on analyzing the variability and predicting the agriculturally important features of rainfall at a single station (Virmani and Huda 1988). It is hoped that concerted efforts by interdisciplinary groups of agricultural scientists and meteorologists will be able to synthesize the different facets to identify optimal agricultural strategies for monsoonal regions.

References cited

Das P K (1986) Monsoons. Fifth IMO lecture. World Meteorological Organization, Geneva.

Gadgil S, Gowri R, Yadumani (1988) Coherent rainfall zones: case study for Karnataka. Proc. Indian Acad. Sci. (Earth Planet. Sci) 97:63-79.

Gadgil S, Hegde M, Gowri R (1987) Agricultural zones of Karnataka. Karnataka state of the environment report 1985-86, India.

Gadgil S, Joshi N V (1983) Climatic clusters of the Indian region. J. Climatol. 3:47-63.

Gadgil S, Yadumani (1987) Rainfall in Karnataka: variability and prediction. Karnataka state of the environment report 1985-86. India.

Hasternath S (1985) Climate and circulation of the tropics. D. Reidel Publishing Co., Dordrecht. 455 p.

Huda A K S (1984) Forecasting the yields of sorghum, pearl millet and groundnut. Paper presented at the Symposium on Crop Yield Forecasting Methodology, Indian Agricultural Research Institute, New Delhi.

Lorenz E N (1969) Predictability of a flow which possesses many scales of motion. Tellus 21:289-306.

Ramage C S (1971) Monsoon meteorology. Academic Press, New York.

Rao Y P (1976) Southwest monsoon. Meteorological monograph. India Meteorological Department, New Delhi.

Shukla J (1985) Predictability. Adv. Geophys. 28B:87-122.

Virmani S M, Huda A K S (1988) Effects on agricultural productivity. Pages 537-556 *in* The impact of climatic variation on agriculture. Vol. 2. Kluwe Academic Publishers.

Notes

Author's address: S. Gadgil, Centre for Ecological Sciences, Indian Institute of Science, Bangalore, India.

Citation information: International Rice Research Institute (1989) Climate and food security. P.O. Box 933, Manila, Philippines.

Soil degradation in relation to climate

R. Lal

Soil degradation—the loss or decline in those properties of soil that affect its life-supporting processes—is affected by climate. That effect is frequently accentuated by the interventions and actions of man. The climatic factors responsible for soil degradation are rainfall (amount, intensity, kinetic energy, distribution), wind velocity, temperature and humidity, and insolation. Important land characteristics that interact with climate include soil texture, structure, and depth; terrain; and internal drainage. Climate-induced soil degradation processes include wind and water erosion, salinization and alkalization, acidification, reduction in soil organic matter content, decline in soil fauna, laterization, and compaction. Soil degradation is a dynamic process and the nature and extent of soil degradation due to a particular factor are difficult to quantify. Important factors in the processes that cause soil degradation include deforestation, expansion of cultivation to steep and marginal lands, intensive use of agrochemicals, irrigation, monoculture, and indiscriminate use of mechanization. The potential arable land area of the world is limited, making it important not only to reverse the degradation trend but also to restore the productivity of degraded lands.

Archaeological evidence suggests that many ancient civilizations vanished because of a decline in the productivity of the soils that supported them. The extinction of the Riparion and Harappan-kalibangan civilizations and the Mayan culture, and the abandoned slopes of the ancient Kingdom of Lydia, as well as the demise of many other civilizations, were the consequence of severe soil degradation (Olson 1981). It has been estimated that, for the 1.5 billion ha of currently cultivated land, an additional 2 billion ha were once biologically productive. It is also feared that each year, some 5-7 million ha of currently cultivated lands are being lost from agricultural production (UNEP 1986). If these estimates are anywhere near correct, the consequences are alarming and the challenge facing mankind is great.

In South Asia, the Himalayan-Tibetan Mountain ecosystem is one of the most severely degraded systems in the world (Dent 1984). In India, annual floods affect about 4.9 million ha of land in Assam, Bihar, Uttar Pradesh, and West Bengal. Dent (1984) estimates that siltation of a reservoir in northern India is about 200% greater than the design flow. In Nepal, it is estimated that 63% of the Shivalic zone, 86% of the Middle Mountain zone, 48% of the transition zone, and 22% of the high Himalayas have been reduced to poor and fair watershed conditions. The levels of

Terai river beds are estimated to be rising 15-30 cm annually. In Pakistan, a 20-yr-old survey of the Upper Indus Basin showed that 84% of the area already had moderate to severe erosion problems (Dent 1984). In China, as much as 46 million ha of the loess plateau that drains into the Yellow River are subject to degradation. Severe degradation problems also exist in the watersheds of the Yangtze, Huaihe, Pearl, Liaolie, and Songhua Rivers.

Arid lands make up about one-third of the world's total land area (Dregne 1976). The deserts around the world seem to be spreading. In India, arid lands occupy about 0.39 million km^2. The desert of Rajasthan has been spreading at the rate of about 1 km/yr, encroaching on about 130 km^2 of fertile land (Singh 1977). Other regions prone to severe degradation are the Ethiopian Highlands, the Andes, Haiti, Dominican Republic, and sub-Saharan Africa. Salinity is a severe problem in southwest and western Asia and in the western U.S. The soil in these and other regions is being rapidly degraded.

Definitions and basic concepts

The World Soil Policy (UNEP 1982) defines soil degradation as, "The decline in soil quality caused through its use by humans. Soil degradation includes physical, biological, and chemical deterioration such as decline in soil fertility, decline in structural condition, erosion, adverse changes in salinity, acidity or alkalinity, and the effect of toxic chemicals, pollutants, or excessive inundation." The FAO-UNEP definition is, "Soil degradation is the diminution of the current and/or the potential capability of soil to produce (quantitative or qualitative) goods or services as a result of one or more degradation processes."

Soil depletion means the eluviation of nutrients by water moving through the soil or removal through the produce harvested. This less drastic process is an initial stage of soil degradation.

To avoid ambiguity, it is important to identify the soil degradation caused by different processes. The critical limits of the soil properties that affect crop production must be quantified to delineate level of degradation. Those limits apparently are different for different soils, previous soil moisture regimes and climatic conditions, land use, crops, and agroecological regions. Without knowledge of the critical limits of organic matter content, water and nutrient status, porosity, and compaction for major soils and crops, it is difficult to judge whether a soil is degraded, and to what degree.

Significant progress has been made in defining the critical limits for salt content of alkaline and saline soils in relation to crop growth, and of toxic levels of Al and Mn for acid soils. However, research data are scarce that delineate certain physical processes of soil degradation in relation to crop growth (e.g. erosion, compaction, effective rooting depth, plant-available water reserves, etc.). For example, the quantity and quality of organic matter necessary to maintain an adequate structural condition is different for different soils and environments, and is not completely understood. The effective rainfall to grow a crop or the amount of lime required to neutralize a unit pH for soils of different properties and for crops of different physiological requirements are also not understood. If neither the critical levels nor the response of different crops to these levels for different soil and management conditions is known, we cannot be sure of the magnitude and trends in soil degradation.

Processes of soil degradation

Major climatic elements responsible for soil degradation are precipitation (amount, distribution, intensity), wetting and drying, wind velocity, temperature regime (including freezing and thawing), and evaporation. The effects are greatly accentuated by the perturbations caused by man. Man modifies the role of climate through many different activities, including deforestation, arable land use, grazing, and irrigation.

Climate-influenced processes that lead to soil degradation include surface and subsurface water flow, climatic aridity and evaporation, and water deficit. Different processes operate in different climatic regions, and more than one process can operate simultaneously in any given region.

Soil erosion by water

Climatic factors responsible for soil erosion by water are rainfall intensity, rainfall amount and distribution, snow melt, wind velocity, and hail. The turbulence caused by an impacting raindrop in shallow overland flow accentuates soil detachment. On steep and undulating terrains with soils of low structural stability, when rainfall intensity exceeds infiltration capacity, runoff rate and amount accelerate soil erosion. Snow melt in spring also causes severe erosion on steep terrains, especially when soil is already saturated. Erosion-induced soil degradation includes decreased organic matter content and plant nutrients, loss of fertile topsoil, exposure of subsoil with unfavorable physical and nutritional properties, and loss of water reserves available to plants.

The relative susceptibility of soils to erosion depends on differences in climatic erosivity. Climatic erosivity is determined by the kinetic energy of rainfall and of the overland flow. A high proportion of detachment is caused by the direct impact of a raindrop. Soil detachment by raindrop impact is influenced by raindrop size and mass, drop impact velocity (speed and direction), and depth of overland flow. Detachment increases with an increase in the depth of overland flow, to a maximum flow threshold approximately equal to raindrop diameter (Meyer 1981, Morgan 1980, Palmer 1963). At deeper depths, the overland flow influences erosion by contributing to runoff erosivity (Onstad and Foster 1975). A raindrop of about 4 mm diameter reaches a terminal velocity of 39.3 cm/s and possesses a maximum kinetic energy of 203 dynes. This energy is sufficient to lift a 1 cm^2 sand layer 1.3 mm thick about 6 cm (WMO 1983).

Rainfall intensity, kinetic energy, and drop size distribution are interrelated. Within certain limits, rainfall intensity increases with an increase in drop size distribution. Soil detachment is proportional to the square of rainfall intensity (Foster 1982) in

$$D_I = a\,i^2 \qquad (1)$$

where D_I = interrill detachment rate (kg/m^2·h), i = rainfall intensity (mm/h), and a is constant. Because soils differ in susceptibility to erosion, Foster (1982) related the interrill soil detachment to rainfall intensity according to the equation

$$D_I = 0.0138\ K_I\ i^2 \tag{2}$$

where K_I = soil erodibility factor for detachment by raindrop impact (kg·h/N·m^2). In general, rainfall intensity of less than 25 mm/h is considered nonerosive (Hudson 1971).

While the relative importance of rain momentum vs kinetic energy has been debated, it is generally accepted that kinetic energy is the major factor responsible for initiating soil splash. Young and Wiersema (1973) observed that lessening rainfall impact energy 89% without lessening intensity decreased soil loss 90% or more.

Kinnell (1981) described two separate forms of kinetic energy in relation to rainfall intensity: a) the rate of expenditure of the rainfall kinetic energy (E_{RR}) (which has units of energy per unit area per unit time), and b) the amount of rainfall kinetic energy expended per unit quantity of rain (E_{RA}) (which has units of energy per unit depth). Thus E_{RA} and E_{RR} are related to rainfall intensity as

$$E_{RA} = C\ E_{RR}\ i^{-1} \tag{3}$$

where C is an empirical constant. The E_{RA} can be calculated from rainfall intensity, as is done in the Universal Soil Loss Equation (Wischmeier and Smith 1978) from an empirical relation of the type

$$E_{RA} = a + b \log_{10} i \tag{4}$$

Wind accompanying rain influences the size and shape of impacting raindrops and their impact velocity. Disrud and Krauss (1971) observed that soil detachment from clods exposed to wind-driven rain was more than that caused by rain without wind. Hail also causes soil detachment. Even low-energy hail can easily increase soil detachment from clods 50%. Hail erosivity depends on size and intensity of hail, wind speed, and surface cover.

Soil erosion by water is often most severe in semiarid and subhumid regions. These are the regions where a prolonged dry season denudes the land of its protective vegetation cover and where intense rains are concentrated in a short period of 3-6 mo. In the subhumid and semiarid regions, accelerated erosion may be more severe in the tropics than in temperate regions because tropical rains are of high intensity and high energy load. The effect of accelerated erosion on crop productivity is more severe in tropical than in temperate zone climates.

Soil erosion by wind

Wind erosion is a severe problem in extremely arid, arid, and semiarid regions where the following conditions prevail: 1) loose, dry, finely divided soil; 2) smooth soil surface devoid of vegetative cover; 3) large fields; and 4) strong winds (FAO 1960). The climatic factors responsible for severe wind erosion are 1) quantity, distribution, and nature of the precipitation; 2) temperature regime; and 3) wind velocity.

Desiccation of soil surface and structural degradation facilitate easy displacement of soil by wind. Wind transports soil particles by suspension, saltation, or mass drift. Wind erosion is caused by turbulent wind with speeds exceeding a threshold of about 20-50 km/h, depending on the history of the soil (Chepil and Woodruff 1963). A wind speed of 20 km/h is considered nonerosive.

Threshold velocity V* differs for soil particles of different sizes. It is described as (WMO 1983)

$$V^* = A\sqrt{\frac{\sigma-\rho\cdot g\cdot d}{\rho}} \tag{5}$$

where σ is particle density (2.65 g/cm^3), ρ is density of air, i.e. (1.2×10^{-3} gcm^{-3}), g is acceleration of gravity, d is diameter of particle, and A is constant (0.1). Since $\rho >> \rho$ the equation is reduced to

$$V^2 = C\cdot A^2, (C = \sigma/\rho) \tag{6}$$

Equation 6 implies a linear relationship between threshold velocity and particle size diameter exceeding 0.1 mm. This relationship, however, is greatly influenced by turbulence.

Chepil et al (1962) also observed that the rate of soil movement by wind (q) is directly proportional to friction velocity cubed. Chepil et al (1962) postulated that q varies inversely to the square of the effective moisture index. The moisture index was defined by Thornthwaite (1948) using precipitation and potential evapotranspiration.

Antecedent soil moisture content plays an important role in the susceptibility of a soil to wind erosion. Wind erosion is significant where the surface soil is very dry. Wet soil has more cohesive resistance than dry soil because water binds the soil particles together. If all other factors are constant, soil is more susceptible to wind erosion following an extended period without rain.

Desertification

Desertification is defined as "the impoverishment of arid, semiarid, and subhumid ecosystems by the impact of man's activities. This process leads to reduced productivity of desirable plants, alterations in the biomass and in the diversity of life forms, accelerated soil degradation, and increased hazards for human occupancy" (UNEP 1977). In simple terms, it means "a change in the character of land to a more desert condition" (Mabbutt 1978).

Desertification is caused by severe wind erosion in arid and semiarid climates (Rapp 1974), and is feared to be spreading in the fringes of arid lands and deserts.

The climatic factors that lead to desertification include low and erratic rainfall, high evaporation losses resulting in low effective moisture content, high runoff losses caused by surface sealing and low infiltration rate, low vegetation cover and denudation, and strong and turbulent winds. The debatable issues have been whether the spread of desert into surrounding regions is caused by temporary drought, long-term climatic change toward aridity, or man-induced climatic change.

Although it is difficult to determine the precise cause-effect relationship, desertification is caused by over-exploitation and misuse of marginal resources (Goudie 1981).

An important climatic index to assess desertification is the Budyko ratio, or the "rational index of dryness." The Budyko ratio is defined as the ratio between annual net radiation and mean annual precipitation at the surface (Budyko 1985). The radiation index of dryness is another possibility for quantifying climatic parameters related to desertification (Budyko 1974), as

$$D = R/LP \tag{7}$$

where L is latent heat of vaporization (600 cal/g^{-1}), P is rainfall (cm), and R is radiation (1 yr^{-1}).

Desertification has not been defined in quantitative terms. The critical limits of effective moisture content in relation to biological productivity of different plant species are not known. It is difficult to estimate reliably the extent of area and the trends without understanding the critical limits of important life-supporting processes of the soil and climatic variables involved. In general, desertification is a problem of the Sahel and other regions lying on the periphery of major deserts (Mabbutt 1978).

Salinization and alkalization

Accumulation of excessive amounts of soluble salts in the root zone to levels toxic to plant growth is caused in arid and semiarid regions by an excess of evaporation over precipitation. The accumulated salts are usually chlorides, sulfates, and carbonates of sodium, magnesium, and calcium. If the predominant cation accumulated in the soil is sodium, the process is also called sodication. High concentrations of Na^+ on the exchange complexes disperse the clay, which may eluviate to the subsoil and form a horizon of massive structure with low water permeability. The source of soluble salts may be indigenous, either from the parent material or from groundwater. The salts also may be brought in with irrigation water and as fertilizers and other amendments. In coastal areas, salt comes either from sea encroachment or is blown inland by wind.

Whatever the source, soluble salts alter the soil profile's physical and chemical properties. A soil is characterized as saline if the conductivity of saturated soil paste exceeds 4 s/cm and as alkaline if the conductivity exceeds 8 s/cm at 25 °C. Alkali soils occur extensively in regions with a mean annual rainfall between 550 and 1,000 mm. In addition to the altered soil structure, plant-water availability is severely curtailed due to high osmotic pressure.

The climatic factors responsible for a salt imbalance include those that reduce or eliminate leaching. Salt accumulation does not occur in humid and subhumid conditions because soluble salts are leached out of the root zone during the rainy season. Climatic aridity, an important factor for salt enrichment, is attributed to low precipitation, high evapotranspiration, high temperature, and low humidity. Saline soils are easily formed in those regions where the ratio of precipitation to potential evapotranspiration (P/PET) is less than 0.75.

Salinity also occurs in irrigated regions of northwestern India and in Pakistan, southern and southeastern Australia, and California. In Pakistan, 1.9 million ha are severely saline and another 4.5 million ha have saline patches (Snelgrove 1967). Alkali soils cover approximately 2.5 million ha in the Indo-Gangetic plains (Narayana and Abrol 1981). Worthington (1977) observed that salt-affected soils amount to 50% of the irrigated area in Iraq, 23% in Pakistan, 50% in the Euphrates valley of Syria, 30% in Egypt, and more than 15% in Iran.

Leaching

Leaching, the reverse of salt accumulation, occurs under very humid conditions in soils with predominantly low-activity clays and free drainage. Leaching is a natural process in soil evolution, similar to soil erosion. However, excessive leaching leads to soil degradation through depletion of bases and clay minerals. Accelerated leaching involves the loss of bases such as calcium, sodium, and magnesium, rendering the soil acidic in reaction, it often replaces bases with exchangeable aluminum.

The most important climatic factor influencing leaching is rainfall amount and spatial and temporal distribution. FAO (1978) proposed a climatic index to assess soil degradation by leaching, for the humid season when P>PET

$$\text{Leaching index} = \sum_{1}^{12} (\text{P-PET}) \tag{8}$$

or, where R is soil moisture reserve,

$$\text{Leaching index} = [\sum_{1}^{12} (\text{P-PET})] - \text{R} \tag{9}$$

The soluble cation may leach out of the root zone in solution or as absorbed/adsorbed cations on the exchange complex of eluviating colloids, mainly clay. Leaching may be vertical and/or oblique, depending on the relative proportion of percolating water moving vertically down the soil profile or horizontally along a gradient. Lateral or oblique drainage often occurs in layered profiles with drastic differences in permeability among the horizons. An eluviated or leached horizon loses bases and a part or most of its clay. The loss of bases and clay alters both physical and chemical properties. Leached soil, therefore, has low chemical fertility but also low plant-available water reserves. The exchange complex is dominated by aluminum.

Leached soils occur in very different climatic regions (WMO 1983): 1) in temperate Atlantic climates in soils with a differentiation of textural B horizon, 2) in humid temperate climates through the process of podzolization caused by acid litter, 3) in subtropical climates where most rainfall is received in winter, 4) in tropical climates with a seasonal humid moisture regime that leads to the formation of Alfisols, and 5) in equatorial climates that form highly leached tropical Ultisols and Oxisols.

Laterization

In semiarid West Africa, a considerable part of the uplands is characterized by an iron and/or manganese-rich hard crust at variable depths. The pedogenesis of this layer is attributed to fluctuating groundwater at this depth, causing alternating oxidation and reducing conditions, thereby solubilizing the iron and manganese present in the soil. This layer is rich in iron and manganese because of the preferential removal of silica during extensive weathering, which leads to the accumulation of sesquioxides. When it lies within the soil body and is protected by vegetation cover, this layer is soft and is called "plinthite." When the layer is exposed to dry and hot conditions (Macfarlane 1976), the soft material hardens into a rocklike crust popularly known as laterite. Ultradesiccation following deforestation causes hardening of soft plinthite into hard crust (Goudie 1973). Laterized horizons are hard and compact and cannot be cultivated. Severe declines in soil productivity are caused by the process of laterization. Gourou (1961) observed that "laterite is a pedological leprosy."

Sanchez and Buol (1975) reported that plinthite occurs in less than 7% of the total land area of the tropics. Van Wambeke (1978) observed that plinthite occurs in less than 2% of the Amazon. The hardened plinthite is, however, widely observed in subhumid and semiarid regions of West Africa. Obeng (1978) estimated about 250 million ha of iron-pan soils in the West African savanna. Similar soils occur extensively in the semiarid tropics of India and Southeast Asia.

The climatic factors responsible for laterization include 1) intense weathering leading to removal of silica and accumulation of sesquioxides, 2) seasonally wet tropical conditions characterized by an intense rainy season followed by a prolonged dry season, and 3) hot and dry environments leading to accelerated soil erosion. Tropical rainforest is the original vegetation of plinthite soils. The savanna cover seen now is the result of forest degradation. In fact, the transformation of forest into savanna vegetation contributes to the formation of extremely hard crusts (Mohr et al 1972).

Physical degradation

Deterioration in soil structure is a severe problem everywhere in the world. Decline in soil structure eventually leads to surface sealing, crusting, compaction, hard-setting, reduction in permeability, poor aeration, and waterlogging. Structurally inert soils, containing predominantly low-activity clays and low organic matter content, are prone to hard-setting and general physical degradation. Lack of organic matter content and a high proportion of silt are responsible for crust formation. FAO (1978) developed this index to characterize soils with these properties:

$$\text{Crusting index} = \frac{\%\ \text{fine silt} + \%\ \text{coarse silt}}{\%\ \text{clay}} \tag{10}$$

This index will exceed 2.5 for soils prone to intense crusting. An index based on soil organic matter content also used by FAO (1978) is

$$\text{Crusting index} = \frac{1.5\ (\%\ \text{fine silt}) + 0.75\ (\%\ \text{coarse silt})}{\%\ \text{clay} + 10\ (\text{organic matter})} \tag{11}$$

This index will exceed 2 for soils prone to intense crusting.

Soils prone to physical degradation occur in seasonally moist regions with intensely hot summers. Hard-setting soils are widespread in the dryland regions of Australia (McDonald et al 1984, Northcote et al 1975), in West African savanna (Charreau and Nicou 1971, Jones and Wild 1975), in Botswana (Sinclair 1985), and in Zambia (Veldkamp 1986). The climatic factors leading to physical degradation are intense rains, high temperatures during the dry season, and ultradesiccation.

Biological degradation

Biological degradation refers to the loss of organic matter, reduction in biomass carbon, and decline in the biotic activity of soil fauna. The climatic factors responsible for biological degradation are water deficit and high temperatures. The decomposition rate of soil organic matter is doubled for every 10 °C increase in mean temperature. The climatic index used by FAO to assess biological degradation is

$$K = \frac{1}{12} \sum_{1}^{12} e^{0.1065t} \times \frac{P}{PET} \text{ (with P<PET)} \qquad (12)$$

where t is mean air temperature during the month, P is precipitation, PET is potential evapotranspiration, and K is rate of humus decay in percent by year. If $P > PET$, $P \div PET = 1$, and if $t < 0$, $t = 0$.

Decline in biotic activity of soil fauna, caused by alterations in soil temperature and moisture regimes and decrease in food availability and diversity, is responsible for physical degradation of soil. A soil devoid of macro- and microfauna is easily crusted and compacted and has poor aeration and impeded drainage. High temperatures, low moisture, and climatic aridity cause a decline in soil biotic activity.

Man-climate interactions

Through judicious management, man can reverse climate-induced soil degradation. But historically, man has exploited natural resources and greatly intensified the rate of degradation. Perturbations caused by man have influenced climate on local, regional, and global scales.

Deforestation

Removing existing vegetation cover alters the soil water balance and energy balance. Deforestation increases surface runoff and decreases soil-water storage. Without the moderating effect of forest on climate, diurnal and seasonal fluctuations in temperature and humidity are greatly accentuated. Notable effects of deforestation on soil are accelerated erosion, soil compaction and physical degradation, biological degradation, and laterization.

Vegetation and soil management technologies exist that permit the productive use of existing forest and its conversion to arable land without attendant adverse effects.

Expansion of arable land area

Soil degradation is increased if marginal lands are brought under cultivation. The thin topsoil is vulnerable and easily eroded, resulting in barren and denuded hills prone to severe gully erosion. Severe problems of accelerated erosion due to land misuse are evident in Nepal, Ethiopia, Haiti, Dominican Republic, and many other lands that should have never been cultivated. Severe wind and water erosion are caused by intensive cultivation.

This is not to say that these lands should be left unused: appropriate land use systems are available that permit economic returns from marginal lands without causing degradation.

Irrigation

Expansion of irrigation into semiarid regions has affected the soil water balance and salt balance. Waterlogging and the spread of salinization and alkalization are directly attributable to the development of canal irrigation in arid and semiarid regions. Development of grandiose irrigation schemes involving creation of large man-made lakes has created other types of ecological problems.

Drainage

Both surface and subsurface drainage have been used to alter soil-moisture regimes and improve aeration. The most impressive drainage system, installed to reclaim land from the sea, is in the Netherlands. Large areas of marshlands and flood lands have been reclaimed. Drainage systems to improve aeration of agricultural land are widely used in Western Europe and in the Midwestern U.S.

Excessive drainage of some soils can cause soil degradation. Drainage of peat soils changes the temperature and water balance and can cause their oxidation and eventual depletion. Drainage of mangrove swamps causes soil acidity to develop to levels toxic for plant growth.

Fertilizers and chemical amendments

Biological degradation, soil acidification, and physical degradation are enhanced by the intensive land use facilitated by the use of fertilizers and other agrochemicals. Chemical amendments are used to manipulate the soil chemistry. Although fertilizers and chemical amendments are essential for high yield, their indiscriminate and excessive use leads to severe problems.

Conclusions

Soil degradation is caused by various processes that are triggered by climatic factors. Soil erosion by wind and water are caused by rainfall amount, rainfall distribution, evapotranspiration, and wind velocity. Salinization and alkalization are caused by insufficient water to leach excess soluble salts out of the soil profile. Chemical degradation and acidification are the results of excessive leaching of the bases and other cations. Removal of tropical rain forest and cultivation has resulted in hardening of plinthite and widespread distribution of laterization in subhumid and semiarid tropics. Biological degradation is caused by the prevalence of year-round high temperatures.

Man's activity can either accelerate the degradative processes or reverse the trend. Indiscriminate deforestation, bringing marginal and steep slopes under arable land use, excessive use of agrochemicals, and irrigation by poor-quality water accentuate soil degradation. Productivity of degraded soils is drastically curtailed.

Soil is a finite, nonrenewable resource. Through judicious land use and adoption of management systems that do not cause gross imbalances in water and energy cycles of natural ecosystems, man can reverse the trends and restore the productivity of degraded lands.

References cited

Budyko M I (1958) The heat balance of the earth's surface. U.S. Department of Commerce, Office of Technical Services, Washington, D.C. 259 p.

Budyko M I (1974) Climate and life. Academic Press, New York.

Charreau C, Nicou R (1971) L'amelioration du profil cultural dans les sols sableux et sablo-argileux de la zone tropicale seche ouest-africaine et ses incidences agronomiques. Agron. Trop. 26:209-255.

Chepil W S, Siddoway F H, Armburst D V (1962) Climatic factor for estimating the wind erodibility of farm lands. J. Soil Water Conserv. 17:162-165.

Chepil W S, Woodruff N P (1963) The physics of wind erosion and its control. Adv. Agron. 15:211-302.

Dent F J (1984) Land degradation: present status, training and education needs in Asia and the Pacific. UNEP investigations on environmental education and training in Asia and the Pacific. Food and Agriculture Organization Regional Office, Bangkok, Thailand. 20 p.

Disrud L A, Krauss R K (1971) Examining the process of soil detachment from clods exposed to wind-drawn simulated rainfall. Trans. ASAE 14:90-92.

Dregne H E (1976) Soils of the arid regions. Elsevier Scientific Publ. Co., Amsterdam. 237 p.

FAO—Food and Agriculture Organization (1960) Soil erosion by wind and measures for its control on agricultural lands. FAO Development Paper 71. Rome, Italy.

FAO—Food and Agriculture Organization (1978) A provisional methodology for soil degradation assessment. Rome, Italy. 84 p.

Foster G R (1982) Modelling the soil erosion process. Pages 297-382 *in* Hydrologic modeling of small watersheds. ASAE, St. Joseph, Micnigan.

Goudie A S (1973) Duricrusts of tropical and subtropical landscapes. Clarenden Press, Oxford.

Goudie A S (1981) The human impact: man's role in environmental change. The MIT Press, Cambridge, Massachusetts.

Goudie A (1984) The human impact: man's role in environmental change. The MIT Press, Cambridge, Massachusetts. 326 p.

Gourou P (1961) The tropical world; its social and economic conditions and its future status. 3d ed. Longmans, London. 159 p.

Hudson N W (1971) Soil conservation. B. T. Batsford, Ltd., London, England.

Jones M J, Wild A (1975) Soils of the West African savanna. CAB Harpenden, United Kingdom.

Kinnell P I (1981) Rainfall intensity-kinetic energy relationships for soil loss prediction. Soil Sci. Soc. Am., Proc. 45:153-155.

Mabbutt J A (1978) The impact of desertification as revealed by mapping. Environ. Conserv. 11:103-113.

McDonald R C, Isbell R F, Speight J G, Walker J, Hopkins M S (1984) Australian soil and land survey field handbook. In Kata Press, Melbourne. 100 p.

Macfarlane M J (1976) Laterite and landscape. Academic Press, London.

Meyer L D (1981) How rain intensity affects inter-rill erosion. Trans. ASAE 24:1472-1475.

Mohr E C J, Van Baren F A, Van Schylenborth J (1972) Tropical soils. Mouton-Ichtiar Barn Van Hoeve, The Hague. 481 p.

Morgan R P C (1980) Field studies of sediment transport by overland flow. Earth Surface Processes 5:307-316.

Narayana V V, Dhruva Abrol I P (1981) Effect of reclamation of alkali soils on water balance. Pages 283-298 *in* Tropical agricultural hydrology. R. Lal and E. W. Russell, eds. John Wiley and Sons, Chichester, United Kingdom.

Northcote K H, Hubble G D, Isbell R F, Thompson C H, Bettenay E (1975) A description of Australian soils. CSIRO Div. Soils, Canberra, Australia.
Obeng H B (1978) Soil, water, management and mechanization. Afr. J. Sci. 5:71-83.
Olson G W (1981) Archaeology: lessons on future soil use. J. Soil Water Conserv. 36:261-264.
Onstad C A, Foster G R (1975) Erosion modeling on a watershed. Trans. ASAE 18:288-292.
Palmer R S (1963) The influence of a thin water layer on water drop impact forces. IAHS Bull. 65:141-148.
Rapp A (1974) A review of desertization in Africa-water, vegetation and man. Stockholm Report 1. Secretariat for International Ecology, Stockholm.
Sanchez P A, Buol S W (1975) Soils of the tropics and the world food crisis. Science 188:598-603.
Sinclair J (1985) Crusting, soil strength and seedling emergence in Botswana. Ph D thesis, Aberdeen University, United Kingdom.
Singh G (1977) Climatic changes in the Indian desert. Pages 25-30 *in* Desertification and its control. New Delhi, India.
Snelgrove A K (1967) Geohydrology of the Indus River, West Pakistan. Sind University Press, Hyderabad.
Thornthwaite C W (1948) An approach towards a rational classification of climate. Geogr. Rev. 38:55-94.
UNEP—United Nations Environment Programme (1977) Desertification: an overview. Paper presented at the United Nations Conference on Desertification, 29 Aug-9 Sep 1977, Nairobi, Kenya.
UNEP—United Nations Environment Programme (1982) World's soil policy. Nairobi, Kenya.
UNEP—United Nations Environment Programme (1986) Farming systems principles: for improved food production and the control of soil degradation in the arid, semi-arid and humid tropics. Paper presented at the Experts' Meeting sponsored by the UNEP, 20-30 June 1973, International Crops Research Institute for the Semi-Arid Tropics, Hyderabad, India. 36 p.
Van Wambeke A (1978) Properties and potentials of soils in the Amazon Basin. Interciencia 3:233-242.
Veldkamp W J (1986) Soil map of Zambia. Soil survey of Mt. Makulu. [Quoted by C. E. Mullins 1986.] Hard-setting soils. Aberdeen University, United Kingdom. 14 p. (mimeo.)
Wischmeier W H, Smith D D (1978) Predicting rainfall erosion losses: guide to conservation farming. USDA Handb. 282. Washington, D.C. 58 p.
WMO—World Meteorological Organization (1983) Meteorological aspects of certain processes affecting soil degradation, especially erosion. Tech. Note No. 178, WHO, Geneva, Switzerland.
Worthington E B, ed. (1977) Arid land irrigation in developing countries: environmental problems and effects. Pergamon, Oxford.
Young R A, Wiersma J L (1973) The role of rainfall impact in soil detachment and transport. Water Resour. Res. 9:1629-1639.

Notes
Author's address: R. Lal, International Institute of Tropical Agriculture, PMB 5320, Ibadan, Nigeria.
Citation information: International Rice Research Institute (1989) Climate and food security. P.O. Box 933, Manila, Philippines.

Coping with climatic variation in plant disease control

J. C. Zadoks

Changes in agricultural practices, sometimes with great phytopathological and economic consequences, are triggered by forces originating within human society or imposed on it by changing climate. Social, economic, and technical changes in society lead to changes in crops and cropping methods, and in the spectrum and pattern of plant diseases. Methods have been developed for short-term warnings and for long-term projections about diseases. For the short term, strategic (preplanting) and tactical (in-season) forecasts have been developed. For long-term projections, three methods are available: 1) time-series analysis, which relates disease and weather phenomena over a number of years; 2) geophytopathology, which analyzes climatic maps and delineates areas where conditions are optimal for a particular disease; and 3) extended surveys of large numbers of fields grown under a variety of conditions, analyzed to predict future trends.

The "climatic system" (Flohn and Fantechi 1984) studied by climatologists, and the "weather machine" (Calder 1974) studied by meteorologists, together determine the long-term and short-term possibilities and constraints of agriculture. Pests and diseases are constraints that are highly responsive to changes in weather and climate. Because climatic change is a fact of life, the questions to be addressed here are straightforward: How did phytopathologists cope with climatic change in the past? How do phytopathologists cope with climatic change today? The answers are less straightforward than the questions.

The history of phytopathology reaches back some two centuries, a period during which climatic changes have occurred in several areas. In fact, many changes of phytopathological interest have been observed, but as a rule they have not been of climatic origin, although climatic effects might have aggravated disease (A. Bourke in Flohn and Fantechi 1984). Changes in human society, sometimes rapid and clearly felt, sometimes slow and nearly imperceptible, have changed crops and cropping methods and therewith the spectrum and pattern of diseases. The word spectrum refers to the species composition of the pathogens (viruses included) and their frequency distribution; the word pattern refers to their distribution in space and time.

Man-made changes

Two examples will be given, one old and one of recent origin. Toward the end of the 18th century, intensification of agriculture was needed in northwest Europe to feed its growing population. In several countries, changing economic and social relations led to a redivision of the land. In Denmark, fencing became compulsory in 1794 (Hermansen 1968). An exotic shrub, the barberry (*Berberis vulgaris*), was recommended by the Royal Agricultural Society of Denmark. Severe epidemics of stem rust (*Puccinia graminis*) on rye, the leading cereal of that time, followed in the period 1800-05. Initially, Danish authorities refused to accept a relationship between fencing by barberry hedges and crop loss by rust. The problem was solved by the gradual eradication of barberry, later enforced by law. Climate had little to do with the problem, which was triggered by social change.

In the Netherlands of today, farmers are under heavy pressure to maintain a decent income. Because land to enlarge their farms is not available, their response has been to specialize in one or two cash crops and one or two filler crops. The rotations have been narrowed, so that potato, the number one cash crop, is grown every second year New problems have appeared. Potato cyst nematodes (*Heterodera rostochiensis, Globodera pallida*) have surged up, and hitherto unknown soil-borne problems have appeared, some of them caused by fungi thought to be rather harmless. In sugar beet, the second cash crop, cyst nematodes and rhizomania (a fungus-transmitted virus disease) endanger production. Economic pressures induced the problems, which have little to do with climatic change.

Plant breeding can have unexpected side effects. There is at least one developing country on record where the problem of stem rust (*Puccinia graminis*) in wheat was solved by resistance breeding. There was no relief from disease, however, because the newly introduced varieties had little resistance against leaf rust (*Puccinia recondita*), which became nearly as damaging as the stem rust had been. Again, resistance breeding solved the problem, but then stripe rust (*Puccinia striiformis*) caused great damage. Over a narrow zone stretching from Morocco and Spain in the west to Thailand in the east, newly introduced high-yielding varieties induced stripe rust problems. Because the temperature preferences of stem, leaf, and stripe rust go from high to low temperature, it is tempting to explain the phenomenon by hypothesizing climatic change. But the real cause of the problem was the introduction of new germplasm. The problem was aggravated by prolonged periods of suitable weather, from 6 to 18 mo, throughout the zone, including the Netherlands (Zadoks 1961), a temperate Atlantic country, and Spain (Nagarajan et al 1984), a Mediterranean country.

Epidemiological theory (Zadoks and Schein 1979) summarizes the man-made changes that increase the risk of devastating epidemics: 1) enlargement of fields, 2) aggregation of fields, 3) increase in the density of host plants, 4) decrease of genetic diversity of host plant populations, 5) increase in specialization leading to short rotations, 6) increase in genetic vulnerability as a side-effect of plant breeding, 7) increase of fertilizer application, 8) widespread application of irrigation, and 9) international exchange of contaminated plant material. Both the spectrum and the pattern of pests and diseases are greatly affected by these changes, which are normal corollaries of development. They took place gradually in western agriculture, and thus were not so destructive. But the Green Revolution in Southeast Asia caused them to occur rapidly and to take society by surprise.

Three points must be stressed: First, plant disease impact is affected by climate and will remain so. Plant disease impact maps are rare. Those published in the European Cereal Atlas (Zadoks and Rijsdijk 1984), and thought to be valid for the 1960s, still seem useful in the 1980s. There are, however, a few notorious exceptions. The pattern of the aphid pests changed completely, probably due to the intensification of wheat growing in Europe, and especially to highly increased N application.

Second, whenever and wherever a crop is grown for maximum yield, after correcting the deficiencies of soil and climate by modern technology, the within-crop microclimate is suitable for the major diseases of that crop. For example, potato late blight caused by *Phytophthora infestans* can ravage the crop in the humid, temperate-atlantic climate of the Netherlands as well as in the semidesert climate of the northern Negev in Israel, where potato is grown under irrigation.

Third, irrigation changes the cropping situation. New crops appear. The spectrum and the pattern of diseases change dramatically (Palti 1981, Rotem and Palti 1969), and diseases flourish.

Climate-induced changes

It is thus difficult to disentangle man-made from climate-induced changes. There is at least one case on record where climatic change is supposed to have aggravated a problem. This is the occurrence of stripe rust on wheat in the Pacific Northwest of the USA. Coakley (1979) argued that slight but significant changes in seasonal temperature and precipitation during periods spanning one or a few decades reduced or enhanced the risk of stripe rust outbreaks. Alas, such meticulous statistical analysis of long-term data sets, difficult in itself, is rarely possible because long-term phytopathological data are scarce.

Phytopathologists cannot yet match the elegance of the analysis by ecologists of the risk of oat crop failure along the slope of a mountain in Scotland. The probability distribution of crop failure in relation to the altitude of the crop was shifted about 100 m downward when the relatively cold period of 1661-1710 was compared with the relatively warm period of 1931-80 (Parry and Carter 1985).

In intensive agriculture, climatic variation does affect the spectrum and pattern of diseases, but the effect of climate becomes less pronounced due to modern technology. Climatic variation should not be confounded by variation in the origin of diseases. Many diseases supposedly originated in the gene centers of their respective hosts. Most crops were transferred to other areas, even other continents, leaving their disease behind them. Human frailty caused reencounters between host and pathogen in the newly cropped areas, often with devastating results. Potato late blight caused havoc in Europe some 200 yr after the introduction of the potato itself, as did southern maize rust (*Puccinia polysora*) in Africa. These human errors have nothing to do with climatic change.

Short-term predictions

Two approaches to short-term predictions can be distinguished: tactical forecast and strategic disease forecast. Tactical forecasts are issued during the growing season. They aim at warning the growers in time to take action to protect their crops. Numerous methods have been developed, based on a variety of principles (Zadoks 1984). Three major principles are 1) predictions based on actual field monitoring, extrapolating the present disease situation to the future; 2) predictions based on weather forecasts, particularly suitable for downy mildews; and 3) various value point systems, where a number of circumstances are given a certain numerical value, and these values are added together to provide a sum, which is handled as a kind of risk assessment. Several of the more successful systems combine elements of different principles mentioned here. Some systems are completely computerized (Rabbinge and Rijsdijk in Gallagher 1984, Spaar and Ebert 1985, Zadoks 1989). Warning systems for fungal diseases have a history of some 70 yr; those for viral diseases are of younger date (Thresh 1985, McLean et al 1986).

The value point system provides a transition to strategic forecasts, made before planting. These may influence the date of planting, the simultaneity of planting dates, the variety chosen, and the timing of fertilizer and pesticide treatments. Typical differences exist between England and the Netherlands; in both countries, advanced growers can easily produce 8 t winter wheat/ha—in England by early planting and up to 12 or more pesticide treatments, in the Netherlands by comparatively late planting and 4 pesticide treatments. Insurance treatments, among which seed dressing is considered indispensable, are in fact strategic decisions. In Europe, computer programs are now being developed for strategic decisionmaking, in part with the explicit objective to reduce insurance treatments.

In developing countries, it was not unusual that planting began only when the village elders had decided that the time was right. There may be great wisdom in such a system, as simultaneity of planting can minimize losses due to pests and diseases (Kiritani and Duffus in Plumb and Thresh 1983). Modern technology includes staggered planting, which provides diseases, disease vectors, and pests the opportunity to breed and multiply continuously. The harmful agents march over a "green bridge" (Zadoks in Gallagher 1984), spanning the gap from one crop season to the next. A crop-free period or break is a cheap way to control rats, insect pests, insect vectors of viral diseases, and fungal diseases. Regulatory action may be needed to make the green bridge long, narrow, and difficult to pass. Climatic considerations determine the timing of the break.

Strategic predictions based on weather data are gradually becoming feasible. Crop management practices in the Sahel might be attuned to precipitation forecasts (Franquin 1984), with crop protection practices included. The time-series analyses of stripe rust in the USA effectively led to strategic warnings about outbreaks, providing lead time to prepare large-scale chemical control (Coakley et al 1983). In

India, the "Indian stem rust rules" for wheat are a beautiful example of strategic warning based on satellite imagery recording the course of tropical cyclones (Nagarajan and Singh 1975, Nagarajan et al 1976). Long-term weather conditions during the off-season often determine the length and width of the green bridge and thus the amount of inoculum available at the beginning of the growing season, as in the stripe rust example. Such knowledge is exploited to estimate the risk during the coming season, especially in the USSR and the German Democratic Republic (Ebert and Poljakov 1981).

Long-term projections

Long-term projections of the disease situation of crops, including mean values and span of variation, may cover periods of 5-25 yr and have a strategic nature. Long-term projections seem to be applied in the USSR (Ebert and Poljakov 1981). Another sector of phytopathology involved in long-term projections is plant quarantine, the objective of which is to prevent harmful agents from following their hosts to new areas. Various national and international plant protection agencies have made long lists of exotic harmful agents and of the damage expected after their introduction. American workers have tried to estimate the relative risks of introducing many exotic harmful agents and the magnitude of the potential losses (MacGregor in Horsfall and Cowling 1978).

I know of no other long-term projections. Fortunately, the methodology is available. Three approaches are envisaged: The first is the time-series analysis already mentioned. The second, which also relies on desk research, is labeled geophytopathology. The third, which relies on extended field surveys, uses a little trick—momentaneous variation is interpreted as temporal variation.

The technique of geophytopathology is rather straightforward (Weltzien 1972, Weltzien in Horsfall and Cowling 1978). The environmental requirements of a particular pathogenic fungus are analyzed in detail, taking into regard all phases of the infection cycle and all relevant environmental aspects, including temperature, humidity, dew, and radiation. Thus, it is possible to identify the climate most suitable for the pathogen. The areas where such a climate exists can be identified using climatological maps. If the host is or will be grown in such areas, the disease under study may be expected there sooner or later. Predictions made for the appearance of powdery mildew (*Erysiphe betae*) of sugar beet in California came true. The technique has been used in the geographical context, but it could also be used in a temporal context to project trends.

The technique is based primarily on desk research, but experimental work may be needed if the available data set is incomplete. If, for a particular geographic area and a particular expected climatic change, the agronomists provide a projection of cropping pattern changes, the phytopathologists should be able to make projections of disease spectra and patterns. The methodology is available in principle; it may need some modification for application to a particular case. It could be complemented by risk analysis, indicating the probability of undesirable events. As yet, such risk analysis is rare (Anderson et al 1985).

The survey technique involves a special application of disease and pest surveys (Teng and Krupa 1980). As many fields as possible are visited during several successive seasons, and relevant agronomical and phytopathological observations are made on the spot, eventually complemented by an inquiry among the growers. If the variation among regions, growing conditions, and years is sufficiently large, some statistically significant trends may be distinguished. Although these trends only reflect the available database, they may be used for long-term projections when the climatologists indicate the expected changes in climate and the agronomists indicate the concomitant changes in cropping pattern.

A good example is a 3-yr survey of groundnut foliar diseases in peasant fields in the Ivory Coast (Savary 1986). The geographical variation ranged from the tropical rain forest to the sub-Sahel savanna. The annual weather pattern ranged from relatively wet to relatively dry years. The agricultural conditions ranged from very low to medium-high inputs. The analysis of the survey provided a statistical "itinerary" from low-input to high-input agriculture, with its phytopathological consequences as follows: With increasing inputs, cercospora leaf flecking disease (*Cercospora arachidicola*) decreased in importance, whereas groundnut rust (*Puccinia arachidis*) increased in importance. The results allow the following long-term projection: Because in due time peasant agriculture will be intensified, making optimal use of the available water by applying suitable tillage techniques and inputs such as fertilizer and herbicides, the crops will be more frequent, more dense, and more lush, and so they will become more vulnerable to groundnut rust. If this long-term projection is accepted by the authorities, it should induce them to organize three lines of defense: 1) breeding resistant cultivars, 2) investigating cultural methods that could avoid or limit rust epidemics, and 3) testing fungicides to keep the rust under control.

For viral diseases in annual crops, the problem is more complex, because there are not only crops and pathogens, but also vectors. A weak point, especially in the tropics, is our limited knowledge of vector behavior. Several insect vectors are polymorphic, some forms spreading only locally but others migrating over considerable distances. Vector behavior may be affected by sudden changes in cultural methods or host germplasm. This is one explanation for the unexpected and often widespread outbreaks of rice viral diseases in Southeast Asia.

Long-distance migration of fungal diseases does occur frequently (Zadoks 1973). The groundnut survey in the Ivory Coast provided circumstantial evidence for long-distance migration of groundnut rust (up to 500 km). Insect vectors of plant viruses can also be dispersed over hundreds of kilometers (Rosenberg and Magor in Plumb and Thresh 1983). Such evidence complicates long-term projections, because first, the long-distance migration possibilities must be detected, and second, the long-term projections can no longer be limited to a "target area" only but must incorporate the "source area" too.

Discussion and conclusions

The foregoing considerations are based on experience with, and are applicable to, annual crops only. They largely refer to airborne diseases. For soilborne diseases of annual crops due to fungi, bacteria, nematodes, or soilborne viruses, there exists a large body of knowledge. The phytopathological principles discussed apply also to perennial crops, but the resulting patterns may be different because of the absence of crop-free breaks and the relatively slow change in the genetic makeup of the crop. Numerical calculation of crop yields under a variety of environmental and cultural conditions, and of crop losses due to harmful agents, have become feasible (Van Keulen and Wolf 1986). It can also be applied to the present issue on the possible effects of climatic change.

The first conclusion is that the past provides us with many examples of phytopathological change. Most of these can be attributed to social and economic changes in human society, several to human error, a few to exceptional weather constellations lasting for several months, and maybe only one to climatic change.

The second conclusion is that phytopathologists have developed methods to deal with the immediate future by short-term forecasts, and with the far future by long-term projections. For long-term projections, the methods available are 1) time-sequence analysis, 2) geophytopathology, and 3) extended surveys.

A third conclusion emerges: Long-term projections, although aimed at a certain target area, have to consider national as well as international developments.

The fourth and final conclusion is that, whenever the responsible authorities set a goal, including target area, expected climatic change, and ensuing changes in crops and cropping practices, phytopathologists are ready, in principle, to produce long-term projections on diseases, their spectra and patterns, and the risks involved. The exercise will require research efforts to be measured in tens of person-years.

Let me finish with a suggestion: I am convinced that international cooperation and funding are needed to establish and maintain a specialized international research unit to study the long-term impact of climatic change on crop diseases. The pathology sections of the international institutes supported by the Food and Agriculture Organization, and consultants from various disciplines could join hands to begin to establish a common methodology. Such an effort will be a worthy preparation for phytopathologists to enter the twenty-first century.

References cited

Anderson W C, Guldin R W, Vasievich J M (1985) Risk assessment of investments in loblolly pine plantations threatened by bark beetles. Pages 328-334 *in* Proceedings of the integrated pest management research symposium. USDA Forest Service General Technical Report 50-56. New Orleans, Louisiana.

Calder N (1974) The weather machine. BBC, London. 143 p. Coakley S M (1979) Climate variability in the Pacific Northwest and its effect on stripe rust disease of winter wheat. Climatic Change 2:33-51.

Coakley S M , Line R F, Boyd W S (1983) Regional models for predicting stripe rust on winter wheat in the Pacific Northwest. Phytopathology 73:1382-1385.

Ebert W, Poljakov I J, eds. (1981) Ueberwachung und Prognose - Grundlagen eines gezielten Pflanzenschutzes. Akademie der Landwirtschaftswissenschaften der Deutschen Demokratischen Republik, Berlin. 288 p.

Flohn H, Fantechi R (1984) The climate of Europe: past, present and future. Natural and man-induced climatic changes: a European perspective. Reidel, Dordrecht. 356 p.

Franquin P (1984) Agroclimatologie et agrometeorologie en zone tropicale seche d»Afrique. Agron. Trop. 39:301-307.

Gallagher E J, ed. (1984) Cereal production. Butterworths, London. 354 p.

Hermansen J E (1968) Studies on the spread and survival of cereal rust and mildew diseases in Denmark. Friesia 8(3):1-206.
Horsfall J G, Cowling E B, eds. (1978) Plant disease. An advanced treatise. Vol. II. How disease develops in populations. Academic Press, New York. 436 p.
McLean G D, Garrett R G, Ruesink W G (1986) Plant virus epidemics. Monitoring, modeling and predicting outbreaks. Academic Press, Sydney, Australia. 550 p.
Nagarajan S, Kranz J, Saari E E, Seiboldt G, Stubbs R W, Zadoks J C (1984) An analysis of the 1978 epidemic of yellow rust on wheat in Andalusia, Spain. Z. Pflanzenkr. Pflanzenschutz 91:159-170.
Nagarajan S, Singh H (1975) The Indian stem rust rules - an epidemiological concept on the spread of wheat stem rust. Plant Dis. Rep. 59:133-136.
Nagarajan S, Singh H, Joshi L M, Saari E E (1976) Meteorological conditions associated with the long distance dissemination and deposition of *Puccinia graminis tritici* uredospores in India. Phytopathology 66:198-203.
Palti J (1981) Cultural practices and infectious crop disease. Springer-Verlag, Berlin. 243 p.
Parry M L, Carter T R (1985) The effect of climatic variation on agricultural risk. Climatic Change 7:95-110.
Plumb R T, Thresh J M, eds. (1983) Plant virus epidemiology. The spread and control of insect-borne viruses. Blackwell, Oxford. 377 p.
Rotem J, Palti J (1969) Irrigation and plant diseases. Ann. Rev. Phytopathol. 7:267-288.
Savary S (1986) Enquete sur les maladies fongiques de l'arachide (*Arachis hypogaea* L.) en Cote d'Ivoire. I. Methodes d'enquete et etude descriptive: les conditions culturales et les principales maladies. Pages 3-29 and 107-128 *in* Etudes epidemiologiques sur la rouille de l'arachide en Cote d'Ivoire. Thesis, Wageningen. 154 p.
Spaar D, Ebert W (1985) Monitoring and forecasting in plant protection in the German Democratic Republic. EPPO Bull. 15:299-310.
Teng P S, Krupa S V, eds. (1980) Crop loss assessment. *In* Proceedings of E. C. Stakman commemorative symposium. Miscellaneous Publication 7. Agricultural Experiment Station, University of Minnesota, St. Paul. 327 p.
Thresh J M (1985) Principles of plant virus disease forecasting. Pages 13-37 *in* Proceedings of the seminar on agrometeorology and pest forecasting, Fulmer Grange, U.K., 24-28 June 1985. CTA, Wageningen-Ede.
Van Keulen H, Wolf J, eds. (1986) Modeling of agricultural production: weather, soils and crops. Centre for Agricultural Publishing and Documentation, Wageningen. 479 p.
Weltzien H C (1972) Geophytopathology. Ann. Rev. Phytopathol. 10:277-298.
Zadoks J C (1961) Yellow rust of wheat. Studies in epidemiology and physiologic specialization. Neth. J. Plant Pathol. 67:69-256.
Zadoks J C (1973) Long-range transmission of phytopathogens. Pages 392-396 *in* Airborne transmission and airborne infection. J. F. Ph. Hers and K. C. Winkler, eds. Oosthoek, Utrecht. 300 p.
Zadoks J C (1984) A quarter century of disease forecasting. Plant Dis. 68:352-355.
Zadoks J C (1989) EPIPRE, a computer-based decision support system for pest and disease control in wheat: its development and implementation in Europe. Plant Dis. Epidemiol. 2:3-29.
Zadoks J C, Rijsdijk F H (1984) Agro-ecological atlas of cereal growing in Europe. III. Atlas of cereal diseases and pests in Europe. Pudoc, Wageningen. 169 p.
Zadoks J C, Schein R D (1979) Epidemiology and plant disease management. Oxford University Press, New York. 427 p.

Notes

Author's address: J. C. Zadoks, Laboratory of Phytopathology, Agricultural University, Binnenhaven 9, 6709 PD Wageningen, The Netherlands.

Citation information: International Rice Research Institute (1989) Climate and food security. P.O. Box 933, Manila, Philippines.

Postharvest losses and food preservation

B. L. Amla

Climate greatly affects the rate of postharvest spoilage and quality deterioration: exponential quantitative and qualitative losses occur within 15-40 °C and at relative humidity above 30%. Small holdings, postharvest handling systems, inadequate storage space, the multiplication potential of pests, high fungal growth, scarcity of fumigants and protectants, and shortage of skilled personnel in the rural sector are the constraints. Grain production of 250 million t in India by 2000 A.D. will pose procurement, distribution, and storage problems. Food supply, food reserve, buffer stock, and food security will assume unprecedented priority. Climatic factors demand specific approaches for preharvest and postharvest grain handling and storage to maintain quality while avoiding pesticide residues and biological contaminants. New technologies of preharvest prophylaxis and postharvest treatment will provide long-term conservation of buffer stock at low cost. Short- and long-term storage systems are needed for food security. The present central grain pool is 18 million t for long-term (at least 3 yr) security; to counter likely climatic aberrations, access to 30 million t is needed. This could be partly achieved by alternate storage methods.

In the last 15 yr, food grain production in India has been increasing steadily, even with wide yearly fluctuations due to climatic factors. From 1970 to 1980, production increased 20% while the population grew by 25%. Current population is roughly 33% higher than in 1971, and food production has increased 35%. During 1986, the country accumulated a food grain stock of 28.4 million t, attaining self-sufficiency in two major crops, wheat and rice. This period has been one of transition, from perennial shortages to a surplus.

This surplus accumulated over several years has brought management and financial problems into focus. The government policy is to maintain a buffer stock of about 10 million t in storage, with another 8 million t of operational stock for distribution. After this requirement was met, the surplus in July 1986 stood at about 10.4 million t. There have been strong signals from concerned organizations that management of these surpluses has posed problems that have resulted in the slowing down of procurement operations. Transportation and storage losses sustained by the sector organized for grain handling alone was estimated to be Rs1500 million in 1985-86 (US$ = Rs12.02), accounting for about 1.6% of the grain handled.

The terms "food security" and "buffer stock" are often used interchangeably, although they have different implications. According to generally accepted definitions, food security is the access by all people at all times to enough food for an active and healthy life. Chronic food insecurity occurs when a population lacks the economic ability to either buy enough food or to produce its own requirements. Transitory insecurity is due to a temporary decline in access to the required quantity of food as a result of instability in food prices or insufficient food production. The purpose of a buffer stock of food grains is to balance market surplus or deficit from season to season, to maintain a continuous supply and stabilize prices in the trade channels. The public distribution system in India claims to cover almost 70% of the population, but often this system is unable to receive grains from the buffer stock in time and in the needed quantities, probably due to transportation and other logistic difficulties.

The four major prerequisites for any meaningful food security system are

- Long-term policy and plans for improving production and productivity of major food crops, with built-in incentives as part of agricultural activity.
- Stability in production based on crop improvement; early warning systems for weather, diseases, and pests; and crop insurance against failures.
- Sufficient reserves to create a sense of security and to maintain a buffer and an operational supply to provide stability in the marketing system. This involves the development of safe and scientific storage facilities, transportation infrastructure, and healthy marketing practices.
- An efficient and equitable distribution mechanism to ensure availability of adequate stocks near areas prone to food shortages.

Climate and storage requirements

Climate plays a critical role not only in crop production, but also in maintenance of postharvest quantity, quality, and nutritional value. The use of pesticides to prevent adverse changes in nutritional content and organoleptic qualities becomes inevitable. Yet the use of chemicals poses problems of pollution and residues. The insects, molds, mites, and enzymes of grain have different optimum temperatures and humidities. Even in temperate regions, grain spoilage can be significant, especially when the pattern of agricultural operations changes. For example, wheat and maize spoilage in the U.S. became much more serious with the introduction of combine harvesters, which thresh grains at a higher moisture content than with traditional hand methods (Christensen and Kaufmann 1974).

Distribution patterns of pests and fungi in temperate, subtropical, and tropical regions indicate taxonomic differences and adaptations to climate. In temperate regions, dormancy in seeds, eggs, spores, and embryos is an adaptation for survival in extreme conditions. Very high temperatures and very low temperatures are inimical to growth and development of storage pests and storage fungi. In the tropics and subtropics, pests and fungi have shorter life cycles, higher fecundity, rapid population growth, and higher physiological activities, and the spoilage inflicted is significantly higher (Fig. 1). Molds and rodents are other major pests in food storage

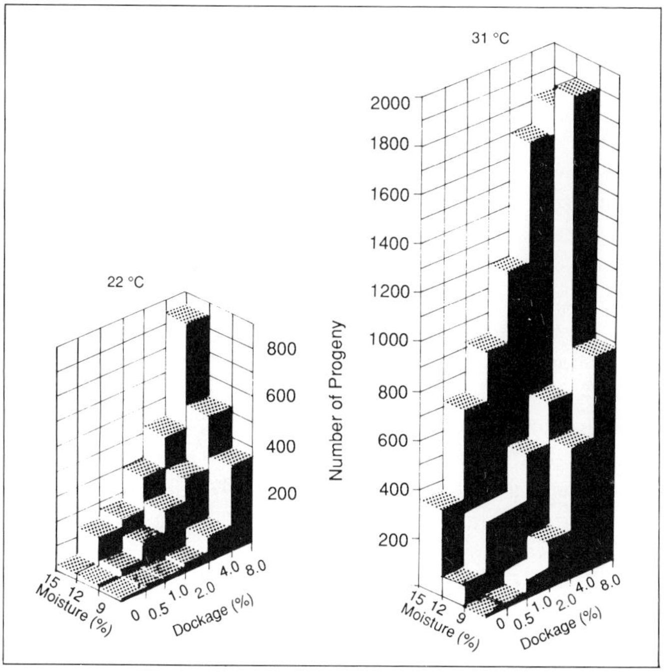

1. Effect of variation in temperature, moisture, and dockage present in wheat on the reproduction of 25 confused flour beetles over a 19-wk period.

and handling installations in the tropics. (A pair of rodents can breed to 800 in a year.) Tropical birds are also problems.

Storage structures, handling systems, and choice of fumigants have to conform to the weather, types of pests, pest diurnal and seasonal cycles, the scale of handling, the distribution pattern, and the end-use of the grains. Bulk storage poses problems of moisture migration and condensation even if grains are adequately dried. Fluctuations in day and night temperatures lead to the development of high moisture pockets, and loss of calories and nutrients. In bulk storage and silo systems, aeration engineering assumes an important role.

Climate and crop infestation

Because crops are climate specific, their geographical distributions are ecological. Human settlements have followed the crop potential of the land. India has about nine agroecotypes, which have different cropping patterns based on rice, wheat, millet, pulses, oilseeds, cotton, jute, sugarcane, tea, coffee, tobacco, and rubber. These cropping systems are bound by soil, moisture, and temperature.

The distribution of pests and diseases is also climate specific. The khapra beetle *Trogoderma* is abundant in the hot and dry climates of northern and central India, but does not thrive well in hot and humid or cool and wet climates. Coastal wet and humid climates provide good conditions for *Oryzaephilus* sp. to grow profusely.

Sitophilus sp., *Tribolium* sp., and moths can live on stored grains for 7 mo in northern India, but grow throughout the year in the southern and eastern parts. Storage fungi also show climate specificity. *Aspergilli, Penicillia, Rhizopus, Mucor, Trichoderma, Fusarium*, and others occur in relative abundance in appropriate climatic regions.

Storage structures

Microclimates are generated by the interactions of storage structure and bulk of stored grains. Aboveground structures are influenced by climatic fluctuations; underground structures provide more stable isothermal conditions. Storage quality of dried grain also varies with the macroclimate of the region. Conduction, convection, radiation, thermal balance, and other meteorological factors play critical roles in the effectiveness of a storage structure. Moisture migration in grain bulk and condensation in some stored grain accentuate spoilage by pests, seed enzymes, and fungi. In the tropics and subtropics, well-aerated structure are required for long-term grain storage (Majumder 1971). Apart from climate, social factors and agricultural land policies also play a role in deciding the type of storage structure and the dimensions of the units.

A rural storage system is dependent on size of harvest, availability of local construction materials, level of skill, and compatibility of the system with climatic factors. A sound storage structure should have low thermal conductivity to prevent moisture condensation; the ability to retain fumigants; resistance to rodent and termite attack; and protection against cross-infestation.

Qualitative losses

In addition to physical or quantitative losses, pests cause significant and serious qualitative losses (in eating and commercial quality and in nutritive value). Food energy loss is a phenomenon of the biodeterioration process. The respiratory activity of grain cells or tissues is initiated by pest infestation, resulting in progressive losses in quality.

There are many loss estimates in the literature, but few consider the qualitative loss that grains sustain during storage, handling, transportation, and other postharvest operations. No loss estimate can be regarded as realistic unless the loss of quality is taken into account (Majumder 1975). Figure 2 depicts physical and quality loss through the accumulation of filth and undesirable metabolites deposited during infestation and biodeterioration.

Weight loss is not a reliable criterion of real food loss (Schulten 1982). A damaged kernel indicates a change in the proximate analysis and infers that the sound grain with its native nutritional, eating, and industrial quality has been altered. Table 1 shows the effect of insect infestation on the uric acid content and quality of wheat flour. The organoleptic quality supersedes all other loss criteria.

Infestation, grain quality/food safety

Rice grain, or for that matter any grain, is an energy capsule. Biologically active tissue in the kernel is degraded by enzymic activity triggered by insects, fungi, and mites when moisture and climate are favorable. Weight:volume ratios and changes

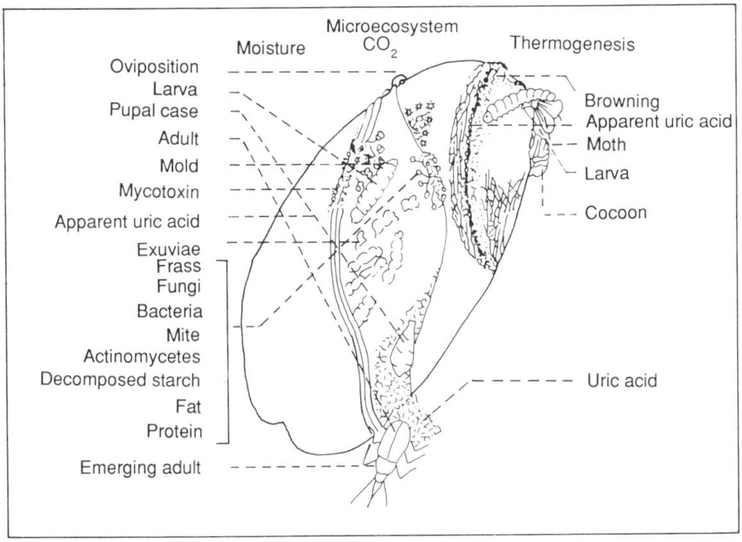

2. Qualitative food losses of an infested kernel. Insect damage: uric acid, moisture, exuviae, chitin, dead insects, infested odor, frass, loss of viability, internal and external microflora. Mold damage: apparent uric acid, moisture, discoloration, mycotoxins, thermogenesis, musty odor, loss of viability. Bacterial damage: deamination, proteolysis, enteric, pathogenic, thermogenesis, toxins, putrefaction. Mite damage: guanine, foul odor, allergens, vector pathogens, debris.

Table 1. Effect of insect infestation on wheat flour (Majumder 1975).

Infestation period (mo)	Insects (no./ 100 g)	Insect fragments (no./ 100 g)	Uric acid (mg/ 100 g)	Acceptability[a]	Gluten (%)	Thiamine (µg/ 100 g)	Loaf vol. (% reduction)	Acceptability[a]
0	0	0	0.1	A	7.4	160	–	A
1	42	370	11.2	NA	6.8	153	7	A
2	110	603	37.7	NA	–	–	19	NA
3	299	850	68.3	NA	4.2	76	23	NA

[a]A = acceptable, NA = not acceptable.

in total weight are deceptive criteria because they do not present an accurate picture of the nutritive and other quality characteristics of the grain.

Basic information on toxicological and nutritional impairments in relation to the energy budget and nature of deteriorative changes is yet to be established. Kaminski et al (1975, 1977) suggested the use of gas chromatography to detect grain deterioration. Gas chromatography can detect volatile metabolites in the head space (3-methyl butanol, 3-actanone, 3 octanol, 1-octen-3-ol). Odors produced by microflora during grain storage have significant relevance to the sensory characteristic of grain (Stawicki et al 1973). These odors have been designated as musty, fungal, urinal, putrid, ammonia, acid, honey, fruity, oily, floral, musky, and other indeterminates. Estimates of these volatiles can serve as reliable and objective criteria for determining loss due to mold and bacteria.

There is a significant relationship between physical criteria and biochemical changes occurring in a grain. Pests create a microclimate by physically removing the bran and depositing fungal inocula, giving rise to accelerated fungal growth. Biochemical changes can be assessed by extractibles such as FFA and by rancidity factors. On the other hand, free amino acids and free reducing sugars provide indices for biodeterioration. An insect that attacks the germ portion causes selective changes in the analytical quality of the grain. The protein-, vitamin E-, and thiamine-rich portions of the germ are lost, thereby reducing the viability and nutritional quality of the grain. A *Sitotroga* attack on rice reduces the grain bulk density. *Sitotroga* feeds on the starchy endosperm, resulting in CO_2 evolution and formation of formozan-like substances with tetrazolium compounds. Calorie loss, impaired nutrition, and accumulation of toxic metabolites become unavoidable.

Health hazards of infestation

During growth and development, insects leave exuviae, fragments, and excreta on the grain. Fragment count has been regarded as one index of quality.

In general, insects are uricotelic and, therefore, produce crystalline uric acid and deposit it on the substrate. Estimation of true uric acid indicates selective insect activity, while Benedict's reaction complex (apparent uric acid) reveals total deterioration caused by moisture, enzymes, insects, mold, and mites (Majumder 1974).

Glandular liquids secreted by *Tribolium* consist of carcinogenic ethyl and methyl-parabenzoquinone (Ladisch 1977). The insect quinones react avidly with amino acids, forming conjugates related to the coal tar dyes that have been proven lethal when fed to mice or painted on their dorsal skin. Approximately 0.025 mg of quinone derivatives can be isolated from the glandular secretion of each insect.

Yet another important factor, drawing increased attention from nutritionists, toxicologists, and health specialists relates to the production of toxic metabolites of fungi. The metabolites or mycotoxins such as islandotoxin, citrinin, patulin, candidulin, rugulosin, notatin, citreoviridin, and aflatoxin have been recognized as hazardous pollutants in cereals, particularly in humid and hot climates.

It is clear that losses based on physical or quantitative parameters alone do not adequately assess true food losses. The importance of deterioration of the energy budget, nutritive value, organoleptic quality, and processing quality cannot be ignored.

Food conservation

In many tropical countries, methods for storing and handling food grains are being improved. Activities, however, have been confined mostly to parastatal storage—the organized sector. But about 70% of the population lives in the rural areas of developing countries, where grain is stored, processed, and consumed mostly in traditional ways. The structures, receptacles, containers, and bags used offer little protection from biodeterioration and quality damage.

Agricultural operations on comparatively small landholdings do not lend themselves to the adoption of the modern techniques of harvesting, transport, and storage evolved in the developed countries. Storage in a silo presents many economic and functional problems. Tropical and semitropical climates with wide diurnal fluctuations aggravate moisture migration, resulting in accelerated enzyme and fungal damage as well as insect infestation.

Rural facilities store 60-70% of the grain production. The following measures could minimize losses without the danger of pollution:

- functional design and development of rural storage containers;
- proper prophylactic measures;
- use of effective nontoxic and selective protectants;
- improved transportation, mill sanitation, inert atmosphere, and hermetic storage techniques; and
- integrated pest control measures using attractant, repellent, chemosterilants, hormones, and parasites.

Rapid progress in food science and technology during the last decade has resulted in the development of dry, dehydrated, desiccated, extrusion-cooked, and freeze-dried packaged products in urban centers. These innovations pose new problems of food sanitation, preservation, storage, and distribution in patterns similar to those of cereals, pulses, and oilseeds. Many of the flexible packaging materials used are vulnerable to insect attacks (Majumder 1970). When stored and transported in tropical climates and marketed with low rates of turnover, these processed products deteriorate rapidly.

Nontoxic chemical compounds used as selective protectants for food products designed for applied nutrition programs not only act as deterrents to insects but also improve the nutritional quality of the food products. These chemical protectants are based on physiological specificities, and the morphology, biochemistry, and immunology of insects. Wherever these highly specific protectants cannot be used effectively, ecologically compatible techniques (Table 2) should be applied. Most of these protectants and technologies do not pose environmental pollution problems.

The principle of integrating preharvest protection with postharvest operations in a prophylactic system needs to be emphasized in storage practices. Integrated pest control measures have to be treated as a cyclical system, from seed sowing to grain harvest to utilization.

The approaches assume significance in the context of large buffer stock maintenance. The present burden of carrying large buffer stocks over a long period calls for innovative approaches to design an economically balanced system between the postharvest storage of native grain and the movement of grain to market, and conversion and retail storage.

Alternate storage options

Maintaining a buffer stock is crucial to the food security system. India has made great strides in establishing such a buffer stock on an almost permanent basis. However, holding these stocks is proving to be an expensive proposition—Rs16,500 million for 1985-86. It is likely that will increase substantially as buffer stocks grow steadily. At the same time, 37% of the country's population remain below the poverty line. This is clear evidence that chronic food insecurity still prevails. The economic feasibility of using part of the grain stock to improve the economic status

Table 2. Measures that provide best control in tropical storage and processing centers (Majumder 1975).

Preharvest prophylaxis to control field infestation by storage fungi.	Malathion, Tricalcium phosphate, *Bacillus thuringiensis,* borate buffer mixtures.
Rodent control.	Burrow fumigation (with liquid fumigation, primate safe, baiting); Rodent repellent.
Sanitation of threshing yard to prevent cross-infestation by resident insect population.	Ecological or habitat control.
Drying to safe moisture content to inhibit mold growth.	Sun drying and mechanical drying; Fungicidal organic acid; Gaseous sterilization.
Storage in bins or rural structures.	*Bacillus thuringiensis* and malathion minifume treatment for fumigation of small bulk with ovicidal effect, sorptive Kaolin.
Storage in bags in warehouses or CAP storage system.	Durofume process and preventive insecticidal and rodent repellent spray on outer aspect of stack; Poison bait in container (trap-cum baiting station treated with quinine hydrochloride [optical attractants] for rodents).
Milled product.	Mill disinfestation, spot fumigation of milling machinery, Entolation, Tricalcium phosphate + vitamin formation incorporation.
Processed products (macaroni, noodles, etc.).	Infrared heat disinfestation, serial fumigation process for in-package disinfestation.

of the population by creating more employment opportunities is worth examining. It is also worthwhile to explore storing grains by recycling them through livestock or poultry, thereby recovering the cost of investment in addition to creating jobs.

In considering any alternative to the existing warehousing methods of managing surplus stocks, the following measures should be kept in mind:

- reducing storage costs,
- arresting grain quality deterioration inherent in conventional storage,
- providing long-term food security.

India is likely (if everything goes well) to have surpluses of 30-40 million t over and above the minimum buffer stock by the turn of the century. Taking present surplus grain availability as 11 million t, we can construct a scenario whereby this overflow is used to produce animal or industrial products such as meat and alcohol, which are much in demand. It should be possible to deploy a part of the surplus grains for other uses and, during any food stress, divert them back again into the mainstream of the food supply or the grain pool for direct consumption.

The present position vis-a-vis grain stock shows a net surplus of 10.4 million t for 1985-86; that could be available for alternate utilization (Table 3). One needs to examine the following options in dealing with this surplus.

Table 3. Grain stock profile.

Item	Stock (million t)	
	1984-85	1985-86
Rice	8.0	9.3
Wheat	21.0	18.9
Coarse grains	0.2	0.1
	29.2	28.4
Buffer stock needed	12.0	10.0
Operational stock	6.0	8.0
	18.0	18.0
Net surplus	11.2	10.4

Table 4. Requirements for storage of 1 million tons wheat surplus.

Capital outlay (assumptions)		
a. Storage space requirement	0.66 million m^2	
b. Labor-days required to procure, store, transport, distribute	13 labor-days/t	
c. Storage and distribution loss over 1 yr	2% on quantity and value	
Financial balance sheet	Input (in million rupees)	Output (in million rupees)
Raw material		
Wheat 1 million t at Rs1,520/t	1,520	0.98 million t 1,489.6
Utilities, manpower, and other overhead costs	535	--
	2,055	1,489.6
		– 565.4
Bioenergy balance sheet		
Wheat 1 million t	318×10^9 kcal	3116.4×10^9 kcal
Labor	--	147.13×10^9 kcal
	318×10^9 kcal	3263.53×10^9 kcal
		$+ 83.53 \times 10^9$ kcal

- storage and subsidized distribution,
- conversion to meat and meat products, and
- conversion to alcohol.

To store 1 million t grain, a fixed investment of Rs682 million on godowns and other infrastructure and a recurring cost of Rs565.4 million (value depletion) is required. It results in a net energy gain of 83.53×10^9 kcal over a period of 1 yr (Table 4), but with an overall loss in management costs. In producing poultry with a mean equivalent to 1 million t grain, the investment required is Rs3,815 million, with a

Table 5. Options for managing 11 million tons of surplus wheat.

Option 1	Option 2	Option 3	Option 4
Storage and subsidized distribution	Conversion to alcohol	Conversion to beef	Conversion to poultry (broilers)
Capital			
Investment in storage godowns and infra-structure: Rs7502 million	Investment in establishing distilleries: Rs27280 million	Investment in establishing beef (buffalo) farm: Rs27005 million	Investment in pens, infrastructure: Rs41965 million
Financial			
	(Recurring, in million rupees)		
Input: 22,605	Input: 31,526	Input: 63,235	Input: 96,833
Output: 16,385	Output: 34,986	Output: 75,328	Output: 117,557
Value added			
−6,220	+3,460	+12,093	+20,724
Bioenergy balance	*Bioenergy balance*	*Bioenergy balance*	*Bioenergy balance*
Input:			
34980 $\times 10^9$ kcal	35519 $\times 10^9$ kcal	56859 $\times 10^9$ kcal	49665 $\times 10^9$ kcal
Output:			
38598.8 $\times 10^9$ kcal	28361.3 $\times 10^9$ kcal	43650.2 $\times 10^9$ kcal	46185.9 $\times 10^9$ kcal
Balance (+)			
+918.8 $\times 10^9$ kcal	−7157.7 $\times 10^9$ kcal	−13209.8 $\times 10^9$ kcal	−3479.1 $\times 10^9$ kcal

value addition potential of Rs1,884 million and a net energy loss of 316.19 $\times 10^9$ kcal (Table 5). Similarly, production of animal meat using 1 million t grain calls for an expenditure of Rs24.55 million on infrastructure and has a value addition of Rs1,099.4 million and an energy loss of 1200.8 $\times 10^9$ kcal. Conversion to alcohol is possible with an investment of Rs2,480 million, which would result in Rs314.6 million value addition and a net energy loss of 650.7 $\times 10^9$ kcal. If feasibility factors rule out complete diversion into one type of industry, options can be exercised to open different routes, as determined by the demand.

A cybernetic model for food handling

An effective management model for food security would include equitable distribution, stock in the pipeline, and adequate quantities in storage to cover climatic failures and emergency requirements and to alleviate chronic nutritional imbalances. Long-term storage for food security should be achieved at as low a cost as possible, through a dynamic balance of procurement, storage, and pipeline holdings. The logistics would have to be planned and operated through a balanced storage, transportation, and distribution infrastructure.

In such a model, the village is taken as a unit of production as well as of consumption. The marketable surplus, about 30% of the total, is available to the organized trade for procurement and marketing. The public sector organization also operates in this surplus pool. India has 557,000 villages, 3,000 small towns, and 300 cities, with about 30 megapolis complexes. By the end of the century, the population

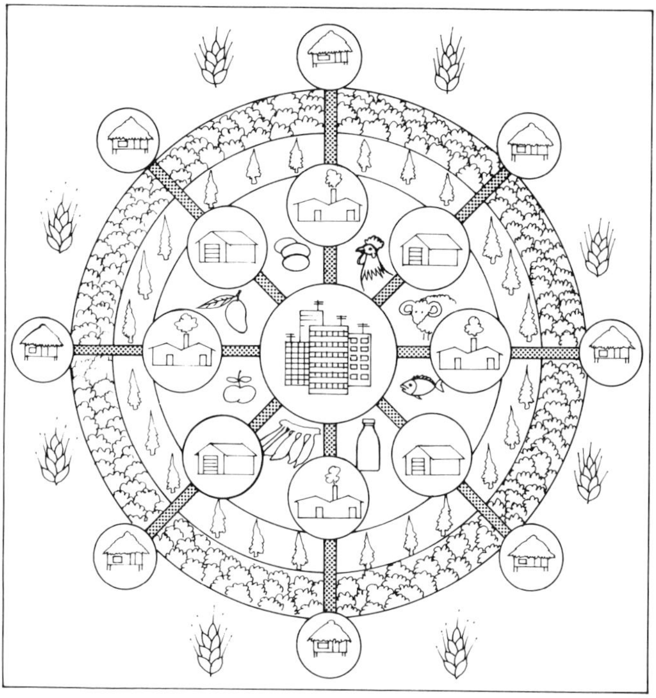

3. Cybernetic model for linking villages with storage-cum-processing centers and urban and rural consumers.

is expected to stabilize at 1 billion. We expect to have a 29.5% urban and 70.5% rural population. To face the problem of food handling and marketing in 2000, we need to examine alternative designs for grain management.

The logistics of distribution start from the primary producer in a village. Postharvest processing and conservation centers are needed near the producing areas, for example, in the town nearest a village or cluster of villages. Some 3,000 such processing-cum-storage centers would be required for drying, bagging, warehousing, and processing. These centers should provide for two-directional movements of the processed products—about 70% to return to the village for the rural consumers and 30% to move forward for urban consumption and industrial uses. Depending on milling capacity and other processing requirements, the centers could provide raw materials at a rated capacity for optimum utilization of the processing equipment. The husk, by-products, chaff, and other waste materials from milling and processing would serve as a fuel or energy resource for the rural sector.

This model would provide inflow and outflow to processing centers (Fig. 3) and balance agricultural production and postharvest operations at farm levels. Processing and storage cost would be kept low due to the proximity of raw materials and a sizable market. Migration of skilled and educated manpower from rural sectors to urban areas could be reversed, giving rise to the development of more active agroindustrial centers or clusters in and around rural areas.

References cited

Christensen C M, Kaufmann H H (1974) Storage of cereal grains and their products. AACC Monograph. Vol. V. American Association of Cereal Chemists, St. Paul, Minnesota.

Cotton R T (1954) Insects in storage of cereal grains and their products. J. A. Andersen and A. W. Alcock, eds. AACC Monograph. Vol. II. American Association of Cereal Chemists, St. Paul, Minnesota. p. 221-274.

Kaminski E, Stawicki S (1977) Characterization and identification of volatile odor substances produced by bacteria in cereal grain. Final technical report. Project No. PL-ARS-21 FG-PO-285, Poznan.

Kaminski E, Stawicki S, Wasowicz E, Przybylski R (1975) Detection of deterioration of grain by gas chromatography. Pol. Pismo 44:441-447.

Ladisch R K (1977) The biology of *Tribolium*. Vol. 3. A. Sokoloff, ed. Clarendon Press, Oxford.

Majumder S K (1970) Packaging and protection against insects in tropical and subtropical areas. Pages 255-269 *in* Principles of food packaging. R. Heiss, ed. Food and Agriculture Organization, Rome.

Majumder S K (1971) Recent progress in the protection of foods. Pages 725-741 *in* Proceedings of the first Asian congress on nutrition. P. G. Tulpule and K. S. J. Rao, eds. International Union Nutrition Sciences, INSA & NSI.

Majumder S K (1974) Control of microflora in stored grains. CFTRI Monograph. Chapter 3. Central Food Technological Research Institute, Mysore. p. 79-107.

Majumder S K (1975) Pre-harvest and post-harvest losses in bread. A. Spicer, ed. Applied Science Publishers. p. 181-199.

Schulten G G M (1982) Post-harvest losses in tropical Africa and their prevention. Food Nutr. Bull. 4(2):2-9.

Stawicki S, Kaminski E, Niewiarowicz A, Trojan M, Wasowicz E (1973) The effect of microflora on the formation of odors in grain during storage. Ann. Technol. Agric. 4:309-336.

Notes

Author's address: B. L. Amla, Central Food Technological Research Institute, Mysore, India.

Citation information: International Rice Research Institute (1989) Climate and food security. P.O. Box 933, Manila, Philippines.

Growing consensus and challenges regarding a greenhouse climate

W. Bach

Fossil fuel combustion, synthetic chemicals production, biomass burning, and forest and soil destruction are the major activities through which humans contribute to the perturbation of the trace gas composition of the atmosphere. An important consequence of this is the modification of the radiation budget (what has become known as the greenhouse climate), with potentially far-reaching impacts on ecosystems, food production, water resources, sea level, etc. There is a growing realization that mankind is faced, not with a CO_2 problem alone, but rather with a major global trace gas problem, which is fueled by many local sources and decisions. The climatic impacts are felt worldwide, especially in the developing countries. It is now clear that a certain degree of climatic change is inevitable, given man's past and present activities. This has led not to inaction but to a growing consensus that only concerted action can help solve mankind's major environmental challenges: climatic change; the threat to the ozone shield; and the destruction of forest, soil, and genetic resources, and hence, the endangerment of food security. The consensus grows that we must follow a two-pronged safety strategy: conduct vigorous research to narrow uncertainties in our knowledge and, at the same time, use the information available to take precautionary measures. This paper briefly reviews the current consensus on the greenhouse climate, the uncertainties, the potential impacts, and the required actions.

Growing awareness

The radiatively active natural constituents of the atmosphere—water vapor (H_2O), carbon dioxide (CO_2), and clouds—by reducing the outgoing longwave radiation to space, help create a comfortable climate on earth with an average surface temperature of 15 °C. This reduction in longwave heat emission to space has become known as the greenhouse effect. Without it, the global mean surface temperature would be an unwieldy −19 °C. The natural greenhouse effect essentially guarantees life on earth.

But, as is often the case, too much of a good thing can cause opposite results. Worldwide monitoring has provided concrete evidence that humans, through such activities as fossil fuel combustion, production of synthetic chemicals, biomass burning, and forest and soil destruction, are significantly changing the chemical composition of the atmosphere, thereby enhancing the natural greenhouse effect. A result of this enhancement will be changes in, for example, air circulation and rainfall patterns, which will have potentially far-reaching impacts on global ecosystems, agriculture, water resources, sea level, etc. In a world where population is increasing rapidly, this is of more than mere academic interest.

A series of troublesome scientific developments has led to a growing awareness of greenhouse gases (GHG) as major agents of climatic change (Bach 1984; Bach et al 1980, 1983, 1984; Bolin et al 1986; MacCracken and Luther 1985a,b; Ramanathan et al 1985; Trabalka 1985; WMO 1979, 1985). These developments include

- Trace gases other than CO_2, such as methane (CH_4), nitrous oxide (N_2O), chlorofluorocarbons (CFC 11, CFC 12), and ozone (O_3), can significantly amplify the greenhouse effect of CO_2. So far, some 30 additional GHG have been identified; the list is still growing.
- The greenhouse effect of the other GHG is already about as great as that of CO_2. Should present trends continue, combined effects could reach a magnitude equivalent to a doubling of the preindustrial CO_2 level, possibly by as early as the first half of the next century.
- Most GHG have greater growth rates, longer residence times, and greater greenhouse effects than CO_2. For example, on a molecule-per-molecule basis, the release of 1 CFC molecule has the same surface heating effect as the addition of 10,000 CO_2 molecules. This means the importance of other GHG should increase relative to CO_2.
- The combination of radiatively active gases (e.g CH_4) and chemically active gases (e.g. carbon monoxide [CO] and nitrogen oxides [NO_x]) can significantly increase tropospheric O_3 and CH_4. This, in turn, enhances the greenhouse effect. It has been shown that the indirect climatic effect (via CO and NO_x) is as large as the direct radiative effect. The radicals OH and HO_2 could moderate the effects of CH_4 and O_3 if they were not themselves significantly depleted by the growing levels of CO, NO_x, and other pollutants.
- Perturbations in stratospheric ozone can noticeably affect tropospheric climate through radiative-dynamical interactions between the troposphere and the stratosphere.
- The present warming (the signal) cannot yet be unequivocably distinguished from natural variation of climate (the noise), because much of the heat is stored within the oceans. This concealment by the oceans means that a certain degree of future warming is inevitable, given past human actions. But it also means that appropriate countermeasures must be taken very soon, if the rate and degree of warming during the next century are still to be mitigated significantly.

Measures taken to reduce the impacts of a greenhouse climate will be more effective if there is an awareness of how closely the greenhouse problem is linked to other environmental issues, such as forest and soil destruction in the tropics, forest dieback in midlatitudes, and the destruction of the earth's ozone shield. For example, reducing fossil fuel combustion through more efficient use would not only reduce a host of climate-influencing GHG, but would also simultaneously decrease the air pollutants and acid deposition responsible for the death of forests and lakes.

Another example pertains to consumer product changes involving the reduction of many CFCs, thereby not only helping protect the ozone layer, but also reducing the potential of a climatic change.

Increasing greenhouse gases — a growing challenge

An assessment of the evolving greenhouse climate requires estimating future trends in the buildup of CO_2 and other GHG. This is a hazardous undertaking, because such trends depend on population, economic, industrial, and sociopolitical developments, as well as on the sources, sinks, lifetimes, and chemical reactions of the trace gases, which are all uncertain. There are, however, differences in the degrees of uncertainty. Therefore, this is not an exercise in futility. Rather, it serves a useful purpose in that it can help identify areas in which there is general agreement or in which more work needs to be done. Information that is based on a broad consensus can then be channeled into the decisionmaking process.

Likely future CO_2 emissions

The annual C emission from fossil fuel combustion, gas flaring, and cement production increased from about 0.1 GtC in 1860 to about 5 GtC in 1980, which corresponds to an average exponential growth rate of 3.4%/yr (Marland and Rotty 1984). After the energy crisis of 1973-74, the C emission growth rate slackened to 1.5%/yr during 1973-82. Projections of future C emissions show a wide range of uncertainty, primarily because of our inability to predict future behavior. There seems to be, however, a downward shift in scenarios projecting future C emission levels. Before 1982, the C emission projections for the year 2050 ranged from 1 to 127 GtC/yr; after 1982, the range was reduced to 1.5-64 GtC/yr (Edmonds and Reilly 1985). For 2050, Keepin et al (1986) give a range of C emission from fossil fuel combustion between 2 and 20 GtC/yr plus 1-2 GtC/yr from terrestrial biota and other sources. Our own estimates for 2030 (Fig. 1) range from 1 to 23 GtC/yr (Bach 1986). An important result from many recent studies is that the future C emission could be low, if the available technology to improve energy efficiency were put to use. Researchers now agree that the annual global C emission could range from 1 to 25 GtC by 2050.

Likely future CO_2 concentrations

Estimates of preindustrial atmospheric CO_2 levels range from 250 to 290 ppm. The first precise measurements on Mauna Loa showed an increase from 315 ppm in 1958 to 345 ppm in 1984, with a mean growth rate of about 1.6 ppm/yr over the past decade (Neftel et al 1985). Future CO_2 levels depend not only on the assumed emission scenarios shown in Figure 1, but also on the transfer processes between the major carbon reservoirs, such as the oceans (with their marine biota and sediments) and the terrestrial ecosystems (with land use changes and soil and forest destruction). In one of the U.S. state-of-the-art reports, Trabalka et al (1985) conclude that, on the basis of their energy-economic CO_2 emissions study, it would be unlikely for a CO_2 doubling (i.e. 550 ppm) to occur before 2025, and it may not occur within the next

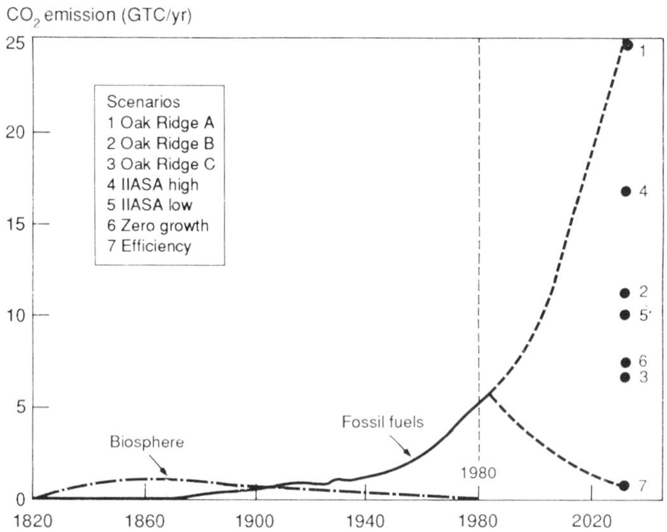

1. Simulation of CO_2 emission. The historical development is based on biospheric sources from 1820 to 1980 and on U.N. statistics for fossil fuel use, cement production, and gas flaring from 1860 to 1980. The projection from 1980 to 2030 is based on the specifics of the listed energy scenarios (Bach 1986).

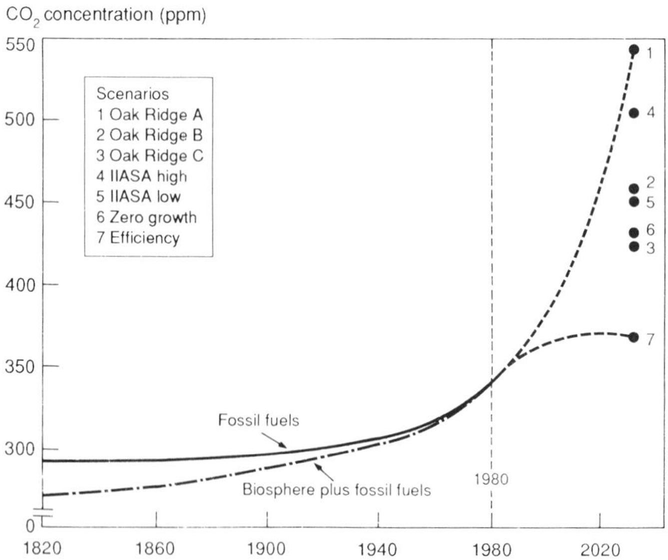

2. Simulation of CO_2 concentration. Calculated with a box-diffusion carbon cycle model (Oeschger et al 1975) using as input the CO_2 emission rates of Figure 1.

century. Based on fossil fuel combustion scenarios, Bolin (1986) gives a range of 370-460 ppm in 2025 and, in agreement with Trabalka et al (1985), concludes that it would be unlikely that CO_2 would double to 550 ppm before 2050. Based on the emission scenarios in Figure 1, our studies (Bach 1986) result in a range of 370-540 ppm in 2030 (Fig. 2). There appears to be general agreement that a CO_2 doubling to 550 ppm may not occur until during the second half of the 21st century.

Likely future levels of other greenhouse gases

Table 1 summarizes the main characteristics of the GHG (Bach 1984, Ramanathan et al 1985). Also shown are the projected concentrations of the various GHG in 2030, with their uncertainty ranges. Because the other GHG are radiatively similar to CO_2, it is common practice to approximate their effects in terms of an equivalent amount of CO_2. Currently, the effect of the other GHG is approximately equivalent to an additional 40-50 ppm increase of CO_2 (Jaeger 1986). During the next 50 yr, the effects of the other GHG are expected to increase relative to CO_2 through, among other factors, their higher growth rates. There is general agreement that this would result in a greenhouse situation equivalent to a CO_2 doubling well before the middle of the next century. Therefore, preventive measures must include reducing emissions of both CO_2 and the other GHG.

Climatic response to greenhouse gases

Assessing the future climatic response to GHG requires mathematical models based on the set of fundamental physical principles that govern the climate system. A hierarchy of one- (1-D) to three-dimensional (3-D) models is available for such assessments. There are two basic approaches: equilibrium response studies (e.g. the study of the climate response to a time-independent doubling of CO_2 or other GHG), and transient response studies (e.g. the study of the response of the model climate to a time-dependent continual increase in CO_2 and other GHG).

Equilibrium response studies

Table 2 shows the equilibrium changes in surface air temperature due to a CO_2-doubling as simulated by a variety of 3-D general circulation models (GCMs) with different characteristics. In the first two experiments with a climatological ocean, the ocean acts as if it were an infinite heat sink: as the atmosphere warms, there is no corresponding warming of the ocean and hence, no return of energy to the atmosphere. The 0.2 K results from such GCMs are typically of an order of magnitude less than those from more realistic models.

In the next three studies, the ocean is treated as a swamp: as a perpetually wet land of zero heat capacity, without any horizontal or vertical heat transport, and with an infinite water vapor source. Such models, with neither a seasonal nor a diurnal insolation cycle, give a surface air warming of 1.3-2.9 K.

The next three experiments, with a fixed-depth mixed-layer ocean model, incorporate some of the surface layer-atmosphere feedback but still have neither horizontal heat advection by ocean currents nor vertical heat transfer, such as in upwelling and downwelling. These models, with a seasonal insolation cycle, yield warmings on the order of 3.5-4.2 K.

The last two models deploy the most sophisticated ocean model schemes. The reason for the relatively moderate warming of 1.5 K may be that these models have not run long enough to have reached equilibrium.

Table 1. Overview of a selection of greenhouse gases influencing climate.

Constituent	Principal anthropogenic sources	Principal sinks and removal processes	Atmospheric residence time	Global average mixing ratio in 1980	Possible increase 1980-2030 (uncertainty range in 2030)	Potential influence on climate and surface temperature change
Carbon dioxide (CO_2)	Fossil fuels, deforestation, soil destruction	Ocean biosphere	6-10 yr	339 ppmv	339-450 ppmv (380-550 ppmv)	Warming in T, cooling in S, 2x $CO_2 \rightarrow 3 \pm 1.5$ K global, 2 to 3 times greater at poles
Methane (CH_4)	Rice paddies, cattle raising, biomass burning, gas leakage, fossil fuels	Photochemical reaction with OH and NO_x; soils in semiarid climates, net CH_4-flux from T to S; reactions with OH and O (^1D) in S	9-10 yr	1.65 ppmv	1.65-2.34 ppmv (1.85-3.30 ppmv)	Direct and indirect greenhouse effect in T; influence on chemistry in S (source of H_2O, reaction with Cl and HCl formation) 2x $CH_4 \rightarrow 0.3$ K
Ozone (O_3)	Indirectly produced through photochemical reactions with other substances	Through oxidation of CO and NO under the influence of NO; catalytic reactions with, e.g., NO_x, Cl_x, HO_x in S	30-90 dy (T) 2 yr (S)	0.02-0.3 ppmv (T) 5-10 ppmv (S) (at 30 km)	12.5% increase in tropospheric O_3	O_3 production in T, leads to warming 2x $O_3 \rightarrow 0.9$ K
Nitrous oxide (N_2O)	Biomass burning, fossil fuels, artificial fertilizers	No major sink in T, photolysis and reaction with O (^1D) to form NO in S	165-185 yr	300 ppbv	300-375 ppbv (350-450 ppbv)	Greenhouse effect in T, impact on O_3-budget in S 2x $N_2O \rightarrow 0.3\text{-}0.4$ K
Chlorofluorocarbons ($CFCl_3$) (CF_2Cl_2)	Propellants, coolants, solvents	No known sink in T, sink in S through photolysis	65 yr 110 yr	0.18 ppbv 0.28 ppbv	0.18-1.10 (0.5-2.0) ppbv 0.28-1.80 (0.9-3.5) ppbv	Increase from 0-1 ppbv $CFCl_3 \rightarrow 0.13$ K $CF_2Cl_2 \rightarrow 0.15$ K

T = troposphere, S = stratosphere, ppmv = parts per million, ppbv = parts per billion, K = degree Kelvin, 2x = doubling of concentration. Sources: Bach (1984), Ramanathan et al (1985).

Table 2. Surface air temperature change (ΔT) induced by a CO_2 doubling as simulated by a selection of 3-D general circulation models.

Model study[a]	Geography[b]	Topography[b]	Seasonal cycle/ diurnal cycle	Ocean	ΔT (K)
OSU Gates et al (1981)	R	R	Yes/Yes	Climatological	0.2
UKMO Mitchell (1983)	R	R	Yes/Yes	Climatological	0.2
GFDL Manabe and Wetherald (1975)	I	None	No / No	Swamp	2.9
OSU Schlesinger (1983)	R	R	No / No	Swamp	2.0
NCAR Washington and Meehl (1983)	R	R	No / No	Swamp	1.3[c] 1.4[d]
NCAR Washington and Meehl (1984)	R	R	Yes/No	50 m mixed layer	3.5
GISS Hansen et al (1984)	R	R	Yes/Yes	65 m mixed layer	4.2
GFDL Wetherald and Manabe (1986)	R	R	Yes/No	68 m mixed layer	4.0
OSU Gates and Potter (1985)	R	R	Yes/Yes	Variable depth mixed layer	1.5
OSU Gates et al (1984) Schlesinger et al (1987)	R	R	Yes/Yes	Complete ocean general circulation model	1.5

[a]OSU = Oregon State Univ., Corvallis, USA; UKMO = Meteorological Office, Bracknell, UK; GFDL = Geophysical Fluid Dynamics Lab., Princeton, USA; NCAR = National Center for Atmospheric Res., Boulder, USA; GISS = Goddard Institute for Space Studies, New York, USA. Adapted from Bach (1988) and updated. [b]R = realistic, I = idealized. [c]Clouds prescribed. [d]Clouds predicted.

Experiments with 3-D models give an average global surface air temperature increase of 1.5-4.5 K for a CO_2 doubling, with an amplification by a factor of 2-3 toward the poles and the largest increases occurring in the winter (CDAC 1983). Recent GCMs with more realistic geography, ocean modeling, and seasonal dependence show a global surface air warming of 3.5-4.2 K and an increase in the global average precipitation rate of about 7-11% (MacCracken and Luther 1985a,b). A number of models also indicate a significant increase in summer dryness in the midlatitude Northern Hemisphere regions; that could markedly affect agriculture and water resources.

So far, very little agreement exists on the magnitude of the changes and their regional diversity. The regional diversity is essential for a useful assessment of the impacts of the greenhouse climate. The realistic allowance for ocean thermal inertia must await the coupling of a 3-D atmospheric GCM with a 3-D ocean GCM. There are still many uncertain features, some of which are discussed later.

Table 3. Surface air temperature change due to trace gas changes obtained from 1-D radiative-convective model calculations.

Study	Gas	Change in concentration	Change in surface temperature (K)[a]
Augustsson and Ramanathan (1977)	CO_2	320-640 ppm	2.9
Lacis et al (1981)		300-600 ppm	2.9
Owens et al (1985)		330-660 ppm	1.7
Bach and Jung (1987).		340-680 ppm	1.9
Wang et al (1976)	N_2O	0.28-0.56 ppm	0.44
Donner and Ramanathan (1980)		0.30-0.60 ppm	0.33
Lacis et al		0.28-0.56 ppm	0.65
Owens et al		0.28-0.56 ppm	0.29
Bach and Jung		0.30-0.60 ppm	0.34
Wang et al	CH_4	1.60-3.20 ppm	0.20
Donner and Ramanathan		1.50-3.00 ppm	0.30
Lacis et al		1.60-3.20 ppm	0.26
Owens et al		1.60-3.20 ppm	0.23
Bach and Jung		1.65-3.30 ppm	0.25
Ramanathan et al (1985)	CFC 11	0-1 ppb	0.13
Bach and Jung		0-1 ppb	0.15
Ramanathan et al	CFC 12	0-1 ppb	0.15
Bach and Jung		0-1 ppb	0.16
Wang et al	CFC 11/	0-2 ppb	0.38
Lacis et al	CFC 12	0-2 ppb	0.68
Owens et al		0-2 ppb	0.33
Ramanathan et al		0-2 ppb	0.55
Bach and Jung		0-2 ppb	0.64

[a]Simulations refer to fixed relative humidity and fixed cloud altitudes. Source: Bach and Jung (1987)

Climate response to GHG is usually investigated with less sophisticated, and hence less expensive, models, such as 1-D radiative convective models (RCMs). Table 3 gives an overview of how well the various model studies agree and the relative contributions of the various GHG to the greenhouse climate. The temperature sensitivity due to a CO_2 doubling is much less in RCMs (1.7-2.9 K) than in GCMs (1.5-4.5 K). The main reason is the neglect in RCMs of such feedback processes as ice-albedo and ocean-atmosphere coupling, which reduce the sensitivity of the troposphere to changes in the infrared radiation fluxes.

Transient response studies

For a climate sensitivity of 4 K for a CO_2 doubling based on GCMs, and for a CO_2 increase from 270 ppm (1850) to 338 ppm (1980), the equilibrium surface temperature increase is 1.3 K (WMO 1985). Reconstruction of the actual Northern Hemisphere surface temperature shows a warming of about 0.6 K from 1880 to 1980 (Jones et al 1986). The most plausible reason for the discrepancy is that the actual response of the climate system lags behind the equilibrium response because of the

Table 4. Computed equilibrium and transient temperature response due to the combined effect of CO_2 and other greenhouse gases (GHG).

Period	Temperature response (K)				
	Equilibrium		Transient[b]		
	Ramanathan et al (1985)[a]	Lal et al (1986)[b]	Wigley (1984)[c]	WMO (1985)[d]	Bach and Jung (1987)[e]
			T_e[f] = 1.5 T_e = 4.5	T_e = 2 T_e = 4	T_e = 1.5 T_e = 4.5
1880-1980	0.79	0.85	0.46 - 1.13[g]	0.4 - 0.8[h]	0.64 - 1.12
1980-2030	1.52	1.64[i]	0.72 - 1.28[j]	0.4 - 2.2	0.70 - 1.89
1880 - 2030	2.31	2.49[k]	1.18 - 2.41[l]	0.8 - 3.0[l]	1.34 - 3.01

[a]CO_2, CH_4, N_2O, CFC 11, CFC 12, O_3, and additional GHG. [b]CO_2, CH_4, N_2O, CFC 11, CFC 12. [c]Ocean diffusion coefficient K = 2 cm^2/s. [d]K = 1-2 cm^2/s. [e]K = 1.3 cm^2/s. [f]T_e = global mean equilibrium temperature for a doubling of CO_2 (from GCMs T_e = 1.5-4.5 K). [g]1850-1985. [h]1850-1980. [i]1980-2050. 1985-2030. [k]1880-2050. [l]1850-2030.

thermal inertia of the ocean. It has been shown that the equilibrium is approached exponentially, with a characteristic "e-folding" time for a variety of atmosphere/ocean models ranging from 10 to 100 yr (WMO 1985). Experiments with a 2-layer atmosphere GCM coupled to a 6-layer ocean GCM showed that the time required for a CO_2-induced climatic change to reach equilibrium was 50-100 yr (Schlesinger 1986).

Thus, a more realistic modeling of the climate as it is evolving (i.e. the transient response of the climate system) requires coupling an atmosphere model with an ocean model in which heat capacity and vertical transport of heat are considered. For reasonable ocean thermal diffusivities of 1-3 cm^2/s, the 1850-1980 transient warming is reduced to about 50% of the equilibrium warming (i.e. about 0.5-0.7 K), which agrees with the observed warming of 0.6 K.

This transient aspect of the climate leads to an important conclusion: even if CO_2 concentrations were to cease increasing, the surface temperature would still continue to increase by about 0.7 K (i.e. 1.3 K – 0.5 to 0.7 K) as it approaches its new equilibrium temperature.

Table 4 summarizes the total GHG effect on the equilibrium and transient temperature response for three periods. The equilibrium studies pertain only to the equilibrium warming of the surface to a step-function increase in GHG. The surface warming ranges of the transient studies reflect the uncertainties in the climate model's sensitivity in the assumed scenarios and in the assumed ocean diffusivity. The assessment suggests, for the period 1850-80 to 2030-50, a total GHG-induced cumulative equilibrium surface warming of about 1.5-6.1 K, of which only about 50% (i.e. about 1-3 K of the equilibrium warming) will be realized by 2030 due to the ocean's inertia.

Figures 3 and 4 show the transient global mean surface temperature increase as it evolves from 1820 to 2030 for the scenarios in Figures 1 and 2 and the GHG changes in Table 1. For the CO_2 effect alone (Fig. 3), the highest scenario could, by 2030, induce a warming similar to that of the Holocene Optimum, about 6,000 yr before

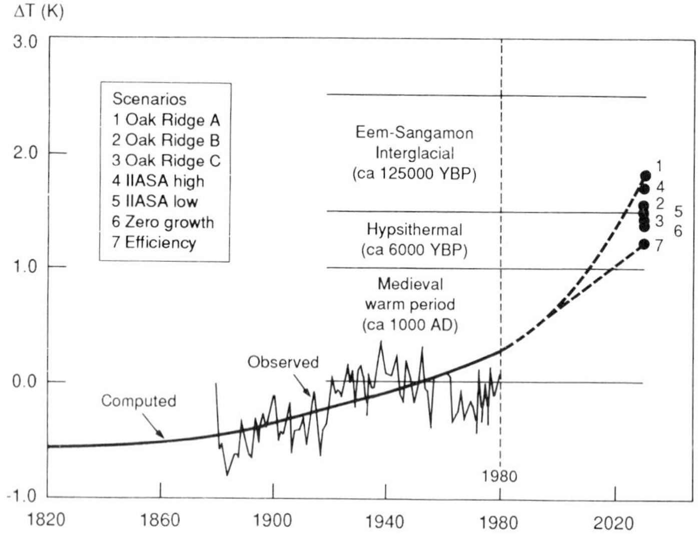

3. Simulation of temperature response to CO_2 effect only. Calculation from 1820 to 2030 with a parameterized form of a 1-D radiative convective model for the atmosphere (Hansen 1983) coupled to a 1-D energy balance model for the ocean (Cess and Goldenberg 1981). Observed temperature for the Northern Hemisphere from 1880 to 1980 (Jones et al 1982). Reference period 1946-60.

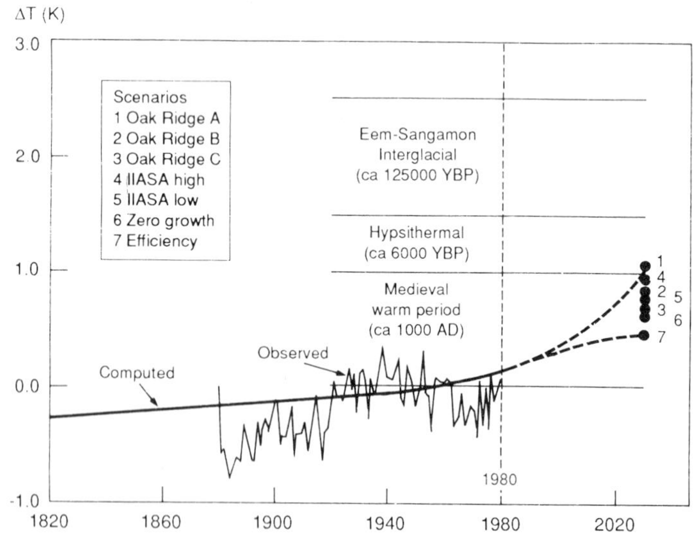

4. Simulation of temperature response to the combined effects of CO_2 and other GHG (CH_4, N_2O, CFC 11 and 12).

the present (yBP). The combined effect of CO_2 and the other GHG (Fig. 4) could even lead to a warming that has not been experienced since the Eem-Sangamon Interglacial (about 125,000 yBP).

Impacts of a greenhouse climate

There are three main approaches to studying the impacts of climatic change on society. The first is to take arbitrary changes in climate parameters (e.g. temperature, precipitation) and deduce crop yield changes. The second considers scenarios based on past climates as analogs of a greenhouse climate. The third is based on climate scenarios derived from model-generated simulations using a variety of GCMs.

Because of the uncertainties in climate projections and a host of other problems, it is not yet possible to assess with any degree of confidence the consequences of global and regional climate change. Nevertheless, it is important to indulge in some first-order impact estimates and, by doing so, to refine the methodology so that it will be available for impact analysis when it is needed (Bach 1988, Kates et al 1985, Meinl et al 1984, Parry et al 1988, Strain and Cure 1985, White 1985).

Agriculture

A number of GCMs indicate that, in a future greenhouse climate, the U.S. maize and wheat belts might become warmer and drier. Simple impact models for this new situation calculate reduced yields of as much as 26% for maize and as much as 10% for wheat. The margins of crop regions would shift poleward by as much as 100 km/degree of warming. Cropping patterns should be expected to be altered at the margins of crop regions. Laboratory experiments and some field tests indicate stimulation of plant growth due to enhanced CO_2 levels. This may be counteracted by the also-increasing levels of such plant and soil pollutants as O_3, SO_2, NO_x, acid rain, heavy metals, etc., as well as by increasing levels of pest and plant diseases in a warmer world.

Industrialized countries may be able to counter some of the impacts through technological developments, genetic diversity, and maintenance of food reserves. But current difficulties in responding to year-to-year climate variability (especially rainfall) are no cause for optimism in the developing world—particularly in light of the increasing population stress.

Water resources

Reduced rainfall and increased evaporation in a warmer world could, in some areas, dramatically reduce runoff, thereby significantly decreasing the availability of water resources for irrigation of crops and hydroelectric power production as well as for industrial or commercial and transport uses. A particular problem is anticipated in providing good-quality drinking water, in light of the increasing uses of artificial fertilizers and pesticides and the expanding production of excrement from feedlots, as well as the increasing numbers of hazardous waste dumps and catastrophes at chemical production plants.

Forests

In a warmer world, the mixed hardwoods and conifers of temperate latitudes would migrate farther north and replace some of the boreal forest. However, by the time the warming became effective, the current rapidly proceeding forest dieback in midlatitudes may have left little forest for migration. Similarly, the warmer and moister climate anticipated for tropical areas could favor expansion of tropical rain forest at the cost of subtropical moist forest—if indeed the ongoing deforestation has left much forest for migration. At any rate, the destruction of forests would feed back on the climate by changing the heat and water budgets as well as the trace gas composition of the atmosphere. Perhaps most worrisome is the speed with which the global warming occurs. There is reason for fear that the already stressed ecosystems may not be able to adapt.

Sea level rise

In a warmer climate, the sea level would rise through the expansion of ocean water and by the melting of snow and ice. With an expected warming of 1.5-4.5 K by the middle of the next century, the sea level could rise between 50 and 150 cm. A sea level rise in the upper part of this range could have very detrimental effects on low-lying coastal areas. These effects might include coastal erosion, increased frequency of storm surge flooding, salt water infiltration (and hence pollution of irrigation and drinking water), destruction of estuarine habitats, damage to coral reefs, etc. A catastrophic 5-6 m sea level rise due to the melting of the West Antarctic ice sheet is not expected to occur within the next century, but it remains a real possibility within the next 200-500 yr.

Uncertainties

The challenges that lie ahead grow out of the existing uncertainties. It is, therefore, useful to summarize the main areas of deficient knowledge (Bolin et al 1986, Wigley 1984, WMO 1985):

- The observed global warming of about 0.6 K over the past 100 yr is uncertain largely because of incomplete data coverage.
- Past trace gas concentrations are uncertain because the sources and sinks, as well as the chemical/climatic interactions, are uncertain. Future projections are even more uncertain, and thus, only scenarios of plausible changes can be devised.
- The time-dependent calculations of the response of the climate system to GHG forcing are very uncertain and must therefore have large error bars.
- The response of the climate system lags considerably behind the instantaneous equilibrium temperature (i.e. between about 10 and 100 yr). This lag is not only time-dependent, it is also sensitive to both the climate sensitivity and the rate of mixing below the mixed ocean layer.
- The sensitivity of the climate system due to the forcing by radiatively active gases is also uncertain, as is reflected in the range of equilibrium temperature change for a CO_2 doubling of 1.5-4.5 K. The same uncertainty is assumed to apply to the other GHG.
- Uncertainty arises from the incomplete knowledge of oceanic mixing processes. In the simple models currently used, these processes are parameterized by an eddy diffusion coefficient which is usually taken to range

between 1 and 3 cm^2/s. The mixing processes, however, are not strictly diffusive.

- Uncertainties come with the models themselves, since they can only be highly simplified replica, both physically and mathematically, of the complex climate system. Principal sources of uncertainty include cloudiness-radiation interactions, ocean and sea ice behavior, and land surface processes (including hydrology).
- The major uncertainty lies in the difficulty of foreseeing future human behavior.

Relief strategies — toward a growing consensus

It is, above all, the industrialized countries which, through their activities in the industrialized and the developing world, are responsible for influencing the world climate. Therefore, it is the industrialized countries which, in their own best interest, have the primary obligation to take countervailing measures—quite apart from the fact that only the industrialized countries have the technical means and the required financial resources to do so.

When should relief measures be initiated? The usual argument is that, until the large existing uncertainties have been reduced, countervailing measures would not be justified. This poses the pivotal question: How much certainty is enough to warrant action? This question involves a difficult value judgment, which has to be viewed against the risks and benefits involved. The danger is, if we wait until we have the certainty that will satisfy all critics, it may be too late for any countervailing action. Moreover, it should be realized that uncertainties always work both ways. This leads to the logical conclusion that it is not complete certainty—which is not achievable anyway—but the inherent uncertainty that supplies the strongest motive for precautionary action. To be on the safe side becomes the main motive for action.

The consensus is growing (e.g. Bach et al 1980, Bolin et al 1986, Trabalka 1985, WMO 1979) that in a world of uncertainty with a high climatic risk, it would be prudent to follow a safety strategy that simultaneously

- advances basic research as well as research on the causes and impacts of climatic change, and
- introduces precautionary measures.

The resulting low-risk policy consists of these relief strategies:

1. A more efficient use of energy resources. This
   - reduces energy demand, and hence fossil fuel use, which
   - results automatically in decreased emission, and hence
   - reduces the impact on environment and climate. It also
   - saves the nonrenewable fossil fuels for purposes less wasteful than mere combustion. The reduced demand
   - permits pollution-free renewable resources to contribute substantially to the energy supply.
2. A more effective control of the production of synthetic chemicals. This includes

- safety clearance of all new chemical substances as to their short-term and long-term effects, individually or in combination, on people, environment, and climate, and
- a replacement of all synthetic chemicals already in use that are detrimental to health and the environment.

3. A more rapid introduction of available abatement techniques. This includes emission reduction
   - before combustion (coal cleaning, coal gasification and liquefaction, desulfurization of liquid fuel),
   - during combustion (burner technology, fluidized bed combustion), and
   - after combustion (flue gas desulfurization, DENOX, catalyzer technology).
4. A more effective regulation of land use. This includes reducing
   - deforestation in the tropics,
   - biomass burning in the subtropics,
   - forest dieback in midlatitudes, and
   - soil destruction worldwide.

Implementation of this strategy would substantially reduce a GHG induced climatic threat.

The international science community has taken further steps to provide the scientific rationale for precautionary measures. On the recommendations of the Villach Climate Conference in October 1985, the Advisory Group on Greenhouse Gases (AGGG) was set up jointly by the International Council of Scientific Unions (ICSU), World Meteorological Organization, and United Nations Environment Programme in July 1986. The AGGG is to conduct

- biennial reviews of global and regional studies on greenhouse gases, and
- aperiodic evaluations of the rates of greenhouse gas increases and their effects.

The annually recurring ozone-hole over Antarctica has led to an initiative among 17 European research centers to establish the European Program in Chemistry of the Atmosphere. Although ozone will be the focal point of this scientific mission, many of the climate-related trace gases also will be investigated.

At its September 1986 General Assembly in Berne, ICSU launched a decade-long International Geosphere-Biosphere Program. Its objectives are

> "to describe and understand the interactive physical, chemical, and biological processes that regulate the total Earth system, the unique environment it provides for life, the changes that are occurring in that system, and the manner by which these changes are influenced by human actions."

In 1985 UNEP initiated a global framework convention for the protection of the ozone layer and the control of global warming, which entered into force in 1988. Its implementation plans, as laid down in the Montreal Protocol in 1987, will be strengthened in 1990.

In 1988 UNEP and WMU set up an Intergovernmental Panel on Climate Change with three working groups to help prepare both the Second World Climate Conference in Geneva in 1990 and the International Convention on the Protection of Global Climate in 1992.

References cited

Augustsson T U, Ramanathan V (1977) A radiative-convective model study of the CO_2-climate problem. J. Atmos. Sci. 34:448-451.

Bach W (1984) Our threatened climate: ways of averting the CO_2 problem through rational energy use. Reidel, Dordrecht.

Bach W (1986) Global studies of the atmosphere: modeling the transient and equilibrium climate response to greenhouse gases - impacts and strategies. Pages 250-291 *in* Proceedings of the international seminar on nuclear war. W. S. Newman and S. Stipcich, eds. Ettore Majoruna Center for Scientific Culture, Erice, Sicily.

Bach W (1988) Development of climatic scenarios from general circulation models. Pages 125-157 *in* The impact of climatic variations on agriculture. Vol. 1. M. L. Parry, T. R. Carter, and N. T. Konijn, eds. Kluwer Academic Publishers, Dordrecht.

Bach W, Pankrath J, Williams J, eds. (1980) Interactions of energy and climate. Reidel, Dordrecht. 569 p.

Bach W, Crane A J, Berger A L, Longhetto A, eds. (1983) Carbon dioxide. Current views and developments in energy/climate research. Reidel, Dordrecht. 525 p.

Bach W, Jung H J (1987) Untersuchung der Beeinflussung des Klimas durch Spurengase mit Hilfe von Modellrechnungen. Munstersche Geographische Arbeiten, Festschrift. 26:45-64.

Bolin B (1986) How much CO_2 will remain in the atmosphere? *In* Greenhouse effect, climatic change and ecosystems. B. Bolin, B. R. Döös, J. Jäger, and R. A. Warrick, eds. John Wiley and Sons, New York.

Bolin B, Döös B R, Jäger J, Warrick R A, eds. (1986) Greenhouse effect, climatic change and ecosystems. John Wiley and Sons, New York.

CDAC—Carbon Dioxide Assessment Committee (1983) Changing climate. National Research Council, National Academy Press, Washington, D.C.

Cess R D, Goldenberg S D (1981) The effect of ocean heat capacity upon global warming due to increasing atmospheric carbon dioxide. J. Geophys. Res. 86:498-502.

Donner and Ramanathan (1980) (incomplete bibliographic data)

Edmonds J A, Reilly J M (1985) Future global energy and carbon dioxide emissions. Pages 216-245 *in* Atmospheric carbon dioxide and the global carbon cycle. J. R. Trabalka, ed. U.S. Department of Energy, Washington, D.C.

Gates W L, Cook K H, Schlesinger M E (1981) Preliminary analysis of experiments on the climatic effects of increased CO_2 with an atmospheric general circulation model and a climatological ocean. J. Geophys. Res. 8:6385-6393.

Gates W L, Han Y J, Schlesinger M E (1984) The global climate simulated by a coupled atmosphere-ocean general circulation model: preliminary results. Climatic Research Institute Report No. 57, Oregon State Univ., Corvallis, Oregon.

Gates W L, Potter G L (1985) The response of a coupled atmospheric GCM and mixed layer ocean model to doubled CO_2. Abstract for Third Conference on Climate Variations: Symposium on Contemporary Climate 1850-2100, Los Angeles, January 5-11.

Hansen J (1983) Can we delay a greenhouse warming? U.S. Environment Protection Agency, Washington, D.C.

Hansen J, Lacis A, Rind D, Russell G, Stone P, Fung I, Ruedy R, Lerner L (1984) Climate sensitivity: analysis of feedback mechanisms. Pages 160-163 *in* Climate processes and climate sensitivity. J. Hansen and T. Takahashi, eds. Geophys. Monogr. 29. Maurice Ewing, Vol. 5. American Geophysical Union, Washington, D. C.

Jaeger J (1986) Climatic change: floating new evidence in the CO_2 debate. Environment 28(7):6-9, 38-41.

Jones P D, Wigley T M L, Kelly P M (1982) Variations in surface air temperatures. Part 1. Northern Hemisphere, 1881-1980. Mon. Weather Rev. 110(2):59-70.

Jones P D, Wigley T M L, Wright P B (1986) Global temperature variations between 1861 and 1984. Nature 322:430-434.

Kates R W, Ausubel J H, Berberian M, eds. (1985) Climate impact assessment: studies of the interaction of climate and society. SCOPE 27, John Wiley and Sons, New York.

Keepin W, Mintzer I, Kristoferson C (1986) Emission of CO_2 into the atmosphere. *In* Greenhouse effect, climatic change and ecosystems. B. Bolin, B. R. Döös, J. Jäger, and R. A. Warrick, eds. John Wiley and Sons, New York.

Lacis A A, Hansen J, Lee P, Mitchell T, Lebedeff S (1981) Greenhouse effect of trace gases, 1970-1980. Geophys. Res. Lett. 8:1035-1038.

Lal M, Dube S K, Sinha P C, Jain A K (1986) Potential climatic consequences of increasing anthropogenic constituents in the atmosphere. Atmos. Environ. 20(4):639-642.

MacCracken M C, Luther F M, eds. (1985a) Detecting the climatic effects of increasing carbon dioxide. U.S. Department of Energy, DOE/ER-0235, Washington, D.C.
MacCracken M C, Luther F M, eds. (1985b) Projecting the climatic effects of increasing carbon dioxide. U.S. Department of Energy, DOE/ER-0237, Washington, D.C.
Manabe S, Wetherald R T (1975) The effects of doubling the CO_2 concentration on the climate of a general circulation model. J. Atmos. Sci. 32:3-15.
Marland G, Rotty R M (1984) Carbon dioxide emissions from fossil fuels: a procedure for estimation and results for 1950-1981. Tellus 36(B):232-261.
Meinl H, Bach W, Jäger J, Jung H J, Knottenberg H, Marr G, Santer B, Schwiesen G (1984) Socio-economic impacts of climatic changes due to a doubling of atmospheric CO_2 content. Research Report to CEC/DFVLR, Dornier-System, Friedrichshafen.
Mitchell J F B (1983) The seasonal response of a general circulation model to changes in CO_2 and sea temperatures. Q. J. R. M. S. 109:113-152.
Neftel A, Moore E, Oeschger M, Stauffer B (1985) Evidence from polar ice cores for the increase in atmospheric CO_2 in the past two centuries. Nature 315:45-57.
Oeschger H, Siegenthaler U, Schotterer U, Gugelmann A (1975) A box diffusion model to study the carbon dioxide exchange in nature. Tellus 27:168-192.
Owens A J, Hales C W, Filkin D L, Miller C, Steed J M, Jesson J P (1985) A coupled one-dimensional radiative-convective, chemistry-transport model of the atmosphere. I. Model structure and steady state perturbation calculations. J. Geophys. Res. 90:2283-2311.
Parry M, Carter T R, Konijn N T, eds. (1988) The impact of climatic variations on agriculture. Vol. 1. Kluwer Academic Publishers, Dordrecht.
Ramanathan V, Cicerone R J, Singh H B, Kiehl J T (1985) Trace gas trends and their potential role in climate change. J. Geophys. Res. 90(D 3):5547-5566.
Schlesinger M E (1983) Simulating CO_2-induced climatic change with mathematical climate models: capabilities, limitations, and prospects. U. S. Department of Energy (U. S. DOE) 021, III.3-III.139. Washington, D.C.
Schlesinger M E (1986) Equilibrium and transient climatic warming induced by increased atmospheric CO_2. Clim. Dyn. 1(1):35-51.
Schlesinger M E, Han Y J, Gates W L (1987) The role of the ocean in CO_2-induced climatic change: a study with the OSU coupled atmosphere-ocean general circulation model. Climatic Research Institute Report No. 60, Oregon State Univ., Corvallis, Oregon.
Strain B R, Cure J D, eds (1985) Direct effects of increasing carbon dioxide on vegetation. U.S. Department of Energy, DOE/ER-0238, Washington, D.C.
Trabalka J R, ed. (1985) Atmospheric carbon dioxide and the global carbon cycle. U.S. Department of Energy, DOE/ER-0239, Washington, D.C. 316 p.
Trabalka J R, Edmonds J A, Reilly J, Gardner R H, Voorhees L D (1985) Human alterations of the global carbon cycle and the projected future. Pages 248-287 *in* Atmospheric carbon dioxide and the global carbon cycle. J. R. Trabalka, ed. U.S. Department of Energy, DOE/ER-0239, Washington, D.C.
Wang W C, Yung Y L, Lacis A A, Mo T, Hansen J E (1976) Greenhouse effects due to man-made perturbations of trace gases. Science 194:685-690.
Washington W M, Meehl G A (1983) General circulation model experiments on the climatic effects due to a doubling and quadrupling of carbon dioxide concentration. J. Geophys. Res. 88(C11):6600-6610.
Washington W M, Meehl G A (1984) Seasonal cycle experiments on the climate sensitivity due to a doubling of CO_2 with an atmospheric general circulation model coupled to a simple mixed layer ocean model. J. Geophys. Res. 89:9475-9503.
Wetherald R T, Manabel S (1986) An investigation of cloud cover change in response to thermal forcing. Climatic Change 8(1):5-23.
White M R, ed. (1985) Characterization of information requirements for studies of CO_2 effects: water resources, agriculture, fisheries, forests and human health. U.S. Department of Energy, DOE/ER-0236, Washington, D.C.
Wigley T M L (1984) Carbon dioxide, trace gases and global warming. Clim. Monit. 13(5):133-148.
WMO—World Meteorological Organization (1979) World climate conference. WMO-No. 537. Geneva.
WMO—World Meteorological Organization (1985) Atmospheric ozone 1985. Vol. III, Chap. 15. Trace gas effects on climate. Rep. 16. Geneva. (see also volumes I and II).

Notes

Author's address: W. Bach, Center for Applied Climatology and Environmental Studies, Department of Geography, Climate and Energy Research Unit, University of Münster, Robert-Koch-Str. 26, Germany.

Citation information: International Rice Research Institute (1989) Climate and food security. P.O. Box 933, Manila, Philippines.

Predictability of weather and climate: long-range forecasting of Indian monsoons

J. Shukla

Space-time variability of monsoon rainfall over India is described, and possible mechanisms for variability on intraseasonal and interannual time scales are suggested. Large-scale persistent anomalies of monsoon rainfall over India are associated with anomalies in planetary scale circulation and boundary conditions, and these relationships are sufficiently strong to be useful in predicting seasonal rainfall.

The sun is the ultimate source of energy for the atmospheric movements that give rise to weather and climate variations. Because the earth's surface is nearly spherical, and the earth revolves around the sun and rotates on its own axis, the amount of solar energy that falls on any part of the surface of the earth-atmosphere system varies with time of day and day of the year. The chemical composition of the atmosphere and the temperature distribution determine the rate at which radiative processes heat or cool different parts of the earth-atmosphere system. In general, the mean climate of the earth is an equilibrium among various physical processes that are directly or indirectly related to the size of the planet, the acceleration due to gravity, and the distribution of ocean, land, and mountains on the earth's surface.

The fact that winds are continuously blowing indicates a permanent source of kinetic energy to compensate for the continuous frictional dissipation. In this respect, the atmosphere can be considered as a heat engine in which the warmer equatorial regions are being heated and the cooler polar regions are being cooled, thus producing energy to sustain the winds.

Within the context of the mean climate of the earth-atmosphere system, two questions are posed: Why does weather change from day to day? Why are monsoons different from one year to the next?

The causes of day-to-day weather changes can be understood by following the concept of hydrodynamic instabilities of fluids, particularly the instabilities of stratified and rotating fluids. The mean climate of the earth's atmosphere is characterized by horizontal and vertical shears of wind, temperature, and moisture. These shears are such that very small disturbances can grow into large aperiodic fluctuations. The day-to-day changes in weather can be understood as manifestations of the growth, decay, and propagation of these unstable disturbances. This is also the principal reason why weather changes cannot be predicted in great detail beyond a few weeks. Because both the growth rate of these instabilities and the state of the atmosphere at any instant are largely uncertain, predictions of the instantaneous state of the atmosphere are useless after a finite amount of time.

During the past 30-40 yr, meteorologists have been greatly interested in determining the limits of weather predictability. Based on a variety of theoretical, observational, and computer modeling studies, it is now generally believed that

- The instantaneous large-scale (a few hundred kilometers) weather in the extratropical regions cannot be predicted beyond 2-4 wk. (We consider the weather unpredictable beyond a period at which the prediction is not better than a randomly chosen weather map.)
- The instantaneous large-scale weather in the tropical regions cannot be predicted beyond 3-7 d.
- Monthly and seasonally averaged weather over large areas (a few thousand kilometers) is more predictable for the tropical regions than for the extratropical regions.

Mechanisms for interannual variability of the Indian monsoon

In contrast to the mechanisms of day-to-day weather changes, which are hydrodynamic instabilities, changes in monthly and seasonally averaged weather (particularly the monsoons) are due to changes in the slowly varying boundary conditions at the earth's surface, which include sea surface temperature (SST), soil moisture, and snow cover. Changes in these conditions can produce substantial changes in atmospheric circulation and rainfall. For example, during certain years the upper layers of the Equatorial Pacific Ocean are warmer by 1-2 °C. This seemingly small change in ocean temperature (referred to as an El Niño episode) produces large changes in the magnitude and the distribution of rainfall over the entire Pacific Ocean and adjoining land. Also, changes in SST in the Equatorial Pacific are significantly related to the Indian monsoon rainfall.

Figure 1 shows the monsoon season rainfall anomaly over India for 116 yr (1871-1986) normalized by its standard deviation. The rainfall anomaly for any year is defined as the departure of rainfall for that year from the long-term mean rainfall. A value of 1 on the ordinate signifies rainfall equal to its mean value (853 mm) plus 1 standard deviation (84 mm), or a total of 937 mm. A value of zero means that the rainfall was equal to the mean (853 mm), and a value of −1 means that the rainfall was the mean $^{-1}$ standard deviation or 769 mm. There were no well-defined trends or periodicities in the rainfall series. The years denoted by black and hatched bars will be referred to as the warm and the cold Pacific years, respectively. Twenty of 24 yr denoted by black bars (warm Pacific years) had below-normal rainfall, and all the years denoted by hatched bars (cold Pacific years) had above-normal rainfall over India. Most of the severe drought years are denoted by the black bars, and only 1 yr (1944) denoted by a black bar had a positive rainfall anomaly of more than one-half the standard deviation. Likewise, most of the excessive rainfall years are denoted by hatched bars.

Figure 2 shows the mean rainfall anomaly (normalized) for different subdivisions of India for warm and cold Pacific years, respectively. The warm Pacific years are

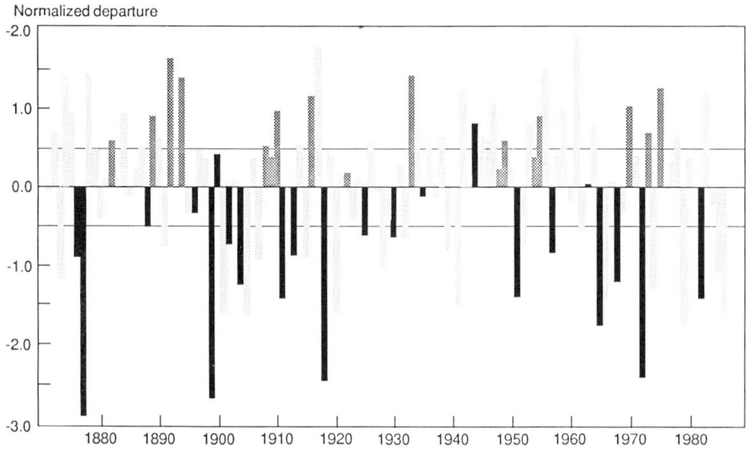

1. Indian summer monsoon rainfall anomaly (departure from long-term mean). Values of 1.0 and -1.0 mean rainfall were 84 mm above and below normal rainfall, respectively. Zero line represents normal rainfall (= 853 mm). Solid and hatched bars denote warm and cold Pacific years, respectively.

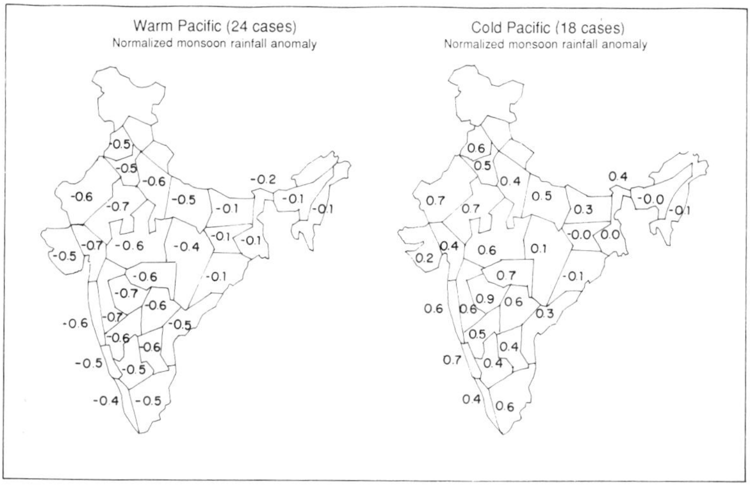

2. Rainfall anomaly (normalized) for Indian subdivisions averaged for years of warm and cold Pacific years.

characterized by widespread droughts over India, and cold Pacific years are characterized by widespread excessive monsoon rainfall. These observational results provide strong evidence for a significant relationship between Equatorial Pacific SST anomalies and Indian monsoon rainfall anomalies.

Figure 3 shows the mean sea level pressure anomaly over Darwin (Australia) for the warm and the cold Pacific years. The warm Pacific years (monsoon drought years) were characterized by large positive anomalies of Darwin pressure, and the

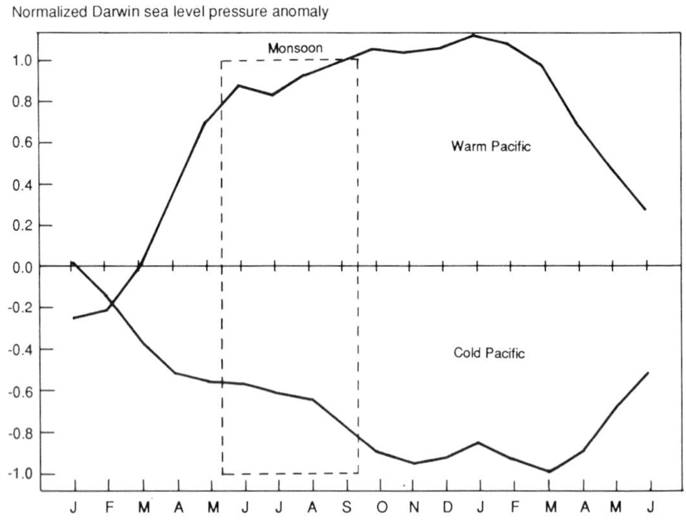

3. Sea level atmospheric pressure anomaly at Darwin averaged for years with warm and cold Pacific.

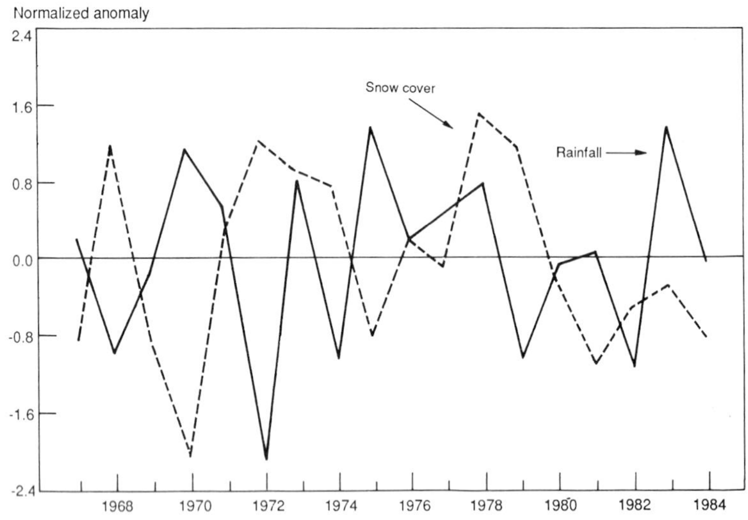

4. Indian monsoon rainfall anomaly and Eurasian snow cover anomaly.

cold Pacific years (monsoon flood years) by large negative anomalies of Darwin pressure. The tendency of Darwin pressure from the preceding winter to spring season is opposite in both cases.

Another example is the apparent relationship between Indian monsoon rainfall and snow cover over Eurasia shown in Figure 4. Winter seasons with a large amount of snow cover over Eurasia were followed by monsoon seasons with deficient rainfall.

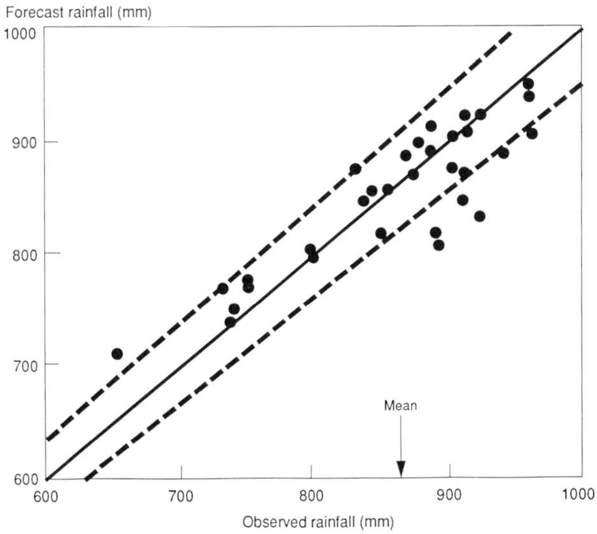

5. Observed and forecast rainfall (mm) for 32 yr of verification. Arrow indicates long-term mean rainfall (853 mm). Dashed lines represent ±5% of the solid line.

The existence of a significant relationship between the slowly varying boundary conditions, global circulation, and monsoon rainfall over India has provided an observational basis for computer modeling studies to simulate these relationships numerically. Several empirical approaches are also being pursued to produce operational long-range forecasting of Indian monsoon rainfall.

Long-range forecasting of monsoon rainfall

Since monsoon rainfall fluctuations are of vital importance to agriculture, drinking water supply, and energy planning in India, numerous attempts have been made to develop techniques to predict monsoon rainfall. These techniques are based on the premise that global changes have a long enough time scale to predict the future state of the atmosphere from the present or from the immediate past. The results presented in the preceding section give some support to the validity of this premise.

Figure 5 shows the verification of a simple regression model developed by D. A. Mooley and the author to predict seasonal mean monsoon rainfall over India. The two predictor parameters used in the regression equation are the tendency of the Darwin sea level pressure, which represents a broader phenomenon referred to as the Southern Oscillation, and the latitudinal location of the 500 mb ridge during April, which represents the mid-tropospheric circulation over India. The regression equation was developed using 30 yr of independent data. The root mean square error of predictions is about 36 mm, which is only about 4% of the mean rainfall, and the prediction of droughts is especially good.

Conclusions

Although there is an upper limit of a few days to a few weeks for deterministic prediction of instantaneous weather, there appears to be some potential for predicting space- and time-averaged circulation and rainfall. This potential appears to be particularly high for predicting seasonal mean monsoon rainfall averaged over large (1,000,000 km^2) spatial scales. Evidence for predictability of monthly and seasonal rainfall over small regions (approximately 100,000 km^2) is yet to be found.

Notes

Author's address: J. Shukla, Center for Ocean-Land-Atmosphere Interactions, Department of Meteorology, University of Maryland, College Park, MD 20742, USA.

Citation information: International Rice Research Institute (1989) Climate and food security. P.O. Box 933, Manila, Philippines.

Stochastic models for rainfall prediction

P. K. Das

In India, long-range forecasts for the onset of summer monsoon rains and total monsoon rainfall are made using multiple regression equations, using predictors selected by a screening procedure. But the association between any given predictor and rainfall changes with time. In addition, there is considerable variation in rainfall within any given meteorological subdivision. When a stable predictor is available, multiplicative autoregressive models for rainfall prediction yield encouraging results. A good predictor for summer monsoon rain is the April position over the Indian peninsula of an upper air ridge at 500 hPA. A model was fitted to the time series of Kabete, near Nairobi in Kenya, and Dinapur, a station with a long time series in Bihar, in eastern India. A nonseasonal component could be discerned in the Kabete time series; it was not visible in the Bihar time series.

In India, the average countrywide rainfall during the monsoon is about 85 cm. In a good monsoon year, the average exceeds 100 cm; a poor monsoon can average only 60 cm. Because of the wide variation between good and poor monsoon rains, long-range predictions are needed.

Long-range forecasts of the summer monsoon (Jun-Sep) rains are prepared each year. They indicate a) the likely date of the onset of monsoon rains over the southern tip of India, b) the total rainfall likely over the Indian peninsula and the northwestern areas of the country, and c) the likely rainfall during the second half of the monsoon for the Indian peninsula and the northwestern areas. Long-range forecasts are made for areas with a high coefficient of variability (Standard deviation ÷ Mean x 100).

Currently, such predictions are made using multiple regression equations that seek a statistical association between monsoon rains and a number of antecedent events or predictors. For example, the predictors for monsoon rainfall over the Indian peninsula are a) pressure differences from normal (in mm) in April and May for Buenos Aires, Cordoba, and Santiago in South America (x_1); b) mean position in April of an upper air ridge at 500 hPA along 75° E and near 12° N, measured in degrees of latitude (x_2); and c) the mean minimum temperature (°C) in March for stations at Jaisalmer, Jaipur, and Calcutta, India (x_3). The rainfall (R) is expressed by

$$R = a_0 + a_1x_1 + a_2x_2 + a_3x_3 \quad (1)$$

where the constants a_0, a_1, a_2, and a_3 are determined by past data. The joint correlation coefficient between R and x_1, x_2, and x_3 is approximately 0.6.

The statistical association between a set of predictors and rainfall changes with time (Rao 1964, Rao et al 1972), and after some years, many predictors lose their utility. Fortunately, the mean position in April of the upper air ridge has remained a fairly stable predictor for a fairly long time (Banerji et al 1978).

Prediction by this means has low resolution in space and time. Ideally, we need forecasts of monthly rainfall over small areas to plan agricultural operations.

Monthly forecasts of rainfall using empirical methods have been tried for 35 meteorological subdivisions of India. But the boundaries of a meteorological subdivision do not necessarily coincide with the political boundaries of a state. Because the density of rain gauges is not the same in each state, monthly rainfall forecasts have not been very successful.

We examined the potential of autoregressive models for rainfall prediction. Specifically, we tried to find the year-to-year dependence of rainfall at a station using a multiplicative seasonal and nonseasonal autoregressive model. The model will indicate to what extent the rainfall of a given month in 1 yr is related to the observation of the previous year.

Two temporal relationships are of interest: between observations of successive months in a particular year and between observations of the same month in successive years. The first could be important for a station that has two peaks of rainfall, a bimodal distribution.

Autoregressive models

An autoregressive model is given by the expression

$$Z_t = \sum_{j=1}^{p} \phi(j) Z_{t-j} + a_t \tag{2}$$

where Z_t represents rainfall at a given time (t), $\phi(j)$ are constants, and p is a finite integer, referred to as the order of the model. Deviations from model-generated values of Z_t are denoted by a_t, a zero mean, uncorrelated, random variable that has a variance σ_a^2 (referred to as noise).

The noise elements may be replaced by a weighted average of earlier values. This leads to the expression

$$Z_t = a_o + \sum_{j=1}^{p} \phi(j) Z_{t-j} + \sum_{j=o}^{q} \theta(j) a_{t-j} \tag{3}$$

where $\theta(j)$ are constants (θ_0 may be taken to be unity). The second term on the right is an autoregressive (AR) process; the last term is a moving average (MA) process.

The variance of Z (σ^2) is related to the variance of noise (σ_2^2) by

$$\sigma_a^2 = \sigma^2 \left[1 - \sum_{j=1}^{p} \phi(j)\, \rho(j) \right] \tag{4}$$

where $P(j)$ is the lag-j autocorrelation coefficient defined by

$$\rho(j) = \frac{1}{\sigma^2} \sum_{j=1}^{p} Z_t Z_{t-j} \quad (5)$$

Different procedures exist for using equation 5 to determine autoregressive coefficients $\phi(j)$ (Box and Jenkins 1976).

To make the time series stationary, it is necessary to difference the series by a backward shift operator (B). This is

$$B = \nabla Z_t = Z_t - Z_{t-1} \quad (6)$$

If the series is differenced 'd' times, then the order of the model is (p, d, q). When the time series has both a seasonal and a nonseasonal component, a multiplicative model is designed to consider both types of fluctuations. The seasonal component is expressed by

$$Z_t = \sum_{i=1}^{P} \bar{\phi}(i) Z_{t-is} + \sum_{i=1}^{Q} \bar{\theta}(i) a_{t-is} + a_t \quad (7)$$

where s is the seasonal period. Considering the 12 mo of the year, $s = 12$. The order of the seasonal model is then $(P, d, Q)_{12}$.

If we express a_t by a nonseasonal model of order (p, d, q), we have

$$a_t = \sum_{j=1}^{p} \phi(j)\, a_{t-j} + \sum_{j=1}^{q} \theta(j)\, a_{t-j} \quad (8)$$

Combining equations 7 and 8, we have a multiplicative model of order $(p, d, q) \times (P, d, Q)_{12}$.

The first component (p, d, q) is nonseasonal. It tells us how the rainfall values in a particular month may be related to rainfall in previous months of the same year. The second, seasonal component tells us how the rainfall in any month is related to the rainfall of that same month in previous years. The important model parameters for rainfall studies appear to be p or P. They indicate to what extent rainfall is influenced by the observations of previous months or years. For a first-order process, the autocorrelation coefficient decays exponentially to zero.

Sometimes analyzing autocorrelation functions (acf) does not clarify what should be the correct order of a model. Akaike's Information Criterion (AIC) (Akaike 1974) is often used to select the correct order. It measures the distance between a model derived from a finite database and a true order model based on an infinite database. If an $AR(p)$ model is fitted to the data, then p usually will be less than the true order. The estimated residual variance will be larger, because additional terms that the model might have missed could explain a larger part of the total variance.

AIC seeks to maximize the probability of getting the true distribution from a fixed sample. The maximizing problem is replaced by the minimizing problem, as

$$\text{AIC (min)} = N / n^{\sigma_a^2} + 2 \text{ (sum of model parameters)} \quad (9)$$

where N is the total number of observations. Model parameters p and q are varied until a minimum AIC is reached. (We do not consider here criteria for model selection that take into account the size of the sample in addition to model parameters.)

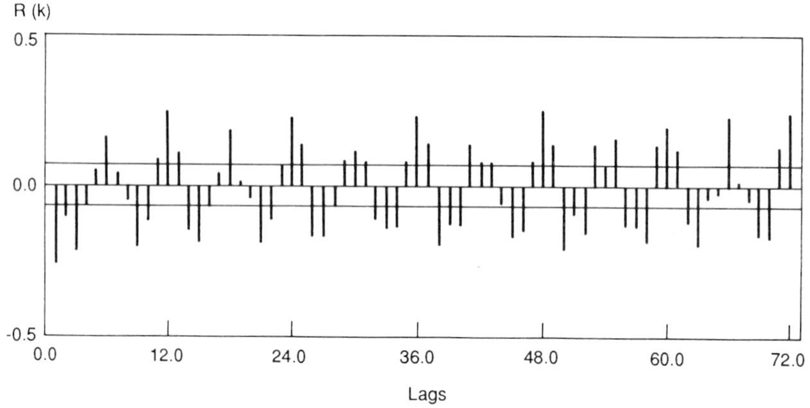

1. Autocorrelation function for Kabete.

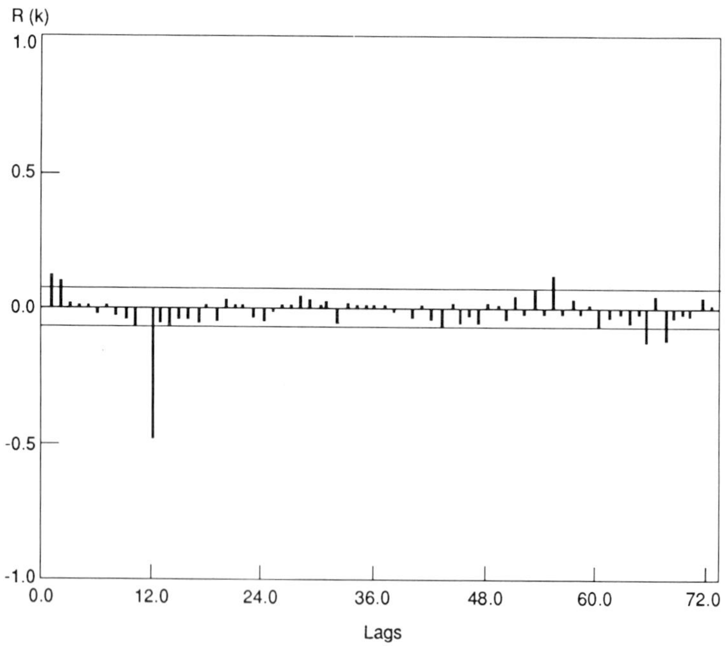

2. Autocorrelation function after seasonal differencing for Kabete.

Results

Data from two stations were considered: Kabete, near Nairobi (0.4° S, 38° E) in Kenya, and Dinapur, in Bihar state in eastern India (24.5° N, 84.8° E). The objective was to compare the model's performance for an African station, where rainfall is bimodal, with its performance in an Indian station, where rainfall has a single annual peak.

Kabete, Kenya

Figure 1 shows the acf for ∇Z_t. The horizontal lines indicate 95% confidence limits. The dominant feature is a sinusoidal pattern representing seasonal variation.

Figure 2 shows $\nabla_{12}Z_t$. This is the seasonal pattern after differencing once. An AR(I) and a MA(I) component may be inferred from the rapid decline of acf (positive values) after lag 1. The MA(I) component coincides with lag 12, which represents a seasonal cycle. The appropriate model is, therefore, $(1, 0, 0) \times (0, 1, 1)_{12}$. The autoregressive coefficient ϕ (1) for the nonseasonal component was 0.4; the

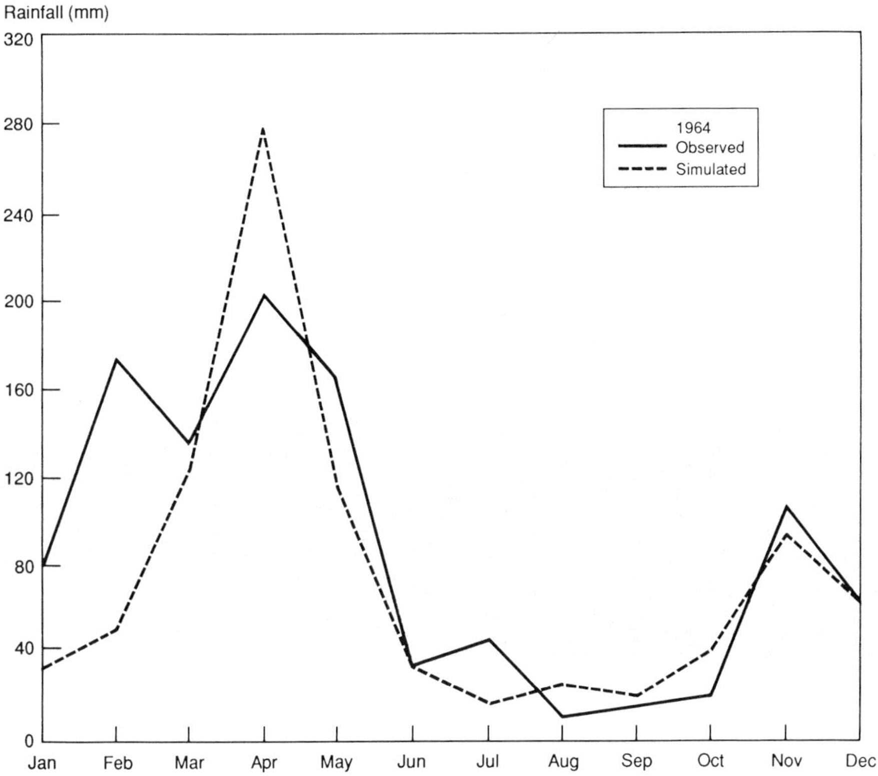

3. Simulated and observed rainfall, Kabete, 1964.

MA(I) coefficient $\bar{\theta}$ (1) was 0.9. These parameter values were verified by finding the sum of squares surface

$$\Sigma\, a_t^2\, (\phi, \theta, \overline{\phi}, \overline{\theta}) \qquad (10)$$

for a range of parameter values and locating the minimum (Salas et al 1980).

Using these values for the coefficients, we compared simulated and observed rainfall for 3 yr (1964, 1965, 1966). Data for 1916-63 were used to fit the model. The results of the comparison are shown in Figures 3, 4, and 5. The model simulates the bimodal character of rainfall fairly well, but the peak rainfall values are underestimated.

If the mean rainfall is used as a forecast, the residual variance is 493 $(mm)^2$. But if the model is used to forecast rainfall, the residual variance is reduced to 185 $(mm)^2$.

Dinapur, Bihar

Data for different time spans were available from 479 stations in 16 districts of Bihar. The coefficient of variation ranged from 15 to 45%. No systematic pattern in the coefficient of variability was found.

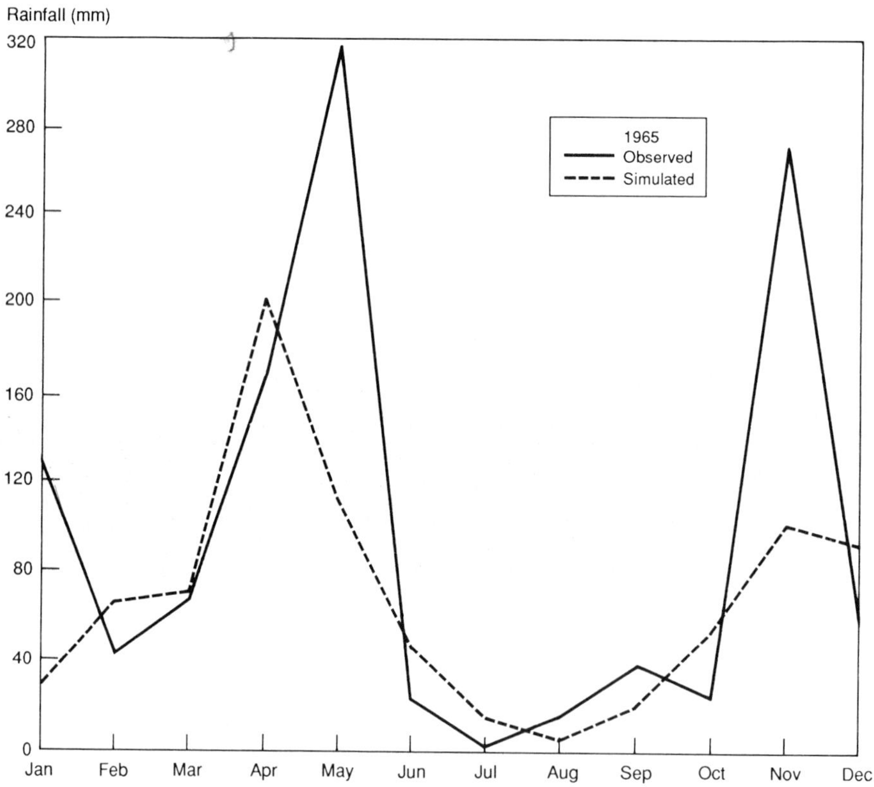

4. Simulated and observed rainfall, Kabete, 1965.

For comparison with Kabete, the time series of Dinapur was used. It had the longest time series (1901-70) of all stations in Bihar. Figure 6 shows the acf values for

$$\nabla_{12} Z_t \tag{11}$$

Only one dominant value at lag 12 is seen: there is no indication of an AR component.

Similar computations were made considering all the stations of Bihar, taking mean values for each month and year. In another series, the stations were clustered according to the coefficient of variability; stations with high and low coefficients of variability were considered separately. But no AR component was found in all these experiments.

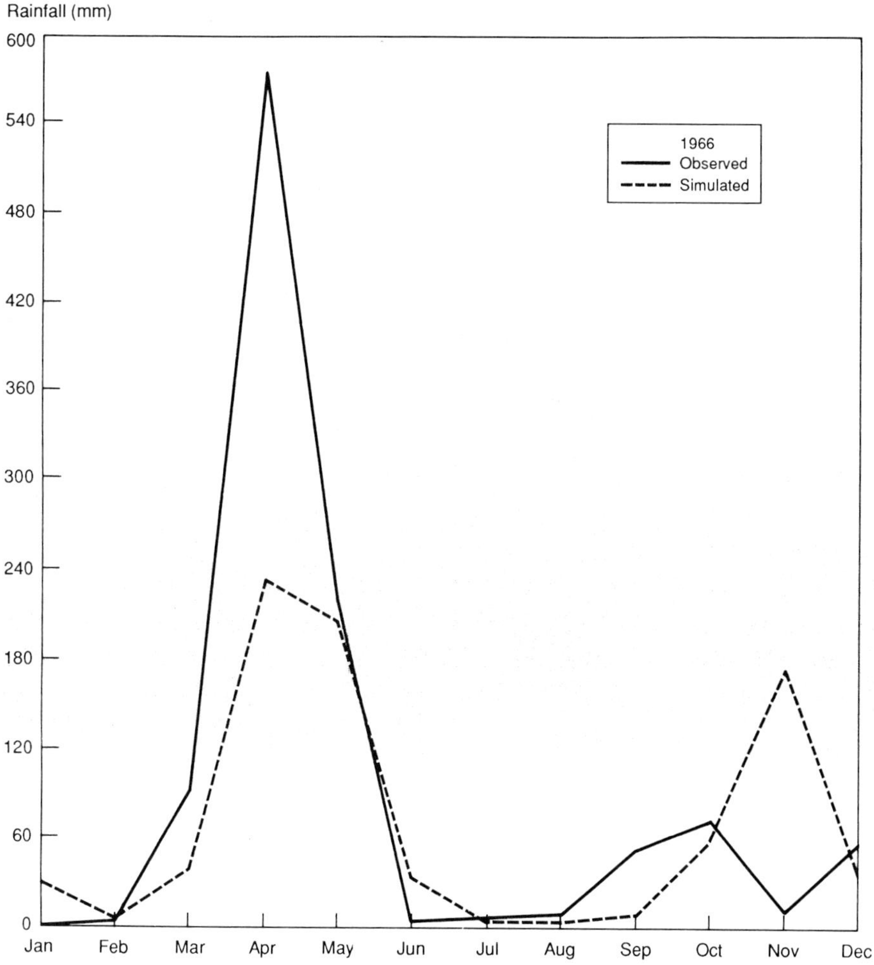

5. Simulated and observed rainfall, Kabete, 1966.

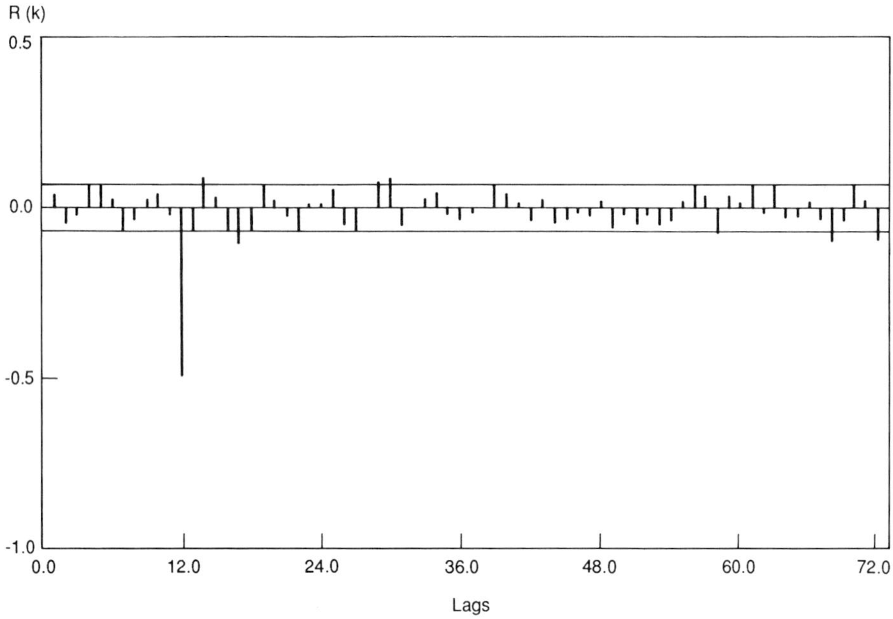

6. Autocorrelation function after seasonal differencing for Dinapur, Bihar.

The absence of an AR component suggests the absence of any nonseasonal fluctuation, such as a low frequency (30-40 d) cycle in rainfall. But it is worthwhile to note that the available data were not for the same time period.

Summary

The main conclusions of the study are

- Information on seasonal and nonseasonal components of rainfall may be obtained through a multiplicative autoregressive model.
- Kabete, near Nairobi, Kenya, shows a nonseasonal component in addition to seasonal variation in rainfall.
- The residual variance is reduced if a multiplicative autoregressive model is used to forecast rain, but the peak values of rainfall are underestimated.
- No nonseasonal component could be found in the rainfall of Bihar, India.

References cited

Akaike H (1974) Markovian representation of stochastic processes and its application to the analysis of autoregressive moving average processes. Ann. Inst. Stat. Math. 26:363-387.

Banerji A K, Sen P N, Raman C R V (1978) On foreshadowing southwest monsoon rainfall over India with mid-tropospheric circulation anomaly over India. Indian J. Meteorol. Hydrol. Geophys. 29:425-431.

Box G E P, Jenkins G M (1976) Time-series analysis, forecasting and control. Rev. ed. Holden Day Inc., San Francisco, USA. 575 p.

Rao K N (1964) Seasonal forecasting - India. World Meteorol. Org. Tech. Note 66. Geneva.
Rao K N, George C J, Ramasastry K S (1972) Agroclimatic classification in India. I.M.D. Met. Monogr. Agric. Meteorol. 4. 18 p.
Salas J D, Delleur J W, Yevjevich V, Lane W L (1980) Applied modelling of hydrologic time series. Water Resources Publications, Colorado, USA. 484 p.

Notes

Author's address: P. K. Das, Centre for Atmospheric Sciences, Indian Institute of Technology, New Delhi 110016, India.
Citation information: International Rice Research Institute (1989) Climate and food security. P.O. Box 933, Manila, Philippines.

Improving response to drought

W. E. Easterling and D. A. Wilhite

Drought, for all of its destructive potential, is a normal, recurrent feature of climate. Deliberate efforts to mitigate the negative consequences of drought too often are the product of ad hoc crisis management when, in fact, prescriptive risk assessment planning is a far more effective strategy. Attempts by governments and donor organizations to respond to drought have proved largely untimely and ineffective. This paper reviews the outcome of a recent international symposium and workshop aimed at improving international capacity to assess and respond to drought. State-of-the-science of drought is reviewed under the rubrics of prediction, detection and monitoring, impact assessment, and response. Major impediments to the formulation of drought assessment and response plans are identified. Actions are proposed to remove these impediments. Recommendations and research needs that would facilitate development of drought contingency plans in both developed and developing countries are presented.

Drought is a normal feature of climate. Recent global experience with this recurring natural hazard has demonstrated the lack of ability on the part of individual citizens, governments, and international organizations to assess and respond in an effective, timely, and coordinated manner. Recent examples include the mid-1970s droughts in the United States and Canada, the 5-yr drought in northeast Brazil from the late 1970s through the early 1980s, the enduring desiccation in the Sahel region of Africa, and the severe drought of 1986 in southeastern United States. When considered in a global context, this lack of preparedness and inability to respond effectively, coupled with projections of increasingly burdened agricultural carrying capacities, is a major concern.

The Institute of Agriculture and Natural Resources of the University of Nebraska and the Department of Energy and Natural Resources of the State of Illinois sponsored an international symposium and workshop on drought in the fall of 1986 at the University of Nebraska-Lincoln to address the issue of drought planning. Funds were provided by the U.S. National Science Foundation, the National Climate Program Office, the United States Agency for International Development (USAID), the U.S. Department of Agriculture, the United Nations Environment Programme (UNEP), the United Nations Development Programme, the World Meteorological Organization, and the University Corporation for Atmospheric Research. Some 150 scientists and policymakers from 25 countries participated.

The symposium program was organized to give participants a broader understanding of the concept of drought and to establish lines of communication among participants. It also provided an excellent opportunity to review state-of-the-art technologies in drought prediction, monitoring, impact assessment, adaptation, and response.

The workshop was the culmination of discussions initiated during the symposium, directed toward the specific question: How can the information presented at the symposium on available and emerging technologies in drought prediction, monitoring, impact assessment, adaptation, and response be used to enhance the drought-coping capacity of state and national governments and international organizations?

Specific workshop objectives were

- To identify information needs and opportunities that could improve national and international capacity to assess and respond to drought.
- To develop an agenda of drought-related research priorities.
- To make recommendations on ways to stimulate the development of drought contingency plans by national governments.
- To suggest ways of involving international organizations in promoting that planning process.

The ultimate goal was progress toward a plan of action for national governments and international organizations. This paper reviews some of the major insights gained in workshop. (Proceedings of the International Symposium and Workshop on Drought will be published by Westview Press, 5500 Central Ave., Boulder, CO 80301, USA, in late spring 1987).

Organization

The workshop, like the symposium, approached the issue of drought in an integrated assessment framework. The building blocks of this approach were drought prediction; monitoring, detection, and early warning; impact assessment; adaptation and adjustment; and planning and response. This approach closely parallels typical models of climate and society interaction (e.g. Kates et al 1985) and thus provides an underlying scientific rationale for organizing the issues of drought.

Some 60 scientists and senior-level policy officials were invited to participate in workshop discussions. The co-organizers sought a balance of scientists and policymakers with representation from many drought-prone regions. Participants included symposium speakers, scientists supported by the USAID/Bureau for Africa, policy officials supported by the UNEP, and some symposium participants who have experience on issues pertaining to drought management. Representatives of 21 drought-prone countries from all continents participated in the discussions.

Workshop participants were divided into five task force groups reflecting the integrated drought assessment components. Participants were placed in a task force on the basis of special expertise; organizers also maintained a relatively equal distribution of policy officials.

A primary goal in dividing participants into groups was to stimulate dialogue among scientists of the many relevant disciplines represented and also between scientists and policymakers. A truly interdisciplinary, science-policy forum was sought: most of the issues concerning drought preparedness reside at the interstices of disciplines and of science and policy. On any spatial scale, the optimal drought policy should be based on scientific understanding and the technological advances that accompany such understanding. At the same time, scientific research aimed at alleviating drought problems should be conducted with an understanding of the realities of policy development and implementation.

Each group was asked to provide written responses to two sets of questions. The first set involved three general questions, the second set was specific to the topical area of the task force. Groups also prepared summaries of the key elements of their discussion. These responses were distributed to participants and summaries were presented during a final plenary session.

Constraints to drought planning

Each group was asked to discuss and respond to the following general questions:

1. What factors are considered to be constraints to the formulation and implementation of drought plans by state and national governments?
2. How can these constraints be circumvented to facilitate the development of drought plans?
3. How can international organizations promote and facilitate drought planning activities?

Only the major themes are identified here. Responses to questions one and two have been combined to facilitate discussion of these issues. The order in which the discussion is presented does not reflect the relative importance of the constraints.

Drought: an interdisciplinary problem

The science of drought spans many disciplinary boundaries, from the physical sciences to the social sciences. Much of the previous drought research has been addressed by several disciplines working somewhat independently of one another. The greatest progress toward effective drought planning will be achieved with several concerned scientific disciplines working together, in an interdisciplinary effort.

Information gaps

A particularly vexing problem in the development of effective drought plans is the general lack of scientific data and information (meteorological, agricultural, economic, demographic, etc.) to track the development of drought and its impact. This problem is compounded by insufficient databases, especially in less developed drought-prone regions.

Coupled with this problem is the ineffective transfer of existing information from the scientific community to appropriate policymakers. Furthermore, policymakers may challenge the usefulness of this information in formulating policies.

It is critical that historical databases and existing reporting systems be protected. It is essential that remote sensing techniques be enhanced and that predictive models be developed through diagnostic and impact studies. Some data acquisition systems could be developed by piggy-backing these programs on other high profile programs.

Impact sensitivities and adaptations

More knowledge of the approximate (secondary) impacts of drought as well as proximate (direct) impacts is needed. Moreover, equal effort should be given to identifying the possible range of adjustment and adaptations available to lessen the impacts of drought. Specifically, detailed assessments of previous droughts could provide guidance for planning for future droughts. In addition, natural analogues to drought, such as persistent drawdown of slow-recharge aquifers in irrigation-dependent agricultural regions, provide opportunities to study drought-like conditions.

Poor policymaker understanding

Policymakers and bureaucrats often view drought as an extreme event and, implicitly, rare and of random occurrence. Thus, droughts are viewed as something outside government control. If droughts continue to be perceived by policymakers as a freak quirk of nature—one that cannot be planned for—they will not be planned for. A related constraint is that when policymakers accept the challenge of planning for drought, they implicitly assume responsibility or blame for the negative effects of drought.

Policymakers, bureaucrats, and the general populace, must be taught that droughts, like floods, are a normal feature of climate. Better understanding of the seriousness of drought and of the advantages of preparedness for drought would facilitate the development of effective drought plans. Awareness of the likelihood of droughts of varying intensities and durations and their probable associated impacts is essential. Appropriate planning implemented during nondrought periods can mitigate some of the impacts, reduce hardships, and improve the government's ability to respond effectively during crisis periods. This, in turn, improves the image of government with its constituents.

The economics of preparedness

Drought preparedness requires resources. The costs of maintaining a high level of preparedness may impede the development of drought plans. Indeed, a major difficulty in assessing the economics of drought preparedness is comparing the benefits of drought planning with the costs of drought. Preparedness costs are fixed and occur now. Given an uncertain discounting, in the future they may be viewed as being too high. Further complicating the issue is the fact that the costs of drought are not solely economic. They must also be stated in terms of human suffering and degradation of physical resources: items with values inherently difficult to estimate.

Postdrought evaluations of governmental responses in the U. S. and elsewhere have shown those efforts to be largely ineffective, poorly coordinated, untimely, and economically inefficient. During the mid-1970s drought in the U.S., for example, the Federal Government appropriated more than $8 billion for drought relief through 40 separate programs administered by 16 different agencies. Although the annual cost of preparedness in the U.S. is not known, the inefficiency of the current approach suggests the need to examine other alternatives. The U.S. General Accounting Office recommended the development of a national drought plan in 1979; no action has been taken. The Australian Government, after the severe 1982-83 drought, examined the need for a national drought policy and has, after considerable study, moved forward with the rudiments of such a policy.

A principal advantage of drought planning is that it is done during noncrisis periods and thus can give due consideration to the complex issues involved. It is essential that drought planning be integrated into existing socioeconomic and political structures at appropriate levels of government and that the process be sufficiently coordinated to avoid duplication. Moreover, drought plans should be incorporated into existing natural disaster plans. All these steps will reduce the costs of preparedness.

A related issue is that drought, like most natural hazards, is not consistent with annual government budgetary cycles. This makes it difficult to dedicate budget lines for drought preparedness, especially given the typical short-term outlook of governments.

Drought plans should be updated annually and the funding required to support continuous activities, such as monitoring, should be allocated each fiscal year. This will ensure a dynamic planning process that reflects changing socioeconomic and political trends.

Once drought plans have been established, a clear path from the drought planning team to high-level government decisionmakers is needed. Streamlining the transmission of a drought response protocol from the planning team to decisionmakers is crucial if plan is to be carried out effectively.

The science of water management

Considerably more scientific knowledge is needed on appropriate and effective water management practices, particularly with respect to drought. Drought should be viewed in the context of a comprehensive system that incorporates all water-dependent biophysical processes and human activities. More research aimed at increasing water-use efficiency (especially focusing on existing technologies such as irrigation and hydroelectric generation) is needed.

Drought prediction

A widely held belief among policymakers is that, because drought cannot be accurately predicted, drought planning is of no value. The result is many facets of drought response planning not necessarily dependent on predictions (e.g. food and/or water storage) are neglected. It is folly to reject planning on this basis.

Although climate predictions may have limited use in present planning, weather forecasts, climate monitoring, and climatic probabilities are indisputably useful tools in the strategic planning and implementation process. Their utility must be clearly defined for policymakers.

Variation in drought vulnerability

There is great spatial and temporal variability in society's vulnerability to drought. For example, some subnational regions are relatively more drought sensitive than others. As crop mixes in an agricultural region change over time, because of factors such as economics (or climate), vulnerability to drought may also change. Superimposed on this spatial and temporal variability is the tendency to view drought as a regional rather than national problem; unaffected regions are apt to be hesitant to commit national resources to manage affected regions.

The traditional response by governments to periods of severe drought has been to share the costs throughout the society. This approach, when undertaken reactively or through crisis management, has been inequitable and untimely. Ultimately, planning can reduce these costs by improving the efficiency of the assessment and response process and by lessening the direct and indirect impacts of drought.

Drought planning: the role of international organizations

The role of international (including donor) organizations in helping develop and implement national and supernational drought response plans involves four critical areas.

Resources

International organizations with charter provisions for drought response should facilitate the development of drought plans by funding international conferences aimed at bringing together scientists and policymakers. Where possible and desirable the organizations should also, provide technical assistance for drought planning.

Information

International organizations should assist governments in obtaining information necessary to develop and sustain drought plans. Primarily this would include climatic monitoring and analysis with emphasis on historical trends; monitoring in near-real time; and, if sufficient skill and capability exist, development of predictions. Attention should also be given to obtaining more (and better) information on drought-related impacts and, equally important, to monitoring changing drought vulnerabilities of alternative production strategies as drought develops.

International cooperation

International organizations should foster the exchange among nations of information concerning drought response actions, particularly transmittal of information from those nations whose efforts have been successful (e.g. India,

Kenya) to those just now facing the problem (e.g. Sudan, the Sahelian countries). Efforts should be focused on defining drought planning regions (multinational or otherwise) on the basis of factors such as similar biophysical characteristics (including climate), economics, and sociopolitical structures.

Drought as part of international development

Drought must be examined in concert with the development process. Some may question the importance of drought planning in developing countries, arguing that as nations develop and general levels of health and well-being rise, vulnerability to drought diminishes. Workshop participants did not agree with this philosophy. They suggested that international organizations should promote drought planning as part of overall development strategy. Perhaps the most effective contribution of international organizations is to use their considerable financial leverage to encourage governments in developing countries to integrate drought plans into long-term development strategies.

Recommendations and research priorities

Space limitations prevent inclusion of the full text of group responses to each question posed during the workshop. Here is a summary of the responses of Task Group 5 to drought planning and drought response. This indicates the types of observations, recommendations, and research priorities articulated by all the groups.

Question: Should state and/or national governments have a drought policy to assess and respond to episodes of severe drought? If so, what should be the objectives of that policy?

> National governments should establish a policy (or strategy) to assess and respond to drought. Droughts of various frequencies and magnitudes should be considered in development plans. National policy should rely on area-wide identification of the existence and severity of drought as well as on assessments of its present and future impact. The national policy must also recognize the need for both coping with and responding to drought. National drought policy should stress the creation of an infrastructure to supply the basic data, analysis, and research needed for assessment and response. The infrastructure should be independent of individual expertise, but cognizant of past experience. A key part of the national policy is the development of a general plan to prepare for and respond to drought episodes, a plan that is effectively integrated with area-wide plans.

Question: To what extent are the products of current impact assessments used in the implementation of drought-response strategies, and how might they be modified or adapted to best augment a national drought strategy? Are the products of those techniques easily interpreted by decisionmakers?

> Responses to drought are always based on some type of assessment of the severity and duration of the event, its current and potential impact,

and the ability of government and other organizations to respond. The most effective impact assessments concentrate on basic aspects of the economy for the affected area, e.g. food, health, housing. Methodologies should be developed to assess the present and future state of these basic needs in some standardized form while establishing an adequate and timely data base for input to these methodologies. Procedures are also needed to synthesize the results of these assessments for persons or organizations responsible for drought response. At the same time, more complete impact assessments should be made to indicate both short- and long-term effects of drought. These more complete assessments must be presented to advisors and planners for consideration in updating assessment and response criteria.

Question: Postdrought evaluations of assessment and response efforts are considered by some to be essential if we hope to improve our ability to cope with drought. How can governments and international organizations arrange for the conduct of ex post facto evaluations of drought assessment and response activities?

The best guide to the future is past experience and the lessons learned from it. This is true of drought occurrence, the physical and biological consequences of drought, and the human response to it. It is important to register the experiences of each major drought so that society can respond more effectively to future events.

Governments should evaluate each major drought episode. This evaluation should include the physical aspects of the drought itself; its impact on soil, groundwater, plants, and animals; its economic and social consequences; and the extent to which predrought planning was useful in mitigating impact, facilitating relief or assistance to stricken areas, and in postdrought recovery.

It is recommended that governments place the responsibility for evaluating drought and societal response to it in the hands of an unrelated agency to ensure an unbiased appraisal of actions taken. For example, the Food for Work Program of India is implemented by state governments, but is assessed by an independent body, the Planning Commission. Evaluations carried out by volunteers and non-governmental organizations can also be invaluable, because they often see weaknesses that official agencies may miss or be reluctant to admit.

International agencies, intergovernmental and nongovernmental, should realize the value of postdrought evaluations, and be prepared to sponsor them in cases where the emergency extends beyond national boundaries, especially where internationally coordinated relief projects have been mounted.

Independent research organizations may see the evaluation of such drought emergencies as valuable to their own programs. Bodies such as Resources for the Future, in Washington, D.C., and the International Federation of Institutes for Advanced Studies already have long-standing experience with such post-hoc analysis of economic and social impact.

Foundations and public research-funding organizations should realize that much of the ability to conduct such studies lies in the world's universities. Funding of postdrought evaluation should be encouraged. Many countries have specialized agencies or corporations capable of analyzing climate impact. The human resources available within these agencies should be tapped, especially in relation to major events (such as the Sahelian desiccation) that occur on a scale too great for individual academic workers.

A postdrought evaluation should seek answers to the following questions:

1. Was the drought plan followed? If not, why?
2. Did the actions taken and measures implemented mitigate the impact of drought effectively? Which actions and measures were effective and which ones were not?
3. Did aid reach all groups in the stricken area? If not, why? How were the target groups for aid identified?
4. Were the measures timely in relation to the events of the drought periods?
5. Was it possible to correct errors during the emergency?
6. How efficient were the actions taken and measures implemented? What financial and human resources were allocated to the relief effort? Where did the resources come from and how were they controlled?
7. How efficient was logistic support and the available infrastructure? Did the aid reach these groups and in a timely manner? Were obstacles encountered that reduced the efficiency of the response? If so, what were they (for example, limitations of personnel, fuel, obstructions by customs officials)?
8. What was the level of cooperation between the agencies involved, both public and private? Did this hinder the flow of information or aid?
9. Was media coverage accurate and realistic in providing details of the event? What media were present? What role did they play in the emergency?

Several aspects of postdrought evaluation suggest that extensive research will be needed on the following topics:

- The effects of prolonged rainfall deficiency on various hydrological quantities, notably the depletion of soil moisture and shallow groundwater. Ground surveys and satellite monitoring will help in the assessment of these effects, which will have a bearing on future stream flow, water use, energy development, and agriculture.
- The effects of drought on land use, vegetation, and soil. The recovery of agriculture in badly affected areas—a process that will be highly relevant to future drought measures—will need to be watched.

- Ex post facto evaluations should not only be conducted, the results should be implemented. Sadly, just as interest wanes with distance from the spark of crisis, so does the urge to regroup and reorganize. Several strategies can be employed to counteract these natural tendencies. First, evaluations can be implemented more easily if they contain specific and clearly articulated recommendations and rationales. Second, decisionmakers may be more willing to authorize evaluations if they have been properly educated or sensitized on the issue. Drought education efforts should specifically address evaluations. Third, recommendations can encourage interest if the planning process specifically calls for an evaluation and implementation sequence. Finally, incentive for implementation can be provided by scheduling various forums for the exchange of information and experiences. Funding agencies should consider making implementation a requirement of participation in their programs.

Question: As participants in this workshop, what courses of action would you recommend to ensure that the momentum for drought planning generated by this meeting will be considered further by governments and international organizations? For example, might a working group under the sponsorship of several United Nations agencies (UNEP, WMO, and FAO) be effective in promoting the concept of drought planning?

The following courses of action were recommended to sustain momentum generated by the workshop:

- Development of model drought plans for nations and states within nations,
- Establishment of a drought planning network, and
- Communication of results of drought planning conference.

Question: What research and information would be valuable in facilitating postdrought evaluation?

Hydrological impact. One of the most important effects of drought is on the hydrology of the stricken region. An assessment of this impact through ground surveys and satellite monitoring is beneficial in assisting future planning efforts, particularly with respect to water use, energy, and agricultural activity.

Agricultural recovery. Depending on the duration, intensity, and spatial characteristics of drought, the agricultural recovery could be handicapped despite average or better rainfall. Input requirements (e.g. seed, fertilizer, pesticides, implements, energy) could be determined on the basis of the magnitude of the drought impact. Assessments of this type would be helpful when drought recurs.

Economic impact. An assessment of losses in the agricultural and agriculture-based industries should be made following each drought episode. The assessment in developing countries must include the condition of livestock, health, and market prices of essential commodities. It is also important to assess the impact of relief measures on the various economic sectors and to determine if there were individual persons, industries, municipalities, or other sectors that were affected substantially but were neglected by available assistance programs. Who should be the target groups or individuals in the future? To what extent was there discrimination against women or female children or children in general?

Decisions during drought. Decisions of governments during periods of drought are made for humanitarian and political reasons. It will be difficult, if not impossible, to change this attitude. Therefore, it is important that evaluations of drought assessment and response efforts be carried out by an organization other than the one with responsibility for implementing program plans.

Social response to drought. In many developing countries, the occurrence and effects of natural disasters are considered inevitable and unavoidable. Events and their effects are considered unmanageable and out of the realm that can be influenced by government. A change in this outlook, based on scientific explanations and approaches, could help mitigate the effects of events such as drought. This could be done by organizing meetings and symposiums in developing countries in which governmental leaders could explain the strategies to the people.

Reference cited

Kates R W, Ausubel J, Berberian M, eds. (1985) Climate impact assessment. John Wiley and Sons, Chichester, United Kingdom.

Notes

Authors' addresses: W. E. Easterling, Climate and Meteorology Section, Illinois State Water Survey, Champaign, Illinois 61820; D. A. Wilhite, Center for Agricultural Meteorology and Climatology, University of Nebraska, Lincoln, Nebraska 68583-0728, USA.

Citation information: International Rice Research Institute (1989) Climate and food security. P.O. Box 933, Manila, Philippines.

The world climate impact studies programme

G. Golubev

The Sahelian drought of 1968-73 and worldwide adverse weather in 1972 contributed significantly to increasing the attention given to weather and climate. Politicians, administrators, and scientists started to contemplate a global climate change, and the stark realities of how vulnerable humans can be to climate and its variability became a general concern. In the Sahel, vegetation disappeared and the land was laid bare. The threat of a Sahara desert extending south became very real.

This was the setting in which the UN General Assembly called for a worldwide conference on the problem of desertification. That conference was held in 1977. Two years later, the World Climate Conference was held in Geneva.

In 1979, the 8th World Meteorological Congress authorized the Secretary General of the World Meteorological Organization (WMO) to establish the World Climate Programme (WCP). Four components were envisaged:

- the World Climate Data Programme,
- the World Climate Applications Programme,
- the World Climate Research Programme, and
- the World Climate Impact Studies Programme (WCIP).

It was agreed that the United Nations Environment Programme (UNEP) would assume responsibility for developing and implementing the impact study program. In accepting that responsibility, UNEP was aware that the science of climate impact assessment was relatively new, and that developing and implementing impact assessments would require an interdisciplinary approach. Steps were taken to establish a multidisciplinary Scientific Advisory Committee, which is responsible for advising the Executive Director on implementation.

The next step was to embark on a project whose main objective was improving our knowledge of climate impact assessment. This project, undertaken in conjunction with the Scientific Committee on the Problems of the Environment (SCOPE), resulted in the comprehensive publication *Climate impact assessment*, SCOPE 27. In an effort to disseminate concepts of climate impact assessment to scientists who might not have access to advanced computer facilities or the modeling capabilities to make full use of the comprehensive program, a simplified version of that publication was produced: *Assessing the social implications of climate fluctuation through climate impact studies* (Riebsame 1989).

The impact of climate and climatic variability on agricultural and food production has been a central theme in the development and implementation of WCIP. Some of the questions that UNEP, in collaboration with other United Nations (UN) agencies and intergovernmental organizations, has included are

- How are different agricultural systems vulnerable to climate and what management policies are required to increase productivity and lessen climate dependency in such systems?
- How does the interaction between different ecological systems and climatic factors adversely affect agricultural production?
- To what extent can improved environmental management increase agricultural production? In particular, how can human activities that modify the environment be minimized?
- What contribution can the international community make toward lessening the suffering caused to millions of people by climatic extremes such as drought?
- How can developing countries build sufficient manpower to handle the complex problem of increased food production by applying effective agrometeorological practices and by introducing climate considerations into agricultural planning?

It is not appropriate, nor convenient, to catalog the projects that UNEP has supported or undertaken to study the impact of climate on food production. The following examples merely illustrate the manner in which UNEP has attempted to find solutions to some of these questions.

During 1982, UNEP embarked on the project *Vulnerability of food systems to climate* in conjunction with the United Nations Research Institute for Social Development and the Centre for Regional, Ecological, and Science Studies in Development Alternatives of India. Two publications were produced during 1986, *Agro-ecological zones of Eastern India* and *Food, nutrition, and environment focus on Gosaba and the Sundarbans.* We expect most of the work to be published in the WCP series.

In collaboration with the International Institute for Applied Systems Analysis (IIASA), UNEP has just concluded a project, Integrated Approaches to Climate Impact Studies. The project focused on food production in three broad zones:

1. High latitudes, where temperature is a constraint;
2. High altitudes, where both temperature and precipitation are limiting factors; and
3. Low latitudes and precipitation-deficient areas.

This study was multidisciplinary, a cooperative effort among geographers, meteorologists, agronomists, economists, and decisionmakers. The methods used were simple. The study is likely to find early application in agricultural planning by various governments. UNEP will certainly use the results in workshops and seminars to improve techniques in climate impact assessment. Two volumes, both titled *The impact of climatic variations on agriculture*, are available.

In Latin America, UNEP initiated a study on the socioeconomic impact of climate on agricultural systems in the humid tropics of South America. The project is being undertaken with the cooperation of the Inter Agencies Group on Agricultural Biometeorology (United Nations Educational, Scientific and Cultural Organization, WMO, and Food and Agriculture Organization). The focus is on social and economic consequences of the agricultural alternatives available within existing climatological conditions in each region. It is expected that the outputs will enable decisionmakers in South America to introduce climate considerations into planning and policy development.

UNEP places great emphasis on training. It has undertaken two training projects in climate and agriculture. The basic aim is to provide developing countries with sufficient manpower to assess the impacts of climate variability on their own agricultural production. The latest course in the series was held in October 1986 in Alma-Ata, USSR.

During the last two decades, climatic extremes, especially drought, have had serious impacts on the social and economic infrastructures of some developing countries. UNEP has joined other international and intergovernmental agencies in the fight against those effects. The Scientific Round Table on Climate and Drought in Africa in early 1984 was part of the UN system's response to the urgent problem of combating the catastrophic consequences of drought in many African countries. Later that year, the Economic Commission for Africa Conference of Ministers adopted the recommendations of the Round Table, including a Plan of Action on Drought. The Plan recommends the establishment of an African Centre of Meteorological Applications for Development. The centre is expected to require substantial international support, particularly in its initial stages. UNEP will join other members of the UN family in extending such support.

As part of the implementation of WCIP, UNEP has cosponsored a number of international symposiums, workshops, and seminars related to drought.

- UNEP participated in the August 1985 Symposium on Drought and Hunger in Africa organized by the National Centre for Atmospheric Research (NCAR), Boulder, Colorado, USA.
- Support was given to the Kenya National Environment Secretariat to organize a regional seminar on drought for eastern Africa. The seminar at UNEP Headquarters in May 1986 focused on experiences during the 1984 drought in Kenya, Uganda, Tanzania, and Sudan.
- UNEP and NCAR cosponsored a workshop, Worldwide Climate Anomalies of 1982-83 and Their Economic and Social Impacts, in Lugano, Switzerland, November 1985. Climate variations in 10 regions were considered and participants constituted themselves into a standing working group with special reference to El Niño Southern Oscillation (ENSO), and teleconnections. The group will focus on the socioeconomic impact of ENSO-related phenomena, identify areas of potential research, and provide continuing reviews of ENSO-related research. The workshop proceedings were published in 1986.
- The Centre for Agricultural Meteorology and Climatology, University of Nebraska, Lincoln, USA, organized a symposium and workshop on drought September-October 1986. The symposium, partly funded by UNEP and cosponsored by WMO, concentrated on drought prediction and government preparedness.

Consideration and management of the environment call for constant monitoring of a number of parameters. A major agricultural resource base is land. A project supported under WCIP—the International Satellite Land Surface Climatology Project (ISLSCP)—has as one of its primary objectives the development of appropriate methods to monitor changes in land surface characteristics. UNEP has provided the means for scientific groups engaged in ISLSCP activities to come together to consider results and to discuss progress in research. A notable activity has been the development of algorithms to convert satellite-derived data into parameters useful for climate impact studies. A meeting to compare and attempt to standardize such algorithms was held in January 1987 in Pasadena, California, USA, with UNEP support.

ISLSCP has been responsible for mounting several major national projects, including the First ISLSCP Field Experiment in the United States, and for undertaking retrospective studies using satellite data archives. A European-African Task Group coordinates ISLSCP activities in Africa, and a number of studies have been carried out in Niger. Several experiments also are under way in Europe, particularly in France, where several ISLSCP projects are supported. An all-Africa meeting is planned later this year, with the cooperation of the Committee on Space Research. The meeting is expected to concentrate on the application of remote sensing to the study of land surface-climate interrelationships in semiarid regions.

To ensure a wider audience for a project which appears to have major potential in assisting climate impact studies that use remote-sensed data, UNEP is cosponsoring the publication of a general audience brochure on the ISLSCP program.

One of the greatest challenges facing mankind today is the socioeconomic impact resulting from a possible climate change due to increased concentrations of carbon dioxide and other greenhouse gases in the atmosphere. Climate, as one of mankind's economic resource bases, faces changes because of mankind's own activities. Within the framework of the WCIP, UNEP and other international organizations, notably the International Council of Scientific Unions (ICSU) and WMO, have continued to assess the possible socioeconomic impact of increased greenhouse gases concentration in the atmosphere and to search for the best scientific advice that can be made available to governments in the area of policy options for coping with a changing climate.

A major event related to this issue was the convening in Villach, Austria, in 1985 of the joint UNEP/WMO/ICSU International Conference on the Assessment of the Role of Carbon Dioxide and of Other Greenhouse Gases in Climate Variations and Associated Impacts. This conference, held with the support of the Government of Austria, drew extensively on the results of a UNEP-funded project, Development of a Scientific Basis for Assessing the Impact of an Increased Concentration of Carbon Dioxide on the Interaction of Climate and the Biosphere, implemented by the International Meteorological Institute, Stockholm. The results of the project were published in SCOPE 29, *The greenhouse effect, climatic change, and ecosystems*.

The conclusions and recommendations of the Villach Conference, published as WMO-No. 661, have been widely distributed within the scientific community and among decisionmakers. During 1987-89, UNEP will place significant emphasis on implementing the Villach recommendations, in accordance with the decisions of the Advisory Group on Greenhouse Gases and the Scientific Advisory Committee. The first group was created by the Villach Conference to review biennially international and regional studies related to greenhouse gases and to assess periodically the rate of increases in the concentrations of greenhouse gases and the effects of those gases.

Among the recommendations was a call for increased public awareness on the issues of greenhouse gases, climatic change, and rising sea level. UNEP has taken the following steps to increase public awareness:

- Greenhouse gases are addressed in volume I of the UNEP/Global Environment Monitoring System Environment Library Series. In this series, scientific topics are explained in a manner easily understood by the layman.
- An audio-visual kit comprising slides, audio, and printed texts on climatic change and people will soon be distributed to educational institutions and libraries.
- A television film on greenhouse gases and climate change has been produced and widely viewed.

It is expected that the project being implemented by the Climatic Research Unit of the University of East Anglia, UK, with UNEP involvement, will soon complete a scientific state-of-the-art report, as well as a simplified version, on sea level rise and increased tropical storm frequency.

The Villach Conference further recommended increased support for the analysis of policy options related to greenhouse gas-induced climatic change. The regions recommended for study include the Amazon Basin, the Indian Subcontinent, Europe, the Arctic, the Zambezi Basin, and the North American Great Lakes. UNEP is directly encouraging governments in these regions to initiate suitable studies as their contribution to a better understanding of the problems. Further, UNEP is planning active collaboration with the governments in Southeast Asia to undertake a socioeconomic assessment and policy exercise related to possible climatic change in Malaysia, Thailand, and Indonesia, and is seeking support to extend support to other developing regions identified by the conference.

With the collaboration of the Beijer Institute, UNEP plans to hold a technical workshop where scientists and decisionmakers will address the issue of greenhouse gas-induced climate change and policy options. The workshop proceedings will be distributed before the end of 1988.

Mankind faces yet another serious threat—the consequences of the depletion of the ozone layer (once again, caused by human activities). The effects of a depleted ozone layer are serious enough to warrant urgent attention by all nations of the world and by the international community. These effects include risks to health, particularly skin cancer and the deadly melanoma; reduction in the quantity and quality of plants, including important food crops; and considerable stress to marine species. There is also growing evidence that certain industrial products, mainly plastics, deteriorate under increased ultraviolet B radiation.

UNEP has played a leading role in sensitizing the public to these effects and in initiating action that could lead to increased protection of the ozone layer. A landmark was the adoption on 22 March 1985 of the Vienna Convention for the

Protection of the Ozone Layer. The convention has now been signed by 26 countries and ratified by 7. These encouraging actions provide impetus to strengthen international legislation to protect the ozone layer from unacceptable damage.

Such legislation must include a definitive protocol to the Vienna Convention to limit the emission of chlorofluorocarbons (CFCs)—the substances now known to be responsible for ozone layer depletion.

For such a protocol to be meaningful, it must be based on scientifically and technically sound information. In this connection, the GEMS/PAC (Atmosphere) subprogram and the UNEP Environmental Law Unit have undertaken a closely coordinated program on legal and technical matters for developing the protocol.

The UNEP Coordinating Committee on the Ozone Layer (CCOL) met in its eighth session at the UNEP Headquarters in February to reassess ozone layer modification. This assessment was based on, and confirmed the findings of, an earlier assessment of the understanding of the processes controlling the present distribution and change in atmospheric ozone undertaken by seven national and international bodies, including UNEP. Both assessments found compelling evidence of increases in the concentrations of gases that control atmospheric ozone. Predictions of a changing atmosphere now being confirmed by observations have implications for a variety of human issues. The assessments provide a warning that humans are conducting a global-scale experiment on the atmosphere without fully understanding the consequences.

Effects of ozone layer modification were not assessed, as is usually done, at the same time as the processes controlling ozone distribution were assessed because of the need to first undertake a major review of the effects of ozone layer modification. This was done at a joint UNEP/United States Environment Protection Agency Conference, Effects of Changes in Stratospheric Ozone and Global Climate, June 1986 in Washington, D.C. At that conference, scientists from 20 countries discussed ultraviolet radiation flux and its likely effects on human health, plants, aquatic organisms, and polymer degradation, and the implications of the greenhouse effect on water resources, agriculture and forestry, and sea level rise. Policy implications of such changes also were considered.

A resumed session of the CCOL in November 1986 at Bilthoven was able to use the results of the Washington Conference to prepare a global assessment of the effects of ozone layer changes.

Executive summaries of the two-part CCOL-prepared assessment were made available to the Vienna Group convened in Geneva in December 1986 to elaborate a protocol on the control of CFCs for the Vienna Convention. CCOL also provided a policy support document, which outlined in nontechnical terms some of the important issues that the Vienna Group needed to take into account in elaborating the protocol.

A report of a two-part workshop on the Control of Chlorofluorocarbons also was submitted to the Vienna Group Meeting. The workshop, convened in response to a Geneva Conference decision GC 13/18 Part 1, was held in Rome in May and in Leesburg, USA, in September 1986. The first part considered current and projected CFC production, production capacity, emissions, use, trade and current regulations of CFCs, and the costs and effects of their application. The range of existing and developing technological options for control also was considered on a sector-by-sector basis, together with their potential costs and effectiveness. The second part of the workshop identified and analyzed against a comprehensive set of criteria various regulatory strategies for CFCs. Among the control strategies examined was establishment of global emission limits and setting of CFC production capacity caps, calculated according to present production and use and subject to revision on the basis of periodic research and assessment.

The future poses great challenges for UNEP, the international community, and mankind as a whole. In UNEP, we are confident that member states will agree to a protocol to the Vienna Convention for the control of CFCs. Great emphasis will be placed on implementing the recommendations of the Villach Conference on greenhouse gases. On climate impacts, we shall encourage the establishment of viable national climate programs and assist those programs in addressing impact assessment problems.

Within the next 2 yr, we shall be celebrating the tenth anniversary of the World Climate Conference. Toward the end of February 1987, a small meeting at the WMO Secretariat in Geneva will be discussing plans for a Second World Climate Conference. When those plans materialize, the opportunity will be taken to take stock of the achievements, failures, and obstacles over the past 10 yr. It will be a period for reflecting and consolidating, and for resolving to improve even more mankind's knowledge of one of the world's greatest economic resources—climate.

Reference cited

Riebsame W E (1989) Assessing the social implications of climate fluctuation through climate impact studies. Pages 451-467 *in* Climate and food security. International Rice Research Institute, P.O. Box 933, Manila, Philippines.

Notes

Author's address: G. Golubev, GEMS/PAC, United Nations Environment Programme, P.O. Box 30552, Nairobi, Kenya.

Citation information: International Rice Research Institute (1989) Climate and food security. P.O. Box 933, Manila, Philippines.

Basic data requirements of an agroclimatic system

S. Unninayar

Agroclimatic systems need specific climatic and agroclimatic information to help achieve sustained food production. This paper reviews the basic data requirements and the types of information products possible from an agroclimatic system. World Meteorological Organization projects to assist members in providing such data and information are outlined. Most social and economic activities either are determined by or are highly sensitive to climate. Relationships between weather and climate and the land surface, soil, and vegetation have been intuitively recognized for thousands of years.

Civilizations have usually prospered during periods of benevolent climatic conditions, but many were unable to optimize agricultural methods to help control the natural system. History documents the disruption or dislocation of socio-economic systems due to their inability to respond to changes in climate or changes in the soil-water resource base caused by inappropriate land use. Even today, because of short-term policies applied under social or economic pressure, many agricultural and land-use practices neglect downstream consequences. While climatic fluctuations affect all economic sectors to some degree, food and fiber production is perhaps the most sensitive and vulnerable to the vagaries of nature.

The impact of climatic fluctuations on agricultural systems relates not only to the actual producers, but also to a large and complex support structure that serves the producer. This support structure includes the development, production, and distribution of seeds, fertilizers, pesticides, and farm equipment, and the provision of rural insurance (government or private sector) as well as financial, farm management, and pest management services. The degree of sophistication of the support service infrastructure varies substantially from country to country. Within countries, agricultural technology ranges from large-scale, mechanized farming systems to small-scale subsistence farming. After years of trial and error, agroclimatic information may have been integrated, albeit in an ad hoc fashion, into farming procedures. However, traditional methods are often unable to adapt to changes easily. Special efforts are needed to introduce mechanisms that are responsive to climatic fluctuations.

Definition of an agroclimatic system

For the purposes of this paper, an agroclimatic system is a system that incorporates the physical properties of atmosphere-land surface-soil and hydrology-vegetation interactions into the planning and management of agricultural (food and fiber) products. The objective of such a system is to achieve a sustainable, optimized production level through the use of weather and climate information, while maintaining environmental integrity and minimizing the degradation of the soil, nutrient, and water resource base. If technological means are available to boost yield—through, for example, the use of chemical fertilizers, new varieties of seed, or mechanical farming procedures—particular care must be taken to ensure that the impacts of these technological factors are not adverse in the longer term.

Data and functional requirements of an agroclimatic system

An agroclimatic system requires

- an observing system to measure climatic elements (usually the most rapidly changing variables);
- a biological and geophysical monitoring system to keep track of the state of the land surface, soil, and vegetation;
- an assessment system that determines both the best land- and water-use strategies and the consequences of existing or planned changes in land use and agricultural practices;
- a data processing and information dissemination system to guide both operational and planning decisions;
- a research component to establish or improve the relations of weather and climate to soil and hydrology for various crop varieties.

Figure 1 is a schematic of the functional units required and the procedures and interactions involved in implementing an agroclimatic system. Essentially, such a system calls for an effective means by which natural resources can be managed. Climatic, water, soil, crop (genetic), and technological (irrigation, fertilizers, pesticides) resources form inseparable components of the system. Today, there is increasing pressure to integrate the scientific expertise required to manage an agroclimatic system with a cohesive structure capable of feeding information to governmental and industrial decisionmakers. Over the last 5 yr, significant developments have taken place in the design of simulation models that attempt to integrate the many factors and variables involved in natural resource management.

Ideally, an extensive range of data is required to implement a meteorologically based agroclimatic system. All data may not be available in any given region. Even so, the system concepts would still apply, although perhaps with diminished efficiency. Even the simplest of systems would result in definite benefits over one that contains none of the necessary components.

Automatic observing stations and satellite remote sensing techniques have improved our capability to monitor basic variables. Advances in crop genetics with bio-engineering and gene splicing techniques promise much more rapid adaptation to changes.

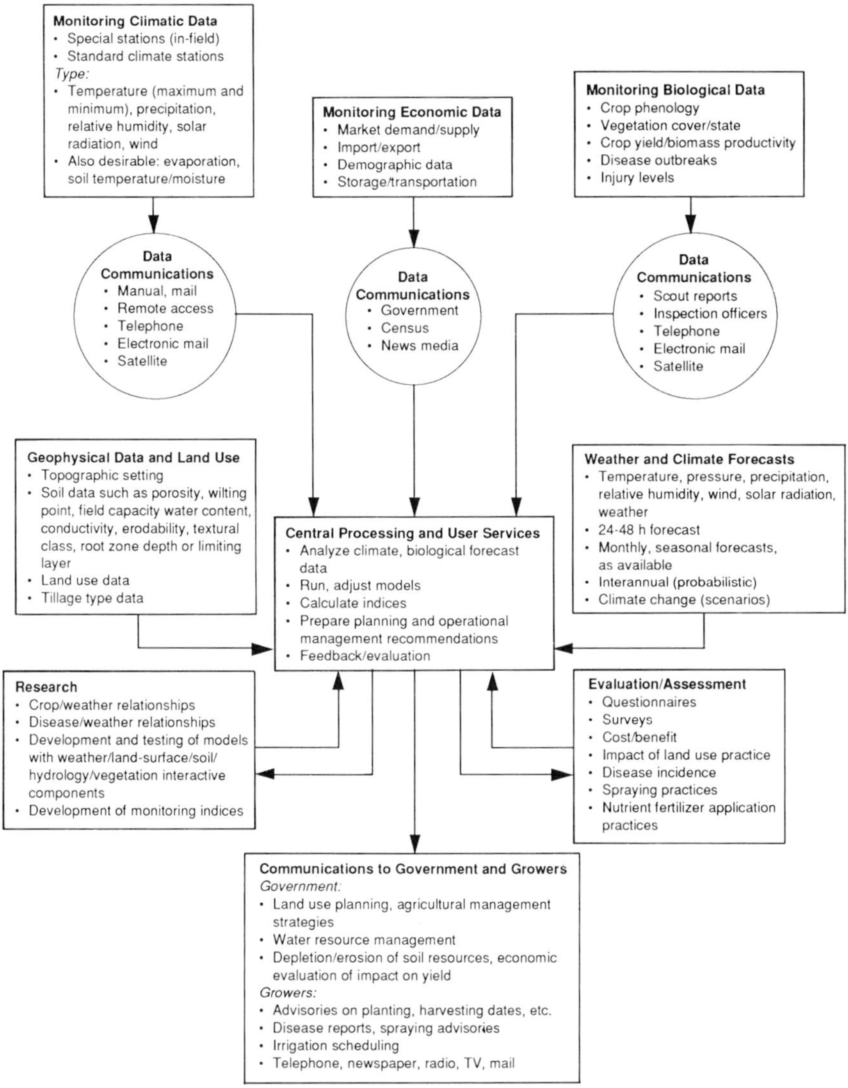

1. Schematic of an agroclimatic system (adapted from Blackburn et al 1986).

A critical component of an agroclimatic system, often lacking in many countries, is a mechanism to disseminate information to the farm producer. A telecommunications infrastructure (at the least radio), through which information can be broadcast, is required. Credibility at the farmer level is important and can be achieved via demonstration projects.

A distinction must be made among the data required for research purposes, those for operational agricultural decisionmaking, and those for weather and climate forecasting. The space and time requirements for data increase (i.e. more stations and more frequent observations) from weather forecasting to agricultural management to research.

For weather forecasting, observations of the basic meteorological variables such as temperature, wind, pressure, precipitation, and relative humidity are required at least twice a day at a space resolution of approximately 1 station/250,000 km^2 for upper air stations and 10 stations/250,000 km^2 for surface stations.

For agricultural purposes, a substantially higher density of stations is necessary, as are additional variables such as evaporation, solar radiation, and soil temperature and moisture. Of the data required, precipitation is often the most spatially inhomogeneous. A network of 20 stations/500 km^2 grid is necessary to measure monthly precipitation with an accuracy of 10-20% (WMO 1985). Higher densities are required to estimate representative areal average precipitation on a daily basis due to the nonuniform scatter of individual cloud cells in, for example, tropical areas or summer convection at higher latitudes. With improvements in satellite remote sensing techniques, it may be possible to use a less dense surface network. Usable information is still obtainable from surface networks of substantially lower space and time resolutions than ideally required.

For an analysis of "the climate of the region," historical time series data are necessary, preferably at least on a daily basis. The requirement for daily data is more stringent in semiarid parts of the world than elsewhere. If, for example, 10-d or monthly totals are used, or are the only data available, serious miscalculations can result when attempting to identify agroclimatic zones suitable for crops. Successful germination may require a certain amount of rainfall with no dry spells exceeding 15 d. Ten-day or monthly rainfall data could miss such dry spells if all of the rain fell on the first and last day of two consecutive periods. Ten-day or monthly rainfall may, however, be sufficient in many regions of the world and even in some semiarid areas, but an analysis of daily rainfall is necessary before this can be determined. For research purposes, hourly or more frequent observations are often required to investigate specific processes. Special observing systems or networks are usually set up for an experimental period to conduct such research studies.

Operational data exchange requirements vary with the application. For weather forecasting, global or at least regional data exchange is mandatory. For this purpose, the World Meteorological Organization (WMO) coordinates, on a voluntary cooperative basis, the Global Telecommunications System (GTS) via which data are internationally exchanged daily from the Global Observing System (GOS) operated by the member states of WMO. The exchanged data are processed at the World Meteorological Centers of WMO, where powerful computers generate forecast products for international dissemination via the GTS. The density requirements correspond to those needed for weather forecasting.

On a national or regional basis, data from a higher density of stations are required to support agricultural operations. Stations located on farms and in water catchment areas are particularly important. Data at the farm level would enable in-situ decisionmaking that is fine tuned to the particular farmsite.

Other types of data required by an agroclimatic system are detailed in Figure 1.

The application of agroclimatic information

Agroclimatic systems must deal with a variety of agricultural and meteorological applications so that decisions of clear short-term and long-term economic benefit could be made. Some examples of applications are

- agroclimatic zoning,
- selection of crop varieties,
- planning of land resources,
- determination of optimum planting dates,
- irrigation scheduling,
- timing of the application of pesticides and fertilizers,
- timing of harvesting and drying operations,
- water-use management,
- microclimatic manipulations to protect crops from adverse weather factors or to improve growth conditions, and
- selection of agricultural methods, land use, and crop geometry.

WMO projects

Several projects under the World Climate Programme and the World Climate Applications Programme of WMO are directed at developing the support structures necessary for agroclimatic systems. A primary need is a computerized climate and agrometeorological database. Over the next 10 yr, the CLICOM project of the World Climate Data Programme plans to deploy microcomputer systems with a comprehensive climate data management and user services package at all national meteorological services that need them (WMO 1986). Being currently developed as a second phase option of the project are software modules with associated training programs to incorporate the user services that directly support planning and decisionmaking in agriculture and natural resource management (WMO, in press). The project is being implemented in coordination with the World Climate Applications Programme and the Hydrology and Water Resources Programme of WMO.

Required national mechanisms

The components required to form an agroclimatic system (Fig. 1) exist in many countries, but usually under the jurisdiction of several government agencies. Integrating the activities of such agencies is difficult. One approach is to establish joint interagency projects and a unit that represents all of the relevant agencies. The scientific disciplines involved can now be integrated via the use of comprehensive, user-friendly "expert-assistance" computer software. The central arm of an agroclimatic data management system should be located within a meteorological service because of the nature of the interactive physical agroclimatic system, i.e. meteorological variables are the fastest changing forcing functions of the system.

The effect of any particular weather or climatic event or series of events is felt on the land surface and in soil moisture, vegetation (biomass), and economic and social activity after lags ranging from a few days to as much as a few years. Thus, recent and current weather events already provide partial information about the future for nonmeteorological components of the system. These facts, generally underutilized in the past, should be recognized and exploited for mankind's benefit. National cooperative mechanisms cognizant of and responsive to fluctuations in the natural environment should be established to achieve a management structure that is capable of viewing nature as an interactive system.

References cited

Blackburn W J, Bootsma A, Gillespie T J (1986) Weather and the control of plant diseases. Chinook Summer/Ete" 1986. p. 48-50.

WMO—World Meteorological Organization (1985) Review of requirements for area-averaged precipitation data, surface-based and space-based estimation techniques, space and time sampling, accuracy and error; data exchange. WCP Report 100. Geneva. 57 p.

WMO—World Meteorological Organization (1986) CLICOM project. WCP Report 119. Geneva. 33 p.

WMO—World Meteorological Organization (1987) WMO regional training seminars on climate data management and user services. World Climate Programme Report Series.

Notes

Author's address: S. Unninayar, World Climate Programme Department, World Meteorological Organization, 41, Giuseppe Motta, Case postale No. 5, CH-1211 Geneve 20, Switzerland.

Citation information: International Rice Research Institute (1989) Climate and food security. P.O. Box 933, Manila, Philippines.

Simulated yields of wheat and rice in current weather and when ambient CO_2 has doubled

F. W. T. Penning de Vries, H. Van Keulen, C. A. Van Diepen,
I. G. A. M. Noy, and J. Goudriaan

Average grain yield and its variability are simulated for wheat and rice crops in four climates. The crops simulated are supplied with ample fertilizer and are weed-, pest-, and disease-free. Situations with continuously optimal water supply and with water shortages are both investigated. Yield predictions are made with two documented and evaluated models. The simulations are repeated for future weather when the CO_2 concentration has doubled. Temperature and precipitation change significantly, but solar radiation, wind speed, and relative humidity are kept constant. The impacts of the higher CO_2 concentration and of future weather on yield are computed. A third set of simulations deals with cultivars adapted to the future weather. The increase in CO_2 level permits potential crop yields to rise 10-50%, but this rise is eroded by the higher temperatures. The results are different for each of the situations considered. Yield variability is low and not much affected in the tropics, but increases in cooler climates. The water use by the crops becomes much more efficient. This boosts water-limited yields, except where precipitation falls. Variability remains high or increases even further. Implications for agricultural research, for breeders, for climatologists, and for agricultural planners are discussed.

Crop yields are strongly influenced by weather. Therefore, climatic changes may have important impacts on agriculture in general and on yields and their variability in particular.

Experimentation to establish the effects of expected climate changes on yield and variability is very laborious, expensive, and slow. An alternative method is systems analysis and simulation, as shown in this paper. We based the simulations on climatic changes predicted by others, and we deal only with the yield variability of single crops. To extrapolate to yield variability on a national scale would also require an analysis of the fluctuations in area used for these crops. This is not attempted.

Wheat and rice were chosen for this study because of their importance to the world food supply and because relatively good models are available for both. Yields and their variabilities are simulated in four geographic regions, characterized by different climatic conditions, for situations with optimum soil water (irrigated crops—potential production) and for situations where soil water availability is dictated by precipitation and soil physical properties (purely rainfed crops—water-limited production). The simulations apply to intensive agricultural systems where nutrients are supplied at a level to obtain maximum yields. Yield reductions due to diseases, pests, and weeds are not considered.

This paper summarizes a report by Van Diepen et al (1987) in which more background and analysis of the results are provided.

Methods

Annual yields of wheat and rice are first simulated for 9-32 yr in different climates. Subsequently, the weather data are modified to reflect climatic conditions under a doubled atmospheric CO_2 level (680 vppm), which is expected to be reached in 70-100 yr. To examine to what extent average yield and its variability change, the same series of simulations is repeated for future weather conditions. Because those conditions are rather different from the present ones, it can be anticipated that new wheat and rice cultivars will be used. An educated guess is made about the characteristics that can be changed in the model to represent such new cultivars. Simulation with those crop properties provides a third set of average yields and yield variabilities.

Yield variability between years is calculated as the coefficient of variation (CV).

Climates

The weather data used are daily values from standard meteorological stations: maximum and minimum temperatures, total global radiation, average wind speed, precipitation, and early morning water vapor pressure. Historical weather data from four locations are used, representing a temperate region (Wageningen, Netherlands), a Mediterranean region (Migda, Israel), a semiarid tropical region with summer rainfall (Hyderabad, India), and a subhumid tropical region (Los Baños, Philippines). The data for Wageningen (1954-85) were obtained from the Department of Meteorology and Physics of the Agricultural University in Wageningen, those for Migda (1962-82) from the Israeli Meteorological Service, those for Hyderabad (1975-83) from the International Crops Research Institute for the Semi-Arid Tropics (ICRISAT), and those for Los Baños (1959-83) from the International Rice Research Institute (IRRI).

Estimates about future weather conditions are based on Schlesinger and Mitchell (1985), and are presented in Table 1. We assume no changes in radiation, wind speed, or relative air humidity. Future weather is "created" by adding or subtracting the projected changes to both the minimum and maximum temperatures; for precipitation, the amount per wet day is changed by a fraction such that the average value given in Table 1 is obtained.

Crop models

The effect of increased CO_2 concentration on crop performance is obtained using relatively simple physiological assumptions. The light-saturated maximum assimilation rate of individual leaves is supposed to have twice the present value (Goudriaan et al 1985) whenever sink limitation is absent. Light-use efficiency at the light compensation point increases by 25% (Goudriaan et al 1984). The effect of high CO_2

Table 1. Projected changes in temperature and precipitation at 4 sites, based on Schlesinger and Mitchell (1985) figures 4.38, 4.39, 4.41, and 4.42. The numbers represent the average increase or decrease in the periods indicated; precipitation change is expressed per day, counting 30 d/mo. The standard deviation (in parentheses) is an indication of the uncertainty.

Temperature change (°C)	Dec-Feb	Jun-Aug
Netherlands	+6 (2)	+3 (1)
Israel	+4 (1)	+3 (1)
India	+3 (1)	+2 (1)
Philippines	+2 (1)	+2 (1)
Precipitation change (mm/d)	Dec-Feb	Jun-Aug
Netherlands	+0.5 (0.5)	–0.5 (0.5)
Israel	0.0 (0.5)	–0.5 (0.5)
India	+0.5 (1)	+1.0 (3)
Philippines	–0.5 (1)	+0.5 (1)

on leaf area development is more difficult to quantify. In agreement with Goudriaan and De Ruiter (1983), the specific leaf area under increased CO_2 is assumed to be 35% lower than under present conditions.

Wheat. The spring wheat model by Van Keulen and Seligman (1987) is applied. It is process-based, executed with time steps of 1 d. Phenological development, dry matter accumulation and distribution, and organ formation are simulated in dependence of weather conditions and water and N supply. In the simulations for this paper, only parameters and functions characterizing the environment and the wheat cultivar are adapted. A fixed sowing date is assumed, specific for each location, irrespective of environmental conditions. For irrigated crops, presowing irrigation is applied in all cases to ensure proper germination. For rainfed crops, the onset of germination is determined by the rainfall pattern. If germination starts but the soil dries out before emergence is complete, crop failure occurs; the model does not allow for resowing.

Rice. The model described by Wolf et al (1986) and Rappoldt (1986) is applied to simulate the rice crop. It follows the same approach as the spring wheat model, but describes the growth processes and soil water balance processes in summary form and in a more generalized manner. The simulations are executed for Los Baños and Hyderabad only, for the cultivar IR8 grown in the rainy season. The simulation starts with successful transplanting into wet soil on a fixed date. Yield reductions due to temporary drought can only result from water deficiency during the mid- and late growing season. The soil used in water-limiting conditions is an upland rice soil with a deep groundwater table and an effective rooting depth of 0.4 m.

Both models are explanatory: results can be fully explained on the basis of physiological, (soil-)physical, and micrometeorological processes.

Results

Current yield level and its variability

Wheat. Average potential yields under the present weather conditions vary from about 3.0 to 6.7 t/ha as a result of climatic differences and are rather stable (CV around 10%). The average values are considerably lower (between about 1.8 and 3.3 t/ha), and the variability is usually much higher (CV up to 84%) in situations with limited water. Results of the simulations for the four locations are summarized in Tables 2 and 3.

Simulated grain yields of irrigated wheat in Migda over a sequence of 21 yr are presented in Figure 1, as an example of the results. (Such results, plotted as a yield-frequency-distribution graph, show an S-shaped curve, indicating that the CV is an appropriate measure of variability.) Yield variability in this situation is due to the combined effects of irradiation and temperature on the various yield-determining processes, including assimilation and respiration on the source side and rates and length of period of organ formation on the sink side.

Figure 1 also presents wheat yields at the same site and under identical weather conditions, but with precipitation as the only water source. There is a dramatic difference in variability between potential yield and water-limited yield in practically all situations, underlining the importance of water availability as a crucial constraint in crop production in many regions of the world. Because of the low and erratic rainfall in Migda, average yields are much lower and the variability is much larger: 84%. This computed variability is probably higher than in reality because of the assumed fixed sowing date and the absence of the possibility of resowing after an early crop failure. Still, it is obvious that variability is large. The CV drops only to 68% if the years with total crop failure are excluded.

Yields in Los Baños and Hyderabad are low because the growth period is relatively short due to the prevailing high temperatures—on average 49 d between emergence and anthesis, and 27 d for the grain-filling period. The relative stability in

Table 2. Predicted yields, in t (dry matter)/ha of wheat and hulled rice, and their variabilities.[a]

Site	Situation[b]	Wheat				Rice			
		Potential production		Water-limited		Potential production		Water-limited	
Wageningen	Current	6.2	11%	3.3	61%	NR		NR	
	Future w.	6.8	26%	3.2	89%	NR		NR	
	Future a.	7.4	22%	3.3	95%	NR		NR	
Migda	Current	6.6	9%	2.1	84%	NR		NR	
	Future w.+	4.1	22%	3.7	68%	NR		NR	
	Future w.–	4.1	22%	1.9	113%	NR		NR	
	Future a.+	6.4	19%	4.2	65%	NR		NR	
	Future a.–	6.4	19%	2.1	112%	NR		NR	
Hyderabad	Current	4.2	8%	1.8	10%	7.8	5%	6.4	27%
	Future w.t0	NC		NC		11.6	7%	9.9	22%
	Future w.	5.7	8%	2.0	11%	10.2	7%	8.6	22%
	Future a.	5.9	9%	2.6	10%	10.3	6%	8.6	23%
Los Baños	Current	3.1	12%	1.8	40%	6.4	9%	6.1	15%
	Future w.t0	NC		NC		9.3	11%	8.3	16%
	Future w.	3.8	10%	2.4	51%	8.1	11%	7.8	18%
	Future a.	4.0	11%	2.5	52%	8.3	12%	7.8	20%

[a]NR = not relevant, NC = not computed. [b]Current = simulation with actual weather data, future w. = future weather, future a. = future weather with adapted cultivars; + or – = highest or lowest estimate of precipitation from Table 1; t0 = CO_2 concentration increase only.

Hyderabad is brought about by the fact that the crop relies entirely on stored soil moisture, combined with the assumption that the profile is at field capacity at sowing, irrespective of preceding rainfall.

The harvest index (HI), the ratio of grain yield (economic yield) to total amount of biomass, is fairly constant when optimum moisture conditions are maintained throughout the season. If not, large fluctuations occur as a result of differences in the

Table 3. The average transpiration coefficient of wheat and rice, in kg H_2O/kg dry matter, and its variability (standard deviation).[a]

Site	Situation[b]	Wheat		Rice	
		Potential production	Water-limited	Potential production	Water-limited
Wageningen	Current	279-46	250-113	NR	NR
	Future w.	121-40	109-60	NR	NR
	Future a.	126-44	108-42	NR	NR
Migda	Current	336-40	322-162	NR	NR
	Future w.+	233-40	166-48	NR	NR
	Future w.–	233-40	154-72	NR	NR
	Future a.+	227-48	178-28	NR	NR
	Future a.–	227-48	149-38	NR	NR
Hyderabad	Current	449-20	383-18	217-6	231-7
	Future w.t0	NC	NC	146-4	142-5
	Future w.	253-19	289-18	156-5	153-5
	Future a.	262-19	279-20	152-6	149-5
Los Baños	Current	360-26	339-40	218-13	217-12
	Future w.t0	NC	NC	149-9	145-8
	Future w.	166-24	160-22	155-10	153-9
	Future a.	175-24	165-20	149-10	147-9

[a]NR = not relevant, NC = not computed. [b]For explanation of situations see Table 2.

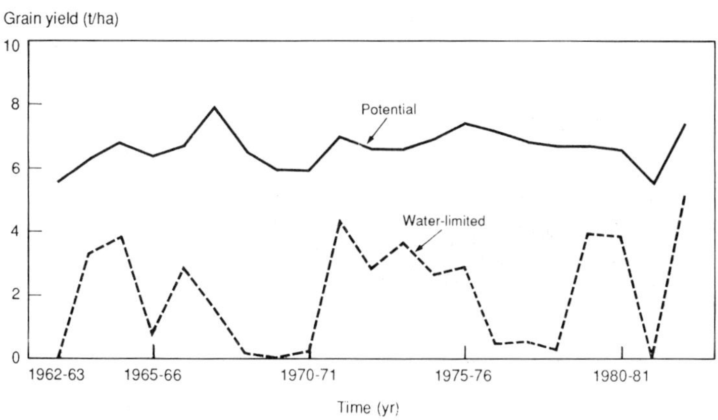

1. Wheat yields in Migda, 1962-82, for the potential growth situation and water-limited growth.

timing of moisture deficiency. The HIs and the range within which they usually fluctuate are presented in Figure 2.

Rice. Average potential and water-limited rice yields and their variabilities under current weather conditions are presented in Table 2, and the corresponding transpiration coefficients in Table 3.

Potential yields vary from 5.1 to 7.6 t/ha for Los Baños and from 7.2 to 8.5 t/ha for Hyderabad. Yield variability is related to the temperature and irradiation regimes, which show stronger fluctuations in Los Baños than in Hyderabad. The higher potential yield in Hyderabad is related to higher irradiation and lower average temperatures during the rice-growing season. Higher irradiation results in higher gross assimilation, and a lower temperature leads to relatively lower maintenance respiration and to longer growth duration (90 d in Hyderabad versus 85 d in Los Baños).

The range in water-limited yields is 3.9-7.6 t/ha for Los Baños and 3.6-8.6 t/ha for Hyderabad, with transplanting dates of 1 Aug and 16 Jul, respectively. These yields should be considered only as indicative, because they depend to a large extent on the rather arbitrarily chosen maximum percolation rate of the soil. Yield variability is related mainly to the rainfall pattern during the grain-filling period: The higher rainfall in Los Baños leads to a lower variability in water-limited yield.

Total dry matter production in the rainfed situation never exceeds total potential production. In some years, however, water-limited production results in a higher HI or even higher grain yield than without water limitation. This phenomenon occurs when drought stress during the preanthesis phase results in less vegetative dry matter at anthesis.

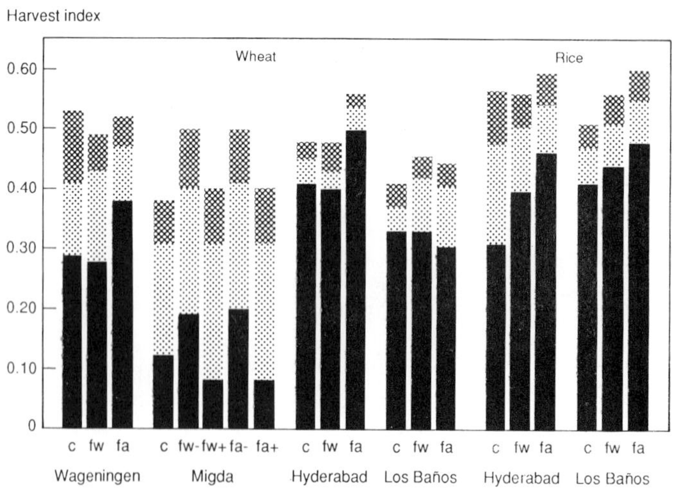

2. Histogram of harvest indices (HI) (kg grain/kg total aboveground biomass) for wheat and rice, water-limited production. The lower part of each bar reflects HI reached or exceeded in 9 of 10 yr; the second part, HI reached or exceeded in 5 of 10 yr; and the upper part, HI reached or exceeded in 1 of 10 yr. C = simulation with actual weather data, fw = future weather, fa = future weather with adapted cultivars, + or - = highest or lowest estimate of precipitation from Table 1.

Future yield level and its variability

Future yields under optimal growth conditions are different from current yields because the high CO_2 level accelerates the rate of assimilation of crops and hence their rate of dry matter production. However, higher temperatures also lead to increased rates of development and consequently to shorter growth durations. The overall effect can be positive or negative, as Table 2 shows.

Wheat. For Migda, with the largest anticipated temperature rise, the effect of the shortened season is by far the strongest. The potential yield decreases by almost 40%. The variability increases substantially to 22% because in more instances sink size is the limiting factor. At the other three sites, the effect on growth rate is the overriding factor, so that potential yields are higher. For Wageningen this is associated with higher variability, but not Los Baños and Hyderabad. The future HI in Migda under optimal growth conditions is substantially lower than under present conditions, and that for Hyderabad substantially higher, while the differences for Los Baños and Wageningen are small (Fig. 2).

An outstanding feature at all sites is the large decrease in transpiration coefficient. Contrary to the effect of other growth-stimulating factors, assimilation goes up considerably without affecting transpiration (assuming a constant relative humidity) at high CO_2. Yet, under rainfed conditions the effects of expected weather changes are more dramatic because variability in precipitation plays an important role. For Migda, we used both extreme values of expected changes in precipitation (Table 1) to investigate the sensitivity for this uncertainty in climate prediction.

Average rainfed yields in Wageningen are hardly affected. The more frequent occurrence of water deficiency during the growing season due to lower precipitation is compensated by higher growth rates under favorable conditions. The variability increases because the difference between dry and wet years is more pronounced. In Los Baños, average yields under rainfed conditions increase by 30% to 2.4 t/ha, but variability rises as well. The combined effect of the higher assimilation rate and more efficient water use result in more favorable growing conditions, although crop failures still occur. In Hyderabad, average rainfed yields go up by 15%. As this crop grows entirely on stored soil moisture, the higher yields are due to more efficient use of available water.

Harvest indices in water-limiting situations show little change, with the exception of Migda, where HI rises substantially (Fig. 2).

Rice. Grain yields increase considerably at high CO_2 level. Without a change in temperature, yields would be 30-50% higher (Table 2). The relative increase in growth rate is not constant over the entire growth period. The greatest gain occurs when the canopy is fully closed. During the first month after transplanting, simulated growth is only very little above the simulated growth for the current CO_2 concentration. This is the consequence of assuming a 35% decrease in specific leaf area.

The temperature increase of 2 °C results in a shortened total growth cycle from transplanting to maturity by 6 d, of which 2 d are at the expense of the grain-filling period. Another consequence is intensification by 15% in maintenance respiration. The resulting yield reduction amounts to 1.1 t/ha for Los Baños and 1.4 t/ha for

Hyderabad. On balance, potential rice yields under future weather are 25-35% higher than current ones, while the variability remains almost the same.

Adapted cultivars in future weather

When the climate changes, adapted cultivars will probably be grown. Such adaptations are simulated by adjusting cultivar-specific characteristics. The effects of changes in many characteristics could have been explored with our models. We have limited ourselves to examining the effects of different relations between temperature and development rate, which lead to different growth durations.

The overall effect on average potential wheat yield is slightly positive compared with that for unadapted cultivars, and much better in the Mediterranean climate. Potential yield is always higher than current potential yield. For the water-limited production situations, the effect is usually modest. The variability, however, remains almost unchanged by choosing other cultivars, although that was what we aimed for. The effect of our choice of better adapted rice cultivars has almost no effect (Table 2,3).

Discussion

The results of these simulation experiments should be considered indicative for the effects of anticipated changes in climate on average yields of wheat and rice and on their variabilities. The results are influenced by the assumptions that, explicitly or implicitly, are incorporated in the models. Many of these are thoroughly evaluated. But some aspects that are important may not have been simulated in sufficient detail. Models aiming specifically at analyzing the effects of ambient CO_2 concentration on crop performance could be developed further. More extensive basic data are also required, particularly on morphological and physiological characteristics.

Another important point is that the simulation results are not accurate under severe water shortage or heat stress. Yields and HIs should not be extrapolated to exceptionally unfavorable years by using the CVs presented.

The computed response of rice yields to future weather is larger than that for wheat. This is a reflection of a constraint in the wheat model, sink size limitation, that is not included in the rice model. Although there is little doubt that this constraint is real, it is conceivable that it can be avoided. The simulated increase in potential wheat yield may therefore be regarded as a pessimistic expectation, while that for rice represents an optimistic view.

The assumptions with respect to agronomic practice used in the models also have a distinct effect on the results. Using a flexible sowing date dependent on the rainfall regime for the rainfed wheat crops—an opportunistic strategy—would probably have resulted in a substantially lower variability in most cases. Also, assumptions regarding fertilizer application in the rainfed wheat crop influence the final outcome.

The considerable effect of high CO_2 on the transpiration coefficient is a major contribution to increased crop production under future weather conditions. Because of induced stomatal closure by increased atmospheric CO_2, transpiration rate on a leaf area basis will decrease. In the wheat crop this reduction is partially offset by increased leaf area formation. Both experimental (Jones et al 1985) and theoretical (Goudriaan et al 1985) studies confirm this effect for seasonal water use. The rice crop model yielded a slightly reduced leaf area. Much experimental evidence exists that demonstrates the phenomenon of stomatal regulation, but it is still uncertain under what conditions it is effective.

The difference in transpiration coefficient between wheat and rice at the same sites (Table 3) is considerable; it is due to different weather conditions because their growing periods are not identical, to the occurrence of growth rate reductions related to sink limitation without feedback on transpiration in the wheat and not in the rice model, and to different concepts in the computation of transpiration in both models.

Whenever a serious shortage of N or minerals occurs during the growing season, as is common in many agricultural systems, the effects of weather on yield and yield variability are more difficult to predict. More efficient water use may then lead to situations where more often, nutrients become the major limitation to crop production. Although higher CO_2 concentrations may slightly improve nutrient-use efficiency, a large part of the favorable effects shown in this study will not express themselves under nutrient constraints. Models for nutrient dynamics in soils and crops are still insufficiently developed, and more research is needed before such simulations are reliable.

Conclusions

Some of the effects on wheat and rice crops computed for climates 70-100 yr from now are large. They will start to emerge in the near future, such as increased water-use efficiency of crops and higher potential yields.

The effect of an increasing CO_2 concentration is positive for yield in the potential and water-limiting situations in all climates. The increase in temperature has a negative effect on growing season duration and hence on total biomass produced, unless adapted cultivars can be used to counteract this effect. However, reduction of preanthesis growth of rice appears to be beneficial for its HI and grain yield. Taken together, average potential yields of wheat and rice will increase 25-50% in the tropics under future weather conditions. For cooler climates the effect is smaller. For rainfed crops in the tropics the yield also increases considerably. The gain appears small in temperate climates, but large if precipitation increases. The variability, higher for water-limited situations than for potential production, generally increases slightly or remains stable.

The effects on yield will probably be smaller for crops with a severe constraint in nutrient availability, but the extent is unknown. Two indicators of soil moisture are computed along with crop growth in the water-limiting situations: average volumetric water content in the 20- to 30-cm soil layer from emergence until maturity, and total soil water in the potential rooting zone at the end of the growing season. They did not show much difference between the three sets of simulations made, and are not discussed further here. However, they should be considered again under nutrient stress. Crops might then use less water and soils may be wetter on average, making the need for fertilizers more explicit.

Physiological research

A major component of the yield increase under future weather is the much more efficient use of water due to high CO_2 concentration. However, a firm description of the conditions under which stomata respond is still lacking. The degree of regulation will affect water-use efficiency, so research in this field is necessary to improve predictions of crop performance.

The potentially large yield increase under high CO_2 concentration in wheat and rice must be accommodated by an increase in sink size, i.e. more and / or larger grains per unit area. Research must quantify to what extent this will be achieved spontaneously in response to the improved carbohydrate supply of the plants, and how much must be achieved by plant breeding.

More experimental work is required to quantify the interactions among high CO_2 level, assimilation rate, and nutrient stress. In many current situations, crop yields are fully determined by nutrient availability. The direct effects of CO_2 and temperature are then probably small. But the indirect effects—lower water use and hence more water in the soil—may be positive and stabilizing. It is uncertain to what extent yields under such conditions will be affected. The comparative advantages of legumes may be explored further, as their capacity to fix N will be reinforced by more assimilates.

To judge the effects of future weather on a geographic scale, experimental research and simulation should expand to other crops and to other plant types such as trees and natural vegetation.

Plant breeding

From the limited exploration of the effects of crop characteristics on yield under future weather, it is somewhat speculative to derive recommendations for plant breeders. Results obtained suggest the need for a wide spectrum of cultivars with respect to growth duration to exploit the relative advantages of various environments. More explanatory research on genotype $\times$ environment interaction is necessary to formulate recommendations for specific cultivars.

The potential growth rate of C_3 crops in the tropics, now considerably below that of C_4 crops, is likely to catch up, even without any genetic engineering of C_4 characteristics. The positive effect of a high CO_2 concentration on growth of C_4 crops is expected to be smaller than on C_3 crops, but the transpiration coefficient of C4 crops will probably also be reduced. Higher temperatures will shorten growing seasons and lower potential yields for existing cultivars.

Maintenance respiration takes an important and increasing share of the total assimilation at the expected higher temperatures, so that further research about cultivars with lower rates holds much promise.

Problems with high temperature stress will probably intensify slowly, particularly in C_3 crops. Breeding will have to identify more tolerant cultivars. Adaptation for increased water stress will probably not be required often.

Climatology

For a more accurate description and prediction of the effects of climatic change on crop production, a better prediction of future weather conditions is indispensable.

Particularly, eventual changes in radiation and humidity are major factors in the response of crops to weather changes. Improvements in general circulation models are necessary to take account of these factors.

References cited

Goudriaan J, De Ruiter H E (1983) Plant growth in response to CO_2 enrichment at two levels of nitrogen and phosphorus supply. 1. Dry matter, leaf area and development. Neth. J. Agric. Sci. 31:157-169.

Goudriaan J, Van Laar H H, Van Keulen H, Louwerse W (1984) Simulation of the effect of increased atmospheric CO_2 on assimilation and transpiration of a closed crop canopy. Wiss. Zeitschr. Humboldt Univ. zu Berlin Math.-Nat. R. 33:352-356.

Goudriaan J, Van Laar H H, Van Keulen H, Louwerse W (1985) Photosynthesis, CO_2 and plant production. *In* Wheat growth and modelling. NATO ASI Series. Series A: Life sciences. Vol. 86. W. Day and R. K. Atkin, eds. Plenum Press, New York.

Jones P, Allen L H, Jones J W, Valle R (1985) Photosynthesis and transpiration responses of soybean canopies to short- and long-term CO_2 treatments. Agron. J. 77:119-126.

Rappoldt K (1986) Crop growth simulation model WOFOST. Documentation Version 3.0. Staff working paper SOW-86-05. SOW, Wageningen, The Netherlands.

Schlesinger M E, Mitchell J F B (1985) Model predictions of the equilibrium climate response to increased carbon dioxide. Pages 82-147 *in* The potential climatic effects of increasing carbon dioxide. M. C. MacCracken and F. M. Luther, eds. Lawrence Livermore National Laboratory, Livermore, California.

Van Diepen C A, Van Keulen H, Penning de Vries F W T, Noy I G A M, Goudriaan J (1987) Simulated variability of wheat and rice yields in current weather conditions and in future weather when ambient CO_2 has doubled. Simulation Report CABO-TT, Centre for Agrobiological Research, Wageningen, The Netherlands.

Van Keulen H, Seligman N G (1987) Simulation of water use, nitrogen nutrition and growth of a spring wheat crop. Simulation Monograph, Center for Agricultural Publishing and Documentation, Wageningen, The Netherlands.

Wolf J, Rijsdijk F H, Van Keulen H (1986) Computer models of crop production. Pages 341-384 *in* Modelling of agricultural production: weather, soils and crops. H. Van Keulen and J. Wolf, eds. Simulation Monograph, Center for Agricultural Publishing and Documentation, Wageningen, The Netherlands.

Notes

Authors' addresses: F. W. T. Penning de Vries, Centre for Agrobiological Research, Bornsesteeg 65, Wageningen, The Netherlands (Present address: IRRI, P.O. Box 933, Manila, Philippines); H. Van Keulen, Centre for Agrobiological Research and Centre for World Food Studies, Bornsesteeg 65, Wageningen, The Netherlands; C. A. Van Diepen and I. G. A. M. Noy, Centre for World Food Studies, Bornsesteeg 65, Wageningen, The Netherlands; and J. Goudriaan, Department of Theoretical Production Ecology, Agricultural University, Bornsesteeg 65, Wageningen, The Netherlands.
Citation information: International Rice Research Institute (1989) Climate and food security. P.O. Box 933, Manila, Philippines.

Potential effects on crop production of carbon dioxide enrichment in the atmosphere and greenhouse-induced climate change

N. J. Rosenberg

Atmospheric concentrations of CO_2 and other radiatively active trace gases are increasing. An enhanced greenhouse effect—warming of the lower layers of the atmosphere, with attendant changes in the regional distribution of temperature, precipitation, and evaporation—is expected in coming decades. Specific regional changes in climate are difficult to predict. Global climate models disagree in significant ways except for a consensus that temperatures will be higher. Additionally, the grid sizes used in such models are too large to assess how specific agricultural systems in smaller areas will be affected. But there is some degree of certainty that crops, particularly C_3 crops, will respond to the increased atmospheric CO_2 concentration with increased rates of photosynthesis. Transpiration rates may be reduced because of stomatal closure caused by high ambient CO_2 concentrations, particularly in C_4 crops, although this response is less certain. This paper reviews some methods used to a) adapt results of general circulation models and other techniques of climate prediction for application to the scale of real geographic crop-producing regions, and b) estimate the regional impact on food production of these climatic changes and of the interactions of climatic change with the direct biological effects of increased atmospheric CO_2 concentration. Suggestions are offered for improvements in methodology for these purposes.

There is growing consensus among climatologists that, as a result of the increasing concentrations of such radiatively active trace gases as CO_2, CH_4, N_2O, chlorofluorocarbons, and others, global climate will significantly change within the next century. These climatic changes will, if large enough, affect agriculture, forestry, and the availability of water resources, hence, global food security; whether for better or worse will depend on the nature and degree of change that occurs in each important food-producing region of the world.

This review briefly describes the mechanisms by which the trace gases affect the global radiation balance and how the so-called "greenhouse-effect" fostered by them may be manifested in changing temperature, precipitation, and evaporation conditions around the world. One of the gases—CO_2—exerts a direct influence on photosynthesis and water use by vegetation that may either counter or augment the effects of climatic change. This subject is briefly introduced in this review and is covered in greater detail in another paper (Gifford, this volume).

An additional purpose of this review is to suggest ways in which researchers and systems analysts may approach the complex problem of predicting how climatic change and the direct biological effects of CO_2 will affect crop production in agriculturally important regions of the world.

The greenhouse effect

Water vapor, CO_2, N_2O, and CH_4 are all essentially transparent to incoming solar radiation but are strong absorbers in the infrared—behaving somewhat like glass in a greenhouse, which lets in the sunlight but traps outgoing thermal radiation. The greenhouse analogy is imperfect, since the gain in heat within a greenhouse is due largely to the suppression of convective transport; nonetheless, the terms "greenhouse effect" and "greenhouse gas" have become commonplace. The term radiatively active trace gas is more accurate.

Water vapor is the most important of the infrared absorbing gases, but CO_2 has strong absorption peaks at about 4 and 15 μm, where water vapor is less absorptive. As a result of human activities, the concentration of CO_2 has increased in the global atmosphere at least 25% since 1750 (280 ppm then, more than 350 ppm today). The increase in CO_2 concentration is expected to double the preindustrial concentration by about the middle of the coming century. Additionally, the concentrations of other radiatively active trace gases, N_2O, CH_4, CCl_4, and the chlorofluorocarbons (CCl_2F_2 and CCl_3F, known commercially as Freon 11 and 12), are also increasing at an even more rapid rate. N_2O and CH_4 may be increasing as a result of human activities, but natural sources and sinks for these gases are known. The Freons and CCl_4 are industrial products—the results of human activities. CH_4, N_2O, CCl_4, and Freons 11 and 12 can, together, generate as strong a greenhouse effect as that caused by CO_2.

The greenhouse effect is a cause for concern. Increasing atmospheric absorption of infrared radiation must lead to a warming of the lower layers of the atmosphere. The capacity of the atmosphere to hold water vapor is an exponentially increasing function of temperature. A warmer atmosphere (all other factors remaining unchanged) leads to greater evaporation. Hence, the CO_2-induced greenhouse effect can "feed back" positively by increasing the quantities of water vapor in the atmosphere unless increased cloudiness compensates for this effect by increasing its reflection of solar radiation to space, thereby reducing the amount that reaches the earth's surface.

Much has been written on the difficulty of identifying the "greenhouse" signal in the noise of natural climatic variation (e.g. Hayashi 1982). Despite the difficulty, a report by Jones et al (1986) based on near-surface temperature data over land and oceans of both hemispheres during the past 130 yr provides evidence of a general warming during this century and a rapid warming since the mid-1970s. Five of the warmest years on record have occurred in the 1980s.

Consequences of the greenhouse effect

Three approaches that have been used to predict how the greenhouse effect will be expressed in terms of global and regional climate changes are discussed by Rosenberg (1987). One approach relies on proxy climatic data from the distant past or paleoclimatic analogues such as tree-ring records, pollen layering in sediments, and isotope ratios in wood or sediments. Another approach is based on instrument records. Sets of warm and cold years are selected from the modern climate record. Differences between specific regions in surface pressure, temperature, and precipitation patterns are determined. Lamb (1986) has reviewed both of these approaches and points out a number of systematic disadvantages which will not be recounted here.

The third approach, which has received most effort, resources, and attention, is simulation modeling. Virtually all components of the climate system including land, ocean, atmosphere, and cryosphere (snow, and land and sea ice) can be described mathematically. The equations can be solved either separately or in concert to study limited interactions or fluctuations of the entire system.

There are many types of climate models ranging in complexity from the one-dimensional, in which some change in the atmosphere (e.g. an increase in dust or smoke) alters the vertical distribution of temperature, to general circulation models (GCMs), which deal with changes in the three-dimensional atmosphere over a fourth dimension—time.

Meehl (1984) pointed out that GCMs are presently the most complex and expensive in terms of computer time but are also the most realistic in the way they explicitly simulate many elements of the climate system. Despite their sophistication, these models are not without problems.

The "state-of-the-art" in GCMs has been assessed by MacCracken and Luther (1985):

> Atmospheric general circulation models are capable of simulating almost all of the observed large-scale features of the climate, and they reproduce the general character of day-to-day variations as well as seasonal changes of the circulation winter to summer. However, these models do not yet adequately represent the observed regional features that are needed for making detailed climate projections and assessments of ecological, agricultural and societal impacts.

Nonetheless, GCMs appear to be the best available tools for predicting in the near future the direction and degree of climatic change. MacCracken and Luther (1985) reviewed results of the then most recent "experiments" involving GCMs in which the increasing atmospheric concentration of CO_2 is allowed to perturb the current atmosphere. Although CO_2 is increasing slowly in the atmosphere (as are the other radiatively active trace gases), it is easier for GCMs to calculate what might happen if a large stepwise change in CO_2 concentration were to occur.

When the models proceed without consideration of feedback processes (changing cloudiness, oceanic capture, or release of heat), the GCMs are in close agreement, predicting a change in global average surface temperature due to a doubling of CO_2 concentration in the range of 1.2-1.3 °C. However, the models that include feedback processes differ considerably in their predictions—1.5-4.5 °C.

The models predict a greenhouse warming of the global average surface air temperature of about 3.5-4.2 °C and an increase in the global average precipitation rate of about 7-11%. The models agree closely with respect to global average surface temperature, but not so well in their projections of the regional patterns of such changes.

Figures 1, 2, and 3 (adapted from Schlesinger and Mitchell 1985) illustrate projections made by three GCMs of changes in climate (temperature, precipitation, and soil moisture) to follow a doubling of the global atmospheric CO_2 concentration. The figures have been redrawn for simplicity and show only Africa and North America in June, July, and August.

Differences between the models of the Geophysical Fluid Dynamics Laboratory (GFDL), Goddard Institute of Space Studies (GISS), and the National Center for Atmospheric Research (NCAR) are shown in these figures. Predictions of temperature increases in the middle of North America range from 2 to 8 °C, depending on the model. Projections for precipitation in this region during June, July, and August range from -1 to +1 mm/d. In West Africa, precipitation changes range from +2 to -1 mm/d. Perhaps most critical is the soil water difference that results from changes in precipitation and evaporation. GFDL predicts as much as 3 cm soil water deficit for the summer season in North America; GISS predicts no change there, while NCAR predicts from zero to -1 cm reduction in soil water. In West Africa the predictions differ widely as well.

Another interesting comparison assembled by Schlesinger and Mitchell (1985) is shown in Figures 4, 5, and 6. Figure 4 compares the GFDL, GISS, and NCAR latitude-time cross sections of zonal mean surface air temperature change for a

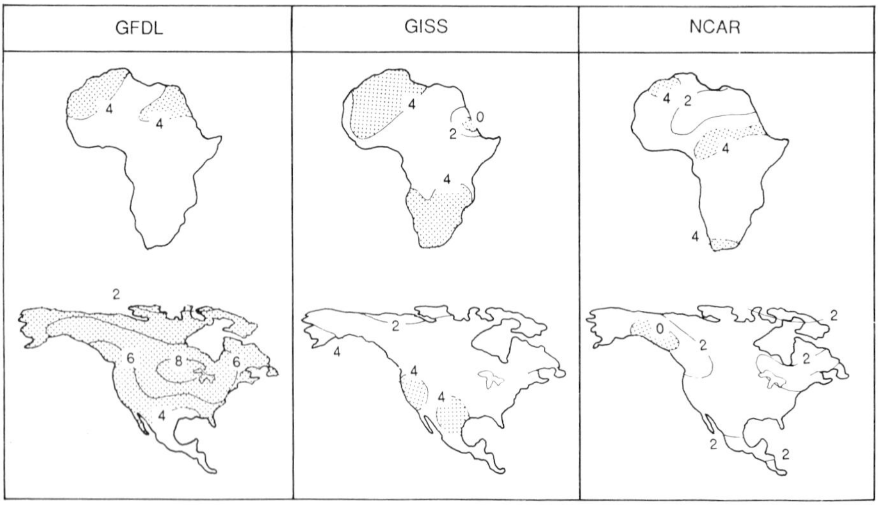

1. Distribution of surface air temperature change (°C) for a doubling of atmospheric CO_2 concentration for June, July, and August simulated by the global climate models of GFDL, GISS, and NCAR. Stippling indicates temperature increases greater than 4 °C. (Redrawn from Fig. 4.39 of Schlesinger and Mitchell 1985.)

doubling of atmospheric CO_2 concentration. GFDL predicts that the region between latitudes 30 and 50° N will be at least 4 °C warmer throughout the year. GISS also predicts that the same zone will be that much warmer, but not from late spring through much of the fall. NCAR indicates that the 30-50° N latitude belt will not be 4 °C warmer at any time.

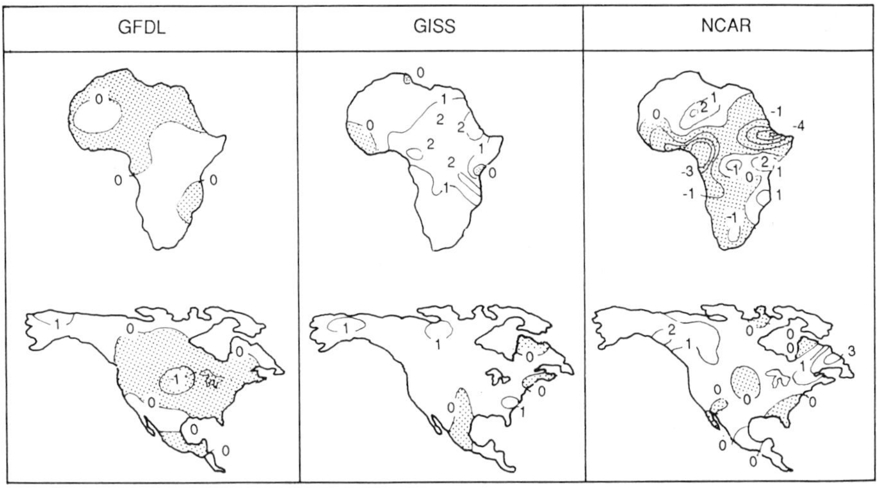

2. Distribution of precipitation rate change (mm/d) for a doubling of atmospheric CO_2 concentration for June, July, and August simulated by the global climate models of GFDL, GISS, and NCAR. Stippling indicates a decrease in precipitation rate (Redrawn from Fig. 4.42 of Schlesinger and Mitchell 1985).

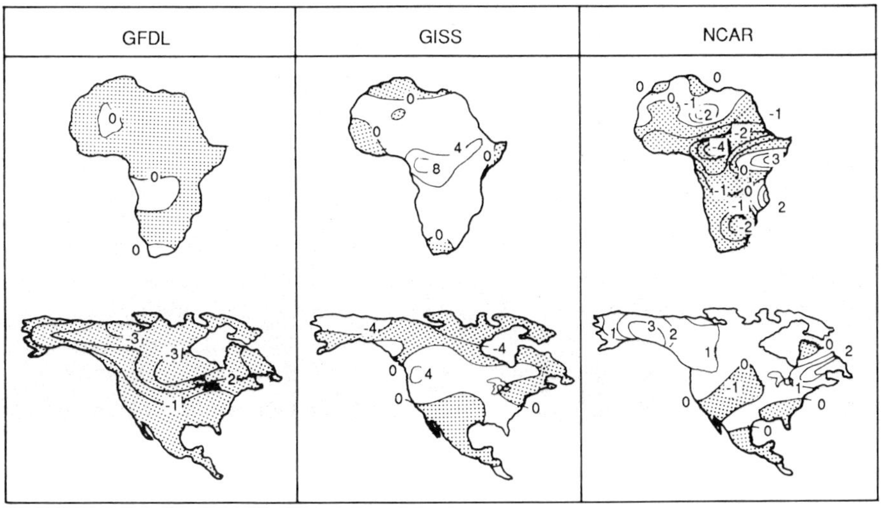

3. Distribution of soil water change (cm) for a doubling of atmospheric CO_2 concentration for June, July, and August simulated by the global climate models of GFDL, GISS, and NCAR. Stippling indicates a decrease in soil water (Redrawn from Fig. 4.45 of Schlesinger and Mitchell 1985).

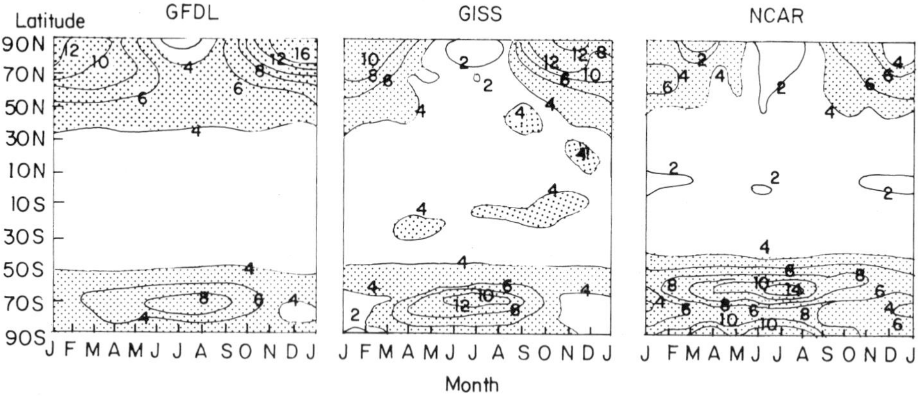

4. Latitude-time cross section of the zonal mean surface air temperature change (°C) for a doubling of the atmospheric CO_2 concentration simulated by the global climate models of GFDL, GISS, and NCAR. Stippling indicates temperature increases greater than 4 °C (From Schlesinger and Mitchell 1985, Fig. 4.40).

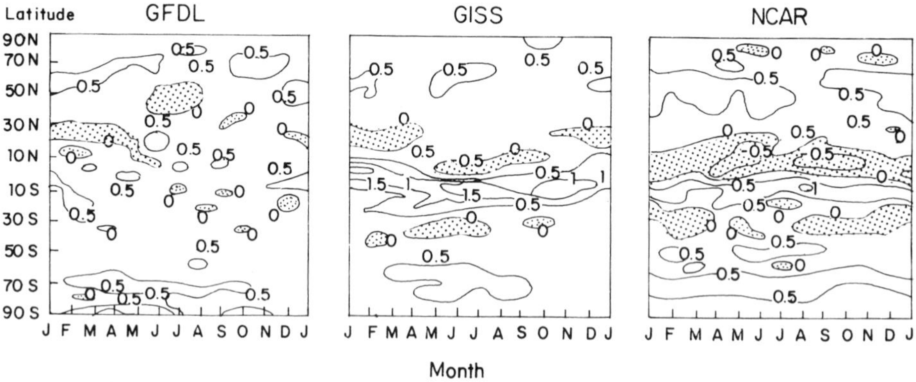

5. Latitude-time cross section of the zonal mean precipitation rate change (mm/d) for a doubling of the atmospheric CO_2 concentration simulated with the global climate models of GFDL, GISS, and NCAR. Stippling indicates a decrease in precipitation rates (From Schlesinger and Mitchell 1985, Fig. 4.43).

Similarly, Figure 5 displays the latitude-time cross section of the zonal mean precipitation rate in millimeters per day for a doubling of the atmospheric CO_2 concentration. For the latitude belt 30-50° N, GFDL predicts a period of summer dryness. The other two models predict no change or a slight increase in precipitation.

Figure 6 displays latitude-time cross sections of zonal mean soil water change in centimeters for a doubling of the atmospheric CO_2 content. This simulation applies only to the ice-free lands. Throughout most of the year, all three models simulate a band of decreased soil moisture near 20° N and a band of increased soil moisture near 10° S. All three models show a moistening of the soil within the belt 30-60° N from October to April. GFDL predicts a drying everywhere from April to September, GISS predicts late summer drying south of 50° N, and NCAR predicts a moistening north of 30° N throughout the year.

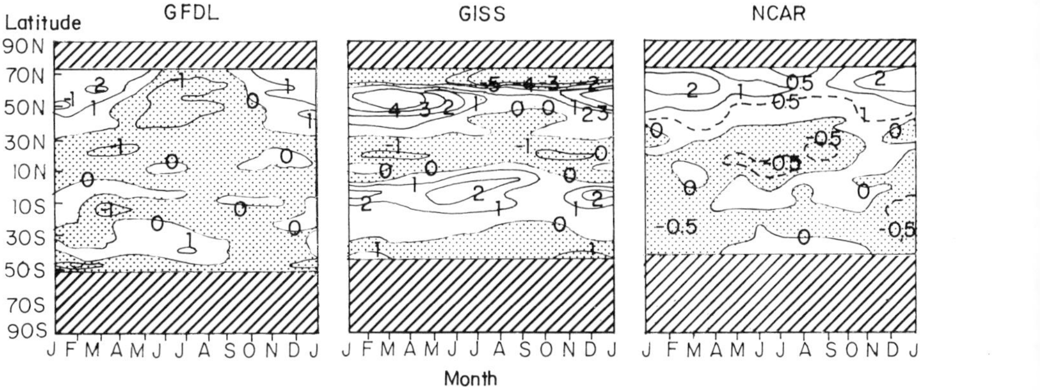

6. Latitude-time cross section of zonal mean soil water content change (cm) over ice-free land for a doubling of the atmospheric CO_2 concentration simulated with the global climate models of GFDL, GISS, and NCAR. Stippling indicates a decrease in soil water; hatching indicates latitudes where there is no ice-free land (From Schlesinger and Mitchell 1985, Fig. 4.46).

Obviously, the "state-of-the-art" in global climatic modeling is not yet sufficient to provide agreement as to what will happen in a doubled-CO_2 world. The detail needed for regional planning purposes is lacking. The models are, in my opinion, good enough at this point to alert us to the probability of significant change in the temperature regime, in the spatial and temporal distribution of precipitation, and in the seasonal availability of soil moisture. They are also good enough to alert us to the possibility that CO_2-induced climatic change might lead to more frequent, more severe, or more protracted droughts. However, careful examination of Figures 1-6 raises the possibility that droughts might actually be moderated by forthcoming climatic change. The issue is complex, and much more information must be brought to bear before we begin calling possibilities by the more commanding term probability.

The most severe predictions stem from the GFDL model, and a trial run with that model (Manabe and Wetherald 1986) produced particularly ominous conclusions with respect to summer dryness in the mid-latitudes. In this scenario, soil moisture is reduced in summer over extensive regions of the middle and high latitudes, including the North American Great Plains, northern Canada, western Europe, and Siberia. With what could be interpreted as a greater than 90% statistical certainty, Manabe and Wetherald predict as much as a 3-cm decrease in mid-continental soil moisture in North America in summer (June, July, August) due to doubling of atmospheric CO_2. This corresponds to as much as a 50% decrease in available soil moisture. Elsewhere, the author (Rosenberg 1987) has raised questions as to the reality of these predictions. But extreme climate changes are considered possible by at least some global climate modelers.

Effects of CO_2 on plant growth and water use

R. M. Gifford (this volume) explains the physiological reasons why plants respond to air enriched in CO_2. In growth chambers and controlled-environment facilities, C_3

plants (small grains, legumes, most trees, root crops, and grasses) demonstrate an increase in the rate of photosynthesis. This is believed to be due to two factors—an increase in the air-to-substomatal gradient in CO_2 concentration, and a suppression of photorespiration. C_4 plants (tropical grasses such as maize, sorghum, millet, and sugarcane) demonstrate a decrease in the rate of transpiration as do the C_3 plants. The C_3 response (increased photosynthetic rate) is well accepted by physiologists; the C_4 response (reduced transpiration) appears to be less certain for both physiological (e.g. Morison and Gifford 1983) and energy-balance (e.g. Jarvis 1986) reasons.

It has yet to be established that either of these responses actually occurs in the field, where CO_2 concentration oscillates considerably on diurnal and seasonal cycles, and where moisture and nutrients are sometimes limiting. Physiological evidence does exist, however, that plants exposed to elevated concentrations of CO_2 suffer smaller reductions in photosynthesis and yield when exposed to moisture stress, soil salinity, and N shortages than do control plants grown at currently ambient CO_2 concentrations (Kimball 1985). It has also been noted that a wide range of terrestrial and aquatic plants grown in open-topped plastic enclosures at about double ambient CO_2 concentration respond positively to increasing temperature (Idso et al 1987).

It appears, then, that elevated CO_2 concentration in the atmosphere will have beneficial effects on plant growth, especially in C_3 species, and may lower transpiration in both C_4 and C_3 species. In both cases, the result is an improvement in water-use efficiency (WUE), where

$$\mathrm{WUE} = \frac{\text{rate of photosynthesis}}{\text{rate of transpiration}} \text{ or } \frac{\text{yield}}{\text{unit of water consumed}}$$

Effect on global food production

Our aim is to find ways to assess the potential impact on food production of the combined climatic and biological effects described above. The task resolves itself into two components:

- adapting the results of global climate models and other techniques of climate prediction for application to the scale of real geographic crop-producing regions (e.g. North American maize belt, Southeast Asian rice zone, etc.); and
- estimating the regional impacts on food production of these climate changes and of the interactions of climate change with the direct biological effects of increased atmospheric CO_2 concentration.

Predicting climatic change on an appropriate scale

While there is strong consensus concerning the global temperature implications of rising concentrations of atmospheric CO_2 and trace gases, at least with respect to the latitudinal distribution of annual means, there is little agreement about related specific changes in regional climates. Some of the models give contradictory results.

How does one approach the problem of predicting the impact of climatic change on actual agricultural systems in regions smaller than the model grid size? One way is to take extremes of change from current regional models as limiting cases and analyze the first-order impacts associated with each case. This is a form of sensitivity analysis, an approach found particularly useful when one has no good basis for selecting one case as more representative of reality than another. To the extent that the first-order impacts are insensitive to the differences among the climate extremes, the fact of the differences loses importance. And the impacts that are sensitive to the differences would provide useful guidelines to the parts of the models most needing refinement. Moreover, methodologies developed in doing this analysis would be valuable for subsequent analysis when climate modelers are more in agreement on the nature of regional climatic change.

Bach et al (1985) made use of GCMs in deriving regional climate scenarios. These they treated not as predictions of regional climatic change nor as estimates of forthcoming events, but rather as "...a set of self-consistent and plausible patterns of climate change." In constructing these scenarios, Bach et al made use of seven GCMs stemming from five separate modeling groups to prepare scenarios of climatic change for western Europe and North Africa. These models differed in their specific characteristics of spatial resolution, temporal resolution, geographic realism, inclusion of ocean-atmosphere feedback processes, and ability to simulate observed climatic conditions, and in the accessibility they afforded to simulated parameters useful for impact analyses (e.g. barometric pressure, vapor pressure). Unfortunately, only temperature, precipitation, and, by derivation, soil moisture changes were actually considered in the work described above.

The range of conditions predicted by the seven GCMs was, as expected, great, leading Bach and his coworkers to conclude that although GCMs can be useful for constructing general scenarios of possible climatic change, they are not yet good enough to provide reliable estimates of change on the detailed geographical and seasonal scales needed for impact assessment.

Another approach appears promising: GCMs available at this time may disagree considerably in estimating equilibrium temperatures as a result of the large increases in atmospheric CO_2 predicted in coming decades. These disagreements are attributable to differences in the models' resolution and parameterization of subgrid scale processes. Gates (1985) pointed out, however, that much more information of possible use in impact analysis can be extracted from GCMs than has been done thus far. The data sets simulated by GCMs permit systematic extraction of a wide variety of statistics important to impact analysis on what Gates terms the "ecosystem scale." Examples of such parameters are length of the growing season, duration of rainless periods, and surface moisture stress.

It also appears likely that the frequency of occurrence of presently extreme events (e.g. cold or hot spells, droughts) may be affected by anticipated climatic changes. These are the events that have important impacts on local vegetative associations or agricultural systems, especially in marginal zones where one or another climatic variable is the major limiting factor.

It should be possible to extract useful information from the files of GCM experiments wherein the climate is perturbed by a doubling (or greater increase) of the atmospheric CO_2. Some examples of extractable data sets are variance in temperature, temperature persistence, and frequency of extremes; storm tracks;

blocking frequencies; cloudiness; solar irradiance of the surface; surface winds; mean jet position; variance and persistence in humidity; and precipitation intensity.

One example of how such data sets may be useful is reported by Mearns et al (1984): An increase in mean annual (or seasonal) temperatures is likely to be accompanied at three midwestern U.S. locations (Des Moines, Fargo, and Evansville) by an increase in the number of days with temperatures exceeding certain high thresholds. Another interesting example was posed by Neild et al (1979): A global temperature increase would make possible the earlier planting and germination of annual crops and the earlier break in dormancy of perennial crops; unless the occurrence of frost is set back in time, such crops become, paradoxically, more vulnerable to destruction by frost.

Despite the richness of GCMs in statistics that may be of possible interest in impact analysis, Gates (1985) pointed out that the effective size or scale of the ecosystem on which climatic impacts actually occur is usually much smaller than the GCM grid size. How, then, can climatic changes be estimated on the local scale from essentially large-scale results of a GCM? Gates proposed one approach that may be useful: to examine the statistical relationships betwween the variations of a particular climatic variable (e.g. monthly mean local or station data) and the corresponding variations in monthly means averaged over an area comparable in size to a GCM grid element. These relationships depend on local factors such as geography, topography, and prevailing circulation.

Gates proposed that variations at each station in a selected area may be quantified in terms of empirical orthogonal functions (EOFs), which define the dominant spatial patterns of local variations that occur in conjunction with variations in the areal mean. This "inversion" of grid-scale results from a GCM simulation permits the assignment of climatic changes to the local scale—giving the statistically most probable local distribution of anomalies when only the area average anomaly is available.

Effects of climatic change and CO_2 enrichment on food production

A number of analysts have used regression techniques to study the potential effects of climatic change and CO_2 enrichment on plant growth. However, far greater emphasis has been placed in these analyses on the effects of climatic change than on the direct effects of CO_2 enrichment. The former is more amenable to treatment, since data sets on crop response to weather and climate are available from real-world experience. Virtually no data exist from outside the growth chamber or greenhouse on plant responses to CO_2 enrichment.

Blasing and Solomon (1982) used regression techniques to predict the impacts of climate change—warming and drying—on maize production in the U.S. maize belt. They concluded that, under certain climatic scenarios, the western portion of the maize belt would be unable to sustain production except under irrigation. A northward movement of the crop was also predicted in the eastern maize belt. Newman (1982), considering changes in evapotranspiration and growing degree days, estimated that the U.S. maize belt would shift in a southwest to northeast direction by about 175 km for every 1 °C increase in regional temperature.

Waggoner (1983, 1984) speculated that in the Midwest about half of a 10% decline in crop production resulting from climatic change would be offset by CO_2 fertilization.

Decker et al (1986) examined the possible impacts of climate change in the U.S. Midwest on animal husbandry as well as on crop and range production. Considering current cropping practices, they analyzed the comparative advantages of specific pairs of crops (e.g. maize vs soybean, maize vs wheat, maize vs sorghum) and predicted certain changes in land use—e.g. maize yielding to sorghum in the western maize belt if summers were to become drier. A weakness of their analysis is its reliance on the prediction of only one of the GCMs (GFDL) (Manabe and Wetherald 1980, Manabe and Stouffer 1980). Nonetheless, their systematic approach to possible change in midwestern cropping patterns, their analysis of possible adaptive techniques (e.g. plant breeding), and their analysis of research needed to develop adaptation are highly instructive.

In regression studies, in general, past yield trends are attributed to changing technology, and annual deviations from these trends are assumed to be due to climatic differences. Strengths and weaknesses of the regression approach as applied to the possible impact of climatic change on food production were discussed by Katz (1977) and Rosenberg (1982).

A different approach to gauge the possible effects of a climatic change in North America was proposed by Rosenberg (1982). Using climatic analogues rather than regression, he was able to show that expansion of the hard red winter wheat zone in North America during 1920-80 occurred over a climatic gradient of temperature, precipitation, and growing season length greater than the changes predicted by GCMs to follow from a doubling of the atmospheric CO_2 concentration (Fig. 7).

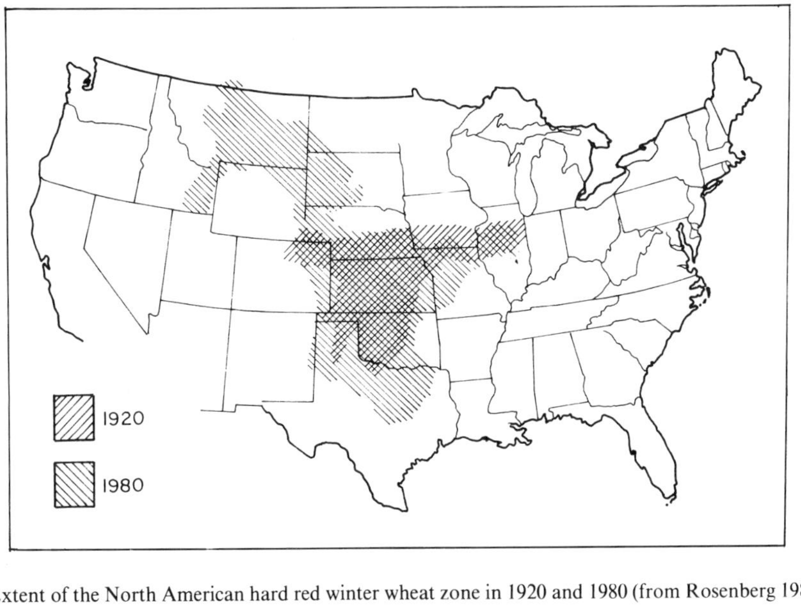

7. Extent of the North American hard red winter wheat zone in 1920 and 1980 (from Rosenberg 1982).

Regression and climate analogues approaches are useful but limited. While they can be applied to a degree to help anticipate how a specific climatic change may affect crop yield and even, ultimately, the crop zonation, no database exists that would permit assessment of the CO_2 enrichment effect or its interaction with climatic change. However, simulation models of plant growth, development, and yield are now available and can be used to anticipate these impacts. Nix (1985) suggests drawing upon well-founded process-type models that deal explicitly with photosynthesis, respiration, evapotranspiration, and other pertinent physical, physiological, and interactive (e.g. soil N level) phenomena.

The literature of simulation models is extensive. My examination of it suggests that the following models may be helpful for an initial look at the possible responses of major food crops.

Maize: CERES-Maize (Jones and Kiniry 1986)
Wheat: CERES-Wheat (Ritchie and Otter 1984)
Sorghum: SORGF (Arkin et al 1976)
Soybean: SICM (Wilkerson et al 1983)
SOYMOD (Meyer et al 1979, 1981; Meyer 1985)
GLYCIM (Acock et al 1982)

These models vary considerably in complexity, input requirements, and the likely ease with which they can be adapted to consider climatic means, variances, persistence, and frequency of critical short-term events such as frosts and extreme winds that are likely to be important in determining actual crop performance in a changed climate.

No useful purpose is served by calculating climate change effects on maize yield with, for example, monthly mean data on precipitation and temperature, as is done in many regression-type models, while hourly data on temperature, humidity, solar radiation, wind speed, etc. are used for the same purpose with, say, soybean. Both CERES-Maize and CERES-Wheat are daily incrementing models for growth, development, and yield. Other models (e.g. SOYMOD) require hourly environmental data. It is a generally accepted strategy in the community of modelers to simulate normal diurnal climatic patterns by hourly increments from daily means.

It should be recognized, however, that the modes of action by which increased CO_2 in the ambient air affects the physiology of photosynthesis and evapotranspiration are not yet fully understood. Although certain models do explicitly consider the effect of CO_2 concentration on photosynthesis in crop plants (e.g. GLYCIM, SOYMOD, CERES), Reynolds and Acock (1985) point out that validation is lacking.

There are impediments to the immediate use of existing crop models for estimating the direct response of plants to increased atmospheric CO_2. Studies in controlled environment chambers show that changes occur in plant morphology, root branching, leaf area index, and other plant characteristics. These may feed back with the environment to increase or temper the direct response of the photosynthetic mechanism to CO_2 enrichment. Additionally, as Reynolds and Acock (1985) point out, while the factors affecting stomatal aperture are known, the mechanisms involved in its regulation are still not understood and "...it is still not possible to predict what the degree of stomatal aperture will be in a given set of circumstances." Direct modeling of the effects of CO_2 concentration increment on transpiration is currently unreliable.

The lack of models of the needed mechanistic specificity suggest that it may be prudent to treat the direct effects of CO_2 enrichment on yield and water differently than the treatment for growth, development, and yield described above. The feasibility of factoring subscenarios of biomass increment (expressed in terms of leaf area or volume), photosynthesis enhancement, and transpiration reduction into the crop growth models should be explored. Sensitivity studies should be made to establish whether or not these direct effects significantly compensate or enhance the effects of climatic change.

Summary

A climatic change is expected to be induced by the increasing concentration of radiatively active trace gases in the global atmosphere. The so-called greenhouse effect, it is generally agreed, will lead to a warming of the troposphere, with greatest temperature increase in the high latitudes and least in the tropics. This change in equator-to-pole temperature gradient will alter atmospheric circulation patterns. Overall, global precipitation will be increased—but its distribution will be altered in ways that are, as yet, far from certain.

A comparison of the results of the three GCMs presented in this paper shows considerable disagreement as to the latitudinal distributions of temperature, precipitation, and soil moisture content annually and seasonally. Additionally, the grid size used in current GCMs is too large to be very useful in analyzing how projected climatic changes will affect current crop-growing regions.

CO_2 enrichment of the atmosphere will increase plant photosynthetic rates and, possibly, reduce transpiration rates. These phenomena must also be considered in assessing the overall impact of forthcoming environmental changes on crop production.

References cited

Acock B, Reddy V R, Whisler F D, Baker D N, McKinion J M, Hodges H F, Boote K J (1982) The soybean crop simulator GLYCIM: model documentation 1982 (PB 85171163-AS). U.S. Department of Agriculture, Washington, D.C.

Arkin G F, Vanderlip R L, Ritchie J R (1976) A dynamic grain sorghum growth model. Trans. Am. Soc. Agric. Eng. 19:626-630.

Bach W, Jung H-J, Knottenberg H (1985) Modeling the influence of carbon dioxide on the global and regional climate: methodology and results. Munstersche Geographische Arbeiten. Schriftleitung: Alois Mayr. Heft. 21. F. Schoningh Publ. paderborn. 114 p.

Blasing T J, Solomon A M (1982) Response of the North American corn belt to climatic warming. Publication No. 2134. Oak Ridge National Laboratory, Environmental Sciences Division, Oak Ridge, Tennessee. 16 p.

Decker W L, Jones V K, Achutuni R (1986) The impact of climatic change from increased atmospheric carbon dioxide on American agriculture. U.S. Department of Energy, Office of Energy Research, Carbon Dioxide Research Division, Report TR031. DOE/NBB-0077. Washington, D.C.

Gates W L (1985) The use of general circulation models in the analysis of ecosystem impacts of climatic change. Climatic Change 7:267-284.

Gifford R M (1989) Exploiting the fertilizer effect of increasing atmospheric carbon dioxide. Pages 477-487 *in* Climate and food security. International Rice Research Institute, P.O. Box 933, Manila, Philippines.

Hayashi Y (1982) Confidence intervals of a climatic signal. J. Atmos. Sci. 39:1895-1905.

Idso S B, Kimball B A, Anderson M G, Mauney J R (1987) Effects of atmospheric CO_2 enrichment on plant growth: the interactive role of air temperature. Agric. Ecosystems Environ. 20:1-10.

Jarvis P G (1986) Stomatal control of transpiration: scaling up from pore to paddock. Pages 28-34 *in* Climate-vegetation interactions. C. Rosenzweig and R. Dickinson, eds. VCAR, Office of Interdisciplinary Earth Studies, Boulder, CO., Report OIES-2.

Jones C A, Kiniry J R, eds. (1986) CERES-Maize: a simulation model of maize growth and development. Texas A&M University Press, College Station, Texas. 194 p.

Jones P D, Wigley T M L, Wright P B (1986) Global temperature variations between 1861 and 1984. Nature 322:430-434.

Katz R W (1977) Assessing the impact of climatic change on food production. Climatic Change 1:85-96.

Kimball B A (1985) Adaptation of vegetation and management practices to a higher carbon dioxide world. Pages 185-204 *in* Direct effects of increasing carbon dioxide on vegetation. B. R. Strain and J. D. Cure, eds. DOE/ER-0238. U.S. Department of Energy, Carbon Dioxide Research Division, Washington, D.C.

Lamb P J (1986) A state-of-the-art review of the development of climatic scenarios. Paper presented at the Task Force Meeting on Policy-Oriented Assessment of Impact Climatic Variations, 30 Jun-2 Jul 1986, International Institute for Applied Systems Analysis, Laxenburg, Austria.

MacCracken M C, Luther F M, eds. (1985) Projecting the climatic effects of increasing carbon dioxide. DOE/ER-0237. U.S. Department of Energy, Carbon Dioxide Research Division, Washington, D.C.

Manabe S, Stouffer R J (1980) Sensitivity of a global climate model to an increase of CO_2 concentration in the atmosphere. J. Geophys. Res. 85:5529-5554.

Manabe S, Wetherald R T (1980) On the distribution of climatic change resulting from an increase in CO_2 content of the atmosphere. J. Atmos. Sci. 37:99-118.

Manabe S, Wetherald R T (1986) Reduction in summer soil wetness induced by an increase in atmospheric carbon dioxide. Science 232:626-628.

Mearns L O, Katz R W, Schneider S H (1984) Extreme high temperature events: changes in their probabilities with changes in mean temperature. J. Climatol. Appl. Meteorol. 23:1601-1613.

Meehl G A (1984) Modeling the earth's climate. Climatic Change 6:259-286.

Meyer G E (1985) Simulation of moisture stress effects on soybean yield components in Nebraska. Trans. Am. Soc. Agric. Eng. 28:118-128.

Meyer G E, Curry R B, Streeter J G, Baker C H (1981) Simulation of reproductive processes and senescence in indeterminate soybeans. Trans. Am. Soc. Agric. Eng. 24:421-429, 435.

Meyer G E, Curry R B, Streeter J G, Mederski H J (1979) SOYMOD/OARDC - a dynamic simulator of soybean growth, development and seed yield. I. Theory, structure and validation. Res. Bull. 1113, Ohio Agricultural Research and Development Center. 36 p.

Morison J I L, Gifford R M (1983) Stomatal sensitivity to carbon dioxide and humidity: a comparison of two C_3 and two C_4 grass species. Plant Physiol. 71:789-796.

Neild R E, Richman H N, Seeley M W (1979) Impacts of different types of temperature change on the growing season for maize. Agric. Meteorol. 20:367-374.

Newman J E (1982) Impacts of rising atmospheric carbon dioxide levels on agricultural growing seasons and crop water use efficiencies. *In* Environmental and social consequences of a possible CO_2-induced climate change. Vol. II, Part 8. Report No. DOE/EV/10019-08. U.S. Department of Energy, Office of Energy Research, Carbon Dioxide Research Division, Washington, D.C. 29 p.

Nix H A (1985) Agriculture. Pages 105-130 *in* Climate impact assessment. R. W. Kates, J. H. Ausubel, and M. Berberian, eds. John Wiley and Sons, New York.

Reynolds J F, Acock B (1985) Modeling approaches for evaluating vegetation responses to carbon dioxide concentration. Pages 33-51 *in* Direct effects of increasing carbon dioxide on vegetation. DOE/ER-0238. B. R. Strain and J. D. Cure, eds. U.S. Department of Energy, Office of Energy Research, Carbon Dioxide Research Division, Washington, D.C.

Ritchie J T, Otter S (1984) CERES-Wheat: a user-oriented wheat yield model. Preliminary documentation. AGRISTARS Publ. YM-U3-04442-JSC-18892.

Rosenberg N J (1982) The increasing CO_2 concentration in the atmosphere and its implication on agricultural productivity. II. Effects through CO_2-induced climatic change. Climatic Change 4:239-254.

Rosenberg N J (1987) Drought and climate change: for better or worse? Pages 317-347 *in* Planning for drought: toward a reduction of social vulnerability. D.A. Willhite and W.E. Easterling, eds. Westview Press, Boulder, Colorado.

Schlesinger M E, Mitchell J F B (1985) Model projections of the equilibrium climatic response to increased carbon dioxide. Pages 81-147 *in* Projecting the climatic effects of increasing carbon dioxide. DOE/ER-0237. M. C. MacCracken and F. M. Luther, eds. U.S. Department of Energy, Carbon Dioxide Research Division, Washington, D.C.

Waggoner P E (1983) Agriculture and a climate changed by more carbon dioxide. Pages 383-418 *in* Changing climate: report of the Carbon Dioxide Assessment Commission. National Academy Press, Washington, D.C.

Waggoner P E (1984) Agriculture and CO_2. Am. Sci. 82:179-184.

Wilkerson G G, Jones J W, Boote K J, Ingram K T, Mishoe J W (1983) Modeling soybean growth for crop management. Trans. Am. Soc. Agric. Eng. 26:63-73.

Notes

Figures 1-3 and the associated discussion were prepared for a paper presented at the International Symposium and Workshop on Drought, Lincoln, NE, USA, 29 Sep-3 Oct 1986, and appear here by permission of the publisher, Westview Press, Boulder, Colorado.

Author's address: N. J. Rosenberg, Center for Agricultural Meteorology and Climatology, 243 L. W. Chase Hall, University of Nebraska, Lincoln, NE 68583-0728, USA.

Citation information: International Rice Research Institute (1989) Climate and food security. P.O. Box 933, Manila, Philippines.

Famine and national and international food prices

J. S. Sarma

Famine—extreme and general food scarcity—unlike a localized disaster, is widespread and may last for an extended time. Often the principal cause is successive crop failures resulting from adverse weather. A sharp rise in food prices is the earliest symptom. High prices severely reduce per capita food consumption; accompanying loss of employment and income intensify that reduction. Disease, starvation, and death follow unless food supplies are moved in time, at affordable prices. Depending on the severity of the crop loss, domestic food supplies may need to be augmented by food aid or commercial imports. Imports depend on available foreign exchange resources and prevailing world prices. International prices are influenced by world crop prospects, including stocks. When these stocks are depleted by large and sudden imports by countries affected by famine, those prices can be expected to rise. National short-term and long-term measures for relief and prevention of famine include arrangements for distribution of supplies, maintenance of buffer stocks to even out supply fluctuations, and programs to reduce instability in crop output. Good transport and communication facilities are needed to move food supplies to areas of scarcity. A responsive and efficient administrative system is necessary, as are information systems that provide advance warnings. Given political will, famine is preventable.

Acute food shortages can be classified according to the nature and duration of the emergency: short-term emergencies caused by earthquakes, floods, cyclones, etc.; medium-term emergencies caused by one or more crop failures due to drought, floods, disease, pests, or mass migration and political upheaval; and long-term disruptions of food supplies caused by war, desiccation, erosion, etc. (Masefield 1974).

Famine is an extreme and general food scarcity. Per capita intake of food grains or other basic foods falls drastically and many people find themselves so deprived of their accustomed food supplies that they suffer considerably unless special efforts are made to provide food. Unlike a localized, short-term disaster, famine is generally widespread and may last over an extended time. Depending on degree, emergencies may affect an entire nation or region and the more vulnerable sections of the population.

The interrelationships among food grain production, grain prices, income, employment, and consumption are complex, particularly during a time of crop failure. When production falls and prices rise, the changes in income from crop production depend on the extent of the fall in output and the rise in prices. The effects also differ depending on whether the production is subsistence or commercially oriented: prices directly relate only to marketed production. For a substantial proportion of producers, marketing may drop far enough that even a very large increase in prices still results in a decline in income; many marginal sellers may become marginal buyers of food. In either case, fall in output reduces the producers' direct consumption of food.

Additional factors affect rural incomes in drought years. Crop failure resulting from drought reduces the demand for labor involved in direct agricultural operations as well as the demand for services relating to input and output marketing and other related occupations, and the income of people employed in these activities is reduced. The overall income of the producers may decline, because food prices do not rise due to availability of supplies from stocks or imports or because the output of nonfood crops also falls. This fall in income reduces the demand for nonfood items and services; this in turn affects the employment and income of people engaged in those activities. In both cases, reduced income results in reduced demand for food. If, in addition, retail prices are high due to hoarding or speculation, low-income people in rural areas are deprived of their accustomed food supplies, leading to hunger and distress. High retail prices also adversely affect per capita food consumption of the urban poor.

Causes of famine

Food shortages can occur because of a sharp decline in food production resulting from adverse climatic factors such as extended drought, floods, or cyclones. These food shortages are intensified when there are successive crop production failures. The decline in per capita food availability may occur in one's own production, exchanged production, or purchased supplies.

Although the obvious deprivation of the family or the individual in a famine is food, the causes leading to the deprivation may be varied: shortfalls in production or supplies, high prices, or falls in income. The producers face reduced availability of food due to crop failure, loss of harvested crops in the field, or harvests kept in home stores. Crop failure also results in reduced market supplies, leading to higher prices. That diminishes consumers' ability to meet their full food requirements through purchase.

An overall deficiency in food supplies may also occur when commercial imports cannot be arranged in adequate quantities because of foreign exchange difficulties or high world prices. In several instances, although the overall food production in a country did not suffer drastically, food supplies in some regions declined sharply because of the breakdown or disruption of internal transport and communications. Frequently the situation is aggravated by panic buying and hoarding. Sometimes a shortage in supplies may be the result of government procurement operations, purchases by military authorities, exports, and even smuggling (FAO 1981). In countries and areas where people are poor and chronically malnourished, even a marginal crop failure, disaster, or disruption can cause large-scale distress and deprivation.

Historically, droughts have greatly affected pastoral societies, such as in Africa, causing considerable hardship. Herdsmen and their cattle must migrate to other areas. Where this is not easy, the cattle perish. Alternatively, cattle must be sold—often at distress prices because of a lack of demand. In these situations, prices of livestock products also fall, causing further disruption to normal economic activities.

Famine symptoms

Prolonged drought, severe floods, or cyclones result not only in reduced food production but also in scarcity of drinking water. People migrate from rural to semi-urban areas, towns, and cities in search of food, water, and employment. In very acute cases, they are forced into relief camps. They are subject to diseases caused by nutritional deficiencies as well as by the unsanitary conditions that develop due to water shortage or excess. Death rates from starvation or disease increase. Alamgir (1980) suggests that "famine is considered to represent a general state of prolonged foodgrain intake deficiency per capita giving rise to a number of sub-states (symptoms) involving individuals and the community that ultimately lead directly or indirectly to excess deaths in a region or a country as a whole." Apart from high prices, these symptoms include increases in interregional migration, crime, and fatal diseases; loss of body weight, malnutrition, and consumption of famine foods; disintegration of customary moral codes; uprooted and separated families; and distress sales of land, livestock, and other productive assets.

Famine relief measures consist of a host of activities to deal with each symptom. This paper focuses on the sharp rise in prices and on measures to prevent this rise in a famine or acute food shortage situation.

Famine and national prices

Various parts of the globe have suffered famines from time immemorial. However, with economic development, particularly the development of agriculture, the incidence of famine in developing countries during the last two decades has declined, except in Africa.

The behavior of prices may be quite different in different famines. For example, there may be an abrupt and sharp rise in food prices that puts food beyond the reach of the poor, who then suffer starvation. In this situation, food may be available in the market, but prices are so high people cannot buy it (Bhatia 1967). In fact, a sudden rise in food prices is an indication of a scarcity of food supplies in the area, and is an early warning of a potential famine. However, occasionally famine conditions may develop in an area without a rise in prices, as occurred in the Ethiopian 1972-74 famine.

For illustration I review here the price situation in a few of the more well-known famines and droughts—the Great Bengal famine 1943, Bihar (India) famine 1966-

67, drought in Maharashtra (India) 1970-73, Indian drought 1979-80, Bangladesh famine 1974, Bangladesh drought 1978-79, Bangladesh floods 1984, Nigeria drought 1972-74, and Ethiopian famine 1972-74.

Great Bengal famine 1943

The Great Bengal famine has been studied by several scholars and the causes and events leading to it have been well documented. In brief, the 1942 winter rice crop in Bengal was adversely affected by an October cyclone, followed by torrential rains in some parts of the province. After the Japanese occupation of Burma in 1942, rice imports were cut off. Also some of the actions taken by the government in relation to the Second World War adversely affected internal transport of food grains.

These and other factors led to a sharp rise in the price of rice in the area. Wholesale prices of rice in Calcutta, which were around Rs 0.17/kg in April, rose to Rs 0.35-0.38/kg by December. By March 1943, prices had risen 50% more to Rs 0.56/kg. By May, rice prices were quoted at between Rs 0.80 and 0.83, and by August they were Rs 0.99/kg. Because of a government order fixing maximum prices, it is difficult to obtain comparable price data beyond September 1943.

In Calcutta, when the food situation became acute consumers had the benefit of supplies from subsidized distribution systems. In other towns, retail prices in the open market rose to extremely high levels. Nonofficial reports indicated prices of Rs 2.14/kg in Chittagong and Rs 2.81/kg in Dhaka in October 1943 (Sen 1981).

These prices most intensely affected vulnerable sections of the population, many of whom also lost income-earning employment opportunities. The distress was much greater in rural areas, where in addition to natural shortages, there was dislocation of normal channels of distribution. Although the overall supply of food grains in the state was not exceptionally low in 1943 compared to 1941, it is likely that availability in the districts declined. It also appears that speculative hoarding of grains by producers and traders contributed significantly to the steep rise in prices.

Estimates of mortality during that famine vary from 1.5 million (based on official figures) to 3 million (including deaths from the aftereffects of famine, from unofficial estimates) (Alamgir 1980, Bhatia 1967, Sen 1981).

Bihar famine 1966-67

India experienced its most severe droughts during two consecutive years, 1966 and 1967. (Average production of food grains in India during the 3 yr 1962-63 to 1964-65 was 83.5 million t.) In 1964-65, it was 89.4 million t. In 1965-66, aggregate food grain production was 72.3 million t—a decline of 17.1 million t, a 19% drop in 1 yr. Output in 1966-67 was only slightly better, 74.2 million t. The extremely good 1964-65 crop year helped to tide over the shortfall in food, with the help of grain imports of 10.3 million t in 1966 and 8.7 million t in 1967.

Nevertheless, the second year of drought aggravated the food situation, particularly in normally food-deficit Bihar State, which was the worst affected in 1966-67. Total production of food grains in Bihar dropped from 7.5 million t in 1964-65 to 7.2 million t in 1965-66 and to 4.1 million in 1966-67. The output of rice fell from 4.3 million t in 1965-66 to 1.6 million t in 1966-67, a drop of more than 60%.

The rise in prices resulting from the shortage of food grains caused widespread migration and incipient starvation. However, the overall impact of high prices was not acutely felt because the public distribution system expanded rapidly; consumers had to depend on the open market for only a very small portion of their food grains needs. Massive relief measures also were taken, by both government and voluntary organizations (Singh 1975).

Maharashtra drought 1970-73

In 1972-73, food grain production in India fell about 9% below that in 1971-72. The impact of drought in Maharashtra State was much larger: food grain output declined by one-third, from 4.95 million t in 1971-72 to 3.05 million t in 1972-73. Even 1970-71 and 1971-72 were not good years; in 1969-70, food grain production was 6.91 million t. The 1972-73 0.75 million t rice harvest was about half of that in 1969-70; the output of coarse grains (including maize) declined from 4.25 million t in 1969-70 to 1.61 million t in 1972-73.

The Maharashtra drought led to a great scarcity of food and water. Between June 1972 and October 1973, prices of rice and sorghum nearly doubled in rural areas of both scarcity and nonscarcity, although the rise in scarcity areas was higher (Table 1).

The Government of Maharashtra undertook a massive relief effort, including both employment and scarcity relief programs. The schemes were aimed at creating productive assets through afforestation, soil conservation, excavation of canals, and construction of drinking water facilities. At the peak of the operations, about 5 million people were employed in scarcity relief works. In addition, food grain distribution was arranged through fair price shops. Consequently, no deaths from starvation were reported in the state, despite the huge shortfall in food production and the sharp rise in prices (Subramanian 1975).

Indian drought 1979-80

The next major drought in India was in 1979-80, when overall food grain production declined 22.2 million t (17%) from the previous year. (Average production of food grains in India in the 3 yr 1976-77 to 1978-79 was 123.2 million t. The shortfall in the

Table 1. Prices of rice and sorghum in scarcity and nonscarcity rural areas of Maharashtra (Subramanian 1975).

Period	Price (Rs/kg)			
	Rice		Sorghum	
	Scarcity areas	Nonscarcity areas	Scarcity areas	Nonscarcity areas
Jun 1972	1.73	1.57	0.98	0.96
Jan 1973	2.12	1.79	1.25	1.10
Jun 1973	3.34	2.69	2.15	1.60
Oct 1973	3.33	2.93	1.95	1.78

1979-80 output was 13.5 million t.) The drop in rice production was 21%, from 53.8 million to 42.3 million t. Because production was poor in several states, government procurement of food grains declined from 13.8 million t in 1979 to 11.2 million t in 1980, while public distribution increased from 11.7 million to 15.0 million t. Domestic production was supplemented by withdrawals of 5.8 million t from stocks. Stocks were drawn down from 17.5 million t at the end of 1979 to 11.7 million t at the end of 1980. No net imports of food grains were made.

Although the withdrawal from stocks was much less than the shortfall in production, some of the worst effects of the drought were prevented because of the public distribution system. The all-India index of wholesale prices of cereals (annual average, base 1970-71 = 100) increased from 167.0 in 1979 to 189.5 in 1980, a rise of 13.5%. Although the 1979-80 production shortfall was large (about 17%), the drought was relatively unknown outside the country.

Bangladesh famine 1974

Unprecedented floods in Bangladesh in 1974 caused severe damage to crops and property and were the immediate cause of famine. The price of rice rose steeply in affected districts. In some, rice prices doubled within the 3 mo Aug-Oct. The situation had to be met through free distribution of cooked food. At one stage, nearly 4.35 million people (more than 6% of the population of the country) were provided cooked food relief. By November, rice prices began going down and the need for relief was less intense. By the end of November, the free distribution centers were closed (Sen 1981). Total mortality in that famine was estimated at 1.5 million persons (Alamgir 1980).

Even before the 1974 floods, Bangladesh had experienced a shortfall in the 1972 winter rice crop, a result of drought. A severe cyclone in December 1973 affected the coastal belt. Thus, although the 1973-74 winter rice crop was good, rice prices had already started to rise. In Dhaka, the average retail price of coarse-quality rice increased 36% between November 1973 and March 1974. In other market centers, price increases were even higher. The overall price situation in Bangladesh between October 1972 and October 1974 can be seen in Table 2. In 1974, prices in the famine districts were more than three and a half times those in 1972.

Bangladesh drought 1978-79

In 1978-79, drought hit the aman and boro rice crops in Bangladesh. In May 1979, rice prices were about 40% higher than the year before. But with international

Table 2. Retail prices of rice, coarse quality (Alamgir 1980).

Area	Price (Taka/kg)		
	Oct 1972	Oct 1973	Oct 1974
Dhaka	1.99	2.62	7.07
Famine districts	2.08	2.57	7.38
Nonfamine districts	1.92	2.66	6.63
Average for Bangladesh	1.95	2.52	6.75

assistance, the government arranged for massive imports of food between July and November 1979. Arrangements were made for shipping, clearance at ports, internal transport, and grain storage. These food imports, together with the opening of 1,500 feeding centers, averted famine (Alamgir 1980).

Bangladesh floods 1984

Although Bangladesh had severe floods in 1984, there was no repetition of the events of 1974 because the government was better prepared to meet the situation. Political commitment to containing the price movements through maintenance of food distribution arrangements was greater. The scope of the Food for Work Program was expanded, and adequate stocks were built up from domestic production and imports. Storage capacity by 1984 was nearly double that in 1980. With these measures, distress was again averted (Clay 1985).

Nigerian drought 1972-74

In Nigeria, 1972-73 and 1973-74 were drought years, particularly in the northern region. Food prices rose continuously following the harvest failure of 1972. Mean prices for food staples in Kano indicate that cereal price levels were several times higher by the hungry season of 1974. The highest price levels and the most violent price fluctuations occurred in Sokoto and Niger states; from a predrought price of 50 naira/t, millet prices soared to more than 200 naira/t in April 1973 and hovered at that level until government relief stabilized the market in March 1974. Millet prices broke the 100-naira/t barrier in the northern zone in February 1973. By June 1973, a majority of markets in the north reported prices of over 150 naira/t; in some, prices in excess of 200 naira/t were reported in May and June 1973 and again in the early months of 1974. "It is quite safe to assume that throughout the twelve northern divisions, grain prices held steadily at double the predrought levels at least for 15 mo" (Watts 1983). As a result, many thousands of Nigerian families had to abandon their homes and migrate to urban areas. There was widespread distress.

The Ethiopian famine 1972-74

The Ethiopian famine in 1972-74 was initiated by a drought that affected the northeastern region, particularly Wollo Province. The rains failed in mid-1972, followed by a near total failure of spring rains in early 1973. The drought was broken in mid-1973, but a new drought situation developed in the south in 1973-74. The effect of the drought on crop output is not clear. Sen (1981) reported that although there was no noticeable decline in food availability for Ethiopia as a whole in the famine year 1973, there was clearly a food shortage in Wollo. (It was reported that for every truck that entered the province to feed famine victims, another truck of food left to feed Addis Ababa.)

Further, despite the disastrous failure of food output, food prices did not go up very much, nor for long, in Wollo. Taking average prices of 1970-72 as the prefamine level, prices in the famine year 1973 on the whole were very close to the prefamine level. But the death toll was estimated between 50,000 and 100,000. People starved to death, even without a substantial rise in food prices. Sen (1981) attributes this to extensive failures of entitlements for different sections of the Wollo population. The Wollo famine is an exception to the general rule that high food prices are a symptom of famine or acute scarcity of food.

Lessons from the price review

Immediately after a crop failure resulting from drought or other natural disaster and the consequent reduction or disruption of food supplies, retail prices rise sharply, particularly causing hardship to low-income families. Further, apart from the overall shortage in supplies, speculation and hoarding can drive prices very high, beyond the reach of the poor.

In some of the cases reviewed, famine was averted and distress mitigated by government steps to augment the food supplies and make them available at affordable prices. Supplies from domestic stocks were augmented from imports or from food aid. Increasing supplies of food and maintaining reasonable prices do not by themselves solve the problem of decreased incomes. Relief feeding or, better still, opportunities to earn income are needed as well as access to supplies at reasonable prices. Food for work, employment, and scarcity relief programs need to be organized to provide income-earning opportunities to the vulnerable sections. With adequate food distribution arrangements from stocks, the tendency to hoard can be discouraged.

Famine and international prices

International prices of cereals are influenced by the levels of world production, trade, and stocks. Any widespread drought in any part of the world affects aggregate production, but the overall effect depends on the behavior of production in other parts of the world. Production shortfalls can be made good by withdrawals from international stocks. (Otherwise, prices are likely to rise.) However, activities in developed countries and in the Eastern Bloc countries, including their stockholding behavior, affect international food grain prices more than developing-country famines.

Table 3 shows world cereal prices, production, trade, and closing stocks from 1961 to 1983. Grain prices were relatively stable in the 1960s. Annual price changes between 1962 and 1969 were less than 5%, with the exception of a 7% increase in 1966. World trade increased to 111 million t in 1965-66, 16 million t more than the previous year, and stocks dipped to 142 million t, the lowest for the decade. In the 1970s, world utilization of cereals increased steadily, but significant changes in production and stocks led to wide fluctuations in prices. The cereal price index reached a peak of 216 in 1974 (base average 1977-79 = 100), and cereal stocks were lowest at the end of 1974-75, around 140 million t, 11.5% of world utilization.

When world prices are high, imports are expensive and cause hardship to poorer countries with foreign exchange constraints. In the early 1980s, stock levels have been high and prices low. The year 1982-83 closed with cereal stocks of 262 million t; preliminary estimates place closing stocks at the end of 1985-86 at 320 million t.

If present world aggregate production and consumption trends continue, the existing levels of stocks are adequate to meet the increased demands arising from droughts in any part of the world. Although famines and droughts cause undue

Table 3. World cereal prices, production, trade, and closing stocks, 1961-83 (World Bank 1984, USDA 1986).

Year	Index of world cereal prices[a] (1977-79 = 100)	World total grains[b] Production	Trade	Closing stocks[c]
		(million tons)		
1961	128	846	72	199
1962	143	806	83	172
1963	143	867	83	174
1964	138	871	98	175
1965	135	924	95	179
1966	144	921	111	142
1967	150	1006	104	169
1968	156	1038	97	189
1969	153	1079	90	222
1970	126	1087	97	207
1971	112	1103	110	165
1972	109	1197	110	183
1973	187	1161	135	143
1974	216	1273	142	149
1975	148	1218	137	140
1976	119	1247	150	148
1977	101	1363	158	201
1978	103	1337	171	201
1979	97	1466	177	231
1980	107	1427	198	207
1981	119	1447	215	191
1982	88	1499	210	227
1983	98	1544	201	262

Sources: World Bank (1984), USDA 1986.
[a]Calculated from prices in constant dollars; cereals include maize, rice, wheat, and grain sorghum. [b]Include wheat, coarse grains, and rice; data relate to 1960-61 (Jul-Jun) and so on. [c]Stocks relate to the end of marketing year. Data are based on aggregate of different local marketing years and exclude those of People's Republic of China and parts of Eastern Europe.

hardship in the affected areas, the additional import demands to meet emergencies in developing countries in general are not very large. Some exceptions were Indian imports of 19 million t in 1966 and 1967 and Russian imports of 25 million t in 1975. Compared with these, the total needs of the 21 African countries that were affected by abnormally high cereal import requirements in 1984-85 were about 11 million t: 5.2 million t of commercial imports and 5.8 million t of aid (FAO/WFP 1985). The increase over normal imports would be even smaller. Scope also exists for diversion of food grains from feed use to food use in drought years when production is low and prices are high, particularly in developed countries. As it is, cattle tend to be slaughtered for food in larger numbers during drought years.

Famine relief vs famine prevention

Although prices play an important role in the prediction of need and subsequent channeling of food to affected areas, maintenance of prices constitutes only one component of a host of other measures for famine relief and prevention. Here, a distinction needs to be drawn between famine relief and famine prevention measures.

In acute famine, immediate steps for relieving distress include organizing relief camps and distributing cooked food through community kitchens run by government or voluntary agencies. Simultaneously, arrangements for rushing food supplies to areas of shortage and for their distribution through fair price shops at affordable prices are necessary. Also, food for work and other employment generation programs need to be undertaken to provide employment and income to the affected people. To be adequately prepared when disaster hits, there is need for contingency planning for famine mitigation (Sarma 1983). Other measures for rehabilitating agriculture and livestock also can be taken.

Role of government policies

Among the long-term measures for famine prevention, the most important are programs to increase agricultural production. While bad weather may precipitate a crisis, the principal causes of mass hunger more often may be the consequences of human acts, such as military conflict and inappropriate government agricultural policies. This is particularly true in Africa, where the 1983-84 drought is estimated by the FAO to have encompassed more than 20 countries and a population of 150 million, with as many as 30 million at risk of food insufficiency (Mellor 1986). Appropriate price policies are a prerequisite to acceleration of agricultural production in Africa. An analysis of the existing situation in this regard shows that many countries in the region lack such policies. In addition, adequate buffer stocks are required to even out supply fluctuations.

Role of infrastructure

The next important measure for both preventing famine and relieving distress, as well as for preventing price increases, is the development of infrastructure. Food grains, whether from imported or domestic stocks, need to be moved to consumption centers in deficit areas. This requires transport facilities, markets, and storage. Transport facilities need to be developed within rural areas, between rural and urban areas, and between urban areas and the rest of the world. Typically, farmers in developing countries store adequate food to manage a single year's crop losses. To store crops at the farm level for 2 yr is more expensive. Hence, when crop production fails for a second year, transport may have to be arranged from distant agricultural surplus areas. Where domestic supplies are to be augmented from imports, adequate facilities at ports or other landing points to handle the imports are needed, particularly when large quantities are to be imported in a short period. For movement within the country, coordinated arrangements between road, rail, and river transport are required.

In Africa, a memorandum prepared by the Overseas Development Administration of the United Kingdom on famine crisis showed that two-thirds of the countries with food shortages or threatened shortages had very poor or poor internal transport (Foreign Affairs Committee 1985). When disasters like floods or cyclones occur and cause food shortages, often the transport facilities are also adversely affected. The development of infrastructure is not only expensive but also takes time. Areas likely to experience periodic food shortages need to be given priority for infrastructure development projects.

A related requirement is the administrative organization needed for handling the various aspects of food distribution at different levels. Warding off famine requires tremendous amounts of coordination and efficiency, and these cannot be organized at short notice (Mellor and Gavian 1987). In countries subject to frequent droughts, advance training of the requisite personnel is a priority.

Famine prediction

Timely famine prediction is as important as famine prevention and relief. Shortfalls in crop production resulting from adverse weather often can be predicted through crop and weather monitoring systems. Any abrupt rise in prices also indicates an actual or potential supply shortfall. Normally, prices fall when the new crop arrives in the market. When this does not happen, it is an indication of the farmers' expectations of the size of the crop. Data on market arrivals are also a useful indicator. Fluctuations in surpluses of food crops that are marketed are wider than those in the crop output. Because in drought years, employment—both agricultural and nonagricultural activities—declines, data on rural wages also need to be monitored.

At the international level, the FAO is already running an efficient Global Information and Early Warning System for Food and Agriculture. In addition, in emergencies it organizes country crop assessment missions, to get a more accurate picture of import needs. As these systems improve, timely prediction leading to famine prevention and relief will be facilitated. However, the processes of famine prediction, prevention, and relief depend to certain extent on the political system and forces operating in a country.

Politics of famine

Famine, like chronic undernutrition and poverty, arises from a complex etiology of political, economic, and environmental factors. In a democratic country where the press is free, food shortages or undue increases in prices in a local area are often brought to light even when local government agencies are not aware of or are slow to react to a developing famine situation. Political pressure from elected representatives often goads the bureaucratic government to take prompt relief measures.

At the other end of the spectrum, examples of government indifference to suffering in areas controlled by or under the influence of rebel or other opposing political groups can be cited. In such cases, even international aid sometimes does not reach those in need. It used to be believed that in a communistic system of government, any shortfall in supplies would be evenly distributed among the different classes of people. Recent evidence of the food crisis in the People's Republic of China in the early 1960s, when millions perished, suggests this is not necessarily so.

Famines are preventable

Year-to-year fluctuations in crop production will continue as long as agriculture depends on rainfall and weather. Recent studies at the International Food Policy Research Institute have shown that the new improved seed/fertilizer-based technologies appear to have contributed to increased production instability. This does not necessarily mean that aggregate consumption should also fluctuate widely. Efforts could be made to reduce some of the output fluctuations through investments in irrigation, propagation of drought-resistant varieties, and other yield-stabilizing measures. Per capita consumption could be stabilized through appropriate trade and stock policies. Donor communities are prepared to assist with food aid where necessary.

The question still remains of variations in the per capita food consumption of different classes of population, particularly the vulnerable groups. Here, the question of access to food also comes up. The range of policies to ensure adequate intakes of food to these sections of the population is wider. They cover arrangements for distribution of food at reasonable prices, food for work, employment guarantee schemes, and supply of cooked food in emergencies. Droughts are recurrent and cannot altogether be prevented. But famines are preventable, provided there is the political will to do so. India and Bangladesh have demonstrated that with careful planning and management and with international assistance, famines and associated distress can be prevented.

References cited

Alamgir M (1980) Famine in South Asia: political economy of mass starvation. Oelgeschlager, Gunn and Hain, Cambridge, Massachusetts.

Bhatia B M (1967) Famines in India 1860-1965. 2d ed. Asia Publishing House, New Delhi.

Clay E (1985) The 1974 and 1984 floods in Bangladesh. Viewpoint, Food Policy, August 1985.

FAO—Food and Agriculture Organization of the United Nations (1981) Agriculture: toward 2000. Rome.

FAO/WFP—Food and Agriculture Organization of the United Nations and World Food Programme (1985) Relief and rehabilitation in food and agriculture of African countries affected by calamities in 1983-85. Situation Report No. 9. Rome.

Foreign Affairs Committee (1985) Famine in Africa. House of Commons, Her Majesty's Stationary Office, London.

Masefield G B (1974) Food and nutrition procedures in times of disaster. Food and Agriculture Organization, Rome.

Mellor J W (1986) Prediction and prevention of famine. Pages 2427-2431 *in* Federation proceedings. Federation of American Societies for Experimental Biology, Baltimore, Maryland.

Mellor J W, Gavian S (1987) Famine - causes, prevention and relief. Science 235:539-545.

Sarma J S (1983) Contingency planning for famines and other acute food shortages: a brief review. International Food Policy Research Institute, Washington, D.C.

Sen A (1981) Poverty and famines: an essay on entitlement and deprivation. Clarendon Press, Oxford.

Singh S K (1975) The Indian famine. People's Publishing House, New Delhi.

Subramanian V (1975) Parched earth: the Maharashtra drought 1970-73. Orient Longman Ltd., Bombay. p. 620-621.

USDA—U.S. Department of Agriculture (1986) Grains: world grain situation and outlook. Foreign Agriculture Circular, June, 1986. Washington, D.C.

Watts M (1983) Silent violence. Food, famine and peasantry in Northern Nigeria. University of California Press, Berkeley.

World Bank (1984) The outlook for primary commodities 1984 to 1995. Staff Commodity Working Paper No. 11. Washington, D.C.

Notes
Author's address: J. S. Sarma, International Food Policy Research Institute, Washington, D.C., USA.
Citation information: International Rice Research Institute (1989) Climate and food security. P.O. Box 933, Manila, Philippines.

Economic adjustments to climatic variability

W. E. Easterling and S. T. Sonka

Research on the interactions of climatic variability and human activities has tended to focus predominantly on impacts of climatic change and less on adaptive/adjustive mechanisms for coping with change. Moreover, most analyses have implicitly assumed instantaneous climatic change. System approaches could be effective means of identifying current adaptive mechanisms in the agricultural sector to changing (as opposed to changed) climate. Those can then be examined using quasi-experimental designs. Decision analysis can provide insight into societal adjustments to changing climate. An approach is proposed that integrates a set of physical and economic models (operating at different levels of aggregation) to provide the capability for assessing adjustments to dynamic climate.

The climate regime in which a society lives and operates is a major determinant of many of the economic characteristics of that society (Clark 1985). That makes the potential for marked changes in climate a matter of concern. In addition to the potential for a changed climate and the likely impacts of that change on economic activities, the effectiveness of reactive measures that could be implemented are of vital interest to decisionmakers (Warrick and Riebsame 1981).

Economic analysis could provide important understanding about the interaction of climate and society. Most analyses to date, however, have been based on some questionable assumptions. In general, analyses have tended a) to address issues of climate variability only during the event itself and, moreover, to implicitly assume an instantaneous climate change; b) to focus on physical processes; and c) to ignore adjustments and adaptations that individuals and society are likely to make as climate changes.

These assumptions have some flaws. Some alternative research approaches could circumvent those flaws and allow a more realistic evaluation of potential economic adjustments to climatic variation. The approaches fall under the categories of quasi-experiments (Kates 1985), expert systems, and decision analysis (Sonka and Lamb 1987). Integrating economic models of different levels of spatial aggregation could give quantitative insight into linkages between climatic change, impacts of that change, and adjustive/adaptive mechanisms for coping with the impacts.

Improving assumptions

As a concept, climate can be mind-numbing—a jumble of statistical moments primarily designed to portray expected weather at a given time and place. In this sense, as Hare (in Kates et al 1985) implies, climate evokes little emotion and is hardly worth the attention of policymakers (who often assume that economics in general operates in quasi-harmony with seasonal rhythms).

Yet in reality, climate is a mixture of entropy; anomalous events such as drought, flooding, heat, and cold waves; subtle fluctuations; and gradual change. Each component is a normal part of climate. Droughts come and go, climates fluctuate and change (as they have in the past), and new entropies evolve. Any adjustment, economic or otherwise, that society makes in response to any one component must be examined against the background of all the other components.

For example, it can be argued that the most rational economic adjustments to drought would be made during both nondrought and drought episodes. A more efficient use of resources is to manage the risks of drought, rather than to react to a crisis. Adjustment is a continuous process, whether or not the climatic event that caused the adjustment persists. Thus, regardless of the peculiar dynamic the atmosphere presents, adjustive response should be evaluated within the context of climate and society interaction, not focused on a single event, such as a drought.

A related issue is how a climatic change (a shift to a new entropy) reveals itself. The prospect of climatic change has several unique characteristics. One of the most important is the relatively long time period over which change can occur. Scenarios for potential alterations in climate patterns over 30, 50, or 100 yr have become relatively common (e.g. Lamb 1979, National Research Council 1979, Palutikof et al 1984, Schlesinger and Mitchell 1985). Given the concern that the climate, at some relatively long time in the future, may well differ from the present climate, what economic analyses would society need?

For the moment, let us pretend that we have perfect knowledge of future climatic change, and that the change will be instantaneous. Would we be satisfied with an economic analysis that simply predicts the future state of the economy, given the altered climate? Probably not. Such an analysis would likely indicate that the altered climate would negatively impact some economic sectors, but have positive effects on others. We undoubtedly would want to extend the analysis to investigate alternatives for mitigating adverse consequences and for exploiting positive potentials of the change. Useful economic analyses would incorporate the effects of potential responses as well as the direct effects (Ausubel 1983). Potential responses include physical reactions to the effects of the changed climate, such as altering production practices or developing new technologies, as well as policy responses by institutions and publics.

Now let us relax our previous assumption of perfect knowledge of the extent and timing of future climatic change. Doing so raises two fascinating research complications. First, it is likely that climatic change will not be instantaneous, but will evolve (Schneider and Thompson 1981, Bryan et al 1982, Hoffert and Flannery 1985). Economic analyses ideally would study the adjustment of economic agents and systems as they responded to the alterations.

Second, climate is not likely to be the only attribute of society that will change across several decades. Population, tastes and preferences, technology, and other factors will evolve as well. Of course, no research effort can accurately predict the future course of factors such as these, but it may be possible to predict how changes in societal parameters will interact with potential climatic change, if at all. Sensitivity analyses describing the effects of such interactions could help decisionmakers understand the potential importance of changes in climate.

This suggests that the purpose of economic analyses of climatic change should be to improve our understanding of the interactions of climate and economic activities, not to predict the future. Given this goal, research must explicitly incorporate potential societal reactions to climatic change. Furthermore, understanding the dynamics of climatic change becomes important as we attempt to discern the directions of potential societal transitions. An important outcome associated with this view is that natural science researchers become important users of the output of such economic analyses. Potential biological and physical innovations could be defined and evaluated against the consequences of their not being available. Gaps in climatological knowledge could be linked explicitly to the decisions that would result from the absence of such knowledge.

Another issue evolving from this concern is the tendency to examine climatic variability apart from the social and political contexts where it is occurring. There is an overemphasis on the physical dimensions of climate. Yet numerous investigations have documented that climatic variability is not solely a physical phenomenon. Heathcote (1973) and, more recently, Wilhite and Glantz (1985) note that it may be useful to define drought in agricultural and hydrological terms as well as in meteorological terms. Garcia (1981) stresses the importance of the economic and social system: "Climatic events per se are not the root cause, in our time, of great disasters, famines, increased misery." Although noting the importance of climate's role, Garcia and numerous others have found that a society's vulnerability to drought is as much a function of human and social actions as it is of climatic events (Garcia 1981, Heathcote 1985, Warrick and Riebsame 1983).

The final common theme is that individuals and societies often tend to act as if they are surprised when a climatic variability event, such as drought, occurs (Heathcote 1973, Hill 1973). Public pronouncements of surprise can be attributed to the desire to attract relief. It is interesting that such relief, particularly for the agricultural sector, is usually forthcoming, even when agriculturalists have not taken action, such as holding excess stocks of feed or purchasing crop insurance, to reduce the adverse effects of drought.

Human actions that make a society more vulnerable to drought sometimes result from viewing climatic variability as a surprise. For example, actions to restrict nomadic movement in the Sahel appear to have made those societies more vulnerable to drought (Garcia 1981). Planting more drought-susceptible crops indicates that sometimes decisionmakers take actions that suggest they do not expect a drought.

Viewed in total, these general themes present a paradox. On the one hand, the knowledge that human actions can make societies more vulnerable to drought and the presence of adaptations suggest an awareness of climate-society interactions. On the other, the element of surprise, whether suggested by pronouncements or by actions, indicates a lack of perception and sensitivity to the potential for climatic variability.

Undoubtedly the truth lies somewhere in between. Some alternative research approaches could provide realistic insights into economic adjustments to climatic variability.

Quasi-experiments and expert systems

Quasi-experiments

Much can be learned about how current and near future economic adjustments to a dynamic climate are being implemented by examining climate fluctuations that have occurred in the recent past. Hare (in Kates et al 1985) suggests that this is guided by Whitehead's dictum, that "how the past perishes is how the future becomes."

The major problem confronting hindsight examinations of recent climate fluctuations has been the limited ability of such studies to distinguish adjustments by society that are climatically induced from responses to nonclimatic factors. An approach to this problem, referred to as a natural or quasi-experimental design (see Kates 1985), is being developed as part of the Climate Impacts, Perception, and Adjustment Experiment (CLIMPAX) (Kates et al 1984).

Natural experiments differ fundamentally from conventional concepts of experimental design. In conventional experiments, such as crop yield trials, biophysical response to climatic change can be studied within the strict controls of a test plot or laboratory. No such laboratory exists for gauging the response of a society or of broader ecosystems to climatic changes. However, in natural experiments it is possible to use matched "before and after" or "with and without" situations as controls. For example, societal behavior during a drought (a "with" situation) is evaluated relative to societal behavior during a normal season (a "without" situation).

In recent years, natural experimental designs have been used to examine, retrospectively, combinations of extreme events. Examples include studies of community disasters (Friesma et al 1979, Rossi and Wright 1982); droughts (Bowden et al 1981); drought, flood, snow, tropical cyclones, and wind in 12 countries at 20 sites (White 1974); and inadvertent climate modification (Changnon et al 1981).

Figure 1 shows a hypothetical example of a quasi-experiment, the CLIMPAX. *A* compares changes in precipitation, indices of wheat yield, and streamflow. Improvements in wheat production technology considerably mask the change between epochs; in the unmanaged watershed, streamflow actually amplifies the change. This longitudinal analysis may illustrate strong impacts and adjustments, but may suffer from other problems, the most serious is the identification of impact in the face of broad secular changes in society or great changes in productivity.

A case-control approach is important in establishing climate effects beyond reasonable doubt. Cases selected in nonfluctuation areas nearby must be similar enough to the fluctuation area to be used as nonaffected controls. *B* of Figure 1 illustrates a hypothetical nonfluctuation series. Comparison of the case (fluctuation)

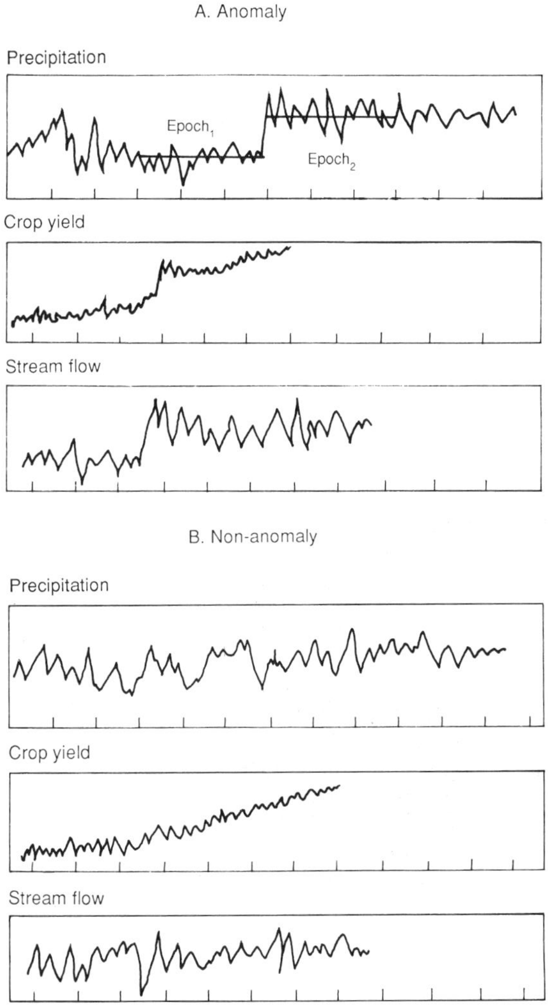

1. Hypothetical paired cases.

and control (nonfluctuation) variables should yield measurable climate impact. Any difference between case and control at discrete time steps over the course of the fluctuation or before, during, and after the fluctuation theoretically should be in a direction and of a magnitude consistent with the fluctuation.

Criteria for selecting control cases (e.g. similarity in crop proportions, urban-rural population profiles, production technologies, natural resource similarities, etc.) are developed. CLIMPAX makes it possible to examine several pairs of fluctuation/nonfluctuation areas, to avoid relying on a single case-control situation. Multiple cases should provide the degree of evidence that makes empirical findings so compelling. Recent work by Karl and Riebsame (1984) suggests that numerous localized precipitation and/or temperature fluctuations in the United States might be candidates for case studies.

Expert systems

One approach for developing a climate-society quasi-experiment begins with an a priori assessment of weather and climate sensitivities within a given economic sector. The questions are: a) which specific processes and activities, as we now understand them, should be affected by climatic variability and b) what is the range of adjustive mechanisms now available for coping with those effects. This information would provide insights into the types of climatic variability (e.g. duration, magnitude, timing, etc.) meaningful to different sectors and would lead to the generation of testable hypotheses about climate impacts and adjustments.

In some sectors, such as in agriculture, an extensive knowledge base exists on the interactions of climate and biophysical processes (Giegel and Sundquist 1984). In others, such as in construction or transportation, knowledge is not nearly so advanced. And in either case, the existing knowledge has not been compiled and synthesized. Moreover, the emphasis in these areas overwhelmingly has been focused on direct impacts; much less knowledge is available on adjustments.

Expert systems have great, largely untapped utility for increasing our understanding of potential adjustments to climatic variability. Arising as they do from the field of artificial intelligence, expert systems embody human expertise within the computer in configurations applicable to real world problem solving (for a fuller discussion of expert systems, see Hayes-Roth et al 1983). Expert systems are a) heuristic—they use judgmental as well as formal reasoning; b) transparent—they can justify a line of reasoning; and c) flexible—they make adding to and updating the knowledge base easy. Expert systems have had some application in such climate-sensitive sectors as agriculture (e.g. Lemmon 1986, McKinnion and Lemmon 1985) and energy (e.g. Electric Power Research Institute 1985).

Stewart and Glantz (1985) argued that expert systems are valid means of synthesizing knowledge in a domain such as climatic change, where the knowledge is less than precise. Unlike previous attempts to bring expert judgment to bear on issues of climatic change, expert systems require logical resolution of differences in judgments from different experts. Stewart (in Murnpower et al, in press) suggests that judgment analysis, in which the judgments of several experts on any one issue are quantitatively modeled, can be used to resolve differences.

Expert systems should not be viewed as substitutes for deterministic modeling. They can serve as a means of integrating a field of expertise in a quasi-objective manner. They are, in a sense, the best guess of the solution to any particular problem. For some classes of problems, including climatic change, we simply cannot afford to wait until deterministic models are sufficiently developed to provide solid answers. In this regard, expert systems provide a bridge from the limits of factual understanding to total understanding of the impacts of and adjustments to climatic changes.

The decision perspective

Analysis of the decisionmaking process used in a given sector by economic agents provides a basis for assessing future climatic variability. Rather than focusing initially on climate and any one of its peculiar fluctuations, the decision framework is addressed in its entirety, and includes production factors (some of which are, some of which are not directly impinged on by climate).

Decision and production processes

Boulding (1985) has stated that, "A decision essentially involves a choice among different images of the future that we conceive to be within our power to achieve." Weather events color those images. In general, we do not have the power to forestall a weather event. Rather, our choices revolve around responding to the possibility that climatic variability may occur or around reacting to climatic variability that has occurred. From the economist's perspective, the challenge is optimal adjustment to the climatic endowment, including its characteristic variability and change (Lovell and Smith 1985).

An example

A simple example will illustrate the issues associated with dynamic decisionmaking in response to climate variability. First, we consider crop production in a commercial agricultural setting, then the example to consider implications for a subsistence agricultural situation and for a public decisionmaker.

Figure 2 presents a schematic of the decision and production process as it might occur in selecting operating inputs for an annual crop, such as cotton, wheat, or maize. The processes are illustrated in a manner quite dissimilar to that typically used to indicate societal adaptations to drought (Kates 1985, Warrick and Riebsame 1983).

The top section, current year's planning, relates to the planning process, where expectations play a major role. Expectations of future weather events, which include some realization that an event such as drought could occur, are one source of uncertainty. Although climate is a source of variability, this approach allows us also to consider climate as a natural resource (Riebsame 1985). Although climate expectations are an important factor in many agricultural situations, they are but one of many economic, social, and physical factors that a decisionmaker weighs in making a production choice.

The potential for climatic fluctuation affects production decisions through expected profit. It is likely, however, that the three factors shown as affecting the production decision are, in reality, not independent but interrelated. For example, a producer in a very secure financial position may have a different attitude toward a fluctuation such as drought than does a producer in a marginal position.

Although it is likely that these interrelationships exist, we know very little about their empirical effect. We might expect the poorly financed producer to adopt more conservative strategies than the well-financed producer, but in reality, the more marginal producer may have to assume that drought cannot occur, because if it does the firm will not survive, no matter what actions the producer pursues. Given that framework, the producer in the weaker financial position may adopt practices that appear to be relatively risky.

The middle section of the schematic relates to actual events, which occur after the planning process is completed and the production decision made. This, of course, is

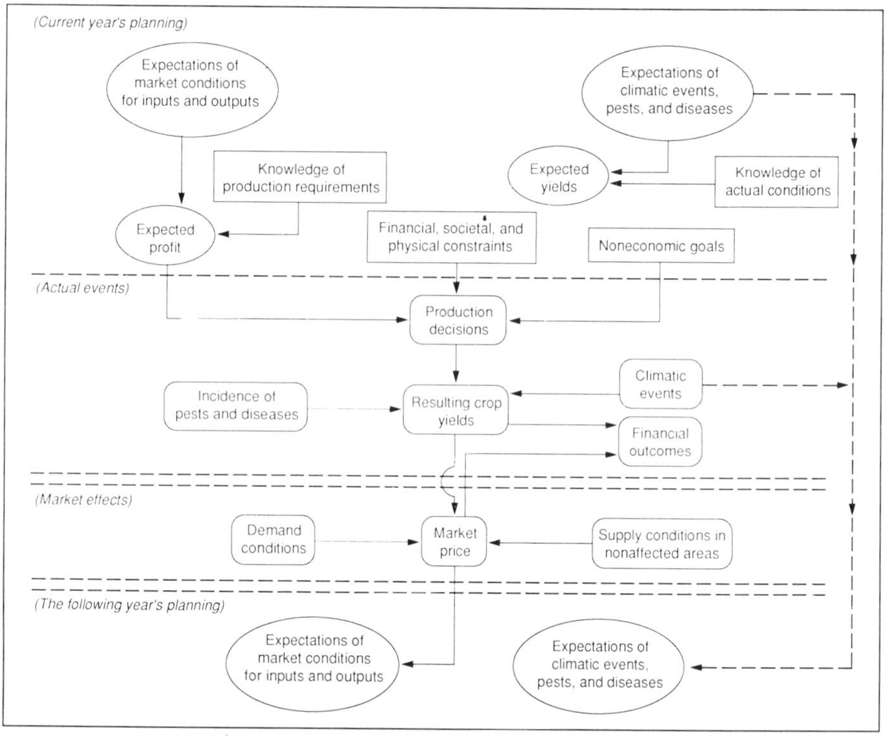

2. Schematic of the interaction of climate with an economic activity (maize production in the central United States), from a decisionmaking perspective.

just another way of saying that uncertainty is a major characteristic of decisionmaking.

Up to this point, we have considered the production process at the level of the individual decisionmaker. But with many producers, financial outcomes cannot be determined until total production is weighed against total demand for the commodity. The third section relates to market effects. Here demand and supply conditions interact to determine market prices. In today's highly interdependent world, market effects are volatile and are affected by some events at the other side of the world from individual producers.

These market effects can mitigate or exacerbate the effects of a drought. For example, the 1983 drought in the central United States was offset by relatively high market prices and government programs. For producers in the southern United States, the 1986 drought was coupled with relatively low output prices, and financial effects were particularly devastating.

The bottom section of Figure 2 emphasizes the continuous nature of decision and production processes. The current year's results are very important in conditioning the following year's processes. These interrelations have many dimensions, including

- physical factors, such as soil moisture and groundwater availability,
- the financial constraints within which the individual firm operates,
- carry-over supplies that affect expected market prices, and
- decisionmaker expectations of possible climatic conditions.

The crucial role of time

Lovell and Smith (1985) assert that three prominent features of climate must be incorporated into frameworks to assess its economic impacts. The stochastic nature of climate and the potential for broad geographical impacts are depicted in the schematic. The third feature is the temporal nature of climate effects. (The annual component of this temporal nature was alluded to previously.)

The sequential nature of decisionmaking is another key component. A significant simplification in the schematic was to present the production decision as a single decision. In reality, a series of interrelated decisions is required. Table 1 shows the structure of a specific example of decisionmaking that is consistent with the previous framework (maize production in the midwestern United States). Within Table 1, operational production decisions are linked to the time periods when choices need to be made.

For midwestern United States maize production, it is well known that drought during the summer can devastate yields. Dramatic examples were experienced in 1974, 1980, and 1983. But notice that essentially no decisions occur during this period (the fertilization decision noted must take place very early in June, at the start of the summer). This underscores the importance of expectations in operational decisionmaking.

The decisions shown in Table 1 are interrelated; choices made early in the production process affect the flexibility of decisionmaking later. For example, in this region, the decision to plant maize often results in the need to apply fertilizer in the preceding fall. Once that occurs, it is very difficult to justify switching to a more drought-tolerant crop, such as oats or soybeans, even if the producer's expectations of drought increase. The option of switching to a more drought-resistant variety of maize remains; that decision might have important financial implications.

Types of decisions and droughts

Another aspect of the role of time in a producer's adaptation to the possibility of drought is the long-term variability of climate. Figure 3 illustrates a time series of streamflow for the Senegal River at Bakel (Faure and Gac 1981). We see three major time periods when drought severely restricted streamflow. Those data suggest two interesting factors.

First, the years preceding the occurrence of drought often had ample or excess rainfall. This has important implications for decisionmaker expectations about future weather events. Also, individuals born in the 1915-20 or 1945-50 periods would not have personally experienced severe drought until they were well into their 20s. Their expectations may not have included severe prolonged droughts, and their decisions may well have reflected that restricted experience. As we learn more about how managers manage and how they make decisions, the hectic, overloaded nature

Table 1. Production choices and time period when climate forecast information could be useful for maize production, Central Illinois, USA (from Sonka et al 1986).

Production choice	Decision time period				
	Fall/winter (Sep-Mar)	Early spring (Apr-mid-May)	Late spring (mid-May-mid-Jun)	Summer (mid-Jun-Aug)	Fall (Sep-Nov)
Crop choice	X	X	X		
Fertilization (timing and amount)	X	X	X	X	
Tillage (timing and amount)	X	X	X		
Planting date		X	X		
Variety selection	X	X	X		
Plant population		X	X		
Harvest date					X

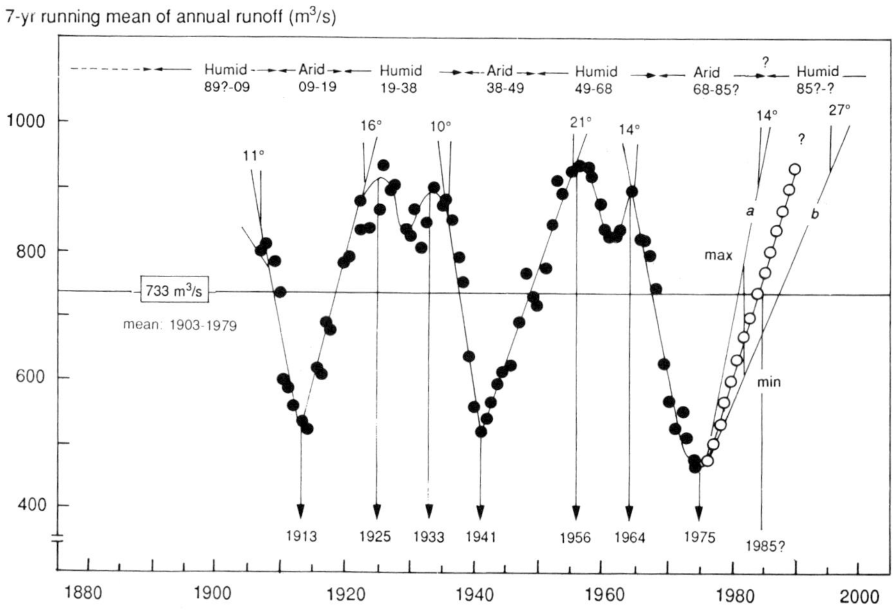

3. Mean annual modulus of River Senegal runoff measured at Bakel (7-yr running mean) (from Faure and Gac 1981).

of decisionmaking becomes more apparent (Van de Ven 1985). Managing the manager's attention has been recognized as an increasingly critical factor. With respect to drought, given its temporal variability, is it likely that a producer who has just seen a run of good rainfall years will expect a drought this year?

The second factor in the uncertainty associated with drought is recognizing that the possibility of drought has an economic effect, even in years of adequate or excessive moisture. The pervasive effect of the possibility of drought does not seem to have received adequate recognition. Often, analyses have focused on the suffering and losses that result when drought occurs. In those instances, human practices that have worsened the effects of the drought typically are identified. Analyses that concentrate only on the extreme event tend to document the adverse effects of such actions during the drought, but discount to zero the benefits of those actions in nondrought years. Where nonreversible physical effects are of interest, this may be appropriate. For economic analyses, however, this approach is not satisfactory.

A recently completed analysis of maize production in midwestern U.S. illustrates this factor (Mjelde 1985). It used a decision analysis framework to estimate outcomes for a hypothetical maize producer over 14 yr, 1970-83. As one part of the analysis, the producer was assumed to make decisions as if he or she believed that drought was going to occur every year. This was compared to an assumption of normal weather. Over the 14 yr, the pessimistic producer would have averaged about $15/ha less in net returns than would someone who had assumed normal conditions.

Of course, in a drought year the pessimistic producer would have done significantly better than the optimistic producer. In the extreme drought of 1983, producers who assumed average conditions might have taken actions that made them more vulnerable to drought. But for the pessimistic producer, the cost of being right in 1983 was reduced over the entire period.

We are not suggesting that actions that are more vulnerable to drought are necessarily preferable. Costs that occur during drought could overwhelm benefits in other periods. The important point is that it is not appropriate to use evaluation methods that consider only a portion of the range of possible climate events.

Some implications

The preceding discussion suggests several implications about the physical or societal impacts of climatic variation. The few mentioned here certainly do not exhaust the possibilities.

Expectations. As we consider means to improve our decision responses to climate, the key role of expanded decisionmaker expectations is highlighted. Several alternatives come to mind. The first is the development of skillful climate forecasts with sufficient lead time to be useful to producers. The scientific problems associated with seasonal climate forecasting are significant. One key issue is to determine the levels of accuracy that would be useful to decisionmakers. Forecasts that do not meet statistical tests of significance may be desired by producers, who have to make decisions no matter what the standard error of the estimate is.

Forecasts are not the only basis for decisionmaker expectations. Knowing the statistical chances for drought may have value, even for producers who have considerable experience in a specific location. For that matter, we know very little about how individuals develop their expectations with respect to weather events and about the information they use in the process. There is considerable evidence that individuals perform poorly in experiments where probabilistic information is important and, probably as significant, consistently have overconfident views of their capabilities (Kahneman and Tversky 1982).

The relevant expectations include economic as well as physical factors. Knowledge of physical conditions in competitive production areas, the size of actual or potential carry-over stocks, and variables that affect demand are some factors for which the producer may want to develop expectations. These factors are important because they indicate the vulnerability of the entire market system to production shortfalls from events such as drought.

Interrelation of climatic variability and decisionmaker constraints. The potential for complex interactions between producers' financial positions and their adjustments or responses to climatic variability has been suggested. Decisionmakers typically face more than financial constraints. Physical factors may also affect the production system every year because of the possibility that drought conditions may develop in any year. For example, planting windbreaks and adopting other agronomic practices, was a response to the Dust Bowl conditions of the 1930s (Worster 1979).

Constraints such as windbreaks are relatively permanent. Drought also can temporarily alter the physical environment with which the decisionmaker works. Annual conditions, such as soil moisture before planting, could materially alter choices.

Lessening the negative effects of climatic variability. A better understanding of the effects of climatic variability on decisionmaking and on society is, of course, desirable. A more compelling goal is to develop mechanisms that will reduce the negative consequences when dynamic climate events occur. Three types of mechanisms come to mind:

- new technologies that are less susceptible to adverse weather,
- societal policies, and
- new management practices.

The decisionmaking framework is helpful in identifying how such mechanisms could assist society.

Focus on climate-human system interaction. The framework shown in Figure 2 concentrates on a drought's impact on a particular aspect of society. This framework explicitly identifies the complexity of the interrelationships between human systems and the range of possible climatic events. In an agricultural setting, the decisionmaker responds continually to the effects of past, current, and future weather events. Individual events, such as drought, obviously are of major importance and occupy part of the manager's attention to climatic events. But drought is not the only event being considered. Understanding the decisionmaker's perspective relative to drought requires a broader perspective than is possible if we only evaluate climate-human system interactions when droughts have occurred.

An interesting extension of this broader focus is to highlight the impact of society's reactions to drought, as well as of the drought itself, on the decisionmaker. The human suffering associated with prolonged drought can be extensive. The effects are often vivid and can be described. Given the availability of modern television communications, news media find that some effects of certain types of climatic variability can easily be transmitted to their audiences.

People respond to these vivid visual reports of suffering. A natural reaction in such situations is to attempt to reduce the suffering. In this decade, that sequence has occurred in the African Sahel drought as well as in this year's drought in southeastern United States. In the first case, worldwide actions were implemented to deliver relief aid. In the second, feedstuffs were delivered to maintain cattle herds.

Such humanitarian actions undoubtedly reduced the immediate suffering of those afflicted. The decisionmaker framework, however, allows us to identify an additional effect of such actions. In Figure 2, expectations are shown to be central to the decisionmaking process. Our expectations, or what Boulding (1985) referred to as images of the future, are also affected by humanitarian actions. One implication of altered expectations could be that it is now plausible for an individual decisionmaker to choose to maintain a smaller internal reserve of food, feedstuffs, or cash.

This comment is not meant to imply that humanitarian relief actions should not occur. The decisionmaking perspective does alert us, however, to the natural human reaction, to expect such actions if similar conditions reappear in the future.

Expanding the analysis

We have been considering a commercial agricultural situation that is, of course, not representative of much of the world's agriculture. In subsistence agriculture, at least two additional factors must be considered: a typically more diverse pattern of production and the need to explicitly include consumption within the decision framework (Jodha and Mascarenhas 1985, Spitz 1980). Fortunately, it is possible to expand the decisionmaking framework to consider subsistence agriculture.

In a commercial-agriculture setting, financial reserves play a major role in mitigating the potential effects of events such as drought. In subsistence agriculture, financial reserves are not available. Physical reserves, in the form of livestock and crop inventories, must be used. The rate at which these reserves can be depleted is a critical decision. The cost of depleting reserves at too fast a rate is severe. In addition to serving as food reserves, grain inventories often are the seedstock for future crops. There are costs associated with maintaining reserves at too high a level just as there are costs for depleting reserves.

Of course, not all important decisions are made by private decisionmakers. Public sector managers make critical choices that can affect society's adaptability to climatic variability. The decisionmaking framework considered here can be altered to accommodate this setting as well. Many of the decision factors are relevant to both private and public sector managers. The role of expectations and uncertainty exists in both cases. Each manager has numerous institutional and physical constraints.

A major distinction is likely to be the way in which good or bad decisions are evaluated. The private decisionmaker measures outcomes in terms of changes in income or revenue. Less direct measures influence the public sector manager. Although the actions of a public sector manager can have major influences on a society's adaptability to drought, the personal income of the manager may not be directly affected. Political processes may be involved. In those circumstances, the attributes of a good decision in the public sector may differ from those of a decision in the private sector.

Integrated models for assessing climatic impacts

A brief presentation of an integrated modeling approach to decision analysis is provided here. The use of models is suggested to extend the body of work related to climatic impacts; it is not suggested as a substitute for more traditional forms of analysis.

The decisionmaking framework in Figure 2 represents a conceptual basis for applying quantitative models to the analysis of climatic impacts on society. Several dimensions of the problem are identified (Warrick and Riebsame 1983). First is the types of processes, which range from microphysical to macroeconomic linkages. The relevant geographic scale varies, from a subunit of a farm to the larger economic region and, by implication, to the world.

The conceptual framework needs to be operationalized if further gains in understanding are to be realized. We need to develop linked quantitative models of the several processes involved to achieve this operationalization.

Plant growth models for the relevant crops are the first component needed. Assuming a regional study area, physical models would be required from several locations within the area. The locations should include marginal as well as central areas. Inclusion of marginal areas would allow identification of important shifts among enterprises (Parry 1985). Data availability, modeling capacity, and resources will determine the number of locations that can be addressed in a specific research effort.

Physical models in themselves cannot completely capture the influence of time and variability in a dynamic climate. The effect of the decisionmaker's constraints, expectations, and attitudes must be included. Considering the sequential nature of decisionmaking, dynamic modeling techniques are required (Katz et al 1982, Winkler et al 1983). These enterprise-level models could be used to evaluate new technologies, alternative management practices, and societal policies. An important point is that an integrated model can be used to perform ex ante evaluations of the mechanisms. Several research teams have successfully developed linked bio-economic models relating to agricultural situations (Ahmed et al 1976, Boggess 1984), including efforts relating to climate and society (Sonka et al 1986, Mjelde et al 1986).

The effects of a dynamic climate and potential adaptations to it are not limited to the individual enterprise. Linkages to regional market economies markedly affect the desirability of alternative mechanisms. The modeling system must include a market-level analytical capability. Such macromodels could exist for different ranges of aggregation, from regional to national or international levels. At whatever level, the model should incorporate effects of exogenous forces outside the study area, such as population changes and production in other regions. In general, efforts to link microprocess and regional models have not been successful.

Conclusion

The time to assess the manner in which economies adjust and adapt to changing (as opposed to changed) climatic constraints and opportunities is at hand. We must extend our thinking about the research approaches that will help us learn how those adjustments take place, both in the recent past, currently, and in the future. Quasi-experiments, expert systems, and decision analyses provide insights into the complex manner in which economies and climate interact. Integration of physical and economic models is the next step in arriving at a calculus of climate-induced adjustments.

References cited

Ahmed J, van Bard C H M, Hiler E A (1976) Optimization of crop irrigation strategy under a stochastic weather regime: a simulation strategy. Water Resour. Res. 12:1241-1247.

Ausubel J H (1983) Can we assess the impacts of climatic change? Climatic Change 5(1):7-14.

Boggess W A (1984) Discussion: use of bioeconomic simulations in production economics. Southern J. Agric. Econ. 16:87-89.

Boulding K E (1985) Human betterment. Sage Publications, Beverly Hills, California.

Bowden M J, Kates R W, Kay P A, Riebsame W E, Warrick R A, Johnson D L, Gould H A, Weiner D (1981) The effect of climate fluctuations on human populations: two hypotheses. Pages 479-513 *in* Climate and history: studies in past climates and their impact on man. T. M. L. Wigley, M. J. Ingrahm, and G. Farmer, eds. Cambridge University Press, Cambridge.

Bryan K, Komro F G, Manabe S, Spelman M J (1982) Transient climate response to increasing atmospheric carbon dioxide. Science 215:56-58.

Changnon S A, Auer A H, Brahm R R, Hales J, Semonin R G (1981) METROMEX: a review and summary. Monograph 18. American Meteorological Society, Boston.

Clark W C (1985) Scales of climatic impacts. Climatic Change 7(1):5-27.

Easterling W E, Sonka S T (1986) An expert system approach to the identification of climate fluctuations for agricultural impacts analysis. Preprint of the joint AMS-National Climatic Data Center Conference on Climate and Water Management-A Critical Era. Asheville, North Carolina.

Electric Power Research Institute (1985) Artificial intelligence: human expertise from machines. EPRI J. 10(5):6-15.

Faure H, Gac J Y (1981) Will the Sahelian drought end in 1985? Nature 291:475-478.

Friesma H P, Caporaso J, Goldstein G, Lineberry R, McCleary R (1979) Aftermath: communities after natural disasters. Sage Publications, Beverly Hills, California.

Garcia R V (1981) Drought and man: the 1972 case history. Vol. 1. Pergamon Press, New York.

Giegel J M, Sundquist W B (1984) A review and evaluation of weather-crop yield models. University of Minnesota, Department of Agricultural and Applied Economics Staff Paper Series.

Hayes-Roth F, Waterman D, Lenat D (1983) Building expert systems. Addison-Wesley, London. 444 p.

Heathcote R L (1973) Drought perception. Pages 17-40 *in* Drought. J. V. Lovett, ed. Angus and Robertson, Sydney, Australia.

Heathcote R L (1985) Extreme event analysis. Pages 369-402 *in* Climate impact assessment. R. W. Kates, J. H. Ausubel, and M. Berberian, eds. John Wiley and Sons, New York.

Hill M K (1973) Farm management for drought mitigation. Pages 195-219 *in* Drought. J. V. Lovett, ed. Angus and Robertson, Sydney, Australia.

Hoffert M I, Flannery B P (1985) Model projections of the time-dependent response to increasing carbon dioxide. Pages 149-190 *in* The potential effects of increasing carbon dioxide. M. C. MacCracken and F. M. Luther, eds. U.S. Department of Energy, Office of Energy Research, Report DOE-ER-0237, Washington, D.C.

Jodha N S, Mascarenhas A C (1985) Adjustment in self-provisioning societies. Pages 437-468 *in* Climate impact assessment. R. W. Kates, J. H. Ausubel, and M. Berberian, eds. John Wiley and Sons, New York.

Kahneman D, Tversky A (1982) The psychology of preferences. Sci. Am. 246:160-174.

Karl T R, Riebsame W E (1984) The identification of 10- to 20-year temperature and precipitation fluctuations in the contiguous United States. J. Clim. Appl. Meteorol. 23:950-966.

Kates R W (1985) The interaction of climate and society. Pages 3-36 *in* Climate impact assessment. R. W. Kates, J. H. Ausubel, and M. Berberian, eds. John Wiley and Sons, New York.

Kates R W, Changnon S A Jr., Karl T A, Riebsame W E, Easterling W E (1984) The climate impact, perception, and adjustment experiment (CLIMPAX): a proposal for collaborative research. CENTED, Clark University, Worcester, Massachusetts.

Kates R W, Ausubel J, Berberian M, eds. (1985) Climate impact assessment. John Wiley and Sons, New York.

Katz R W, Murphy A H, Winkler R L (1982) Assessing the value of frost forecasts to orchardists: a dynamic decision making approach. J. Appl. Meteorol. 21:518-531.
Lamb P J (1979) Some perspectives on climate and climate dynamics. Prog. Phys. Geogr. 3:215.
Lemmon H (1986) COMAX: an expert system for crop management. Science 233:29-33.
Lovell C A K, Smith V K (1985) Microeconomic analysis. Pages 293-322 *in* Climate impact assessment. R. W. Kates, J. H. Ausubel, and M. Berberian, eds. John Wiley and Sons, New York.
McKinnion J M, Lemmon H E (1985) Expert systems for agriculture. Comput. Electron. Agric. 1:31-40.
Mjelde J W (1985) Dynamic programming model of the corn production decision process with stochastic climate forecasts. Ph D dissertation, University of Illinois at Urbana-Champaign, Illinois.
Mjelde J W, Sonka S T, Dixon B L, Lamb P J (1986) Integration of dynamic programming and simulation models to value lead time of information forecasting schemes. Department of Agricultural Economics Staff Paper DIR 86-1 SP-4, Texas A & M University, College Station, Texas.
Murnpower J, Phillips L, Renn O, Uppuluri V C R (1987) Expert judgment and expert systems, Springer-Verlag. (in press)
National Research Council (1979) Carbon dioxide and climate: a scientific assessment. National Academy of Sciences, Washington, D.C. 22 p.
Palutikof J P, Wigley T M L, Lough J M (1984) Seasonal climate scenarios for Europe and North America in a high-CO_2 warmer world. U.S. Department of Energy, Carbon Dioxide Research Division, Technical Report TRO12, Washington, D.C. 70 p.
Parry M L (1985) The impact of climatic variations on agricultural margins. Pages 351-368 *in* Climate impact assessment. R. W. Kates, J. H. Ausubel, and M. Berberian, eds. John Wiley and Sons, New York.
Riebsame W E (1985) Research in climate-society interaction. Pages 69-84 *in* Climate impact assessment. R. W. Kates, J. H. Ausubel, and M. Berberian, eds. John Wiley and Sons, New York.
Rossi P, Wright J (1982) Social science and natural hazards. Abbott Books, Cambridge.
Schlesinger M E, Mitchell J F B (1985) Model projections of the equilibrium climatic response to increased carbon dioxide. Pages 81-147 *in* The potential climatic effects of increasing carbon dioxide. M. C. MacCracken and F. M. Luther, eds. U.S. Department of Energy, Office of Energy Research, Report DFOE/ER-0237. Washington, D.C.
Schneider S H, Thompson S L (1981) Atmospheric CO_2 and climate: importance of the transient response. J. Geophys. Res. 86:3135-3147.
Sonka S T (1987) Adaptation and adjustments in drought-prone areas: research directions. Pages 351-367 *in* Planning for drought: toward a reduction of societal vulnerability. D. A. Wilhite and W. E. Easterling, eds. Westview Press, Boulder, Colorado.
Sonka S T, Lamb P J (1987) On climate change and economic analysis. Climatic Change 11(3):291-311.
Sonka S T, Lamb P J, Hollinger S E, Mjelde J W (1986) The economic use of weather and climate information: concepts and an agricultural example. J. Climatol. 6:447-457.
Spitz P (1980) Drought and self-provisioning. Pages 125-147 *in* Climatic constraints and human activities. J. Ausubel and A. K. Biswas, eds. Pergamon Press, New York.
Stewart T R, Glantz M H (1985) Expert judgment and climate forecasting: a methodological critique of "climate change to the year 2000". Climatic Change 7(2):159-184.
Van de Ven A H (1985) Central problems in the management of innovation. Strategic Management Research Center Paper #21, University of Minnesota, Minnesota.
Warrick R A, Riebsame W E (1981) Societal response to CO_2-induced climate change: opportunities for research. Climatic Change 3:387.
Warrick R A, Riebsame W E (1983) Societal response to CO_2-induced climate change: opportunities for research. Pages 20-61 *in* Social science research and climate change. R. S. Chen, E. Boulding, and S. H. Schneider, eds. D Reidel Publishing Company, Boston, Massachusetts.
White G F, ed. (1974) Natural hazards: local, national, global. Oxford University Press, New York.
Wilhite D A, Glantz M H (1985) Understanding the drought phenomenon: the role of definitions. Water Int. 10:111-120.
Winkler R L, Murphy A H, Katz R W (1983) The value of climatic information: a decision-analytic approach. J. Climatol. 3:187.
Worster D (1979) Dust bowl. Oxford University Press, New York.

Notes

Portions of this paper were taken from Sonka and Lamb (1987), Easterling and Sonka (1986), Sonka (1987), and Kates et al (1984).

Authors' addresses: W. E. Easterling, Climate and Meteorology Section, Illinois State Water Survey, 2204 Griffith Drive, Champaign, Illinois 61820, USA; S. T. Sonka, Department of Agricultural Economics, University of Illinois, Urbana, Illinois 61801, USA.

Citation information: International Rice Research Institute (1989) Climate and food security. P.O. Box 933, Manila, Philippines.

Agronomic adjustments to climatic variation

N. S. Jodha

Traditional farming practices in dry tropical regions of India evolved to adjust to climatic variability. The traditional food security system has included farm-level reserves and recycling devices, collective risk sharing arrangements, and diversified and flexible production systems based on integrated use of the highly heterogeneous natural resource base. Because of rapid changes, especially in population growth, market forces, and technological and institutional interventions in the rural sector, traditional agronomic adjustments are becoming increasingly ineffective. An integration of traditional and modern technology offers scope for salvaging the situation, with the help of appropriate institutional measures.

The dry tropics, owing to their low and highly variable rainfall, are one of the most unstable environments for agriculture. Farmers in these regions have evolved their own food security systems, which include farm-level commodity reserves, food and fodder recycling devices, flexible consumption patterns, self-provisioning strategies in consumption and production, collective risk sharing, and diversified and flexible production and resource use practices (Chambers 1984, Jodha 1978, Jodha and Mascarenhas 1985, Spitz 1980, Walker and Jodha 1986). Traditional practices maximize productivity during good rain years and minimize losses during poor rain years. The production patterns, often termed folk agronomy, strengthen the farm-level food security systems.

Various elements of traditional farming practices help the farmer to adjust to the risky production environment of the dry tropics. However, despite their scientific rationale and stability-promoting attributes, a number of traditional practices are becoming ineffective in the face of the rapid changes affecting rural areas. These changes include population growth, the increasing role of market forces, and the side effects of institutional interventions in agriculture. There is potential for salvaging the situation through modern technology and appropriate institutional measures.

Rainfall variability is the only climatic variation discussed here. The dry tropics, for our purpose, include arid and semiarid tropical regions characterized by rainfed agriculture.

Traditional farm practices

Traditional farming systems cover practices ranging from choice of cultivar to diversification of land use in adjusting to climatic variations. They can be grouped into the following interrelated categories:

1. Adaptation to long-term agroclimatic patterns
   Some traditional farming practices may be regarded as adaptations to the long-term pattern of rainfall and its interactions with soil types (Kates 1985). Well-adapted cultivars and animal species and diversified patterns of resource use are in this category. These structural features facilitate operational adjustments to short-term (intrayear and interyear) fluctuations in rainfall.
2. Adjusting the environment
   A few measures designed to manipulate spatial and temporal availability of moisture to plants, termed water security measures, help in adapting environmental factors to farm enterprises. Moisture conservation measures such as bunding, mulching, and weeding are in this category. They help establish the effective availability of moisture to plants.
3. Diversification and flexibility
   A number of crop and resource management practices are used by the farmer to meet perceived uncertainties in moisture availability. These practices relate to spatial and temporal cropping decisions, input and resource use practices, and harvesting strategies.

A selective inventory of farming practices that directly or indirectly contribute to adjustment to climatic variations is presented in Table 1. Variants of these practices are called by different names in different parts of India. The ultimate goal is to generate the maximum number of options for a farmer operating in a production-constraining environment. The multiple options help stabilize agricultural production and the farmer's food supply.

Interregional differences

The dry tropics are characterized by considerable heterogeneity. Folk agronomy, as a response to natural resource specifics, differs in the extent of adjustment mechanisms in different regions. More measures to achieve diversity and flexibility are found in regions with a higher degree of rainfall variability, or weather risk, than in regions with relatively stable rainfall.

Table 2 illustrates this situation with data from three areas with very different degrees of climatic risk (taken from studies conducted by the International Crops Research Institute for the Semi-Arid Tropics [ICRISAT]). The extent of diversification and flexibility is lowest in the Akola villages with stable rainfall and highest in the Jodhpur villages with the lowest and most unstable rainfall. Sholapur falls in between. The only two exceptions to the general pattern of positive association between the degree of risk and diversity are the smaller number of crops planted in the more risky Jodhpur area and the low extent of intercropping in Sholapur area. The former is due to overall limitations on the number of possible crops one can grow in the sandy arid area of Jodhpur. The Sholapur situation can be attributed to rabi (postrainy) season cropping in deep Vertisols where, due to the

Table 1. An inventory of traditional farming practices with their potential role in adjustment to climatic variations.

Farm practices by their primary focus	Adjustment/adaptation processes: Long-term adaptation to unstable environment	Adjusting environment to crops, etc	Short-term adjustment through: Diversity	Flexibility
1. Key components of farming system				
• Crop-livestock integration	√		√	
• Well-adapted cultivars, animal species	√			√
• Diversified farming	√			√
• Common property resources	√		√	
2. Water security measure				
• Limited access to irrigation	√	√		
• Moisture conservation		√		
• Summer plowing/harrowing/crack filling		√		
• Weeding/thinning/intercropping		√		
• Mulching		√		
3. Crop planting: spatial and temporal arrangements				
• Intercropping			√	√
• Plot splitting			√	√
• Toposequencing			√	
• Staggered planting			√	√
• Relay cropping			√	√
• Sequential cropping			√	√
• Crop rotation/fallowing			√	√
• Resowing/gap filling				√
• Postrainy season cropping			√	√
4. Input/resource use				
• Limited ex-ante commitment				√
• Self-provisioning				√
• Divisibility				√
• Organic recycling				√
5. Harvest activity				
• Harvesting in installments				√
• Salvage operations				√

known status of the soil moisture and the limited scope for interspecies competition, intercropping as a diversification strategy is not required (Jodha 1980).

Interyear differences

In addition to interregional differences, the extent of adjustment to climatic variations in an area changes from year to year, depending on the moisture situation in the planting season. In fact, such interyear changes are the very essence of short-term adjustments to weather variability. Table 3 presents some details from ICRISAT's village-level studies in three agroclimatic zones (Jodha et al 1977). In these studies, plot data were collected from panels of farmers for the 10 yr ending in 1984. Daily rainfall was recorded in each village studied. During low soil moisture

Table 2. Indicators of the farmer's long-term strategies against weather risk in three districts with different degrees of weather risk in the dry tropical regions of India.[a]

Details of degree of risk and adaptation	Akola	Sholapur	Jodhpur
A. Characteristics of weather-risk[b]			
Annual average rainfall (mm)	820	690	382
Probability of favorable soil moisture conditions for rainy season cropping	.66	.33	.21
Length of growing season (d)	200	155	60
Incidence of crop failure (average of 3 yr)			
– plots with complete crop failure (%)	4	17	33
– plots with partial crop failure (%)	7	24	–
B. Indicators of spatial diversification			
Scattered land fragments (no.) per farm	2.8	5.8	7.3
Split plots (no.) per farm	5.0	11.2	–
Fragments (no.) per farm by distance from village			
– Up to 0.8 km	1.5	1.4	1.2
– 0.8-1.6 km	1.1	3.4	4.3
– More than 1.6 km	0.2	1.0	1.8
C. Indicators of crop-based diversification			
Extent of intercropping (%)	83	35	100
Total sole crops planted (no.)	20	34	12
Total combinations of mixed crops planted (no.)	43	56	30
D. Crop-livestock based mixed farming[c]			
Ratio of land and livestock values	93:7	91:9	63:37
Ratio of crop and livestock incomes	80:20	71:29	69:31
E. Occupational and institutional adjustments			
No. of occupation (sources of income) per household	1.5	2.3	2.9
Households with more than 2 occupations (%)	14	18	39
Households with incidence of seasonal out-migration (%)	2	24	33
Cases of land tenancy induced by risk sharing/management considerations (no.)	9	66	–

[a]Table reformulated from Jodha (1967), Binswanger et al (1980), Walker and Jodha (1986), Singh and Walker (1982), Walker et al (1983), and other unpublished data from ICRISAT's village level studies (Jodha et al 1977). [b]Information under first three items is based on district-level details of soils and long-term average rainfall of the concerned districts. The rest of the information under the table relates to 2 villages in each of the district. Number of sample households was 120 each in Akola and Sholapur and 140 in Jodhpur. Reference periods cover 1975-76 to 1977-78 for Akola and Sholapur, 1963-64 to 1965-66 for Jodhpur. [c]Assets (livestock/land) and income valued in terms of prices for 1976-77 for all regions.

years, cropped area per household declined in all the areas. The extent of intercropping increased during low soil moisture years (except in Sholapur), as well as the use of drought-tolerant and short-duration crops. Input-use practices during high and low soil moisture situations were influenced accordingly. These interyear differences in cropping decisions represent agronomic adjustments to rainfall variability.

In villages in the Jodhpur area (Table 4), during a drought year emphasis increased on salvage operations, such as collecting the fodder if not the grain from the withering crops, to minimize losses. Similarly, during a drought year,

Table 3. Changes in cropping decisions in response to soil moisture situation during planting period in three districts in the dry tropical regions of India.[a]

Soil moisture and cropping decisions	Changes in the proportion of area under crops during the planting period in					
	Mahabubnagar		Sholapur		Akola	
	I[b]	II[c]	I	II	I	II
A. Soil moisture (mm) during (6 wk) planting period[d]						
Mean per week	56.3	21.6	129.2 (175.2)[e]	36.0 (58.5)	68.6	22.7
Range during 6 wk	48-99	21-28	67-150 (150-220)	28-41 (44-77)	63-74	12-59
B. Cropping decisions[f]						
Cropped area (ha) per household	4.2	3.9	7.5	6.0	5.4	5.9
Proportion of the following in gross cropped area (%)						
• Intercropping	83	85	38	25	94	90
• Crops with low moisture needs[g]	29	36	54	62	21	23
• Crops with high moisture needs[h]	5	3	11	6	48	43
• Long-duration crops[i]	8	7	13	8	16	11
• Short-duration crops[j]	1	4	2	5	3	8

[a]Table based on data from ICRISAT's village-level studies conducted from 1975 to 1985 in three regions. Data relate to 30 farm households in one village in each of the three districts. [b]Years with high soil moisture. [c]Years with low soil moisture. [d]Based on daily rainfall recorded in each of the villages. Weekly soil moisture (cumulative moisture availability in soil) for successive weeks is worked out on the basis of computer system, which takes into account precipitation, evapotranspiration, soil type, etc. For details see Keig and McAlpine (1977). The rainy season crop planting period coincides with the 23rd-28th week of the calendar year. [e]For Sholapur area where the bulk of the area is planted in the postrainy season, the main crop of the area is planted in the postrainy season, the main crop planting period covers the 41st-46th wk. The relevant soil moisture information for this period is indicated in parentheses. [f]Table covers only rainfed crops. [g]Include caster seed in Mahabubnagar and sorghum in other areas. [h]Include groundnut, some other oilseeds, and minor pulses. Also include rainfed wheat in Sholapur and cotton in Akola. [i]Include mainly pigeonpea. [j]Include mungbean and other minor pulses planted only when sufficient moisture is available at the beginning of the season.

dependence on noncrop production, such as fodder from trees and bushes, increased. Efforts to save as much moisture as possible for promising, surviving crop plants through thinning and frequent weeding also were made during drought. Resource commitment to the withering crop was reduced to save cost.

Decline in effectiveness of adjustments

Irrespective of interregional differences, traditional farming systems contain several elements of adjustment to climatic variations that have performed reasonably well in the past. However, now there are clear indications that most of those adjustment

Table 4. Loss-minimizing practice adopted by farmers during a drought year and a nondrought year in the study villages in Jodhpur area in the dry tropical region of India (Jodha 1967).[a]

Item	Drought year (1963-64)	Normal year (1964-65)
Characteristics of weather risk		
Rainfall during the year (mm)	159[b]	377
Total rainy days (no.)	8	21
Plots covered by risk/loss-minimizing farm practices (no.)		
Collecting weeded material as fodder	53	5
Harvesting field borders for fodder	68	66
Harvesting premature crops	27	–
Harvesting crop by-product only	49	2
Harvesting mature crop	16	144
Interculturing	7	65
Weeding more than once	18	–
Thinning	37	–
Abandoning postsowing operations	36	–
Using hired resources for postsowing operations	2	24
Harvesting premature *Z. nummularia* (bush) for fodder	92	–
Lopping trees for fodder/fuel	53	4

[a]Table based on data from Jodha (1967). Data relate to 160 plots belonging to sample households, in 2 villages. The information is based on physical monitoring of the plots during the reference years. [b]Rains received only at the beginning of the crop season.

measures are losing their effectiveness. Traditional farming systems were shaped by such factors as low population pressure, subsistence orientation of agriculture, and the absence of technological and institutional interventions. Low population pressure made them resource extensive in character. Subsistence orientation led to domination of self-provisioning in both production and consumption spheres. Traditional technology, which involved elements of organic farming, low input use, and some resource conservation, led to low but stable productivity.

However, increased population pressure, the increased role of market forces, and technological and institutional interventions in rural areas have changed the whole environment. A number of traditional farm practices have become ineffective or unfeasible. Resource extensive farming is no longer possible. The increased role of market forces has not only generated an additional source of risk (e.g. price risk), but has curtailed the role of self-provisioning as a source of flexibility in dryland agriculture. The new technologies, which emphasize modern input-responsive, high-yielding cultivars of major crops and precision in management practices, are also fairly insensitive to the diversity and flexibility that in the past helped sustain farming in the dry tropics.

Market forces and public institutions (such as extension services) also support high-intensity, modern-input-based farming over low-intensity diversified farming. Institutional interventions, particularly land distribution programs, have caused a rapid decline in the area of common-property lands and have helped push submarginal lands under the plow (Jodha 1986a). There is less concern for heterogeneity and diversified use of drylands. These issues (Jodha 1986b) are summarized in Table 5.

Population growth, market forces, imbalances in new technological options, and institutional interventions have jointly or separately helped to reduce diversity and flexibility in dryland agriculture. The net result is a reduced range of options through which the farmer can adjust to climatic variations. Other consequences are reduced competitiveness of dryland crops, particularly low-value coarse cereals like sorghum and millet, and an alarming rate of ecological degradation in dry tropics. Extrapolation of these tendencies raises questions about the long-term sustainability of farming in many parts of the dry tropics.

Possible way out

In the context of rising pressure on land and the dynamics of development, there is little scope for sustaining resource-extensive farming or for insulating dry tropics agriculture from the various processes of change. Under such circumstances, the best approach is to adapt the processes of change to the specificities and requirements of dryland agriculture, and thereby restoring its capacity to adjust to climatic variations. However, to do so, one will have to go beyond not only folk agronomy, but also the modern scientific measures directed toward production stability.

A few steps, elaborated elsewhere are indicated here (Jodha 1986b). They can be divided into four categories. The best mix has to be determined by the specificities of homogeneous tracts demarcated within the very heterogeneous dry tropics.

1. High-intensity, high-productivity, arable dry tropics farming. New high-yielding, input-responsive, short-duration cultivars, in association with new techniques of water harvesting (such as broad beds and furrows) are more relevant in the relatively high and dependable rainfall zones. The spillover effects should cushion the impact of drought years.
2. Higher priority to high-value crops (e.g. oilseeds, less perishable vegetables, and dryland horticulture). Because of their low value status, growth in productivity of the major dryland coarse cereals may not ensure income growth for the farmer. However, changes in currently low-value crops could make them suitable for alternative uses, including as processed foods and animal feeds.
3. Incorporate the rationale of folk agronomy into a new approach to stabilize dryland agriculture. The elements conducive to diversity, flexibility, and hence a range of multiple options include promotion of certain non-conventional crops for drylands; higher priority to minor crops, some of which fetch higher market prices than major coarse cereals; alternative land uses, especially for submarginal lands; and strengthening of the livestock support system.
4. Complement production or production technology with appropriate institutional measures. Market and price supports are required to sustain

Table 5. Indicators of decline of selected farm practices which helped adjustment to climatic variations.

Traditional farm practices underlying adjustment mechanisms	Factors and processes reducing the effectiveness of adjustment mechanisms[a]			
	Increased demographic pressure	Increased role of market forces	Technological interventions	Institutional interventions
Use of cultivars with Long duration, indeterminate character, high stalk-grain ratio, low but stable yield	Reduced feasibility of resource-extensive practices (A)	Input, output market largely geared to requirements of modern high-yielding cultivars (A, D)	Promotion of new high-yielding, short-duration, determinate-type cultivars suited to intensive farming (A, D)	Planning and policies conducive for resource-intensive farming systems (A, C)
Diversified cropping systems through Intercropping, crop combinations, emphasis on minor crops, varied planting arrangements, rotations/fallowing	Reduced feasibility of resource-extensive practices, diversity, etc. (A, C, D)	Push for narrow specialization, disregard of self-provisioning needs, discrimination against low-value crops, increased role of price risk (C, D)	Emphasis on precision, standardization, sole cropping, etc. reducing flexibility/diversity and range of options (C, D)	Emphasis through research and extension services, on standardized management practices and major crops, etc. (C, D)
Input management involving Organic recycling, self-provisioning, flexibility in input use	Reduced opportunity for organic recycling, flexibility, etc. (D)	Emphasis on market inputs reducing operational flexibility (D)	Emphasis on modern high-cost inputs and rigidity in rate and date of input use, disregard of organic farming (D)	Heavy emphasis on modern inputs for agricultural development (D)
Integrated use of heterogeneous environment and livestock support system through Common property land, integrated use of crop-tree-bush, conservative use of submarginal lands	Push for resource-intensive system by plowing on common and submarginal lands, disregarding diversity and use capability considerations (A, C)	Inducement to overuse of fragile resources (A)	Imbalances in research and development generating options less conducive to integrated, diversified use of resource base (A, C)	Insensitivity of land policies toward heterogeneity and fragility of lands and their use potential (A, C, D)

[a] Components of adjustment processes (Table 1) adversely affected by different factors are indicated as follows: A = long-term adaptation to unstable environment, B = adjusting environment to crops, C = diversity, D = flexibility.

production gains. Some protection against production and market risk through formal measures like crop insurance or regulated markets could be considered. Research and development facilities, in terms of a wider network of research to match the heterogeneity of the dry tropics and to induce research on processing technologies for alternative uses of dryland crops, are other major areas where institutional support is crucial. Infrastructural facilities and overall development of largely less developed dry tropical regions can integrate them with the rest of the economy. This can help in reducing pressure on the land and in withstanding the consequences of climatic variability.

References cited

Binswanger H P, Virmani S M, Kampen J (1980) Farming system components for selected areas in India; evidence from ICRISAT. Research Bulletin 2. International Crops Research Institute for the Semi-Arid Tropics, Patancheru, Andhra Pradesh, India.

Chambers R (1984) Rural development: putting the last first. Longman, London. 246 p.

Jodha N S (1967) Capital formation in arid agriculture: a study of resource conservation and reclamation measures applied to arid agriculture. Ph D thesis, University of Jodhpur, India.

Jodha N S (1978) Effectiveness of farmers' adjustment to risk. Econ. Polit. Weekly (Q. Rev. Agric.) 13:A38-A48.

Jodha N S (1980) Intercropping in traditional farming systems. J. Dev. Stud. 16:427-442.

Jodha N S (1986a) Common property resources and rural poor in dry regions of India. Econ. Polit. Weekly 21:1169-1181.

Jodha N S (1986b) Research and technology for dryland farming in India: some issues for future strategy. Indian J. Agric. Econ. 41:234-247.

Jodha N S, Asokan M, Ryan J G (1977) Village study methodology and resource endowments of the selected villages. Economics Program Occasional Paper 16. International Crops Research Institute for the Semi-Arid Tropics, Patancheru, Andhra Pradesh, India.

Jodha N S, Mascarenhas A C (1985) Adjustment in self-provisioning societies. Pages 437-468 *in* Climate impact assessment: studies of the interaction of climate and society. R. W. Kates, J. H. Ausubel, and M. Berberian, eds. John Wiley and Sons, New York.

Kates R W (1985) Interaction of climate and society. Pages 3-36 *in* Climate impact assessment. R. W. Kates, J. H. Ausubel, and M. Berberian, eds. John Wiley and Sons, New York.

Keig G, McAlpine J R (1977) A computer system for the estimation and analysis of soil moisture regimes from simple climatic data. 2d ed. Techmemo, Division of Land Research, Commonwealth Scientific and Industrial Research Organization, Melbourne, Australia.

Singh R P, Walker T S (1982) Determinants and implications of crop failure in the semi-arid tropics of India. Economics Program Progress Report 40. International Crops Research Institute for the Semi-Arid Tropics, Patancheru, Andhra Pradesh, India.

Spitz P (1980) Drought and self-provisioning. Pages 125-147 *in* Climatic constraints and human activities. J. Ausubel and A. K. Biswas, eds. Pergamon Press, Oxford.

Walker T S, Jodha N S (1986) How small farm households adapt to risk. Pages 17-34 *in* Crop insurance for agricultural development: issues and experiences. P. Hazell, C. Pomareda, and A. Valdes, eds. The Johns Hopkins University Press, Baltimore.

Walker T S, Singh R P, Asokan M, Binswanger H P (1983) Fluctuations in income in three villages of peninsular India. Economics Program Progress Report 57. International Crops Research Institute for the Semi-Arid Tropics, Patancheru, Andhra Pradesh, India.

Notes

Author's address: N. S. Jodha, International Crops Research Institute for the Semi-Arid Tropics, Patancheru 502 324, Andhra Pradesh, India. (Present address: Farming Systems Division, International Centre for Integrated Mountain Development [ICIMOD], GPO Box 3226, Kathmandu, Nepal)

Citation information: International Rice Research Institute (1989) Climate and food security. P.O. Box 933, Manila, Philippines.

Global climatic change, water resources, and food security

P. H. Gleick

Alterations in climate due to increasing atmospheric concentrations of carbon dioxide and other trace gases must be viewed with concern, if not alarm, because of the risk of adverse impacts on the availability and quality of water resources. Hydrologic changes may lead to alterations in the pressure on food resources in many regions of the world. This paper explores the mechanisms by which food security—defined here as the ability of a region to produce or otherwise provide food without threatening the well-being of either the population or the existing political system—could be affected by climate-induced changes in water resources. Shifts in the timing, magnitude, or location of major hydrologic phenomena, such as runoff, soil moisture, or monsoons, are discussed, together with the important vulnerabilities of agricultural productivity to changes in water availability. Some of the concomitant political and socioeconomic implications are also presented, with suggestions for ways to enhance the reliability of food resources dependent on uncertain water supplies.

Widespread improvements in agricultural productivity have been achieved over the last century using a wide range of technological advances. Future improvements, however, are likely to be constrained by the lower quality of new lands brought into production, by growing limitations on capital for crop expansion and mechanization, and by increasing population pressures. On top of these constraints are new uncertainties about future climatic conditions and the effect of anthropogenic climatic changes on water availability. To better understand some of the impacts of climatic change on food security, plausible changes in water supply and their possible effects on food production are explored. The cases discussed here include increases and decreases in both the average and the variability of water-resource availability.

Of all human activities, agriculture is considered the most sensitive to climatic conditions and to climatic variability. History suggests that climatic fluctuations of sufficient severity and extent to affect agricultural production cause pressures not only on the economy and society of the regions in which they occur, but also on a wide range of political and security considerations (Glantz and Wigley 1986, Gustafson 1981, Rosenberg 1981). This makes it important to study evolving climatic changes that alter the productivity of agriculture—either for better or for worse.

Hydrologic conditions play a major role in agricultural productivity—too little or too much water at critical times can reduce yields, damage soils, or cause crop failures. In many regions of the world, water availability is a more important factor in limiting reliable crop production than is temperature. Because of the value of information on the availability and reliability of water resources, the plausible effects of future climatic changes on water supply must be explored before major new water resource programs are formulated or extensive new agricultural developments planned.

Unfortunately, information on future changes in water resources often is either unavailable or unreliable, because of the complexities of global and regional climate systems and the difficulties of modeling such systems. The possibilities developed here are only a subset of the many climate outcomes possible. Nevertheless, these cases represent a wide range of possible changes in water resource conditions. Their likelihood, plausibility, and effects should be studied. As more information on actual climatic change becomes available, more accurate and detailed scenarios can be refined and studied.

Water availability and food security

Agricultural production in any given region is highly dependent on the climatological and geophysical conditions in that region (precipitation patterns, soil conditions, temperature regimes, etc.). Although remarkable advances have been made that reduce the vulnerability of crops to natural climatic fluctuations, we are still reminded regularly of our sensitivity to extreme climatic events, even those of relatively short duration. Although long-term changes in hydroclimatic variables (such as precipitation and evapotranspiration) are not, in themselves, necessarily a cause for alarm, there are a number of reasons why such changes should be viewed with concern.

In many areas of the world, particularly semiarid and subhumid regions, the most important limitation to increased agricultural production is limited water availability during certain parts of the year. Yet these are the regions whose production must meet growing agricultural demands during the next several decades. Table 1 lists regions where crop production might be expanded, were it not for limitations on water availability in certain forms or at certain times. In these regions, there is often little or no access to irrigation water, either from groundwater resources or through the transfer of water from water-rich to water-poor regions. Agricultural development in these regions is, therefore, dependent primarily on the prevailing patterns of rainfall. Changes in hydroclimatic conditions may ease or exacerbate the availability of water in these regions.

Even in areas with irrigation or groundwater resources, total water demands vary with evapotranspiration and with competing demands from other sectors of society. In times of water shortages, agricultural productivity may be adversely affected. If climatic conditions were to permanently reduce freshwater availability or increase overall demand, some regions that are now productive might have to be taken out of production.

Table 1. Water-limited regions with agricultural potential.[a]

Region	Water/Climate limitation
Southwest and south-central United States	Net water quantity Soil conditions
The Great Plains of Canada	Length of growing season Soil conditions
The Steppes of the Soviet Union	Net water quantity Length of growing season Soil conditions
Parts of India, Australia, Southern Africa, Mexico	Seasonality of water availability Net water quality Soil conditions
Northeastern Brazil	Seasonality of water availability Soil conditions
Coastal South America	Seasonality of water availability Soil conditions
Dry savannas of Africa	Net water quantity Soil conditions

[a]Major semiarid and subhumid regions with potential for expansion of crop production, given appropriate hydrologic conditions. Expansion of crop production in these regions depends on a wide range of socioeconomic and political conditions.

Similarly, regions dependent on seasonal water supplies may experience a shift in the timing of the onset of the rainy season, or a shortening or lengthening of that season. In India, for example, rainfall and runoff are concentrated within a few months of the monsoon season. Any delay in the onset of the monsoon can have severe consequences for agriculture. Because of the existing variability of monsoons on the Indian subcontinent, India has taken action to improve the reliability of its food reserves, to protect them from 2 yr of inadequate monsoons (Slater 1981). A change in climatic variability would lead to the need to reevaluate the management or design of such a program.

Future hydrologic changes

Much of what we know about possible changes in water availability due to future anthropogenic climatic change comes from work using complex, three-dimensional general circulation climate models (GCMs). General circulation models incorporate hydrologic parameters with varying degrees of complexity. Details on these parameterizations can be found in Hansen et al (1983), Manabe (1969a,b), Manabe et al (1981), Schlesinger and Gates (1980), and Washington and Meehl (1983, 1984). GCMs attempt to reproduce climate dynamics as accurately as possible, given the limitations on our ability to mathematically describe complex meteorological phenomenon and our ability to develop adequate global-scale data on oceanic, atmospheric, and terrestrial variables. The greatest limitations to using GCMs to assess changes in water availability are their coarse resolution and their simplified hydrologic parameterizations (Gleick 1986).

Because of these limitations, research to develop other methods for evaluating the hydrologic implications of climatic change has increased (Beran 1986; Gleick 1986, 1987a; Nemec and Schaake 1982; Revelle and Waggoner 1983; Schwarz 1977; Stockton and Boggess 1979; EPA 1984). Although accurate predictions of future regional hydrologic changes cannot yet be made, these authors identify a number of important hydrologic sensitivities. Substantial changes in regional hydrologic normals and moments may result from the types of temperature and precipitation changes anticipated by state-of-the-art GCMs. To make reasonable estimates of these hydrologic changes, certain types of information would be needed from GCMs and regional models (Table 2).

When GCMs are perturbed by artificially doubling the atmospheric concentration of carbon dioxide (CO_2), the models predict global warming. The warming is often expressed as an average increase of 3.0 °C, with a 95% probability that it will be between 1.5 and 4.5 °C (Dickinson 1984). Such a warming would be accompanied

Table 2. Hydrologic effects of climatic change.

Hydrologic variable of interest	
Useful precipitation	
Surface runoff	
Available soil moisture	
Groundwater	
Temperature	
Monsoonality (onset, ending, intensity, location)	
Storm events	
Temporal scale of interest	
Long-term (greater than annual)	
Annual	
Seasonal	
Monthly	
Daily	
Spatial scale of interest (political)	
Global	10^8 km^2
Continental	10^7 km^2
Country/Region	10^6 km^2
Local	10^3-10^5 km^2
Spatial scale of interest (hydrologic)	
Global	10^8 km^2
Continental	10^7 km^2
Regional	10^5-10^6 km^2
Watershed	10^2-10^5 km^2
Statistical scale of interest	
Mean	
Variance	
Persistence	
Skew	
Higher moments	
Hydrologic impact of interest	
Quantity	
Quality	
Peak events (high and low)	

by an increase in the average evapotranspiration rate and an overall intensification of the global hydrologic cycle.

A more intense global hydrologic cycle will result in both increases and decreases in regional precipitation rates. These meteorological changes may be accompanied by changes in the seasonality and variability of precipitation, which in turn would affect seasonal water shortages or excesses. Even though global annual average precipitation is generally expected to increase as a result of doubled concentrations of atmospheric CO_2, there is concern that the additional water is less than the total that would be evaporated by the increase in temperature. Mather and Feddema (1986) studied 12 regions around the world; none showed a shift to wetter soil conditions after an increase in atmospheric CO_2 concentration, and seven showed significantly drier conditions. The water deficit (the difference between the water available from precipitation and lost to evapotranspiration) has been shown to be closely related to agricultural yields (Mather and Feddema 1986).

In another example, Mediterranean-style climates, with distinct rainy and dry seasons, may experience intensifications of both, with concomitant problems of flooding and soil moisture deficits (Gleick 1987b). Soil moisture deficits also may be exacerbated during summer months in midcontinental regions (Manabe et al 1981, Manabe and Wetherald 1986, Mitchell 1983). Although the spatial resolution of GCMs is poor, the physical mechanisms for such drying appear plausible: a faster snowmelt during spring months (Gleick 1987b, Manabe and Wetherald 1986), a shift in the midlatitude rainbelt (Manabe et al 1981, Manabe and Wetherald 1986), a decrease in the ratio of snow to rain in winter precipitation due to increased average winter temperatures (Gleick 1987b), and a decrease in the overall transfer rates for latent heat.

Each of these hydrologic effects is uncertain. Rather than attempting to squeeze more information out of GCMs and regional hydrologic models than can safely be derived from them, a set of generalized cases will permit us to speculate on the nature and likelihood of beneficial or detrimental changes in water availability.

Implications of change in water availability

For agricultural purposes, the climate of a region can be defined in many ways. Here we explore how food security might be affected by changes in climate that alter the availability of water, either through changes in soil moisture available to plants or in changes in the amount of groundwater or surface runoff that can be applied as irrigation. Food security is defined as the ability of a region to produce or otherwise provide food without threatening the well-being of either the population or the existing political system.

Although many hydrologic variables could be explored, the number of cases expands very quickly with the number of variables and the number of degrees of freedom (x^n, X = degrees of freedom, n = number of variables). To assist water resource planners, we limit our efforts to those cases with either significant impacts or high probabilities. Less critical cases would be those that are interesting, easier to mitigate, or harder to distinguish from existing hydrologic noise.

Four cases involving changes in the mean and the frequency of water availability are explored here (Table 3). The cases chosen represent a set of hydrologic changes that are important because they may generate impacts beyond our ability to easily adapt. Each of these cases is, if not more probable than any other, at least not improbable. They would affect many temporal and spatial scales described in Table 2. These examples assume no drastic changes in the patterns of global or regional water transfers and management. Although some such changes are possible—indeed many are desired, even today—such changes typically come in response to changes in the characteristics of the geophysical or socioeconomic environment (Golubev and Biswas 1986).

For each case, the change in water availability is compared to a base case of theoretically unperturbed availability. The base case, plotted in Figure 1, represents a hypothetical time series of water availability—runoff, groundwater, soil moisture, or some similar measure. Particular attention is paid to the frequency and severity of water shortages and excesses, identified here as the number and intensity of droughts and floods (defined simply as water availability below or above a specified level).

Table 3. Changes in the mean and variability of water availability: four plausible cases.

	Mean	*Variability*
Base:	Historical	Historical
Case 1:	Increase	Increase
Case 2:	Increase	Decrease
Case 3:	Decrease	Increase
Case 4:	Decrease	Decrease
Forms such changes could take[a]		
Case 1.	Increased precipitation, decreased evapotranspiration, increased storm frequency, increased storm severity, decreased drought severity, increased drought frequency, increased flooding, increased overall or seasonal runoff and soil moisture.	
Case 2.	Increased precipitation, decreased evapotranspiration, decreased storm frequency, increased storm severity, decreased drought severity, decreased drought frequency	
Case 3.	Decreased precipitation, increased evapotranspiration, increased storm frequency, decreased storm severity, increased drought severity, increased drought frequency	
Case 4.	Decreased precipitation, increased evapotranspiration, decreased storm frequency, decreased storm severity, increased drought severity, decreased drought frequency, decreased flooding, decreased overall or seasonal runoff and soil moisture	

[a]Water availability may take a number of forms, including available surface runoff, available groundwater, and available soil moisture. Changes in each of these variables must be evaluated. These impacts could appear singly or in combination. The list is not exhaustive, given the wide range of possible hydrologic effects.

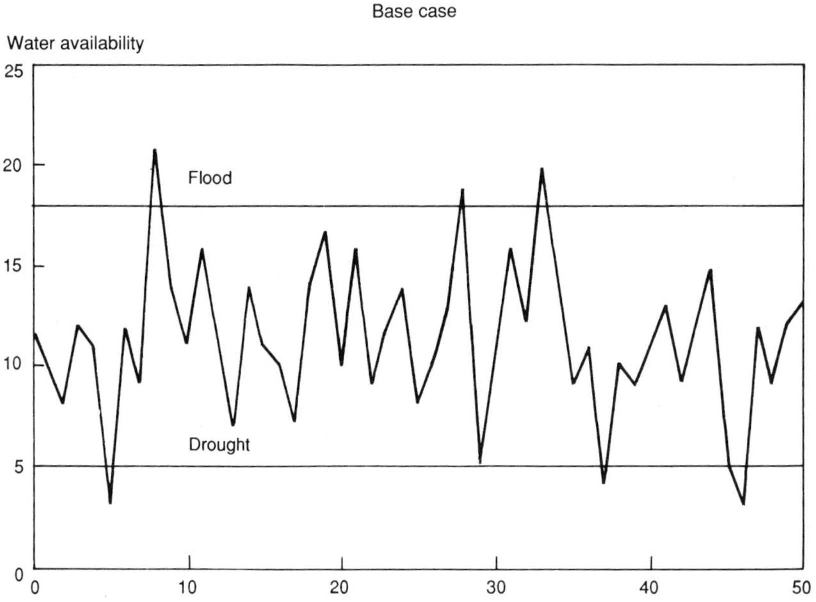

1. Water availability for a hypothetical watershed. Length of record is 50 time periods. During the time period, 3 droughts and 3 floods (defined as water availability below or above a specified level) occurred.

This hypothetical scenario of 50 yr or seasons has three instances of drought and three instances of floods.

Case 1. Increased mean and increased variability

An increase in both the mean and the variability of water resources can occur in a number of ways—increased precipitation rates, decreased evapotranspiration rates, increased storm frequencies and intensities, decreased drought severity, increased drought frequency, increased flooding, increased overall or seasonal runoff and soil moisture. Several of these responses, such as an increase in both precipitation rates and in seasonal runoff and soil moisture, are favorable for agricultural productivity in marginal, water-short regions.

Unfortunately, responses to these hydrologic changes in productive agricultural regions may not all be positive. In particular, increases in storm frequency and intensities will mean an increase in the frequency of large climatic events, such as severe drought, large floods, and high precipitation rates during water-sensitive periods of the crop cycle. Figure 2 plots the new water availability in an existing record following an increase in both the average and the variability of water availability. As the figure shows, the frequency and severity of flooding increased dramatically, from three to seven events. The frequency of drought also increased. While an increase in mean water availability could be beneficial to water-short regions, each of these extreme events can have adverse effects on crop productivity.

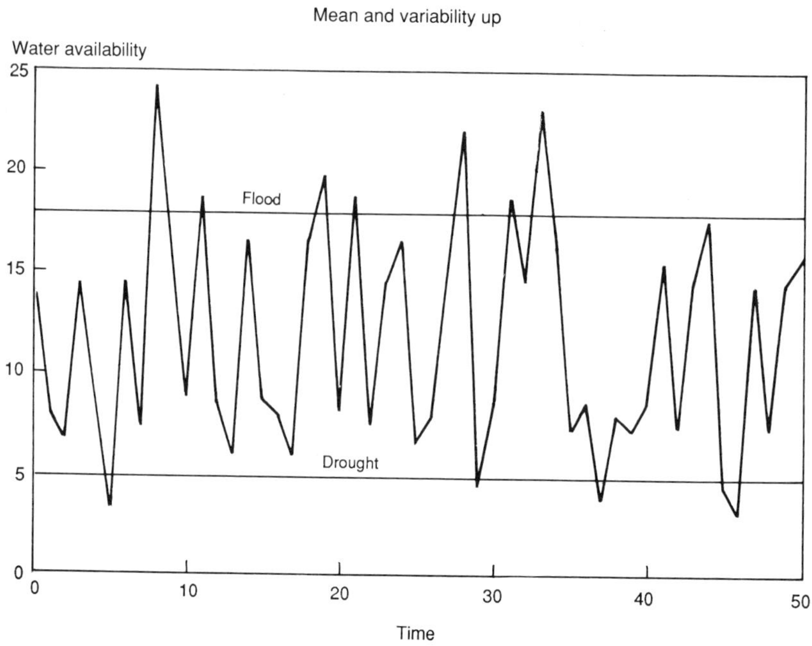

2. Mean and variability of water availability in the hypothetical watershed have increased. Note the increase in severity and frequency of flooding.

Case 2. Increased mean and decreased variability

Some adverse impacts of increased storm frequency, increased drought frequency, etc., of case 1 are mitigated with reduced variability. In particular, this suggests the possibility of increased precipitation rates, decreased evapotranspiration rates, decreased storm frequencies, increased storm severity, decreased drought severity, and decreased drought frequency. Although an increase in storm severity is a possibly adverse outcome from an overall increase in mean, the hydrologic advantages from a decrease in drought frequency and severity could be significant. Figure 3 shows a considerable reduction in both frequency and severity of droughts and an increase in intensity of flooding events, compared to Figure 1.

The case appears to offer the greatest overall advantages for water-limited regions. A note of caution, however: it has been observed that increases in annual average precipitation may not result in either increased soil moisture availability or increased runoff during the parts of the year when water is most needed (see, for example, Gleick 1987b).

Case 3. Decreased mean and increased variability

This example is at the opposite end of the spectrum. A decrease in mean water availability accompanied by an increase in overall variability is perhaps the worst scenario for water-limited regions. Such an outcome could include decreased precipitation rates, increased evapotranspiration rates, increased storm frequencies, increased or decreased storm severity, increased drought severity, and increased

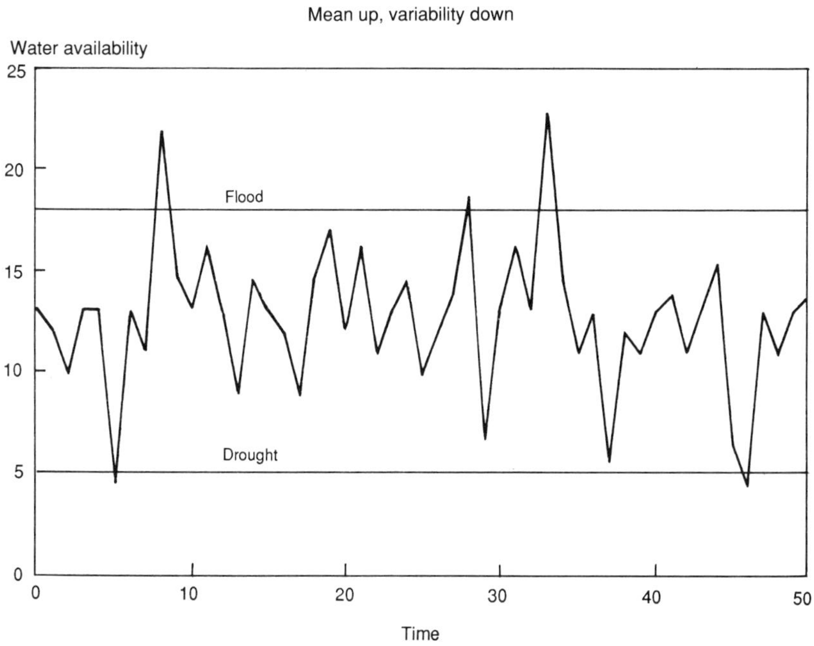

3. Mean water availability has increased, while variability has decreased. During this period, severity and frequency of drought have decreased significantly. Flooding frequency has not changed, although the severity may increase.

drought frequency. With the exception of a possible decrease in the severity of storms, these hydrologic changes would not permit much (if any) increase in food production on marginal lands. At the same time, they could lead to greater pressures on existing production by increasing both the frequency and severity of drought. Figure 4 shows the effects of reduced mean water availability and increased variability on both drought and flood. The number of droughts in this hypothetical record increased by more than a factor of two, and the severity of the droughts also increased. Similarly, the frequency of high runoff (or other net water availability measures) has increased.

Case 4. Decreased mean and decreased variability

This situation might offer some relief to flood-prone areas, but at the cost of adversely affecting other sensitive hydrologic regions, particularly water-short regions. Decreases in the mean and variability of water availability might take the form of decreased precipitation, increased evapotranspiration, decreased storm frequency, decreased storm severity, increased drought severity, decreased drought frequency, decreased flooding, and decreased overall or seasonal runoff and soil moisture. Water-short regions could suffer increased drought severity (due to the decreased mean availability), but fewer droughts—a mixed blessing. Similarly, decreases in mean soil moisture or runoff would decrease crop productivity in many regions; a decrease in storm frequencies and intensities would be beneficial. Figure 5

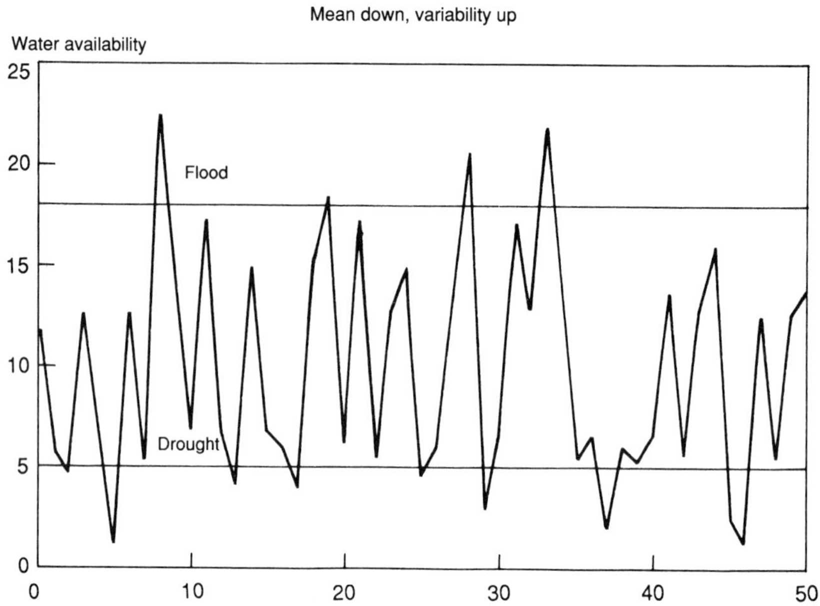

4. Mean water availability decreases and variability increases. The frequency and severity of drought have increased dramatically, the frequency of floods has increased slightly because of the increased variability.

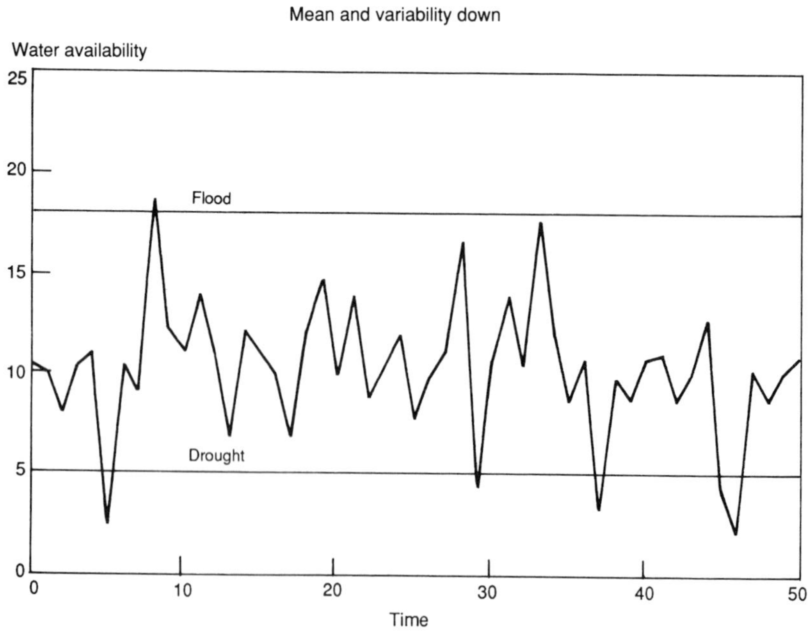

5. Mean and variability of water availability decrease. Note the decrease in flooding frequency and severity and increased frequency of drought due to a decrease in the mean availability of water.

shows an increase in drought severity and a decrease in both the frequency and severity of floods.

Effects on agricultural production

Each of the cases described could have positive or negative consequences for agriculture. For the net agricultural effects of hydrologic changes to be entirely (or almost entirely) beneficial to agricultural production, very specific spatial and temporal distributions of the hydroclimatic changes would be required. The region-specific changes that would be needed to improve the reliability of agricultural production are

- Increased annual average net precipitation (available precipitation after evapotranspiration losses) in arid, semiarid, and subhumid areas;
- Increased seasonal average net precipitation in arid, semiarid, and subhumid areas;
- Decreased or constant annual average net precipitation in humid areas subject to flooding;
- Decreased or constant seasonal average net precipitation in humid areas subject to flooding;
- Decreased variability of water availability in flood-prone agricultural areas;
- Decreased variability of the onset, ending, duration, and location of monsoonality;
- Decreased frequency and persistence of droughts in arid, semiarid, and subhumid areas;
- Increased useful seasonal runoff (runoff available during periods of soil moisture deficit) in arid, semiarid, and semihumid regions;
- Increased mean soil moisture availability in regions with net soil moisture deficits during sensitive growing periods; and
- Decreased mean soil moisture availability in regions with net soil moisture excesses during sensitive growing seasons.

As even this incomplete list shows, many of these changes would be mutually contradictory or highly improbable (intensification of the overall hydrologic system, decreases in the net precipitation in flood-prone regions, increases in net precipitation in subhumid regions).

Another factor is rarely considered. Many of the changes that might be good for agricultural productivity in the long term might be bad for natural plant and animal communities, or for flood control and municipal water supply. Such contradictory outcomes must be considered in assessing possible climatic impacts.

Conclusion

Given the stochastic nature of many hydrologic variables, it is reasonable to expect distinct agricultural winners and losers, even if no readily apparent hydrologic trend is yet evident. Who wins and who loses may well depend on the ability of different regions to adapt to altered hydrologic conditions, which in turn may depend on the flexibility and resilience of existing water resource institutions and facilities.

Those regions where agricultural productivity is already low because of hydrologic limitations may be especially sensitive to such climatic changes as length of growing seasons or availability of moisture in the soil. Even regions in which climate or soils do not limit agricultural productivity would be affected by changes in water availability. In those regions, even slight decreases in overall yields would have important ramifications for world food supplies, if such changes resulted in large changes in the availability of foods on the world market.

The tendency for water resource planners to develop projects for transferring water from water-rich to water-poor regions, or to construct additional reservoir, flood control, or irrigation facilities, will continue. In both developed and developing regions of the world, major new hydrologic facilities carry high economic and environmental costs. Reliable information on the distribution in space and time of rainfall and runoff in a river basin is essential for designing irrigation, drainage, and flood-control systems, as well as for operating and managing these systems (Revelle 1981). Yet, as the four cases discussed here demonstrate, climatic changes are likely to replace the known statistical behavior of existing hydrologic basins with unpredicted or unpredictable behavior.

Uncertainties about future climatic conditions will increase the costs of new hydrologic facilities by increasing the range of conditions for which such facilities will have to be designed. It would be premature to plan and build such facilities if simple institutional, organizational, or economic adjustments can make existing water supply systems more flexible and reliable.

With no changes in the existing agricultural structure (i.e. no changes in irrigation practices, no water conservation programs, no changes in crop genetics, etc.), only slight changes in the timing or magnitude of water availability would alter the productivity or the quality of important crops in most regions. Some reductions in water availability could be withstood if certain structural changes are made. The changes could help reduce the vulnerability of society or segments of society to climatic events, despite the fact that such vulnerability may increase as populations increase and more marginal lands are brought into production, and as costs for agricultural improvements rise.

References cited

Beran M (1986) The water resource impact of future climate change and variability. Paper presented at the U.S. Environmental Protection Agency International Conference on Health and Environmental Effects of Ozone Modification and Climatic Change, 16-20 Jun 1986, Arlington, Virginia.

Dickinson R E (1984) Modeling evapotranspiration for three-dimensional global climate models. Pages 58-72 *in* Climate processes and climate sensitivity. Am. Geophys. Union Monogr. 29. J. E. Hansen and T. Takahashi, eds. American Geophysical Union,

EPA—United States Environmental Protection Agency (1984) Potential climatic impacts of increasing atmospheric CO_2 with emphasis on water availability and hydrology in the United States. Strategic Studies Staff. Office of Policy Analysis. Office of Policy, Planning and Evaluation, Washington, D.C.

Glantz M H, Wigley T M L (1986) Climatic variations and their effects on water resources. *In* Resources and world development. McLaren and Skinner, eds. Springer-Verlag, Berlin.

Gleick P H (1986) Methods for evaluating the regional hydrologic impacts of global climatic changes. J. Hydrol. 88:97-116.

Gleick P H (1987a) The development and testing of a water-balance model for climate impact assessment: modeling the Sacramento Basin. Water Resour. Res. 23(6):1049-1061.

Gleick P H (1987b) Regional hydrologic consequences of increases in atmospheric CO_2 and other trace gases. Climatic Change 10(2):137-160.

Golubev G N, Biswas A K (1986) Large scale water transfers: emerging environmental and social experiences. United Nations Environmental Programme. Tycooly Publishing Limited, Oxford, England.

Gustafson T (1981) Reform in Soviet politics: lessons of recent policies on land and water. Cambridge University Press, Cambridge. 218 p.

Hansen J, Russell G, Rind D, Stone P, Lacis A, Lebedeff S, Ruedy R, Travis L (1983) Efficient three-dimensional global models for climate studies: models I and II. Mon. Weather Rev. 3(4):609-662.

Manabe S (1969a) Climate and the ocean circulation. I. The atmospheric circulation and the hydrology of the earth's surface. Mon. Weather Rev. 97(11):739-774.

Manabe S (1969b) Climate and the ocean circulation. II. The atmospheric circulation and the effect of heat transfer by ocean currents. Mon. Weather Rev. 97(11):775-805.

Manabe S, Wetherald R T (1986) Reduction in summer soil wetness induced by an increase in atmospheric carbon dioxide. Science 232:626-628.

Manabe S, Wetherald R T, Stouffer R J (1981) Summer dryness due to an increase of atmospheric CO_2 concentration. Climatic Change 3:347-386.

Mather J, Feddema J (1986) Hydrologic consequences of increases in trace gases and CO_2 in the atmosphere. Paper presented at the International Conference on Health and Environmental Effects of Ozone Modification and Climatic Change, 16-20 Jun 1986, Arlington, Virginia.

Mitchell J M Jr. (1983) The seasonal response of a general circulation model to changes in CO_2 and sea temperature. Q. J. Roy. Meteorol. Soc. 109:113-152.

Nemec J, Schaake J (1982) Sensitivity of water resource systems to climate variation. Hydrol. Sci. 27(3):327-343.

Revelle R R (1981) Uses of climatic knowledge in the food systems of developing countries. Pages 39-44 *in* Climate's impact on food supplies: strategies and technologies for climate-defensive food production. AAAS selected symposium 62. L. E. Slater and S. K. Levin, eds. Westview Press, Boulder, Colorado.

Revelle R R, Waggoner P E (1983) Effects of a carbon dioxide-induced climatic change on water supplies in the western United States. *In* Changing climate. National Academy of Sciences, National Academy Press, Washington, D.C.

Rosenberg N J (1981) Technologies and strategies in weather-proofing crop production. Pages 157-180 *in* Climate's impact on food supplies: strategies and technologies for climate-defensive food production. AAAS selected symposium 62. L. E. Slater and S. K. Levin, eds. Westview Press, Boulder, Colorado.

Schlesinger M E, Gates W L (1980) The January and July performance of the OSU two-level atmospheric general circulation model. J. Atmos. Sci. 37:1914-1943.

Schwarz H E (1977) Climatic change and water supply: how sensitive is the Northeast? *In* Climate, climatic change, and water supply. National Academy of Sciences, Washington, D.C.

Slater L E (1981) Three approaches to reducing climate's impact on food supplies. Pages 99-130 *in* Climate's impact on food supplies: strategies and technologies for climate-defensive food production. AAAS selected symposium 62. L. E. Slater and S. K. Levin, eds. Westview Press, Boulder, Colorado.

Stockton C W, Boggess W R (1979) Geohydrological implications of climate change on water resource development. U.S. Army Coastal Engineering Research Center, Fort Belvoir, Virginia.

Washington W M, Meehl G A (1983) General circulation model experiments on the climatic effects due to a doubling and quadrupling of carbon dioxide concentration. J. Geophys. Res. 88(C11):6600-6610.

Washington W M, Meehl G A (1984) Seasonal cycle experiment on the climate sensitivity due to a doubling of CO_2 with an atmospheric general circulation model coupled to a simple mixed-layer ocean model. J. Geophys. Res. 89(D6):9475-9503.

Notes

Author's address: P. H. Gleick, Energy and Resources Group, University of California, Berkeley, California (Present address: Pacific Institute for Studies in Development, and Security, 1681 Shattuck Ave., Suite H, Berkeley, CA 94709, USA). This work was supported by a Social Science Research Council-MacArthur Foundation grant.
Citation information: International Rice Research Institute (1989) Climate and food security. P.O. Box 933, Manila, Philippines.

The effect of climatic variations on food production

M. L. Parry and T. R. Carter

Between 1983 and 1987, the International Institute for Applied Systems Analysis and the United Nations Environment Programme co-funded studies on the effects of climatic variations on agriculture. About 80 scientists and planners in 11 case study regions participated. Studies were undertaken in two types of climatic-sensitive regions: in high latitude, cool regions (studies in Saskatchewan, Iceland, Finland, northern USSR, and Japan) and in semiarid regions (studies in the central sierra of Ecuador, northeast Brazil, Kenya, southern USSR, central India, and Australia). Some of the results of that work are summarized in this paper.

Methodology

The research strategy

Studies of impact and interaction. A distinction has been drawn between two approaches to the study of the effect of climatic variations on agriculture: impact approaches and interaction approaches (Kates 1985).

An impact approach is based on an assumption of direct cause and effect, where a climatic event (e.g. a drought) operating on a given exposure unit (e.g. agriculture) may have an impact (Fig. 1a). In reality, of course, so many intervening factors operate that it is misleading to treat these three elements (drought-activity-impact) in isolation from their environmental and societal milieu. Very few studies (if any) have followed this method.

An interaction approach assumes, first, that a particular climatic event is merely one of many processes (both societal and environmental in origin) that may affect the exposure unit and, second, that the impact is not separate from the exposure unit (as suggested in Fig. 1a) but is one of the many processes that constitute it (Fig. 1b).

Types of interaction models. Interaction models achieve greater degrees of realism by considering the cascade of impacts through physical and social systems as orders of interactions (Fig. 2a). The cascade of effects is traced as it works its way through a hierarchy of scales (from local to global) and through a network of systems (agricultural, economic, social, political).

Additional complexity may be introduced by studying interactions of the same order between different sectors. Figure 2b shows interactions between the concurrent effects of drought on agriculture, forestry, water resources, and transportation.

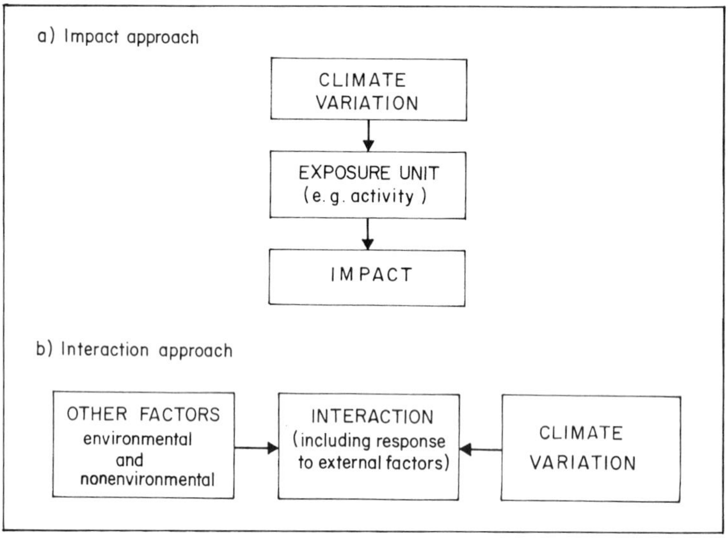

1. Schema of simple impact and interaction approaches in climate impact assessment (Parry and Carter 1988).

Only recently have interactions between climatic variations themselves and other physical systems received attention (Fig. 3). Two sets of such interactions can be distinguished:

- Those in which the effects of the climatic variation are transmitted through other physical systems (e.g. by pests and diseases or by changes in soil structure, soil nutrients, soil erosion or salinization).
- Those in which the effects of the climatic variation are themselves affected by other concurrent environmental trends (such as acid deposition and groundwater depletion).

Finally, we can introduce greater complexity into our conception of the interactions between climate and society by considering different types of response. Figure 3 distinguishes between two orders of response: adjustments at the enterprise level, which might include changes of crops, increased irrigation, changes in fertilization, etc., and policy responses at the regional, national, and international level. This schema broadly represents the conceptual framework adopted in the International Institute for Applied Systems Analysis/United Nations Environment Programme (IIASA/UNEP) project (Parry et al 1988).

An integrated approach to climate impact assessments

One advantage of distinguishing between types of explanation while considering the complexity of interactions is that it discourages acceptance of causal chains, which while conceptually tidy, massively oversimplify processes of climate impact. But some form of simplification is necessary to make sense of the enormous number and complexity of pathways by which the effects of climatic variations are passed through biophysical, economic, and social systems. One aid to comprehension is a

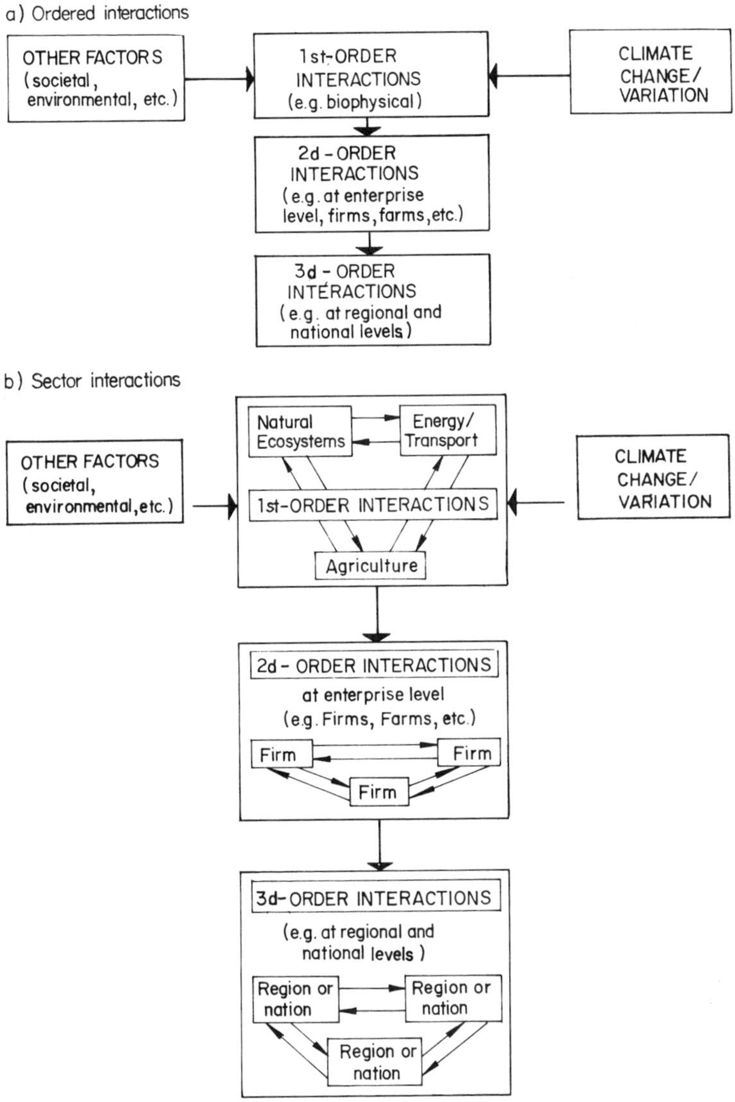

2. Schema of interaction approach in climate impact assessment with ordered interactions, and ordered interactions and interactions between sectors at each level (Parry and Carter 1988).

hierarchy of models (Fig. 4), to conceptualize the linkages between the various systems and enable an integrated assessment of climate impacts. Such hierarchies of models normally include

- Models of climatic variation (based on outputs from global climate models and/or analysis of the instrumental record).
- Biophysical models of first-order relationships—those between certain climatic variables (e.g. temperature, precipitation, insolation, wind speed) and biophysical supply or demand (e.g. biomass productivity, energy demand). In

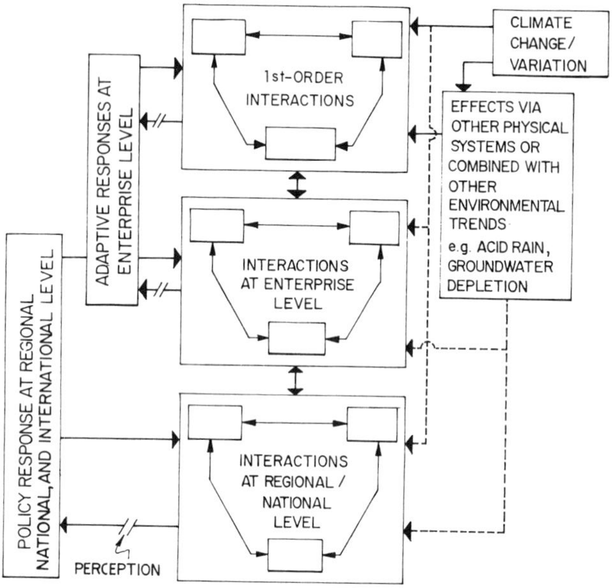

3. Schema of the IIASA/UNEP project study method: an interactive approach to climate impact assessment with ordered interactions, interactions at each level, and social and physical feedbacks (Parry and Carter 1988).

agriculture, these would consider effects on crop growth, livestock productivity, health, etc.

- Economic models of second-order relationships—those at the enterprise level (e.g. farms, firms, agencies, institutions). In agriculture, these would consider the effects (*inter alia*) of changes in farm-level production on regional and national output, etc.
- Economic, social, and political models of third-order interactions at the regional, national, or international level.

This approach yields the following types of integrated assessments: biophysical, enterprise-level, sectoral, and regional.

Types of analytical methods

The direct method. The scientific method most frequently adopted in climate impact assessment is the direct method. In this method, for example, the effects of a change in an input variable (such as a change in precipitation) are traced, in a number of steps, along a number of pathways (e.g. precipitation - crop biomass productivity - forage level - carrying capacity - livestock production - meat and milk supply) (Fig. 5a). The question is thus posed: "For a given climate change, what is the impact on, for example, ecosystems, economy, and society?" An advantage of this approach is that assessments can be made even if the number of climatic scenarios is restricted (as is the case with projections of possible future climates). The analysis is conducted on the basis of the character of the climatic changes rather than on their likely impacts.

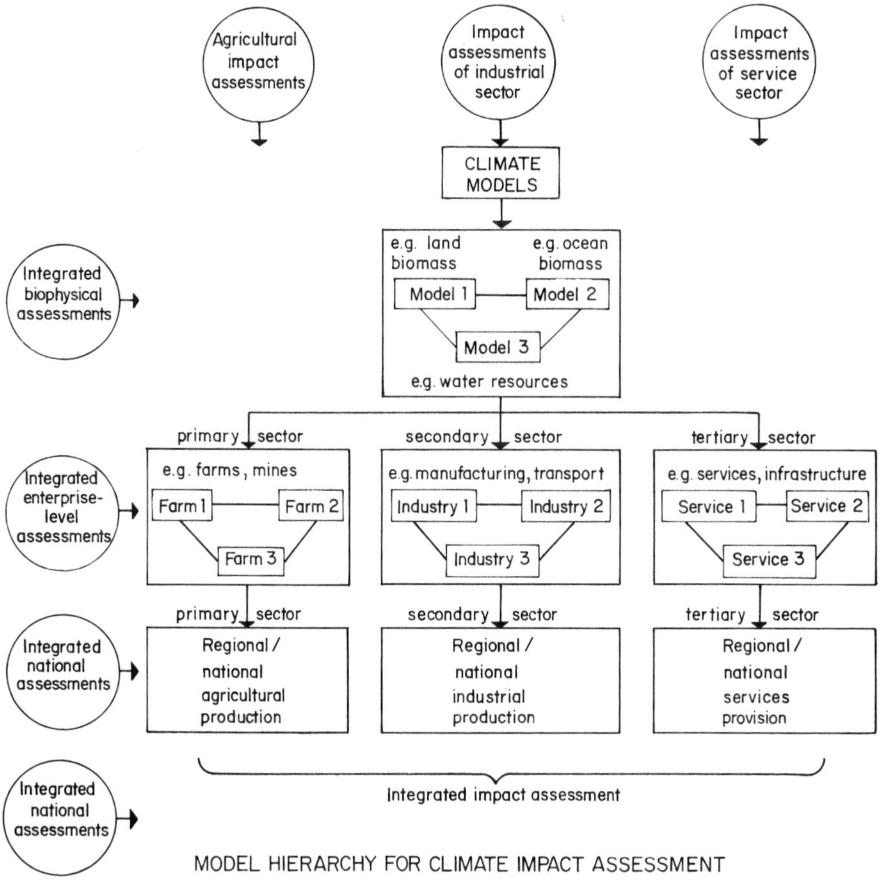

4. A hierarchy of models for integrated climate impact assessment (Parry and Carter 1988).

The adjoint method. An alternative or adjoint method (Parry and Carter 1984) focuses first on the sensitivity of the exposure unit. It addresses the following questions:

1. To what aspects of climate is the exposure unit especially sensitive?
2. What changes in these aspects are required to perturb the exposure unit significantly?

Climatic scenarios can be characterized partly on the basis of these detected sensitivities and partly along lines adopted in the direct approach. The steps are illustrated in Figure 5b. An advantage of this approach is that it can help identify sensitivities independently of state-of-the-art climatic scenarios and can allow climate changes to be expressed in the form that has direct meaning for the exposure unit. In Saskatchewan, for example, recent work has characterized climate impacts on agriculture in terms of the frequency of days of blowing dust. Previous sensitivity studies of agriculture had identified such events as critical to the sustainability of cereal production in the region (Williams et al 1988).

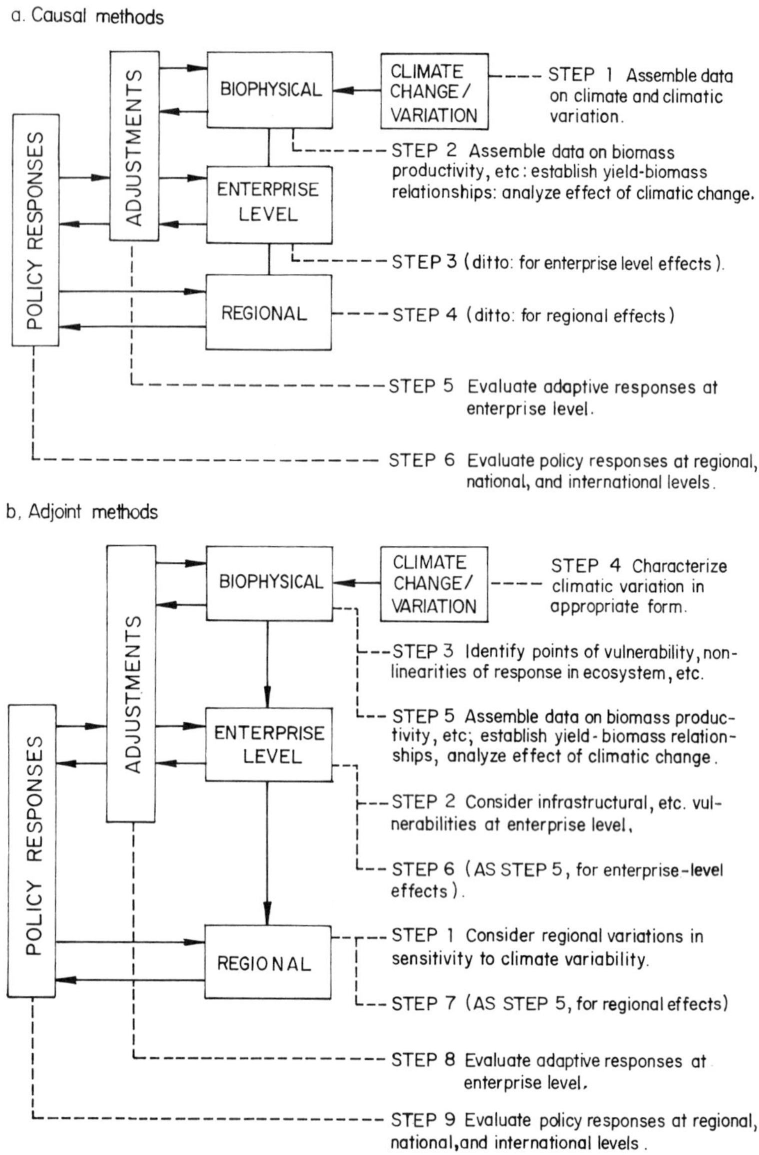

5. Causal (top-down) and adjoint (bottom-up) methods for climate impact assessment. The adjoint approach was adopted in the IIASA/UNEP project (Parry and Carter 1988).

Types of experiments

For either the direct or the adjoint analytical method, two broad types of impact experiments can be conducted (Fig. 6).

Impact experiments. In impact experiments, each set of models in the hierarchy simulates a limited number of feedbacks within its own subsystems. The form of analysis is essentially sequential, estimates of effects are based largely on assumed and essentially static sets of agronomic and economic responses.

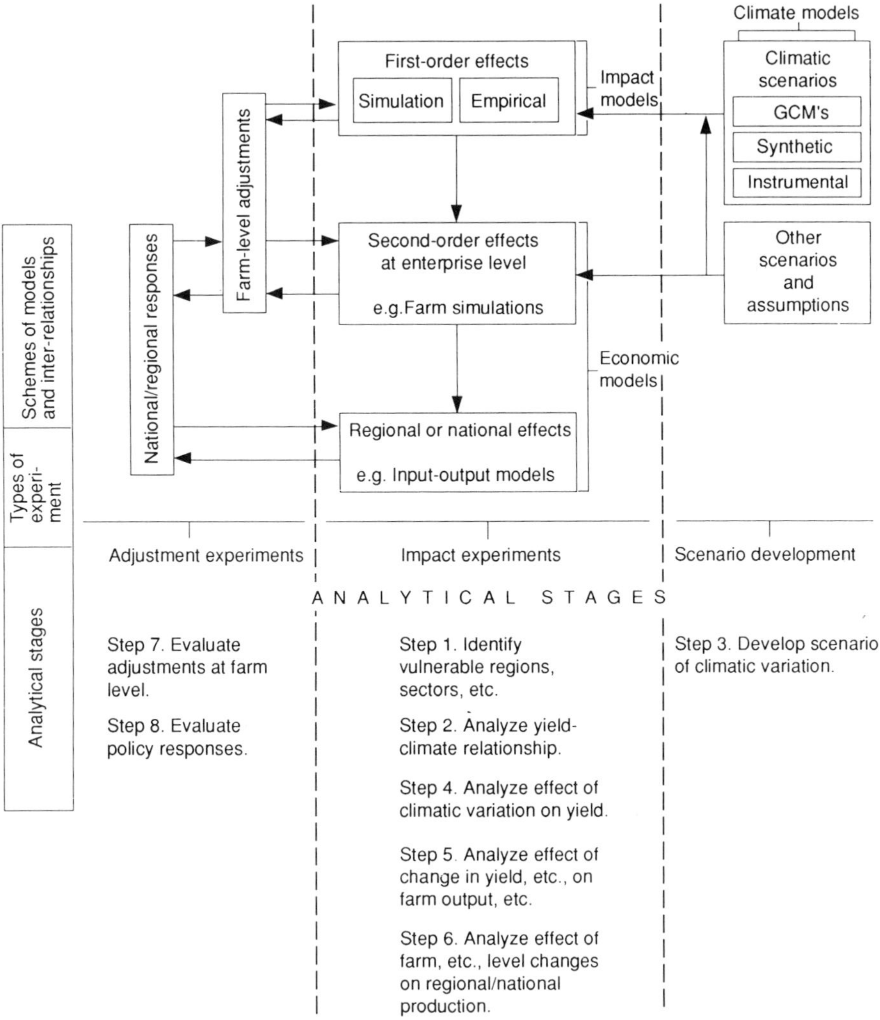

6. Conceptual framework and stages of analysis adopted in the IIASA/UNEP project (Parry and Carter 1988).

Adjustment experiments. By altering some assumptions about management and technology, it is possible to evaluate various options available to offset or mitigate the effects of climatic variations. These responses might occur at the enterprise level (for example, a farm-level decision to switch to a different crop) or through a change in government policy (such as a shift in farm subsidies). Experiments for different crops, varying amounts of fertilizer, etc., can enable a new set of impact estimates to be generated. Those can be compared with the initial estimates, to help evaluate appropriate responses to climatic variations.

In some instances it has proved useful to perform these experiments in sequence: first, evaluating impacts on the (unrealistic) assumption that the economic and social systems will undergo no change and second, adjusting experiments to incorporate an increasingly complex pattern of assumed responses (Parry et al 1988).

Techniques

Development of climatic scenarios

To assess the sensitivity of social and economic systems to climate, we need some realistic methods for quantifying the characteristics of the present-day climate and of likely climatic changes. One technique is to use climatic scenarios. In the integrated analytical system, these form the upper tier of the model hierarchy in Figure 4.

Instrumental meteorological observation data offer the basis for scenario development by providing information on medium- and short-term climatic fluctuations that have occurred in the past. They are valuable because a) they can provide a reference against which to compare different climatic scenarios (e.g. by adopting a standard baseline period for computing climatic averages and variability), and b) they can be related to their recorded impacts (e.g. on crop yields) in developing or testing impact models.

Three types of climatic scenario are particularly useful in analyzing climate impacts.

1. Instrumental scenarios allow us to specify a plausible future climate by examining the instrumental record for climatic anomalies. We can use these anomalies (which we know have occurred in the past and presume could occur in the future) as scenarios of future climate. The impacts would probably differ from the original events due to changed social and economic conditions. The IIASA/UNEP project included scenarios of anomalous individual years, anomalous successive years, and 1-in-10 year anomalies.
2. Synthetic scenarios composed of climatic data generated arbitrarily to simulate a climatic change are of particular use for testing the sensitivity of impact models. For instance, the effects of changes in growing season temperature and precipitation on spring wheat yields in the Saratov region of the southern European USSR were investigated in the IIASA/UNEP project. Each variable was altered systematically (e.g. growing season precipitation was adjusted by 5% increments relative to 1951-80 mean values) (Pitovranov et al 1988b).
3. Scenarios based on outputs from general circulation models (GCMs), which are physically based numerical models of the climate system. GCMs have been used to estimate the climatic changes that might be expected, given an increase in the concentration of atmospheric carbon dioxide (CO_2) and other radiatively active gases. Most experiments so far have considered the effects of a doubling of CO_2. The models produce outputs for a network of grid-points over the whole globe, to show the simulated change in seasonally averaged climatic variables (e.g. temperature, precipitation rate, cloud cover) between $1 \times CO_2$ (present) and $2 \times CO_2$ (future) equilibrium conditions. While GCMs have been reasonably successful in reproducing the continental-scale features of present-day ($1 \times CO_2$) temperatures and pressure patterns, other climatic

elements, including precipitation, are not simulated as well (Gates 1985). Nevertheless, GCM outputs were used in a wide range of impact experiments in the IIASA/UNEP project to demonstrate the feasibility of linking climate models and impact models while awaiting more accurate GCM predictions in the future.

Modeling first-order impacts

To estimate the impacts of the types of climatic scenarios defined on a particular biophysical system, we move to the next level of the model hierarchy (Fig. 5). This requires using models of biophysical response to climate.

Estimating biophysical responses to climate. Three types of agroclimatic models were used in the IIASA/UNEP project.

The simplest method of relating agroclimatic resources to climate is to combine or manipulate meteorological variables into an agroclimatic index. Such an index can be used for identifying areas suitable for different crops because it can incorporate, within a single term, the climatic variables that have the most influence on plant growth. In the IIASA/UNEP project, an index of agricultural potential in Kenya (Akong'a et al 1988) and one of hydrothermal potential in the European USSR (Nikonov et al 1988) were used.

A second type of model is of the empirical-statistical variety, developed by relating a sample of annual crop yield data to weather data from the same time period and area, using such statistical techniques as regression analysis. These models can have a high practical value for large-area yield prediction. They usually require only modest quantities of data and computational time. However, the approach does not provide a causal explanation of the relationship between climate and crop yield, but tends to identify only those variables that show a strong association with crop yield on short time scales. Empirical-statistical models are probably most valuable for estimating the impacts of short-term climatic anomalies, the magnitude of which are within the range of conditions under which a model was constructed (i.e. not requiring extrapolation of model relationships). Assessments using these models are perhaps best conducted in areas where crop yields are highly sensitive to a single climatic variable, such as precipitation for bean and cotton yields in northeast Brazil (Magalhaes et al 1988).

Simulation models can be used in assessing plant sensitivity to climate. These are based on an understanding of the relationships between the basic processes of plant and crop growth and environmental factors (such as water supply, temperature, solar radiation, and soil fertility). Despite their general requirement for detailed meteorological and physiological data (first for validation, then for applications in impact assessment), simulation models are more firmly based on experimental observation than are statistical models, and provide probably the best opportunity to conduct useful climate impact experiments in agriculture (particularly when considering longer term effects of changes in climate). In the IIASA/UNEP project, simulation models were used to assess drought impacts on maize in Kenya (Akong'a et al 1988), sorghum in India (Jodha et al 1988), barley in Ecuador (Bravo et al 1988), and spring wheat in the Saratov region of the USSR (Pitovranov et al 1988b).

Linking models of climate to models of biophysical response. One intention of using the hierarchical approach to climate impact assessment in the IIASA/UNEP project was to link models at different levels. To do this, data from the climatic scenarios should be used as inputs to biophysical models. For practical purposes, this linkage is a major criterion for selecting scenarios and impact models. The scenario determines whether the climatic data that are available for impact assessment are suitable for operating a particular impact model; the impact model's data requirements influence the type of scenario that can be constructed. For example, the temporal and spatial resolution of GCM-derived scenario data is often too coarse to be input directly into an impact model. In these circumstances, a synthetic data set may be substituted. If the climatic data for a scenario period are missing from the instrumental record and are required as an input to a biophysical model, a sensitivity study of the model may indicate that those data can be approximated, either by substituting data from a neighboring meteorological station or by substituting long-term averages.

Integrating the assessments of impact

The IIASA/UNEP study also sought to integrate biophysical effects with their economic and social impacts, both within single sectors (such as agriculture and water resources) and within a region.

Integrated sectoral assessments. Integrated sectoral assessments can be used to trace the impacts of climatic variations through biophysical responses, to the enterprise level, and then to the social and political implications of both sets of impacts. For example, reduced rainfall on forage and feed crops in consecutive years (1983-84) caused sharp increases in calf mortality in the southern rangelands of Ethiopia (Cossins 1986). In past centuries, this would have led to high death rates among pastoral people. This has not been the case in recent decades because of the availability of alternative means of livelihood and the work of famine relief operations. Similar investigations elsewhere in Kenya have considered the relationship between climate, forage, and meat and milk output (Akong'a et al 1988).

Integrated regional assessments. To some extent, sectoral studies are conducted for methodological and disciplinary convenience, because climate impacts are rarely restricted to a single sector of a regional economy. An integrated regional assessment should consider most or all of the significant interactions between sectors and between levels of impact. We can illustrate two approaches to assessments of this type, using a single example for each.

1. Assessments by observation represent an empirical examination of all available information concerning a climatic event and its biophysical, economic, social, and political impacts. An example is the analysis of the impact of the 1979-80 consecutive drought in northeast Brazil, in reducing local food and cash crop production, in reducing aggregate regional production, and in decreasing regional participation in the national economy (Magalhaes et al 1988).
2. Assessments by simulation represent attempts to model many of the interactive aspects of a climate impact described in the empirical approach.

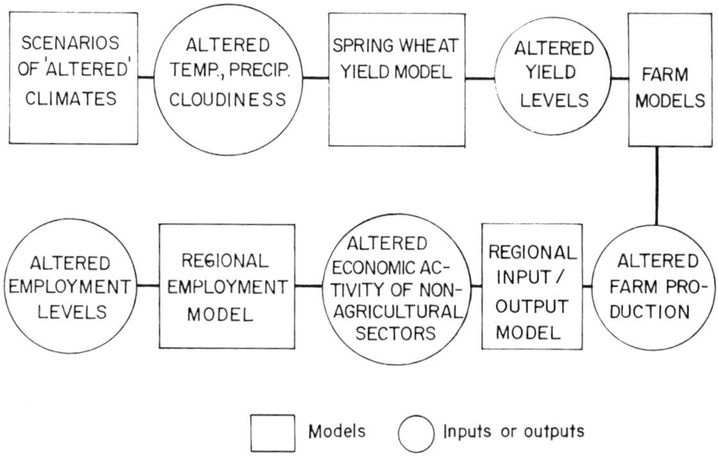

7. Hierarchy of models for assessing impacts of climatic change in the Saskatchewan case study, IIASA/UNEP project (Williams et al 1988).

An example is the Canadian case study. Climatic scenarios of altered temperature and precipitation were used in a simulation model to estimate altered yield levels of spring wheat (Fig. 7). The altered yields were input to farm-level models, converted to production figures, then aggregated by farm size and soil zone to give provincial production and commodity changes. These were used as inputs to a regional input-output model, which considers impacts on both the agricultural and nonagricultural sectors. Finally, using a third, employment model, changes in output levels for various economic sectors were translated into changes in employment (Williams et al 1988).

Results

Samples of results are presented here, selected from a wide range of experiments for both impacts and adjustments. They are drawn from the studies conducted in Canada, USSR, and Japan (Parry et al 1988). In each region, a series of three or four impact experiments were conducted along the lines outlined and in a broadly compatible manner. These were followed by a series of adjustment experiments to evaluate the appropriateness of various responses. Figure 8, which illustrates results for Japan, gives some indication of the estimates made.

In the examples summarized here, the emphasis was on the potential effects of climatic changes due to increased atmospheric CO_2. Scenarios of these changes were based on outputs from the $2 \times CO_2$ experiments with the Goddard Institute for Space Studies (GISS) GCM (Hansen et al 1983). To compare these effects with the effects of short-term climatic variations, the results were matched with those for an extreme decade and extreme year. Because different crop-weather models require different data (e.g. 10-d, monthly, annual), this array of scenarios varied somewhat among the case studies. The USSR study, for example, includes synthetic scenarios that vary temperature and precipitation by arbitrary increments. In general,

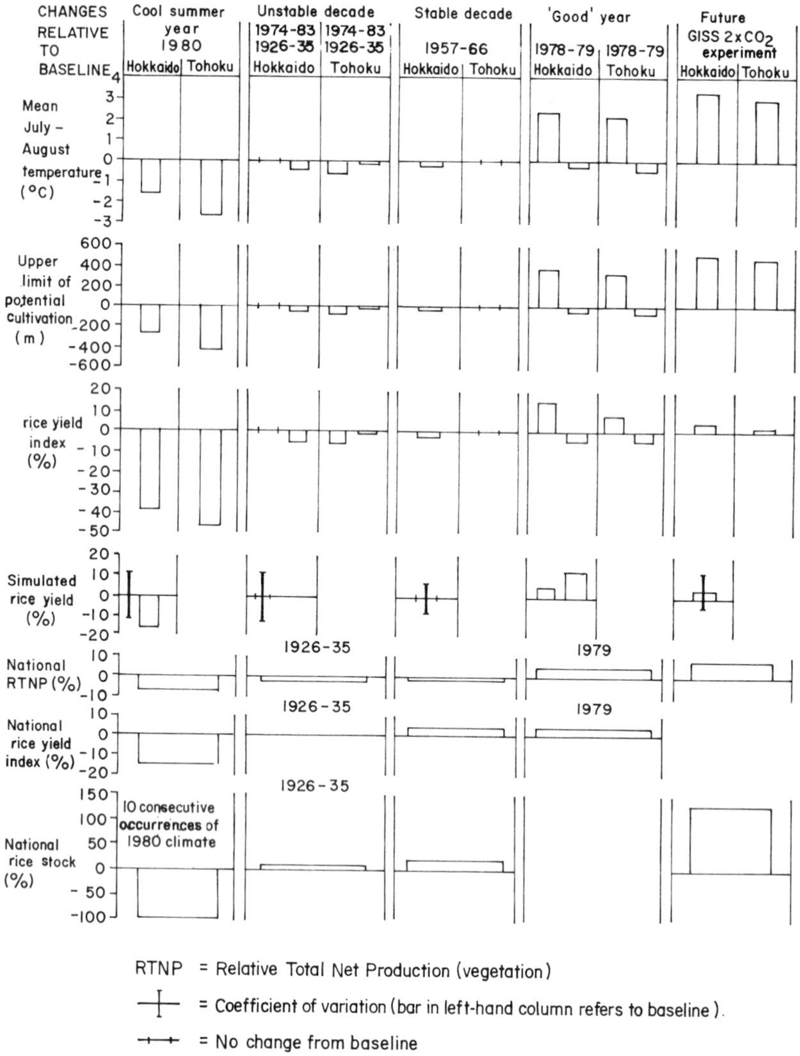

8. Estimated effect of climatic variations on agriculture in Japan (Parry and Carter 1988).

however, the results permit comparison both of the potential effects of a $2 \times CO_2$ climate in different regions and of the effects of a long-term climate change vs the effects of short-term climatic variability. The results are, of course, not predictions. A high level of uncertainty is attached to GCM predictions of regional climatic change and to the estimations of their effects on agriculture.

Effects on crop yields

A recent review of results from a number of studies (Warrick et al 1986) suggests that, despite the diverse modeling methods and scenarios adopted, there is a remarkable degree of unanimity in the expected direction of effects of climatic changes on cereal yields in the core wheat-growing areas of North America and Europe. With no change in precipitation or radiation, slight warming (+1 °C) might increase average yields by about 5 ± 4%; a 2 °C increase might reduce average yields by about 10 ± 7%. Reduced precipitation also tends to decrease yields, implying that both higher precipitation and warmer temperature could have offsetting effects on yields.

We should add that different effects on the same crop may occur in different regions. Also, different cultivars may respond differently to similar changes in climate. These variations can be important in indicating the kinds of technological adjustments that may be most appropriate for responding to a given climate change.

If we ignore the adjustments in agriculture, such as a switch of crops, that are likely to accompany or at least to follow a long-term climatic change, we may conclude that, given the kind of changes in climate most likely to result from increased CO_2 and other greenhouse gases (particularly that warming will be enhanced in higher latitudes, with summer dryness becoming more frequent over the continents at middle latitudes in the Northern Hemisphere), then decreases in yields in the order of 10% might occur in the core wheat production areas of North America and the USSR.

Spatial effects on crop location

One of the major adjustments most likely to occur is the spatial shift of cropping areas, a shift somewhat akin to the shift of biomes that has occurred as a response to long-term climatic changes in the past. We can map this effect through a sequence of steps (summarized by Parry 1985):

1. Identification of the climatic variables that constrain crop growth in the region and limit its spatial extent.
2. Identification of critical levels of the variables that limit the spatial extent of the crop.
3. Characterization of climatic changes as changes in the critical levels.
4. Mapping these as a shift in extent of the crop.

A logical development of this approach is to consider the shift in cropping potential for the climatic scenarios predicted by GCM 2 × CO_2 experiments. Taking the example of wheat growing in the United States, under a 2 × CO_2 climate generated by the GISS GCM, three changes occur: a great extension of the winter wheat belt into Canada, a switch from hard to soft wheat in the Pacific Northwest due to increased precipitation, and an expansion of areas in fall-sown spring wheat in the southern latitudes due to higher winter temperatures (Rosenzweig 1985).

Potential effects on the regional agricultural economy

Hierarchies of linked models, similar to those discussed, have been used in the IIASA/UNEP project to assess the potential effects of climatic changes on the regional economy (Parry et al 1988). Of course, these impact experiments record the impacts that would occur if there were no changes in the farming system. However, farming systems have been remarkably adaptive to climatic changes in the past, and we can reasonably expect them to be as adaptive in the future. It is possible to evaluate the various options available for either mitigating the negative effects or exploiting the positive ones by altering some of the assumptions in our agroclimatic and economic models (for example, by experimenting with a switch to a different crop, or with different amounts of fertilizer, or with different amounts of irrigation). For each set of these adjustment experiments, we can generate a new set of impact estimates that can be compared with the initial impact estimates (or the unadjusted system). Thus, we can begin to identify the more appropriate types of adjustment to various types of climatic change.

Regional variations in effects on crop yields. The higher temperatures that may be expected to prevail under conditions of increased atmospheric CO_2 tend to favor higher yields of cereal crops currently grown in regions where temperatures now limit the growing season. For example, if we assume that these warmer conditions will not create problems of water supply and that Japanese rice production will remain fully irrigated, we may expect average rice yields in central Japan (Tohoku) to increase by perhaps 5% (Yoshino et al 1988). However, where cereal production is already drought prone, increased rates of evapotranspiration may well place a brake on output. In Saskatchewan, simulations of wheat yield point to severe early summer drought stress on young spring-sown wheat, with consequent 20-35% yield reductions, depending on soil type (Williams et al 1988). One way to mitigate these negative effects would be to switch from spring- to winter-sown wheat. However, for a number of reasons the regional pattern of crop yield responses can be expected to be extremely varied.

First, complexity in the spatial pattern of climate change is likely, which is not simply the result of varying degrees of absolute change in different regions, but also stems from the magnitude of these changes in relation to the existing climate. Thus, while effective temperature sums (ETS) in central Japan (Tohoku) increase by 27% in the GISS $2 \times CO_2$ experiment, in the north at Hokkaido (where they are already 25% lower than in Tohoku) the increase is 37%. Thus, much of the variation in yield responses estimated for irrigated rice in Japan is a function of the geography of existing agroclimatic potential.

Second, in the estimates presented we have assumed the same terrain, soils, management, etc. When regional variations in these are introduced, localized variations in responses to climatic change can result. Overlaid on these variations are differences in infrastructure (farm size, etc.) and management (levels of fertilizer application, pesticides, etc.). Even if we assume the same management for all farm sizes, the effects of similar yield changes on farm income vary according to farm size, partly because of varying yield-income functions and partly because different size farms are found on different soil types.

Third, because different crops have different growing requirements, they frequently respond differently to changes in their environment. In the central European USSR, for example, a moderate warming is estimated to increase winter wheat yields (which at present are limited by a short, relatively cool growing season) but decrease barley yields (spring-sown barley is suited to cool conditions but susceptible to the early summer moisture stress that might accompany increased temperatures).

Resultant changes in cropped area. Three consequences can be identified from the geographic complexity described: spatial shifts of crop potential, spatial shifts of comparative advantage, and spatial variations in crop yield sensitivity.

Areas that, under present climatic conditions, are judged to be most suited to a given crop or combination of crops or to a specified level of management will change location. In essence this means that a change will occur in the range of crops that can be profitably grown at a particular place, the result of the shift in the physiological limits for growth of different crops. We might expect, for example, that under a warmer climate, both winter and spring wheat might expand northward, assuming that terrain and soils permit. The point is also well illustrated in northern Japan, where the limits of the safely cultivable area for irrigated rice fluctuate markedly between warm and cool periods. Those limits could be expected to expand under a climatic warming induced by increased atmospheric CO_2 or other greenhouse gases (Fig. 9).

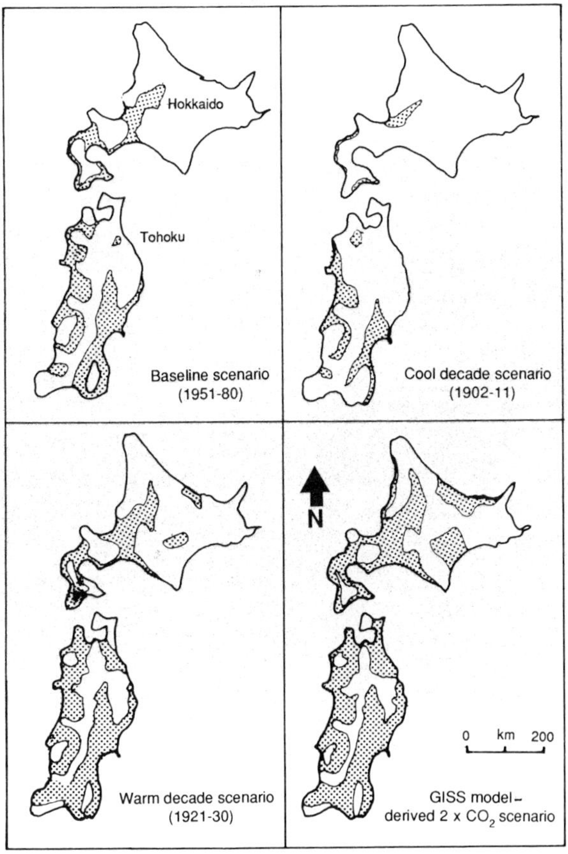

9. Safely cultivable area for irrigated rice in northern Japan under four climatic scenarios. The safely cultivable area is defined by the minimum level of accumulated temperature during the growing season required for the crop to complete its normal life cycle (Yoshino et al 1988).

Likely to be much more important are the changes in area under different crops resulting from differential changes, either in their relative profitability at a particular place or in the comparative advantage that one crop may hold over another.

The picture is further complicated by the fact that the same crop grown in different regions will respond differently to the same change in climate. In northern Finland under a GISS $2 \times CO_2$ climate, barley would relish a higher ETS without moisture shortage. In southern Finland, yields could actually decrease due to early summer moisture stress (for details, see Kettunen et al 1988).

Effects on crop responses to management. One of the difficulties in estimating effects of climatic change on agriculture is that the sensitivity of yield to inputs such as fertilizers and pesticides also varies with climate. Generally speaking, the closer the climatic conditions are to the ideal conditions for plant growth, the greater the plant response to fertilizer applications. This means that adjusting levels of fertilization can be an effective means of stabilizing yield variability resulting from short-term climatic changes.

Effects on variability of production. Even if we disregard those changes in interannual variability of temperature and rainfall that may occur as part of a transition to a warmer climate and assume that climatic variability will remain unchanged, the effect of a change in mean climate on mean yield can be different from its effect on the above- or below-average yields. Figure 10 shows that in central and southern Finland, simulated changes in the lower (95 percentile), average, and higher (5 percentile) spring wheat yields are different for recent cool (1974-82) and warm (1966-73) periods. Similar (although perhaps less pronounced) differences are likely to occur with wheat varieties having higher thermal requirements under a GISS $2 \times CO_2$ climate.

Downstream economic effects

We can use the results of experiments with farm simulation models and input-output models to estimate the effects of climatic changes, via crop yields, on farm incomes, employment, and levels of economic activity in nonagricultural sectors. These allow an estimation of the effects of a specific climatic event such as an extremely dry year or dry decade, if that event were to occur now. But the background factors are constantly changing and, indeed, would almost certainly change in response to a longer term transition in climate resulting from, for example, increases in atmospheric CO_2 and other greenhouse gases.

With these caveats in mind, we can report that experiments in Saskatchewan indicate that, under the changes in temperature and precipitation indicated by the GISS $2 \times CO_2$ experiment, total provincial farm income decreases 26%, on-farm employment 3%, and provincial gross domestic product 12%. Considerable margins of error embrace these estimates. For details see Williams et al (1988). In this experiment the change in climate is treated as a sudden, step-like event: no adjustment is allowed for changes in technology, management, prices, harvested area, etc. In reality, we can be sure that these would change substantially between now and the time by which levels of atmospheric CO_2 double. (Current estimates of the CO_2 doubling time are 2050 to 2100 [Bolin et al 1986].)

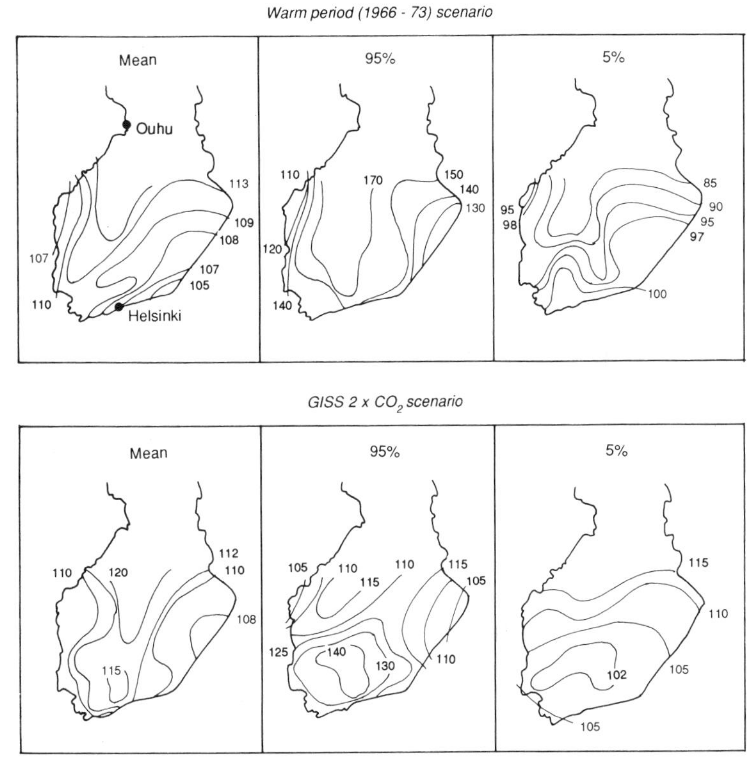

10. Effect of climate on mean, lowest (95 percentile), and highest (5 percentile) spring wheat yields in Finland for warm period (1974-82), with present-day variety and for GISS 2 x CO_2 climate with adapted variety having a thermal requirement 120 growing degree-days (GDD) greater than present varieties. Isolines indicate percentage of yield simulated for 1959-83 (Kettunen et al 1988).

Some technological responses to climatic changes

We refer here only to adjustments that could be put in place now. This enables us to parameterize and input them to the linked models. Vague assumptions about future changes in technology, demand, and prices are much less easy to specify. The adjustments considered are of three types: crop variety, soil management, and land allocation.

Changes in crop variety. Three aspects of this have emerged. First, our simulations in Canada, Finland, and northern USSR indicated that some spring-sown crops (e.g. wheat, barley, and oats) would experience reduced yields under the GISS 2 × CO_2 climate, due to increased frequency of moisture stress early in the growing period. A switch to winter wheat or, in some areas, to winter rye might reduce the effects of high evapotranspiration in the early summer as well as take advantage of the longer potential growing season (assuming that snow cover remained sufficient to protect the crop against winter kill).

Second, a logical way to exploit longer and warmer growing seasons at high latitudes is to use long-duration varieties with higher thermal requirements. In some cases the yields of present-day varieties, such as spring wheat in northern USSR, tend to be reduced under the GISS 2 × CO_2 climate. However, spring wheat varieties with thermal requirements 50 or 100 growing degree-days (GDD) greater than present-day varieties exhibit increased yields.

Third, we know little about what changes could occur in the interannual variability of temperature and precipitation. But, even if we assume no change in variability under an altered climate, the effects on crop yields are not insignificant. It is possible to test a number of varieties for stability of yield. Figure 11 illustrates the effects of the GISS 2 × CO_2 climate on two different varieties (for further details, see Yoshino et al 1988).

Changes in fertilizing and drainage. The IIASA/UNEP project includes two types of fertilizer experiments: the first increases levels of fertilizer application to optimize yields under a GISS 2 × CO_2 climate. The second varies applications to maintain, for example, 1980 production levels. Variable applications of fertilizer to stabilize yields by offsetting the effects of anomalously cool or warm summers are being tested for feasibility by the Icelandic government (for details, see Pitovranov et al 1988a and Bergthorsson et al 1988).

Increased precipitation predicted in the GISS 2 × CO_2 experiment might be expected to lead to increased soil erosion, which might offset the beneficial effects of a warmer climate and technological improvements. Improvements in soil drainage are therefore a feasible adjustment, tested in the IIASA study of the Leningrad region, northern USSR. The impact indicated is of slightly reduced winter rye yields, presumably a result of the leaching of soil nutrients. This effect would have to be

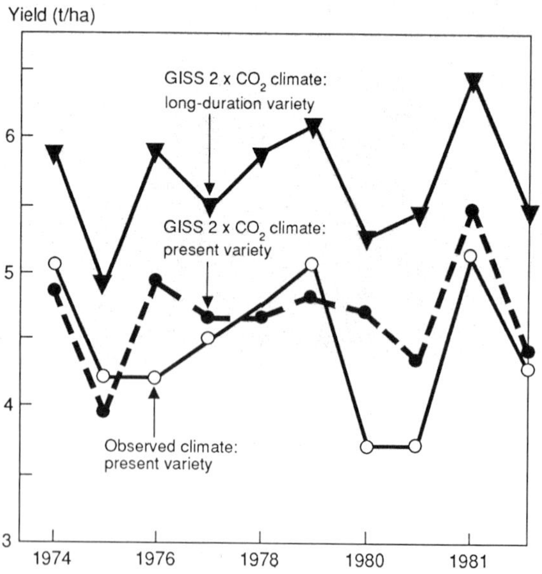

11. Simulated rice yields under observed (1974-83) and GISS 2 × CO_2 climates in Hokkaido, northern Japan. Responses of two rice varieties to doubled CO_2 climate are shown (Yoshino et al 1988).

weighed against reduced erosion and more efficient disposal of nitrate pollutants in the region to fully assess the consequences of such a measure (Pitovranov et al 1988a).

Changes in land allocation. Different crops respond differently to changes in climate and to various levels of fertilizer application under different climates. Any attempt to maximize output of each crop while minimizing production costs is likely to identify different allocations of land to alternative crops under different climates. Experiments in the Central Region, European USSR, for a 1 °C arbitrary increase in mean annual temperature indicate an optimal land use that increases the area under winter wheat, maize, and vegetables while decreasing the area allocated to spring-sown barley, oats, and potatoes. This pattern of land use begins to resemble that now found farther south in the USSR and points to the value of using regional analogues to identify possible responses to climate change.

Experiments in Saskatchewan have tested the efficacy of removing marginal cropland from production as a means to mitigate the effects of drought. Wheat crops on this land tend to be profitable in years of normal or above-normal rainfall but major losses can occur in dry years. The current work has identified previously unimproved land brought into wheat production over 1951-81 (about 14% of all current cropland). In the simulations, this is converted back from wheat to pasture for beef cattle. Total provincial output from the new mix of land uses is compared with that from current land use for a variety of drought and nondrought years (R.A. Fautley, 1985, pers. comm.).

Research priorities

Four needs for future research emerge from this summary.

- First, we need more specific and user-oriented information regarding climatic change (its likelihood, nature, magnitude, areal extent, duration, and [most important] rate of onset). It is also important that information on variability be available, to provide estimates of possible changes in those extreme conditions we have seen to be important in agricultural decisionmaking.
- Second, an important path of climatic impact on the most climate-sensitive sectors of our society (farming, fishing, and forestry) is indirect, through changes in other physical systems (soil chemistry, ocean currents, agricultural pests and diseases and their vectors). Although we have barely begun to grasp these interactions, they clearly have a major influence on some production systems. Climatic changes that affect the frequency of these conditions could alter the incidence of such outbreaks.
- Third, we need to specify with greater precision the interaction between climate and crop growth. It should then be possible to trace, with greater confidence, the downstream effects of first-order impacts on other sectors of economy and society by referring to a hierarchy of the three types of models we have considered: climatic, impact, and economic.
- Fourth, and perhaps most important, we need to explore in greater detail the range of technological and policy adjustments available in agriculture, to evaluate their efficacy in mitigating negative impacts or exploiting new options offered by climatic changes.

References cited

Akong'a J, Downing T E, Konijn N T, Mungai D N, Muturi H R, Potter H L (1988) The effects of climatic variations on agriculture in central and eastern Kenya. Pages 121-270 *in* The impact of climatic variations on agriculture. Vol. 2. Assessments in semi-arid regions. M. L. Parry, T. R. Carter, and N. T. Konijn, eds. Kluwer, Dordrecht.

Bergthorsson P, Björnsson H, Dyrmundsson O, Gudmundsson B, Helgadottir A, Jonmundsson J (1988) The effects of climatic variations on agriculture in Iceland. Pages 381-509 *in* The impact of climatic variations on agriculture. Vol. 1. Assessments in cool temperate and cold regions. M. L. Parry, T. R. Carter, and N. T. Konijn, eds. Kluwer, Dordrecht.

Bolin B, Döös B R, Jäger J, Warrick R A, eds. (1986) The greenhouse effect, climatic change and ecosystems: a synthesis of present knowledge (SCOPE 29). Wiley, Chichester. 542 p.

Bravo R E, Cañadas Cruz L, Estrada W, Hodges T, Knapp G, Ravelo A C, Planchuelo-Ravelo A M, Rovere, Salcedo Solis T, Yugcha P T (1988) The effects of climatic variations on agriculture in the central Sierra of Ecuador. Pages 381-493 *in* The impact of climatic variations on agriculture. Vol. 2. Assessments in semi-arid regions. M. L. Parry, T. R. Carter, and N. T. Konijn, eds. Kluwer, Dordrecht.

Cossins N J (1986) The impact of climate variation on animal production: commentary on a paper by J. J. Burgos. Draft paper presented at the Task Force Meeting on Policy Oriented Assessment of Impact of Climatic Variations, 30 Jun-2 Jul 1986, International Institute for Applied Systems Analysis, Laxenburg, Austria.

Gates W L (1985) The use of general circulation models in the analysis of the ecosystem impacts of climatic change. Climatic Change 7(3):267-284.

Hansen J, Russell G, Rind D, Store P, Lacis A, Lebadeff S, Ruedy R, Travis L (1983) Efficient three-dimensional global models for climate studies: models I and II. Mon. Weather Rev. 111:609-662.

Jodha N S, Gadgil S, Huda A K S, Singh R P, Virmani S M (1988) The effects of climatic variations on agriculture in dry tropical regions of India. Pages 495-578 *in* The impact of climatic variations on agriculture. Vol. 2. Assessments in semi-arid regions. M. L. Parry, T. R. Carter, and N. T. Konijn, eds. Kluwer, Dordrecht.

Kates R W (1985) The interaction of climate and society. Pages 3-36 *in* Climate impact assessment: studies of the interaction of climate and society (SCOPE 27). R. W. Kates, J. H. Ausubel, and M. Berberian, eds. Wiley, Chichester.

Kettunen L, Mukula J, Pahjonen V, Rantanen O, Varjo U (1988) The effects of climatic variations on agriculture in Finland. Pages 511-614 *in* The impact of climatic variations on agriculture. Vol. 1. Assessments in cool temperate and cold regions. M. L. Parry, T. R. Carter, and N. T. Konijn, eds. Kluwer, Dordrecht.

Magalhaes A R, Filho H C, Garagorry F L, Gasques J G, Molion L C B, Neto M da S A, Nobre C A, Porto E R, Rebouças O E (1988) The effects of climatic variations on agriculture in Northeast Brazil. Pages 271-380 *in* The impact of climatic variations on agriculture. Vol. 2. Assessments in semi-arid regions. M. L. Parry, T. R. Carter, and N. T. Konijn, eds. Kluwer, Dordrecht.

Nikonov A A, Petrova L N, Stolyarova H M, Lebedev V Yu, Siptits S O, Milyutin N N, Konijn N T (1988) The effects of climatic variations on agriculture in the semi-arid zone of the European USSR: A. The Stavropol Territory. Pages 57-62 *in* The impact of climatic variations on agriculture. Vol. 2. Assessments in semi-arid regions. M. L. Parry, T. R. Carter, and N. T. Konijn, eds. Kluwer, Dordrecht.

Parry M L (1985) Assessing the impact of climatic change on marginal areas. Pages 351-367 *in* Climate impact assessment: studies of the interaction of climate and society (SCOPE 27). R. W. Kates, J. H. Ausubel, and M. Berberian, eds. Wiley, Chichester.

Parry M L, Carter T R, eds. (1984) Assessing the impact of climatic change in cold regions. Summary report SR-84-1. International Institute for Applied Systems Analysis, Laxenburg, Austria. 42 p.

Parry M L, Carter T R (1988) The assessment of effects of climatic variations on agriculture: aims, methods and summary of results. Pages 11-95 *in* The impact of climatic variations in agriculture. Vol. 1. M. L. Parry, T. R. Carter, and N. T. Konijn, eds. Kluwer, Dordrecht.

Parry M L, Carter T R, Konijn N T, eds. (1988) The impact of climatic variations on agriculture. Vol. 1. In cool temperate and cold regions. Vol. 2. In semi-arid regions. Kluwer, Dordrecht.

Pitovranov S E, Iakimets V, Kiselev V I, Sirotenko O D (1988a) The effects of climatic variations on agriculture in the subarctic zone of the USSR. Pages 615-722 *in* The impact of climatic variations on agriculture. Vol. 1. Assessments in cool temperate and cold regions. M. L. Parry, T. R. Carter, and N. T. Konijn, eds. Kluwer, Dordrecht.

Pitovranov S E, Maximov A D, Sirotenko O D, Abashino E V, Pavlova V N, Carter T R (1988b) The effects of climatic variations on agriculture in the semi-arid zone of the European USSR: b. The Soratov Region. Pages 629-664 *in* The impact of climatic variations on agriculture. Vol. 2. Assessments in semi-arid regions. M. L. Parry, T. R. Carter, and N. T. Konijn, eds. Kluwer, Dordrecht.

Rosenzweig C (1985) Potential CO_2-induced effects on North American wheat producing regions. Climatic Change 7:367-380.

Warrick R A, Gifford R M, Parry M L (1986) CO_2, climatic change and agriculture. Assessing the responses of food crops to the direct effects of increased CO_2 and climatic change. Pages 393-473 *in* The greenhouse effect, climatic change and ecosystems: a synthesis of present knowledge (SCOPE 29). B. Bolin, B. R. Döös, J. Jäger, and R. A. Warrick, eds. Wiley, Chichester.

Williams G D V, Fautley R A, Jones K H, Stewart R B, Wheaton E E (1988) Estimating effects of climatic change on agriculture in Saskatchewan, Canada. Pages 219-379 *in* The impact of climatic variations on agriculture. Vol. 1. Assessments in cool temperate and cold regions. M. L. Parry, T. R. Carter, and N. T, Konijn, eds. Kluwer, Dordrecht.

Yoshino M M, Horie T, Seino H, Tsujii H, Uchijima T, Uchijima Z (1988) The effects of climatic variations on agriculture in Japan. Pages 723-868 *in* The impact of climatic variations on agriculture. Vol. 1. Assessments in cool temperate and cold regions. M. L. Parry, T. R. Carter, and N. T. Konijn, eds. Kluwer, Dordrecht.

Notes

Authors' address: M. L. Parry and T. R. Carter, Atmospheric Impacts Research Group, P.O. Box 363, Edgbaston, Birmingham, UK.

Citation information: International Rice Research Institute (1989) Climate and food security. P.O. Box 933, Manila, Philippines.

Assessing the social implications of climate fluctuation through climate impact studies

W. E. Riebsame

Approaches and methods useful for assessing and mitigating the impact of climate fluctuations on human activities include sensitivity analysis, empirical case studies, integrated and comprehensive assessments, and projective analysis. Suggestions for establishing impact assessment capabilities in climate-sensitive regions are offered, along with recommendations for future efforts to improve our ability to evaluate and mitigate climate impacts on social well-being.

Human experience and simple prudence dictate greater efforts to incorporate climate considerations into the way we manage natural resources and the way we structure certain human activities, such as food production and distribution. This paper presents an overview of approaches and methods useful to government officials and researchers who must measure, evaluate, and mitigate the social effects of climate fluctuations. It is directed to the resource manager or researcher who must take climate into consideration in planning and analysis, but who may be unfamiliar with the methods available for climate impact assessment. It is hoped that this review can provide a broad blueprint for establishing climate impact assessment capabilities, especially in regions where little is known about climate sensitivities and potential impacts. The material is abstracted from a monograph to be published under the auspices of the United Nations Environment Programme (UNEP) as part of the World Climate Impact Program (WCIP).

WCIP was established to provide governments and other decisionmakers with credible indications of the impact of climate variations, both natural and man-made, on economic and social activities. The UNEP took responsibility for implementing WCIP in three areas: 1) improving the methods available for climate impact assessment, to produce more credible assessments for decisionmakers; 2) investigating the potential impact of human-induced climatic changes, with emphasis on the greenhouse effect; and 3) studying the impact of climate variations on the world's food systems, with the goal of increasing food security.

Early in the development of WCIP, it was recognized that a major obstacle to increasing our understanding of the influence of climate on society was the lack of proven methods for conducting climate impact studies and the need to attract more researchers, especially social scientists, to the study of climate-society interactions. Climate has often been neglected in contemporary studies of people-environment relationships, partly because it is a subtle and shifting component of the natural environment. Yet it is precisely that shifting character of climate that emphasizes the importance of our ability to measure and predict its influence on human activity. Credible impact assessments are needed to discern the role that climate has played, and may play in the future, in the development of resources for human well-being.

UNEP's program for improving climate impact studies has included a wide-ranging review of theory and research (Kates et al 1985), support for impact case studies and projective assessments (Parry et al, forthcoming), and preparation of a guide to impact study, to assist the development of impact assessment capabilities in climate-sensitive areas (Riebsame, forthcoming). Future efforts will include regional seminars to provide impact assessment expertise and guidance to resource managers and researchers and integrated impacts and policy option studies in selected regions.

Climate impact assessment goal

Climate impact assessment can be defined simply as the empirical measurement of climate's effects on natural and social systems (early climate impact studies were just this). Modern climate impact assessment is a multifaceted endeavor to understand the climate vulnerabilities of human populations and activities, to ascertain the magnitude and distribution of impacts from climate fluctuations, and to formulate and help implement procedures for mitigating future climate impacts. Impact studies also can provide guidance for the application of climate data to improve the resilience of resource management activities and to aid long-range planning of resource management schemes.

The practice of climate impact assessment can also have several secondary benefits.

- Improving communications among disciplines and governmental units that have not worked closely together in the past (e.g. meteorological services and economic development agencies), but must now integrate their efforts to solve complex development problems.
- Broadening the overall sensitivity of planners and policymakers in various decisionmaking settings to the problems that can be caused by climate and other environmental variables.
- Illustrating the value of monitoring climate and social variables, leading to improved data collection.

Approaches to climate impact assessment

Climate impacts are studied for a wide variety of reasons and from many different perspectives. A government official may need to assess the climate vulnerability of a particular region or economic activity to formulate development plans, or may be called upon to assess current impact to direct emergency relief more efficiently. Decisionmakers responsible for managing certain natural resources (e.g. cropland, forests, grazing lands, water) need to consider climate variability to assess the reliable flow of these resources. University researchers might examine the relationship between the climate and society to develop procedures for more efficient use of natural resources, or even as an opportunity to learn more about how society functions or how climate links to other biophysical systems. Climate impact studies might help us better understand some of the natural factors in long lasting, devastating resource shortages common to some regions. Of course, projected impacts of potential climate changes are crucial to better long-term resource policy and planning.

Several approaches and methods are applicable to climate impact studies. Assessment types include 1) a priori sensitivity analyses conducted to identify specific vulnerabilities in social systems, 2) empirical impact studies focused on certain places or defined climate fluctuations, 3) integrated assessments that link climate effects to broader socioeconomic structures and associated comprehensive assessments that evaluate the multiple impacts of climate fluctuations across multiple economic sectors, and 4) projections of the likely impacts of future climatic changes. Projections (often based on statistical models that link climatic changes to outcomes in crop yields, runoff, fish populations, etc.) are especially important now because of the need to anticipate the effects of climatic changes that might result from human-caused increases in carbon dioxide and the other greenhouse gases.

No single approach can fulfill the many different knowledge needs. Flexible methods are necessary because of large differences in the assessors' goals, units of analysis, data availability, physical linkages between climate and other environmental factors, and socioeconomic settings. A wide range of approaches are suggested here, with examples of how they might be applied to climate impact assessment, especially in agriculture. However, this short paper cannot include the detailed methodological discussions found in the monograph from which it is abstracted.

Sensitivity analysis

Early climate impact research showed that climate-society linkages are interactive rather than one-way. Climate fluctuations may lead directly to negative impacts on, for example, crop yields or freshwater runoff. At the same time, social development and human response produce a continuing adjustment that moderates or exacerbates such impacts. The sensitivity and flexibility of human systems help determine the nature and extent of impacts from specific climate fluctuations.

Some social systems exhibit special vulnerability to climate impacts at various stages in their development. For example, water supply systems tend to alternate between periods of large excess capacity and periods of a closer match between supply and demand, as new facilities are built and use increases. This pattern produces "windows of vulnerability," during which climate fluctuation may be particularly damaging. An example of this alternating vulnerability is shown in Figure 1a. Large storage capacity increments in the Denver, Colorado, water supply system in 1957 and 1965 initiated periods of excess capacity. Those were eroded by steadily increasing demand and occasional drought (e.g. 1963-64 and 1977-78), when shortages occurred. The goal of impact assessment is to anticipate and mitigate such periods of heightened sensitivity.

The challenge to impact assessors is to identify trends that make systems more or less sensitive to impacts before the impacts accrue. What does the continued

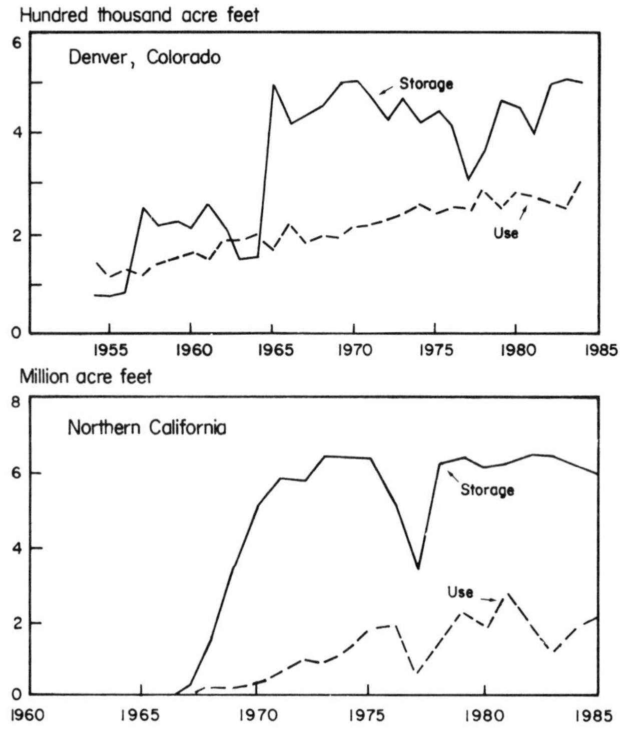

1. Water storage and use records for two water supply systems in western USA: Denver, Colorado and northern California.

development of new crop cultivars mean for future agricultural sensitivity? How do market trends for commodities, energy, and fertilizers alter our food system's susceptibility to climate disruption? Questions like these, which are being addressed vis-a-vis food security at this conference, should be asked of all major resource management systems (e.g. water, energy, forests, grasslands).

Sensitivity analysis, perhaps by applying a priori measures of system reliability (e.g. ratios of storage to demand in water systems, comparisons of crop yield and climate variability, or measures of system resiliency as used in flood control analysis), can improve the choice of case study areas, analytical units, and data collection methods. This approach is being applied in the Climate Impacts, Perception, and Adjustment Experiment (termed CLIMPAX) in the USA. For example, comparisons of water system storage and demand values are being used to identify particularly sensitive water systems in the western USA.

Water systems with a close match between supply and demand are more vulnerable to climate stress. Compare Figure 1a for the Denver water system with Figure 1b, the record of storage and use for the State Water Project in California. The California system maintains a larger supply buffer, and is thus less vulnerable to drought-induced shortages.

Approaches to agricultural sensitivity analysis range from simple field surveys to detailed simulation analyses, depending on the data available. One outcome of the growing concern for global food supplies is more and better research on agricultural practices throughout the world. Thus, the climate impact assessor now rarely faces a total lack of information on a region's agricultural history, current practice, recent climate, and crop productivity. Although data may not be sufficient to create statistical crop-climate models for all areas, at least the broad outlines of the climate sensitivity of locally and regionally important crops can be discerned from government reports and research literature.

In a region poorly documented by agricultural research or crop statistics, an impact assessor might begin a sensitivity analysis by interviewing farmers, agricultural extension agents, marketing experts, and others in touch with that region's agricultural activities. The goal is to build a roster of the climate conditions that lead to agricultural impacts. The assessor should also seek information on the most sensitive phenological periods of current crops, recent yields, farm income and well-being, and rate of such inputs as fertilizers and pesticides.

A typical agricultural sensitivity analysis might include statistical yields analysis and examination of standard crop calendars for matches between phenological sensitivity and episodes of climate extremes. Two often-neglected approaches include assessing the existing roster of farmer adjustments and transferring lessons from analogous agricultural cases.

Assessing the range of agricultural adjustment. The range of adjustments or alternatives available to farmers and other resource managers faced with climate stress is an important determinant of sensitivity, and a factor in mitigating climate impacts. In theory, the more options an agricultural society faced with climate stress has, the less sensitive that society is to severe impacts. Theoretically, adjustments might range from small changes in practices (e.g. intercropping) to large-scale alterations (e.g. building new water storage and irrigation facilities). A roster of potential responses to climate fluctuations, created by evaluating previous research or through field surveys, not only provides some idea of an area's sensitivity to climate, but can identify particularly vulnerable areas or groups (those with few available adjustments) where government action is necessary to ameliorate impacts.

Heijnen and Kates (1974) used field surveys conducted along a moisture gradient in the Usumbara Mountains of Tanzania to develop a roster of adjustments (Table 1). They found that physical and social factors in the dry and moderately moist sections of their survey transect work against the storage of surplus food supplies. In the dry area, this gap in the adjustment repertoire of smallholder farmers is exacerbated by their inability to change location. This indicates a particularly vulnerable area and population.

In a similar study, N. S. Jodha (Jodha and Mascarenhas 1985) compared the activities of farmers in a semiarid zone of India during a normal rainfall year and during a drought year. He found that practices that offer some salvage value from crops, protect resources, and augment supplies of even inferior products became more widely practiced during the dry period. If impact assessors can ascertain the viability of such adjustments before climate fluctuations occur, they can better anticipate impact magnitude as well as better plan for amelioration.

Jodha and Mascarenhas (1985) surveyed the literature on smallholder, mostly self-provisioning farmers in parts of Africa and India, to develop a comprehensive

Table 1. Farmer adjustments in the Usumbara Mountains, Tanzania (Heijnen and Kates 1974).

Type of adjustment	Available moisture classification		
	High	Moderate	Low
Accept self-insured loss			
Work for wages to buy food	+	+	+
Sell cattle to buy food			
Use savings to buy food	+	+	+
Store more than one season's food when crop is good	+		
Distribute and share loss			
Move to another farm	+	+	
Ask help from friends and relatives		+	+
Ask help from the government		+	+
Eliminate moisture waste			
Weed plots	+	+	+
Stop planting when rains are not enough	+	+	+
Change moisture requirements			
Plant drought-resistant crops	+	+	+
Affect source			
Pay for rainmaker			+
Pray	+	+	+
Change location			
Have plots in different places	+		
Plant in wet places	+		
Move cattle			
Improve moisture storage and distribution			
Irrigate	+	+	+
Schedule for optimal moisture			
Plant without rain			
Plant only when enough rains come	+	+	+
Stagger planting	+		

list (Table 2) of farming features that decrease sensitivity to climate and other impacts. It is reasonable to assume that the presence of a greater number of such features decreases overall sensitivity, although some conflict may exist between adjustments (e.g. irrigation may be expensive, lowering the financial reserves of a farmer, and may lead to monoculture as opposed to diversified operations). Such studies of adjustment to climate stress can be useful to policymakers.

Agroclimatic analogue approaches. Another approach to sensitivity analysis involves the use of agricultural analogues. Agricultural impact assessments often must be conducted in the absence of good yield data. For example, data on regional maize yields in most areas of Kenya are based on very few years and infrequent spot checks. But, while conducting a broad climate impact assessment for the country, Downing et al (1985) found it reasonable to infer yield sensitivities between similar agroecological zones. They transferred impact potentials from a few selected areas where more data were available to less well-observed regions. Such rough yield estimates, matched with new field studies and improved climate information, can provide critical indications of climate sensitivity.

Table 2. Farmer adjustments in self-provisioning societies (Jodha and Mascarenhas 1985).

Adaptive features of farming systems	Short-term adjustments through										Features observed in	
		(A) risk/loss minimization				(B) risk/loss management						
	Long-term flexibility or reliability	Salvage operations	Midseason adjustment	Cut back on hired resources	Change in techniques	Cut in current commitments (consumption, etc.)	Resource augmentation (conservation/ recycling)	Supplementary earning (migration)	Inventory depletion (assets)	Dependence on others	India	Africa
1. Diversified production strategy												
Crop/stock mixed farming	x	x	x				x	x	x	x	x	x
Mixed cropping	x	x	x								x	x
Combining crops of varying maturity, drought tolerance, input needs, and end uses	x	x	x								x	x
2. Operational diversification												
Toposequential planting		x	x								P	x
Staggered planting			x	x							x	x
Plot scattering/ splitting		x	x								x	x
Varied plant spacing			x	x	x						x	x
Input use skipping/ splitting			x	x	x						x	x
3. Flexible resource-use patterns												
High dependence on own resources	x		x	x	x			x			x	x
Limited ex-ante commitment to a current production			x	x	x						x	x
Accretionary process asset buildup	x			x		x	x		x		x	x
Recycling the resources	x		x		x			x	x		x	x

Table 2 continued.

	Short-term adjustments through:										Features observed in	
		(A) risk/loss minimization				(B) risk/loss management						
Adaptive features of farming systems	Long-term flexibility or reliability	Salvage operations	Midseason adjustment	Cut back on hired resources	Change in techniques	Cut in current commitments (consumption etc.)	Resource augmentation (conservation/ recycling)	Supplementary earning (migration)	Inventory depletion (assets)	Dependence on others	India	Africa
4. Flexible consumption patterns												
Close link between consumption and production	x	x	x			x	x				x	x
Recycling the products	x					x	x				x	
On-farm storage	x					x			x		x	x
Flexible long-term commitment	x					x		x			x	
5. Adapting the environment												
Irrigation	x	x	x								x	x
Moisture conservation	x	x	x								x	P
Perennial crops	x		x					x			P	x
6. Traditional forms of rural cooperation	x							x		x	x	x

P = partial.

Agroclimatic zoning is useful in evaluating the crop production potential of a region, and broad agroecological zones based on climate data have been defined for most of the tropics. Table 3 is essentially a matrix of temperature constraints (rows) and moisture constraints (columns). Average temperature indicates the growing season and accumulation of heat by the plant; the likelihood of frost reflects the possibility of singular events that will damage a crop. Moisture limitations essentially reflect the ratio of precipitation to evaporation and set a limit on crop yield. The cells of the matrix are filled in with actual crops that prosper in a zone under normal conditions. Climatic changes and their effects on yields can be represented by motion along rows and across columns. An expanded matrix might indicate the crops that are marginal in each zone which could replace current cultivars if temperature or moisture conditions change.

Empirical impact studies

No single set of empirical impact assessment methods can be prescribed for all cases. The questions asked by impact assessors are too varied and complex to yield to a fixed set of approaches. But climate impact assessment is typically oriented toward assessing past or potential effects of defined climate fluctuations. Thus, the "case study" becomes a key general approach. Candidate cases of climate fluctuations might be selected from the climate record of an area or from well-known and memorable episodes of obvious climate impact. Greater weight can be given the results of such case studies if some control is applied to data collection and analysis.

In the CLIMPAX project, researchers are choosing case studies on the basis of the opportunity to compare the situation of a region affected by a climate fluctuation to a nearby, similar region that did not experience the climate fluctuation (Karl and Riebsame 1984). This case-control approach can isolate climate impacts from the multitude of other physical and socioeconomic factors (e.g. market variations, soil erosion, pests, war) that might affect the indicators (e.g. crop yields, farm income, nutrition) selected for impact analysis.

Where such case-control analysis is not feasible, the assessor is urged to take a longitudinal approach that allows comparison of impact indicators before, during, and after the climate fluctuation of interest. Temporal concurrence of fluctuations and impacts is not proof of a causal connection, but when supported by verified links in a logical cascade of cause, effect, and feedback, the concurrence of climate fluctuation and socioeconomic changes points to a climate-society interaction. Impact assessments can be strengthened if researchers examine other nonclimatic explanations for the effects observed. For example, farm production and income logically might be depressed during a spell of drier climate, but the same effects could have resulted from increased energy and fertilizer costs or decreased crop prices. Testing such alternative explanations lends credence to the attribution of cause.

Case studies also might focus on especially vulnerable regions rather than on specific climate fluctuations. This was the approach in a series of projective case studies at the dry and cold margins of world agriculture conducted by the International Institute for Applied Systems Analysis and UNEP (see Parry et al, forthcoming). The Working Group on Socio-Economic Impacts, convened at the 1985 Villach Conference on the Greenhouse Effect, proposed particular focus on

Table 3. Agroclimatic zones in the tropics (Nix 1985, after Jaetzold 1983).

Belts of z. \ Main zones	0 (perihumid)	1 (humid)	2 (subhumid)	3 (semihumid)	4 (transitional)	5 (semiarid)	6 (arid)	7 (periarid)
TA Tropical Alpine zones Ann. mean 2-10 °C	Glacier Mountain swamps	I. Cattle-sheep zone			II. Sheep zone			High altitude deserts
UH Upper highland zones Ann. mean 10-15 °C Seasonal night frosts	F	Sheep-dairy zone	Pyrethrum-wheat zone	Wheat-barley zone	U. highland ranching zone	U. H. nomadism zone[4]		
LH Lower highland zones Ann. mean 15-18 °C M. min. 8-11 °C norm. no frost	o r e	Tea-dairy zone	Wheat/maize[2]-pyrethrum zone	Wheat/(M)[2]-barley	Cattle/sheep-barley zone	L. highland ranching zone	L. H. nomadism zone[4]	
UM Upper midland zones Ann. mean 18-21 °C M. min. 11-14 °C	s t	Coffee-tea zone	Main coffee zone	Marginal coffee zone	Sunflower-maize[3] zone	Livestock-sorghum zone	U. midland ranching zone	U. midland nom. zone[4]
LM Lower midland zones Ann. mean 21-24 °C M. min. > 14 °C	• Z o n	Lower midland sugarcane zone	Marginal sugarcane zone	Lower midland cotton zone	Marginal cotton zone[6]	L. midland livestock-millet zone	L. midland ranching zone	L. midland nom. zone[4]
L Lowland zones **IL** Inner lowland zones Ann. mean > 24 °C Mean max. > 31 °C	• e s	• Rice-taro zone	• Lowland sugarcane zone	• Lowland cotton zone	• Groundnut zone	Lowland livestock-millet zone	Lowland ranching zone	Lowland nom. zone[4]
CL Coastal lowland[5] Ann. mean > 24 °C Mean max. < 31 °C	•	• Cocoa-oilpalm zone	Lowland sugarcane zone	Coconut-cassava zone	Cashew nut-cassava zone	Lowland livestock-millet zone	Lowland ranching zone	Lowland nom. zone[4]

[1] Inner Tropics, different zonation towards the margins. The T for Tropical is left out in the thermal belts of zones (except at TA), because it is only necessary if other climates occur in the same country. The names of potentially leading crops were used to indicate the zones. These crops can also be grown in some other zones, but they are then normally less profitable. [2] Wheat or maize depending on farm scale, topography, a.o. [3] Maize is a good cash crop here, but maize also in LH 1, UM 1-3, LM, and L 1-4. [4] Nomadism, seminomadism and other forms of shifting grazing. [5] An exception because of the vicinity of cold currents are the tropical cold Coastal Lowlands cCL in Peru and Namibia. Ann. mean there between 18 and 24 °C. [6] In unimodal rainfall areas growing periods may be already too short for cotton. Then the zone could be called Lower Midland Sunflower-Maize Zone. •Not in Kenya.

three broad areas: 1) the tropics, with special emphasis on water resources and the effects of high temperatures on forests and health; 2) arctic areas, which are likely to experience the largest temperature changes associated with the greenhouse effect; and 3) coastal zones vulnerable to rising sea level. The working group also proposed regional sensitivity and impact assessment in the Baltic, Zambesi Basin, and Amazonia. Large-scale comparative studies also have been proposed between the North China Plain and the North American Great Plains, and the Baltic and the North American Great Lakes. These areas exhibit broad similarities in climate and human activity.

Integrated and comprehensive assessments

In an integrated assessment, the investigator follows the effects of a climate perturbation through and beyond a particular sector (such as agriculture), attempting to evaluate the impact on multiple aspects of the agricultural economy and the broader economy. There are no generally accepted guidelines for conducting integrated impact assessments, but a few good examples exist. Anderson (1978) used aggregation to examine the impacts of climate fluctuation on the Australian agricultural economy. He examined such factors as the characteristics of farmers, the likelihood of satisfactory farm performance, which depends on input decisions, yields, prices, or other measures of returns (e.g. nutrition), and government interventions. These functional relationships were applied to recent droughts that affected large areas of Australia, beginning with farm-level impacts and aggregating upward to national impacts.

Anderson traced the impacts of drought using established multiplier coefficients of changes in agricultural productivity as they affect other sectors. As his analysis proceeded, impacts became less distinct because of the canceling influences of good and bad effects in other sectors of the economy. Nevertheless, he estimated that climate is responsible for 38% of the variation in gross value of Australian agricultural production and a similar proportion of the variation in farm income. This translates into 1-2% changes in gross national product. However, if adverse agricultural conditions were to continue for several years, the net effect on gross output might be more than 10%.

A comprehensive view of climate impacts can be developed by linking several integrated assessments. Comprehensive assessments might focus on a particular climate event or scenario, tracing the impacts through multiple economic sectors to evaluate, as fully as possible, the numerous ramifications of climate in a region or nation. But good comprehensive assessment requires a level of interagency and interdisciplinary cooperation that is rarely found. Most so-called comprehensive assessments produce few new insights or measurements, but are simply lists of impact types.

Projecting future impacts

Policymakers are turning increasingly to climate impact assessment for information on potential impacts of climatic changes. Governments must be concerned with environmental trends that may reduce the availability of natural resources. The looming possibility in the near-future of climatic changes associated with the greenhouse effect provides the preeminent reason for attempting to anticipate future climate impacts. However, even without human perturbations, it is reasonable to expect that the future climate will exhibit roughly the same amounts of variability as climate has in the past. Thus, climate impact studies are needed whether or not human activity is changing the climate. Theories of human-induced climatic change are compelling, however, because they offer some indication of the likely direction and magnitude of future climate fluctuations. Using even the uncertain projections of future climate currently available, impact assessors can begin to identify the broad range of possible costs and benefits, as well as the policy options that are likely to help society cope with, or even benefit from, climatic change.

Projecting future climates

The first concern in conducting projective impact assessments is the choice of future climate scenarios. It is prudent to explore more than one type of climate projection in assessing potential future impacts. Three approaches can be used to develop scenarios for projective studies:

- historical analogue, where past climatic conditions are assumed to recur in the future;
- modeled climatic conditions based on statistical or dynamic models with altered boundary conditions; and
- convenient increments.

Each has weaknesses and strengths that must be weighed in choosing an approach applicable to the impacts problem at hand. Historical analogues can be very detailed but may not capture the full range of potential climatic changes. Models vary greatly in how well they simulate the climate: the popular models produce similar results, but differ enough that impact projections may vary with choice of model. In addition, modeled climate data are too coarse to be applied to detailed regional or local impact projections.

As an alternative to model data, simple convenient values (e.g. 10% changes in precipitation and 1-3 °C changes in temperature) consistent with past fluctuations at a certain place or with predictions based on model simulation (or both) are somewhat arbitrary but have several useful characteristics. Impact assessments based on ranges of such values are insensitive to swings in projections due to changes in modeling technology and they have the special capacity to facilitate intuitive, imaginative estimates of impacts. The obvious convenience of the numbers indicate that they are simply rough estimates. As such, in policymaking exercises they are less intimidating to resource managers. However, the greater detail of analogue or model scenarios may suggest more reliability than is warranted.

Translating climate projections into impacts

Once credible climate projections are selected, many of the approaches to impact assessment can be used to translate the climate scenario into projected impacts. Relationships between climate and natural resources, expressed as simple qualitative statements of the direction of impacts, indications of the rough ordering and ranking of different impacts, or more formal statistical or physical simulation models can be applied to extrapolate impacts.

The great problem here, of course, is that projections often must assume a rather static social and technological state, because of the difficulty of predicting future technological innovations and economic and social changes that will underlie, or perhaps even negate, future impacts. The most that can be accomplished, given our current ability to predict future technological and social change, is to lay out the broad trends apparent now and attempt to evaluate their implications. For example, is continued crop genetics research likely to produce agricultural systems more or less capable of handling future climatic change? Are there indications now on the probable future trends in energy systems, water management, health care, population dynamics, settlement and urban development, and national or international economic and political relationships that will increase or decrease society's ability to cope with projected climatic changes? At the least, the climate impact assessor should attempt to identify current trends which, if continued, would lead to large future problems, given climatic change. We should urge a policy of "least regret," where at a minimum, decisionmakers monitor trends that might worsen climate sensitivities or make segments of society vulnerable to climate-induced catastrophe.

Policy implications of projections

Because of the great uncertainty surrounding current estimates of future climatic changes, it is premature at this time to urge decisionmakers to modify drastically climate-sensitive policies, in anticipation of future impacts. The impact assessor's goal should be to use projections of climate and its impact chiefly to 1) suggest variables that should be monitored to identify and measure impacts in various sectors; 2) sensitize resource managers and policymakers to the nature and scale of potential impacts; and 3) encourage decisionmakers to avoid decisions that lock them into risky activities and to consider a broader range of alternative responses.

The latter use of impact projections may be the most critical. Decisionmakers' reluctance to consider uncertain forecasts in planning future activities is at least partly related to their eagerness to avoid initiating new policies that might later be shown to be incorrect or that would be made inappropriate by subsequent change. When some change is needed, decisionmakers often respond to the risk by making only incremental changes in past activities rather than by sharply altering policies. It may be possible for impact assessors to take advantage of this by pointing out that the goal of considering future climate possibilities in planning today is to offer an increased repertoire of small policy adjustments that might mitigate future impacts in an evolutionary manner, rather than eventually requiring large, more drastic policy changes.

Building impact assessment capabilities

Because governments must plan, monitor, and manage natural resources for the good of their people, it is important that they evaluate the potential for negative climate impacts and be prepared to alleviate associated social disruptions. This is particularly important in the developing world, where governments are making decisions now on water resource projects, agricultural development, forest management, energy systems, and infrastructural investments that will profoundly affect the course of their future development. Many of these investments extend across time-frames of decadal and longer climatic changes, such as those that might accompany the greenhouse effect.

Collaborative approaches

Climate impact assessment demands a multidisciplinary approach. No single investigator or area of expertise is broad enough to encompass the many and varied aspects of climate-society interaction, even in a local setting or within a single economic sector. At a minimum, climate impact assessment requires knowledge of the climate itself and how climatologists describe and monitor it, the socioeconomic activity or sector affected, and the techniques to trace effects as they ripple beyond the initial area of impact and impinge on different socioeconomic sectors. In addition to experts in, for example, climate, hydrology, agronomy, nutrition, economics, anthropology, and even psychology, an impact assessment team should, at various points in the assessment process, also include decisionmakers who must consider climate impacts in their plans. Their feedback and guidance can help assure that impact studies are addressing critical issues and making progress toward their solution. By involving policymakers in the process at key points, assessors can improve the credibility of their results and recommendations.

Creating multidisciplinary, multiagency research and assessment capabilities is difficult. Depending on the resource or economic activity affected, it is common for impact assessment responsibilities to be assigned to one particular government department. For instance, the agricultural bureau typically will be given the task of monitoring and ameliorating climate impacts on crop and livestock production; meanwhile, a different agency will assess impacts on water or energy resources.

Framers of the World Climate Program, with its emphasis on national climate programs, recognized the need to forge linkages between multiple research fields and government agencies. The program was designed to be broadly multidisciplinary and to encourage collaborative research on climate itself, data management, applications, and impact. Suggestions for establishing national programs include the concept of an intra-agency office that would coordinate the research efforts of all agencies with some climate components in their terms of reference. Or climate impact assessment capabilities might be developed within a mission agency, with provision for interagency consultation and sharing of data and personnel. Climate programs with impact assessment components exist in Hungary, Indonesia, and Japan. Other countries conduct coordinated climate impact studies under the aegis of the national meteorological service.

Developing assessment capabilities

Several steps apply to the establishment of any type of climate impact assessment structure by governments or scientific groups:

1. Identify the relevant experts and agencies that should be involved in a multidisciplinary assessment. This step may include forming an impacts

assessment network that can be called on for special studies, advice, and objective review of methods and findings. It also includes identifying the agencies and persons who should take part, perhaps building a roster of personnel who can be assigned to climate impact assessment projects as needed.

2. Assess the full range of data needs and availability. This critical step is often neglected. The data necessary to climate impact studies (e.g. climate, crop yield, water use, energy production, income, and health) are often collected and stored by several different agencies or research groups. Disparate data sets will have uneven coverage, quality, and accessibility. Impact assessors need to catalog the data, noting its coverage and availability, or build a central archive or clearinghouse for key data sets. A wide ranging impacts data catalog can be compiled relatively easily for distribution to impact assessors. Minimum information should include data type, archive location, form (e.g. printed records, computer tapes), coverage (dates and areas), and some indication of collection techniques and quality.
3. Develop links between assessors and policymakers. It is critical that decisionmakers in a position to reduce future climate vulnerabilities place credibility in the assessment process and its products. Assessors and decisionmakers must work together to assure that the process is aimed at critical problems and that it produces realistic and implemental solutions. This requires that assessment results be conveyed to decisionmakers in a succinct, timely, and understandable manner.

With these steps taken, ad hoc assessments can be initiated quickly and carried out efficiently. The expertise and data will be more easily brought to bear on problems as they arise. There is also a need to encourage ongoing climate and society research focused on enduring sensitivities and recurring problems. One example is a government-university partnership that combines research with impact assessment and the application of impacts knowledge to improve decisionmaking.

Maintaining an assessment capacity

In areas particularly sensitive to climate impacts, it would be useful to establish a standing impact assessment system to operationally assess the impact of climate on various economic activities. This might be accomplished within a national climate program structure, in an existing mission agency, or by establishing a new office with interagency responsibilities. A standing assessment capacity could have a relatively narrow focus. For example, following the 1982-83 El Niño, the Philippine Meteorological Service established a drought-watch office. The office's main goal is to monitor drought before it becomes severe, and to estimate actual and potential impact as it develops.

One of the most important roles a standing assessment office can fill is in briefing government officials and others involved in climate-sensitive policy formation. The continuing experience of climate impacts worldwide demonstrates the need to take climate into consideration as regions and nations develop. Better and more credible impact assessments can aid the development of more climate-proof planning, raising the security level of human activities dependent on climate-sensitive resources.

Applying assessments to decisionmaking

The impact assessor's goal should be to provide credible measures of climate sensitivities and estimates of the likelihood, magnitude, and distribution of current and future impacts. Such assessments can

- Raise the climate consciousness of decisionmakers,
- Improve planning for resources management in climate-sensitive areas,
- Provide early warning of developing impacts,
- Identify ways to ameliorate impacts and improve the efficiency of aid and relief programs, and
- Better prepare society for future climate impacts.

Recommendations

Many climate-sensitive regions lack adequate impact assessment capabilities. Our goal should be to extend climate impacts research and mitigation to these critical regions. Three recommendations that can further this goal are

1. Regional seminars focusing on how government officials and researchers can apply impact assessment methods to particular climate problems should be organized in selected, critical areas. Experts in impact studies would work with local investigators to establish assessment capabilities, assess overall climate sensitivities, and outline a program for monitoring and mitigating climate impacts.
2. A series of case studies should be developed. Actual climate fluctuations from the recent record of sensitive regions should be selected and multidisciplinary teams should be organized to assess the full range of impacts, perceptions, and adjustments associated with those fluctuations. Climate fluctuations will act as natural experiments, to shed light on how regions might respond to future climatic changes. Fluctuations affecting international resources or less well-studied regions (e.g. Southeast Asia) should be given priority.
3. We should consider establishing a program of post-impact mitigation, wherein government officials and researchers respond to specific climate-induced problems (e.g. the droughts associated with the 1982-83 El Niño), with the goal of making recommendations for lessening future climate impact sensitivity. Post-impact mitigation teams would have the benefit of a recent impact experience, heightened awareness on the part of policymakers, and knowledge of other postdisaster factors that offer a window of opportunity for mitigating future impact. The teams should be international, multidisciplinary, apolitical, and innovative in their suggestions for lessening future climate impacts.

References cited

Anderson J R (1978) Impacts of climate variability on Australian agriculture. Pages 1-36 *in* The impact of climate on Australian society and economy. Commonwealth Scientific and Industrial Research Organization, ed. Mordilliac, Australia.

Downing T E, Akon A J, Mungai D N, Muturi H R, Potter H L (1985) Climatic variability and agricultural production in central and eastern Kenya: introduction. National Environment Secretariat, Nairobi.

Heijnen J, Kates R W (1974) Drought in northeast Tanzania: comparative observations along a moisture gradient. Pages 105-114 *in* Natural hazards: local, national, global. G. F. White, ed. Oxford University Press, New York.

Jaetzold R (1983) Agro-ecological zonation in Kenya. Network Environ. Dev. 3 (December):1-4.

Jodha N S, Mascarenhas A C (1985) Adjustment in self-provisioning societies. Pages 437-464 *in* Climate impact assessment. R. W. Kates, J. Ausubel, and M. Berberian, eds. John Wiley and Sons, New York.

Karl T R, Riebsame W E (1984) The identification of 10- to 20-year temperature and precipitation fluctuations in the contiguous United States. J. Climatol. Appl. Meteorol. 23:950-966.

Kates R W, Ausubel J, Berberian M, eds. (1985) Climate impact assessment. John Wiley and Sons, Chichester, New York.

Nix H A (1985) Agriculture. Pages 105-130 *in* Climate impact assessment. R. W. Kates, J. Ausubel, and M. Berberian, eds. John Wiley and Sons, New York.

Parry M L, Carter T R, Konijn N T, eds. (?) Assessment of climate impacts on agriculture. Reidel, Dordrecht, Holland. (forthcoming)

Riebsame W E (?) Approaches to climate impact assessment. To be published by John Wiley and Sons, under the auspices of the United Nations Environment Program. (forthcoming)

Notes

Author's address: W.E. Riebsame, Department of Geography and Natural Hazards Research and Applications, Information Center, University of Colorado, Campus Box 482, Boulder, Colorado, 80309, USA.

Citation information: International Rice Research Institute (1989) Climate and food security. P.O. Box 933, Manila, Philippines.

Deforestation and climatic changes in the Amazon Basin

E. Salati

Brazil has no natural desert areas, but recent changes in the ecosystem of the Amazon indicate a tendency to desertification, especially where dense forest cover has been removed from sandy soils. A strong recirculation of water vapor from the Atlantic Ocean, brought in by year-round trade winds from the east, is evident. This vapor produces rains, of which as much as 75% is restored to the atmosphere through evaporation and exudation processes, depending on specific characteristics of each ecosystem of the Amazon region. Analysis has indicated that the Amazon region is a water vapor source for neighboring regions, including the Brazilian Central Plateau and some areas of the Andean Elevated Plain. Deforestation of an area as big as the Amazon (about 6 million km^2) might introduce serious changes in climatic conditions of adjacent areas and influence agricultural production programs. Reforestation, whether natural or artificial, is difficult in poor soil areas, especially in sandy soils, because topsoil has been severely degraded and many species of natural vegetation lost. Methods used as well as possible schemes and consequences are discussed.

Because of its geographic situation and geomorphology, Brazil has no desert regions. However, depending on the definition of a desert or on the concept of desertification, it can be said that in some regions the situation is changing, making continuous, profitable agricultural activity difficult.

Some authors working in Brazilian areas have indicated various changes in the original ecosystem that may lead to desert-like conditions (Ab'Saber 1977, Fearnside 1979, Nimer 1980, Salati and Marques 1984, Vasconcelos 1978).

- Sand dune movement following deforestation in areas close to coastal regions.
- Sand or sandy soil movement over agricultural areas when soil that was naturally covered with underbrush has been subjected to yearly cultivation.
- Deforestation in the Amazon region, especially in those areas covered with forests over sandy soils. Two to three million ha are deforested each year.

When cultivation is practiced in areas of sandy soils previously covered with dense forest vegetation, the nutrients in the top layer of the soil are quickly lost. Savannas are created, despite precipitation of more than 2,000 mm/yr. Some natural ecosystems with savanna type vegetation exist in the interior Amazon forest, others are beginning to emerge in recently deforested areas.

Climatic alterations are possible as a consequence of intensive deforestation and the introduction of agricultural practices incompatible with the humid tropics. These changes can occur within the Amazon Basin or extend to neighboring regions, such as the Brazilian Central Plateau and areas of the Andean Elevated Plain. The Brazilian Central Plateau, with a yearly rainfall of 1,200-1,300 mm, already is covered with natural savannas. Slight changes in rainfall intensity or distribution could have great effects on agricultural production in that region. Agriculture there is already limited due to low soil water retention capacity and short drought periods (veranicos) which occur during summer vegetative growth (December-March).

- Natural biomass reduction in the semiarid regions of northeast Brazil, where precipitation can be less than 400 mm/yr.

  This change (a consequence of the natural vegetation destruction) is characterized by xerophilous plants, accelerating erosion processes, and lixiviation of the thin soil layer existing over crystalline emerging rocks.
- Mineral exploration activities. Even though the Brazilian Legislation recommends soil replacement after mineral exploration, this is not always done, either because the law is not carried out or because the soils often are composed of a thick layer of organic matter over sandy soil. Once the organic matter layer is destroyed, natural soil recovery is practically impossible.

Here I deal with the possibility of changes in the Amazon region and their influence in neighboring areas. I also explain why agricultural techniques should be adapted to the humid tropics. The necessity for reforestation to reestablish water and thermal balance in densely deforested areas is evident.

The Amazon

The Amazon Basin, with approximately 6 million km^2, has one geomorphological characteristic that distinguishes it from other great hydrographic basins of the world. Its main channel formed by the Solimoes and Amazon Rivers has an average discharge of 175,000 m^3/s, 1/5 to 1/6 of all the water released into all the earth's oceans. This area has an extremely low declivity: a difference in level of about 100 m in 4,000 km. This central strip constitutes the Amazon Plain, which is covered with a forest extremely rich in vegetative species and that shelters varied fauna.

Surrounding this plain are, on the west, the Andes Mountains with average altitudes of more than 4,000 m; on the north, the Guiana Plateau with an average altitude of 1,000 m; and on the south, the Brazilian Central Plateau with average altitudes of about 700 m. The Amazon Basin, which is bisected by the Equator, has a horseshoe shape with the opening toward the Atlantic Ocean. The water vapor that gives it its humid tropical characteristics comes from the Atlantic and is brought into the region by trade winds blowing from the eastern quadrant (Salati and Marques 1984, Salati and Vose 1984).

Colonization of the Amazon started shortly after the discovery of the Americas. Both the Portuguese and the Spaniards explored the resources of the region. The Portuguese went up the Amazon River, east to west; the Spaniards came from west to east, after Orellana descended the Amazon in 1541.

During the 400-yr colonization period, the Amazon ecosystem did not change greatly in its ecology. However, more recently, especially since 1960, ever greater pressure has been exerted on the dynamic balance of the region.

In a general way, we can say that up to modern times, the Amazon system maintained a dynamic soil-plant-atmosphere balance. During the past 15 yr, many researchers have tried to establish the dependency relationships among the plants, animals, soil, and the hydric system of the region. Others have tried to characterize the various stages of ecosystem evolution to reach the current balance. Five recent books reflect and summarize current scientific information: Salati et al (1983), Sioli (1984), Prance and Lovejoy (1986), Fearnside (1986), and Dickenson (1987).

Changes following colonization

The changes introduced in the region during the various stages of colonization can be summarized as follows:

Changes in native communities

When the Portuguese and Spaniards started their South American colonization, they found various peoples with different cultures. In the Amazon area, the Incas, who had greater social development, lived in the elevated parts of the Andean region and apparently exerted some influence even over the lower altitude areas of West Amazon. Different civilizations existed in the lower Amazon: The most advanced were the inhabitants of the Marajo' Island, in the mouth of the Amazon, who settled there from 400 B.C. to 1350 A.D.

With the European colonizers' arrival, the native communities of the lower Amazon, who were living in an almost Neolithic development stage, were easily conquered. It is hard to estimate the size of the population of the lower Amazon at the beginning of colonization; however, according to some surveys there were more than one million inhabitants. The pressure exerted on the native communities, as well as the gradually increasing entry of colonists, reduced the number of Indians in the Brazilian Amazon to less than 50,000 within five centuries. This pressure still persists, and the future of the native communities depends primarily on the land possession policies established in the region.

Change of fauna

Almost all species of the Amazon region fauna have been subjected to continual pressure. At the beginning of colonization, the high value of the skins of animals (such as jaguars and manatees) and the great beauty of bird plumage led to their devastation. As time went by, in addition to the ever-increasing demand for animal skins, a market developed for the meat of some species (alligators, tortoises, tapirs). Thus, the killing continued.

Nowadays, even though hunting and trading of animal skins are forbidden, the practices continue, and the pressure exerted on the fauna of the region is constant. (During a trip in March 1986, on the highway connecting Cuiaba to Porto Velho I came across several trucks and cars carrying live wild animals.)

To protect wildlife, special reserves have been founded by the IBDF (Brazilian Forest Development Institute) and by SEMA (Environment Special Secretariat). However, countless invasions in those reserve areas persist. Until serious control and repression measures are adopted, the fauna will continue to be threatened.

Changes in soil cover

A great change in the ecological balance of the Amazon system began recently, with the building of roads crossing the region from north to south and from east to west. With this construction, the Amazon "firm lands" were made accessible to colonizers. Colonization incentives by INCRA (Colonization and Agrarian Reform National Institute) and by SUDAM (Amazon Development Superintendency) have led to countless investments in attempts to develop agriculture in the humid tropics. As a result of such intensive colonization, some areas of the Amazon have been greatly deforested (Fearnside and Salati 1985). In some communities of Rondônia, more than 50% of the forests were cut down in just 10 yr; it is estimated that about 20% of the vegetative cover in Rondônia and Mato Grosso has already been replaced by other forms of soil use.

In Amazon as a whole, exponential deforestation already must have altered more than 10% of the dense forest area. Yet in most cases, it has not been possible to establish agricultural production. Often, the land is simply abandoned and repopulated by secondary forest, of little ecological and no economic value.

Concerns about the possibilities of ecological alterations occurring on a regional, and eventually on a global basis began in response to this intensive colonization, especially during the last 15 yr.

Researchers acknowledge that changes are introduced in biogeochemical cycles on local and regional levels when forests, primarily heterogeneous ones, are replaced by yearly or perennial monocultures. However, the great doubt is whether those changes might induce climatic alterations in the region itself and in adjacent areas.

Through satellite data, estimates are that 45 million ha have been deforested during the last 2 decades, and that the current deforestation rate is about 2.5-3 million ha/yr. It is most important to stress that it has been very difficult to establish sustainable agriculture in deforested areas (Fearnside 1986). The agricultural ecosystem decays, especially as a result of decreased soil fertility.

Deforestation and climatic changes

The Amazon Basin hydric balance has been studied by several authors (Leopoldo et al 1982b, Lettau et al 1979, Villa Nova et al 1976). The general conclusion is that the Amazon Basin receives $11.87 \times 10^{12} m^3/yr$ of rainwater and loses $6.43 \times 10^{12} m^3/yr$ through evapotranspiration and $5.45 \times 10^{12} m^3/yr$ through the Amazon River discharge. The average mass of water vapor in the region is 24×10^{10} t. In some places, such as in dense forest areas, the loss of water through evapotranspiration can be higher.

Systematic measurements in a model basin have shown a hydric balance such that 75% of the rainwater is returned to the atmosphere through evaporation and transpiration, while just 25% is discharged through the small river formed in that hydrographic basin (Leopoldo et al 1982a).

Two distinct methods have been used to evaluate water vapor recycling.

The first method to evaluate water vapor recycling in the atmosphere used values of water vapor flux obtained through radiosonde. Studies of the vertical distribution of water vapor flux as well as its deviation, summarized by Salati and Marques (1984), show that

- Water vapor entering the region comes from the Atlantic Ocean, and is of the same order value as the vapor produced by evapotranspiration within the Amazon Basin itself—approximately 6.4×10^{12} t/yr.
- Precipitation, in amount and distribution, cannot be explained simply by the vapor condensation coming from the ocean.
- A water vapor flux exists from the Amazon to adjacent regions, especially to the southwest and west.

Another method to assess the extent of water vapor recycling used ^{18}O & D concentrations in rain and river waters of different regions (Salati et al 1979). This isotopic method estimates how much water vapor leaves the region and its isotopic concentration.

Systematic measurements of isotopic concentrations in water vapor, rainwater, and river water have been carried out in different years. The decrease in ^{18}O concentration was less than that expected in the process of continuous removal of water vapor through precipitation. This could be a consequence of water vapor produced by evapotranspiration mixing with oceanic vapor, so that rain in a certain place is formed by a mixture of the two.

The hypothesis of water vapor recycling leads to ^{18}O values closer to those observed in nature.

This suggests that water vapor recycling must exist in the atmosphere within the Amazon Basin, and must be relevant to the existing water budget. It is evident that the changes the Amazon forest is enduring will alter the water and energy balance.

In deforested areas of poor soils, the establishment of a sustained agriculture is difficult and involves a change of hydric and energy balance. Two situations have occurred:

1. After deforestation and 2 or 3 yr of cultivation, the soils are abandoned, supposedly to regenerate naturally.

   What is happening, however, is new growth of herbs and shrubbery, forming a savanna, on the cleared land. The degree of that change depends on the type of soil and the size of the deforested area. When large areas are deforested, the natural germplasm bank for regeneration no longer exists, and many plant species are lost.
2. Pasture is planted for extensive cattle raising.

   In many cases, this agricultural practice has not given the expected results. What does happen is pasture degradation, with an invasion of plants that are sometimes poisonous to the cattle. Efficient pasture are difficult to establish, mainly due to lack of phosphorus.

   The most recent experience is an explosive deforestation rate in Rondônia, as a consequence of colonization programs. At the start, the programs had some purposes and dimensions; in reality, what occurred was an uncontrollable devastation.

In just 15 yr, almost 30% of the state, 6 million ha, has been deforested, without any control through government actions. The situation is chaotic, and the environmental degradation will be very hard to control. The ecosystem biomass has decreased. Degraded savannas are replacing the forests (SEMA 1986).

Based on the results of past experiences and an assessment of potential problems, the following steps are suggested for rational use of land in the humid tropics:

1. Preference should be given to forest handling projects.
2. "Extractive reserves" for forest products such as rubber, chestnut, and medicinal herbs should be established.
3. Agro-silviculture projects should be encouraged.
4. Extensive cattle raising should be avoided.
5. Crops requiring yearly cultivation should be established only in richer soils, with cropping accompanied by soil conservation measures.
6. Fruit tree cultivation, both for the fresh-fruit trade and for processing, should be stimulated.
7. Cutting forests in the poorest sandy soils should be prohibited.
8. Large-scale monocultures should be avoided.
9. Ecological and economic planning, taking into consideration potential climatic changes in the Amazon Basin and adjacent areas, should be organized immediately.
10. Wood trade, both national and international, should be rationalized and controlled.

References cited

Ab' Saber A N (1977) Problemática da desertificaçao e da savanizaçao no Brasil intertropical. Geomorfologia 53.

Dickinson R E ed. (1987) The geophysiology of Amazônia - vegetation and climate interactions. John Wiley and Sons, New York. (in press)

Fearnside P M (1979) O processo de desertificaçao e os riscos de sua ocorrencia no Brasil. Acta Amazônica 912.

Fearnside P M (1986) Human carrying capacity of the Brazilian rain forest. Columbia University Press, New York. 293 p.

Fearnside P M, Salati E (1985) Rondônia: sem florestas na proxima década. Ciência Hoje 4(19):92-94.

Leopoldo P R, Franken W, Matsui E, Salati E (1982a) Estimativa da evapotranspiraçao da floresta Amazônica de terra firme. Acta Amazônica 12(3):23-28.

Leopoldo P R, Matsui E, Salati E, Franken W, Ribeiro M N G (1982b) Composicçao isotopica das preciptaçoes e da água do solo em floresta Amazônica do tipo terra firme na regiao de Manaus. Acta Amazônica 12(3):7-13.

Lettau H, Lettau K, Molion L C B (1979) Amazônia's hydrologic cycle and the role of atmosphere recycling deforestation effects. Mon. Weather Rev. 107(3):227-238.

Nimer E (1980) Subsidio ao plano de açao mundial para combater a desertificaçao. R. Bras. Geogr. Rio de Janeiro 42(3):612-637.

Prance G T, Lovejoy T E, eds. (1985) Amazonia. International Union for Conservation of Nature and Natural Resources, Pergamon Press, New York. 442 p.

Salati E, Dall'Olio A, Matsui E, Gat J (1979) Recycling of water in the Amazon Basin; an isotope study. Water Resour. Res. 15:1250-1258.

Salati E, Marques J (1984) Climatology of the Amazon Region. Pages 85-126 *in* The Amazon. Limnology and landscape ecology of a mighty tropical river and its basin. H. Sioli, ed. Dr. W. Junk Publishers, Boston.
Salati E, Shubart H R, Junk W, Oliveira A E (1983) Amazônia, desenvolvimento integraçao e ecologia. Editora Brasiliense, Sao Paulo. 327 p.
Salati E, Vose P B (1984) Amazon basin: a system in equilibrium. Science 225:129-138.
SEMA—Secretaria Especial do Meio Ambiente (1986) Diretrizes ambientais para o Estado de Rondônia. Brasilia, Brasil. 222 p.
Sioli H, ed. (1984) The Amazon. Dr. W. Junk Publishers, Boston. 763 p.
Vasconcelos S J (1978) Processos de desertificaçao do Nordeste. Sudene/Sema-12. Recife Brasil.
Villa Nova N A, Salati E, Matsui E (1976) Estimativa de evapotranspiraçao na Bacia Amazônica. Acta Amazônica 6(2):215-228.

Notes

Author's address: E. Salati, Salim Farah Maluf Foundation, Rua Tagipuru, 235—2° andan CEP 01156, Sao Paulo-SP, Brazil.

Citation information: International Rice Research Institute (1989) Climate and food security. P.O. Box 933, Manila, Philippines.

Exploiting the fertilizer effect of increasing atmospheric carbon dioxide

R. M. Gifford

High CO_2 concentrations fertilize plants by stimulating photosynthesis, suppressing photorespiration, and reducing transpiration per unit leaf area. CO_2 enhancement of growth occurs at both optimal and nonoptimal levels of other environmental variables (light, water, temperature, nitrogen nutrients, salinity). Severely phosphate-deficient plants may not respond to higher CO_2 concentrations. The globally increasing CO_2 concentration, therefore, represents an improving component of the fitness of the environment for secure food production. This will partially counter any deteriorating aspects of agricultural environments (e.g. adverse climatic change, soil loss and deterioration, acid precipitation). Because yield increase percentages in response to high CO_2 are larger for drought- and salt-stressed plants than for nonstressed plants, some marginal cropping sites (e.g. on arid boundaries) may show less year-to-year yield variation. This would be an improvement in the stability of food production from such sites. Because C_3 species will benefit more than C_4 species, substitution of C_3 for C_4 crops may become more worthwhile. Communities with access to fertilizer may be better able to exploit higher C_2 atmospheres. Cropping boundaries may move onto more saline and drought-prone soils, although this would probably be bad policy in the long term. Genetic variation in within-species responsiveness to high CO_2 may enable the breeding of cultivars to take greater advantage of a high CO_2 atmosphere.

The increasing atmospheric carbon dioxide (CO_2) concentration has three clearly identified primary effects that might have an impact on food production:

- Stimulation of plant net photosynthesis,
- Reduction in stomatal aperture, and
- Increase in the absorption of thermal back-radiation from the earth (i.e. the greenhouse effect).

Most public discussion of CO_2 effects is related to climatic implications of the greenhouse effect. The other two primary responses, which together constitute the CO_2 fertilizer effect, may be equally important to global food production and security.

Only infrequently do agroclimatic models of crop yield incorporate atmospheric CO_2 concentrations, even though many deal explicitly with photosynthesis and stomatal conductance (Reynolds and Acock 1985). CO_2 concentration has been increasing for a century, from a pre-industrial level of about 270 to the current 350 ppm (Gammon et al 1985). Some of the yield increases that we ascribe to cultural and genetic improvement may in fact be due to the change in atmospheric composition. With the potential for a sixfold increase in atmospheric CO_2 by the time the world's fossil fuels are exhausted (Gifford 1982), the direct effects of change in atmospheric CO_2 should be combined with the indirect effects of climatic change in yield-predictive models (Penning de Vries et al, this volume).

Two aspects of the security of food production are considered here. The first concerns whether the trend of food crop productive potential of any region will hold constant, increase, or decrease with time. This is important to long-term environmental planning, population planning, and international trade. The second concerns the size of year-to-year variation in agricultural output owing to variable weather. This is important to those concerned with food storage, international aid, and food crisis management. How does increasing atmospheric CO_2 affect these two aspects of food security?

Direct impact of CO_2 concentrations on crop yield

Biochemically, higher CO_2 concentrations are expected to both increase CO_2 fixation and decrease photorespiration rates of C_3 species (e.g. temperate climate cereals, rice, sugar beet), while having little or no effect on C_4 species with biochemical CO_2-concentrating mechanisms (e.g. maize, sorghum, sugarcane) (Tolbert 1983, Warrick and Gifford 1986). These expectations are borne out in CO_2 response curves of whole leaf photosynthesis (Gifford 1979b.) Despite the occasional downward acclimation of leaf photosynthesis after several days' exposure (Cure 1985), continuous CO_2 enrichment usually increases growth and yield of C_3 species. Even in C_4 species, some growth increase is sometimes observed, but the mechanism is unclear. In one analysis, high CO_2 did not increase the net photosynthetic efficiency of spaced C_4 plants but did increase leaf area, light interception, and hence growth (Morison and Gifford 1984). For C_3 species, the most common response is a 20-40% yield increase for a doubling of CO_2 concentration. However, the responses reported range from –20 to +600% (Kimball 1983). While those extremes are probably erroneous, much of the variation probably represents true species or genotypic differences. Cotton seems noteworthy for its high CO_2 responsiveness. Some variation may reflect interactions between CO_2 concentration and other growth-limiting environmental variables.

It is a common misconception that other growth-limiting factors will eliminate CO_2 responsiveness. High CO_2 concentration can increase the efficiency of use of growth-limiting resources (Warrick and Gifford 1986). For example, at low light intensity, CO_2 enrichment can have a pronounced effect on plant growth. Although the absolute responsiveness to high CO_2 is generally less for plants grown at low than at high light intensity, the percent yield increase can be greater for plants grown at low light intensity (Gifford 1979a). There are at least two reasons for the high CO_2 sensitivity of light-limited plant growth. First, for C_3 species, quantum efficiency (i.e. the initial slope of the light response curve of photosynthesis) is increased by high CO_2 because of competitive inhibition of the oxygenase activity of the carboxylating enzyme (Rubisco) (Ehleringer and Björkman 1977). Second, the light compensation point, at which photosynthesis just balances whole-plant true respiration, is decreased because of the increased initial slope of the photosynthesis v light curve. Thus, a plant that is not growing because of its being at the light compensation point will be moved into positive C-balance if CO_2 is increased, thereby experiencing infinite percent growth enhancement. There may be other causes in some species, such as the recently reported partial suppression of dark respiration by high CO_2 concentration (Gifford et al 1985).

This means that for crops that grow in poor light, such as rainy-season tropical crops, percent yield increase at high CO_2 levels should not be limited by low light. Broadly speaking, the efficiency (ϵ) of conversion of intercepted radiation into plant dry matter in normal atmosphere is a fairly conservative property of vegetation over a wide range of species and incident radiation levels (Monteith 1977). In our studies, doubling the present CO_2 concentration of the atmosphere caused an approximately one-fifth increase in ϵ of wheat canopies (Morison and Gifford 1984, unpubl.).

Yields of crops grown under drought conditions also respond to higher CO_2 concentrations. The greater the drought stress, the greater the relative response to CO_2 (Gifford 1979c) even in C_4 species (Gifford and Morison 1985). Reasons for this might include

- Reduced rate of water loss. The lower stomatal conductance in high CO_2 is more significant to yield of a crop suffering water deficiency (i.e. under water deficits the stomatal effect of high CO_2 as well as the photosynthetic effect is significant),
- Improved osmoregulation (Sionit et al 1980), and
- Increased dark respiration due to stress, and hence increased impact of a photosynthetic increment on the diurnal carbon balance of the crop.

Thus, crops growing in semiarid environments would be expected to have as great or greater percent yield increase under high CO_2 as do well-watered crops.

Crops with nutrient deficiencies can also respond positively to higher CO_2 concentrations, although that general statement and the mechanisms that explain it await further investigation. Nitrogen-limited plants do respond with increased growth to high CO_2 during growth (Goudriaan and de Ruiter 1983, Wong 1979). In some cases the percent response was for deficient plants as much as for nitrogen-sufficient plants. Part of the mechanism is that the C:N ratio of the vegetation is greater for high-CO_2-grown plants. If this C:N ratio changes in the commercially harvested part of the crop, nutritional quality may be less. If the protein content of food products is to be sustained, any CO_2-induced increase in production will require a commensurate increase in nitrogen fertilizer use.

While the interaction with phosphate has been little studied, indications are that phosphorus deficiency substantially reduces the CO_2 responsiveness of growth (Goudriaan and de Ruiter 1983).

Another important stress in semiarid areas is soil salinity. What little research has been done on CO_2 effects on salt-affected plants indicates that high CO_2 reduces the severity of yield reduction by salt (Gale 1980). This might be because reduction in stomatal conductance reduces the transpirational need to take up brackish water and, as for other stresses, the presence of stress-induced respiration leads to a relatively larger effect of CO_2 enrichment on the plant's diurnal carbon balance.

Warm environments seem more likely to foster CO_2-induced yield increase than cool ones. This arises from a combination of the relative solubilities of CO_2 and O_2 as a function of temperature (Jordan and Ogren 1984) and from the temperature sensitivity of the kinetic parameters of the primary carboxylating enzyme of C_3 photosynthesis. At the single-leaf level, the optimum temperature for net photosynthesis in fact increases as a function of CO_2 concentration (Berry and Raison 1981). At the whole-plant level, one might expect the less favorable balance between gross photosynthesis and whole-plant respiration also to lead to a greater responsiveness of growth to high CO_2 concentrations. In practice, however, quantitative responses in controlled environments are more complex and not always easy to interpret (Warrick and Gifford 1986).

Overall it seems safe to conclude that the fertilizer effect of globally increasing CO_2, acting alone on food production and hence on long-term food security, is positive everywhere. The magnitude of the stimulus to production will vary with region, genotype, and inputs, but will everywhere tend to counter any loss of food security due to climatic change, soil deterioration, acid precipitation, etc. When fossil fuel resources are exhausted and atmospheric CO_2 concentrations begin very slowly to decline again, over millenia, there will be a progressive loss in food security. But if the fossil fuels are depleted on a plausible time scale, that is several centuries away.

CO_2 and yield variability

To the extent that year-to-year yield variations at specific sites are due to rainfall variation, salinity stress, or episodes of high or low temperatures, higher atmospheric CO_2 concentrations may reduce the coefficient of variation of yield. Because high CO_2 concentrations cause greater percent yield increase in drought- and salinity-stressed plants than in nonstressed plants, higher CO_2 will probably reduce the yield difference between stressed and nonstressed crops. Similarly, high CO_2 will attenuate the impact of a high temperature spell on yield, since high CO_2 increases the optimum temperature for net photosynthesis, although the optimum temperature range is narrower than it is at low CO_2 (Acock and Allen 1985, Berry and Raison 1981). Because enhanced maintenance respiration is a component contributor to the decline in growth at high temperature, I would expect atmospheric CO_2 to improve the diurnal carbon balance more at high than at optimum temperatures, especially where maintenance respiration is reduced by high CO_2 itself (Gifford et al 1985, Silsbury and Stevens 1984).

Although there are good reasons to believe that crops growing in low temperatures will respond only weakly to high CO_2, Sionit et al (1981) found the opposite effect with the tropical chilling-sensitive species okra. High CO_2 made some yield possible at a temperature too cool for survival in normal air. Thus, it is possible that yield variability due to cool weather may be less for some crops in high-CO_2 environments.

Pests and diseases also cause considerable yield variation. The impact of high CO_2 concentrations on pest and disease incidence has barely been investigated. In general, populations of herbivorous insects are held in check by their ability to acquire enough nitrogen compounds from healthy unstressed plants (White 1984): plant matter is just too poor in available nitrogen and too rich in carbon. In 24-h feeding experiments, soybean looper larvae ate much more leaf on plants grown at 650 ppm CO_2 than on normal plants. The low leaf N content of the high-CO_2 leaves not only increased 24-h consumption, it also decreased larval growth rate, increased frass production, and reduced conversion efficiency (Lincoln et al 1984). These nutritional effects presumably would reduce fecundity. Thus, in the longer term, vegetation grown in high CO_2 concentrations would be unable to support large insect populations. Butler (1985) found this in flea beetle, pink bollworm, leafhopper, and predaceous fly on CO_2-enriched cotton, with low available nitrogen.

It is difficult to predict how the balance among rates of individual insect feeding, population size, and number of species able to thrive on a crop growing in high CO_2 atmospheres will work out in terms of crop damage and frequency of crop failure. The situation with plant disease is even more difficult to document.

Lodging is a common source of yield variability; high CO_2 concentrations may exacerbate this. In some species, plants grown in high CO_2 concentrations are taller than control plants (Acock and Allen 1985). Consequently, it is possible that some of the potential yield increase due to high CO_2 may be lost by increased lodging. Nevertheless, height does not necessarily increase as much as stem dry weight (Acock and Allen 1985), and strength of the stalks might well increase in high-CO_2 atmospheres.

Overall, it is not possible to make a straightforward statement about the direct effect of high CO_2 on interannual yield variability, even when changes in weather patterns are ignored. It depends on the source of the yield variability. In some places, yield would be more variable; in others, it could be less variable.

Maximizing CO_2 responsiveness of crop production

There are both managerial and breeding opportunities to amplify the potential benefits of higher CO_2 concentration on crop yields. Because high CO_2 benefits the yield of C_3 species more than it does that of C_4 species, C_3 species will become increasingly more competitive economically, all else being equal. For example, sugar beet, a temperate C_3 crop grown by several major sugar-consuming nations, will gain some advantage over sugarcane, a tropical C_4 crop often grown by sugar-exporting nations. Plans over the long term to find alternatives to cane sugar for export might be advisable for countries whose sugar operation is economically marginal. For staple starchy C_4 crops like maize and sorghum, gradual partial substitution by CO_2-responsive C_3 alternatives like sweet potato, cassava, rice, or wheat bred for high-temperature areas might be possible.

Farmers who are able to buy fertilizers will be in a better position to maximize the fertilizer effects of CO_2. Nitrogen and phosphate deficiency can reduce the responsiveness of yield to CO_2. In the face of increased nutrient demand by CO_2-

stimulated crops, fertilizer application would maximize the increased food potential as well as maintain soil mineral status.

The increase in crop biomass is likely to also increase crop residue, since harvest index is barely altered by high CO_2 (Cure 1985, Gifford 1977). This increase in crop residue could, if managed to increase soil organic matter content and surface litter, bring further increases in yield via greater soil fertility, better water absorption, and reduced erosion. But that requires long-term investigation to confirm.

The greater lodging propensity possibility could be accommodated by introducing shorter-straw varieties. Dwarf varieties are already available, although in general, they yield lower than the semidwarfs. It could be that in attempting to overcome lodging-related yield variability engendered by high CO_2, some of the potential yield increase might have to be foregone, especially if CO_2-stimulated weeds were more difficult to control.

Broadly speaking, high CO_2 concentrations have little direct impact on crop longevity. Observations of slightly accelerated senescence (Chang 1975, St. Omer and Horvath 1983) are not universal. Flowering can be hastened in some species but retarded in others, as has been found in sorghum (Marc and Gifford 1983). A managerial option of changing the maturity-type of the cultivar might be worthwhile for fine tuning a species to a higher CO_2 atmosphere.

Higher CO_2 concentrations also open the option to move cropping onto more marginal and stressful sites, particularly at hot, arid, and saline boundaries. This is because high CO_2 enables plants to better withstand such stresses. Although exercising such an option would secure more food in the short term, it would also undermine the possible CO_2-derived benefits of more stable soil and less season-to-season variability. Nevertheless, where population growth continues to be paced by the ability of the community to acquire subsistence food and/or money, the same socioeconomic factors that drive people to produce food and cash crops on highly variable, low-yielding, and edaphically vulnerable cropping margins would presumably continue to drive them to exploit the variable margins established under high-CO_2 atmospheres.

The great range in growth responsiveness to CO_2 enrichment reported in the literature suggests genetic variation. This has had little investigation. One analysis of data on lettuce that had been selected for horticultural attributes in either cool and ambient CO_2 conditions over six generations, or warm plus 3× ambient CO_2 conditions, did not reveal any effect of the CO_2 environment on the CO_2 sensitivity of fresh-weight yield (Maxon-Smith 1977). However, there has been no selection for dry weight growth at low or high atmospheric CO_2 alone.

The primary photosynthetic carboxylase enzyme (ribulose biphosphate carboxylase, cor Rubisco) in higher plants appears to have evolved to the maximum affinity for CO_2 relative to its competitive inhibitor, O_2, which gives rise to photorespiration. All attempts to increase the affinity for CO_2 of the higher plant carboxylase have failed (Gifford 1987). In normal air, Rubisco is not saturated with CO_2. As atmospheric CO_2 increases, the intracellular concentration of CO_2 will move closer to the substrate-saturated plateau of the enzyme.

It seems that a price of high affinity of an enzyme for its substrate may be a relatively low maximum (substrate-saturated) velocity of reaction. Apparently evolution has adjusted the kinetics of the enzyme to arrive at a compromise, between tight CO_2 binding (to compete with O_2 binding) and slow turnover of the reaction because of the tight CO_2 binding. The optimum position of this compromise is presumably a function of the prevailing atmospheric CO_2 concentration. In more primitive plant species which evolved in early high-CO_2 atmospheres, there are many versions of Rubisco that have a lower specificity for CO_2 relative to O_2. It may be possible to breed plants with photosynthetic CO_2 fixation enzymes that can take better advantage of high atmospheric CO_2 levels by having a higher saturated velocity of action, at some expense of specificity for CO_2 binding. Such a breeding objective is the opposite of the present objective of that research field.

The inhibitory effect of high CO_2 concentrations on true respiration, at least in some species (Gifford et al 1985), might offer another possibility for breeding plants better adapted to high CO_2. But so little is known of this unexpected response, little can be said about its prospects.

CO_2 fertilizer effect in relation to climatic change

It is believed that a doubling of CO_2 concentration will cause an approximately 3 °C (±1.5 °C) increase in global mean surface temperature. Using ecological data from the International Biological Program for permanently vegetated sites in diverse ecosystems at various latitudes around the world, Lieth (1972) derived an overall expression for the temperature dependence of annual net primary productivity (NPP, t/ha) of natural plant communities:

$$NPP = 30/\ [1 + e^{(1.315 - 0.119\ T)}] \quad (1)$$

where T is the annual mean temperature (°C).

The equation smooths over the considerable variation in productivity attributable to the presence of other limitations to plant growth (water, nutrients, light, etc.). Using equation 1, the effect of a 1 °C temperature increase on productivity ranges from an 8% increase at 5 °C to a 1% increase at 30 °C. Thus, in the middle temperature range, where most vegetation and crops grow, a 3 °C temperature increase attributable to a CO_2 doubling would lead to approximately 15% higher annual productivity of perennial vegetation such as forest or perennial grassland, only half as much as the yield increase expected from the direct CO_2 fertilizer effect itself. Acting together, these two effects might increase productivity of perennial vegetation 40-50%. These calculations must be seen for what they are—pro tem rules of thumb—but they do give a heuristic insight into approximate magnitudes.

Unfortunately, most human food is not produced from perennial vegetation but from annual, mostly cereal, crops which senesce and die after seed formation. Warmer conditions have a much more powerful effect on increasing the rate of plant development than on increasing biomass productivity per unit time. Thus, for a given annual genotype, grain yield is negatively related to seasonal average temperature because of a shortened period from sowing to maturity at warmer temperatures. For example, for wheat yield in the British East Midlands, Monteith (1981) noted a 6% decline per °C increase in mean summer temperature. By how much would mean temperature need to increase to annul the direct CO_2 fertilizer effect on the grain yield of a given genotype? For wheat grown with water and nutrients adequate to produce a leaf canopy that intercepts all the incident radiation by the time spikelet growth begins, warm temperature reduces yield by reducing the number of grains per m^2 (N), according to

$$N = (11 \times 10^3 R_f)/(T_f - 4.5) \tag{2}$$

and by decreasing mean kernel weight (W mg) according to

$$W = 64 - (1.6\ T_g) \tag{3}$$

where R_f is mean radiation (MJ/m^2 per d) and T_f is mean temperature (°C) during spikelet formation, 4.5 °C is a "base temperature" for development, and T_g is mean temperature (°C) during grain filling. These relations, developed by Fischer (1985) from diverse experimental data, can be used to predict yield potential of semidwarf wheats as a function of temperature and radiation. If one assumes as a broadly generalized rule of thumb (Gifford 1977, 1979c; Kimball 1983) that for all relevant conditions of temperature and radiation, a 100% increase in CO_2 concentration causes a 30% increase in wheat yield, the following composite equation allows comparison of the greenhouse effect and the CO_2 fertilizer effect on wheat yield:

$$\text{Yield } (g/m^2) = \frac{R_f\,(700 - 17T_g)^{[1 + 0.3\,(C_a - 340)/340]}}{(T_f - 4.5)} \tag{4}$$

where C_a is the atmospheric CO_2 concentration in ppm.

From data presented by Nix (1975) for conditions pertaining to experimental wheat crops at various latitudes, I have used equation 4 to calculate the temperature increase that would be required to reduce yield by an amount equal to the yield increase due to a CO_2 concentration of 680 ppm (Table 1). It ranges from 1.5 °C to 2.4 °C, averaging about 2 °C.

Table 1. The increase in mean temperature (ΔT) during both spikelet development and grain filling of semidwarf wheats required to annul an assumed 30% yield increase attributable to a doubling of CO_2 concentration. The temperature/radiation model used is that of Fischer (1985); examples of wheat environments are from Nix (1975).

Site	Mean temperature		Mean radiation (MJ/m^2 per d)	"ΔT" (°C)
	Before anthesis	Grain filling		
Swift Current, Canada (50° N)	17.5	18.0	23.0	2.2
Horsham, Australia (37° S)	10.9	16.7	13.8	1.5
Dubbo, Australia (32° S)	12.0	18.0	16.3	1.7
New Delhi, India (28° N)	14.0	22.0	16.7	1.9
Clermont, Australia (23° S)	15.8	21.1	17.6	2.1
Kimberley, Australia (15° S)	23.0	24.2	19.7	2.4

This first-order analysis indicates that the negative effect of a 3 °C temperature increase would slightly exceed the positive direct effect of a CO_2 doubling on wheat grain yield, assuming all other factors being equal. However, all other factors would not be equal. A farmer could choose cultivars that mature more slowly to counteract the hastening effect of warmer conditions on plant development, or he could sow and harvest earlier, or, in year-round growing environments, he could grow more crops per year. I suspect that with appropriate managerial changes, selection of adapted genotypes, and specifically breeding, there should be opportunity to bypass the first-order yield-reducing effect of warmer conditions on annual crops and, indeed, to capture in grain yield some of the potential net primary productivity indicated by equation 1. Such potential for yield increase would be in addition to the increase from the direct CO_2 fertilizer effect. The effect of any change in rainfall could also be substantial. Apparently this could be positive or negative, depending on the location.

Conclusions and research recommendations

To predict the quantitative effect of higher atmospheric CO_2 concentration on particular species in specific regions will require crop yield modeling combined with experimental investigation of numerous aspects. Many crop models are now being used or developed for yield prediction, agricultural management, and agrotechnology transfer. However, at present there is minimal interaction among crop modelers, those concerned with CO_2 fertilizer effects, and those studying climatic change. We need to find a means to facilitate such contact, to design models valid for future atmospheric and climatic conditions, and to guide CO_2 research to produce data in forms pertinent to existing models.

Some broad areas of currently deficient knowledge of direct CO_2 effects on plants are

- Explanation of all primary mechanisms of the action of CO_2 concentration on plant growth. Possible primary effects on phenology (e.g. senescence), morphology (e.g. branching and leaf area), and dark respiration have only been cursorily investigated, as has the downward adjustment of leaf photosynthesis rate during continuous exposure to high CO_2.
- Quantification of the interaction of CO_2 and other yield-limiting factors. Nutrient deficiency and nonoptimal temperature are outstanding gaps here, although further work on water and light interactions is also needed.
- Investigation of genetic variation in CO_2 responsiveness of yield and of opportunities in breeding programs to capitalize on high CO_2 concentrations.
- Quantification of secondary effects of high CO_2 concentrations on crop yield, especially with regard to weeds, pests, and diseases.

In coming to grips with the effects of increasing CO_2 on crop yield and food security, it must be explicitly recognized that the atmospheric CO_2 increase is a continuous, gradual amelioration of the global agricultural environment; climate, climatic change, and year-to-year production stability are statistical phenomena involving shifting probabilities and risks (Warrick and Gifford 1986). With appropriate socioeconomic organization and population control, it should be possible to capitalize on CO_2 fertilizer effects to reduce the effect of changing climatic and weather variability on regional food security. At the same time, the CO_2 fertilizer effects may decrease yield variability from some sources (e.g. drought) while increasing it from other sources (e.g. lodging).

References cited

Acock B, Allen L H Jr. (1985) Crop responses to elevated carbon dioxide concentrations. Pages 53-97 *in* Direct effects of increasing carbon dioxide on vegetation. B. R. Strain and J. D. Cure, eds. United States Department of Energy DOE/ER-0238, Washington, D.C, USA.

Berry J A, Raison J K (1981) Responses of macrophytes to temperature. Pages 277-338 *in* Encyclopedia of plant physiology. Vol. 12A. O. L. Lange, P. S. Nobel, C. B. Osmond, and H. Ziegler, eds. Springer-Verlag, Berlin.

Butler G D (1985) Populations of several insects on cotton in open-top carbon dioxide enrichment chambers. Southwestern Entomol. 10:264-267.

Chang C W (1975) Carbonic anhydrase and senescence in cotton plants. Plant Physiol. 55:515-519.

Cure J D (1985) Carbon dioxide doubling responses: a crop survey. Pages 99-116 *in* Direct effects of increasing carbon dioxide on vegetation. B. R. Strain and J. D. Cure, eds. United States Department of Energy DOE/ER-0238, Washington, D.C, USA.

Ehleringer J, Björkman O (1977) Quantum yields for CO_2 uptake in C_3 and C_4 plants: dependence on temperature, CO_2 and O_2 concentrations. Plant Physiol. 59:86-90.

Fischer R A (1985) Physiological limitations to producing wheat in semitropical and tropical environments and possible selection criteria. Pages 209-230 *in* Wheats for more tropical environments. Proceedings of the international symposium. International Maize and Wheat Improvement Center, Mexico.

Gale J (1980) Use of brackish and solar desalinated water in closed system agriculture. Pages 315-323 *in* Biosaline research: a look to the future. A. San-Pietro, ed. Plenum Press, New York.

Gammon R H, Sundquist E T, Fraser P J (1985) The history of carbon dioxide in the atmosphere. Pages 25-62 *in* Atmospheric carbon dioxide and the global carbon cycle. J. R. Trabalka, ed. United States Department of Energy DOE/ER-0239, Washington, D.C., USA.

Gifford R M (1977) Growth pattern, CO_2 exchange and dry weight distribution in wheat growing under differing photosynthetic environments. Aust. J. Plant Physiol. 4:99-110.

Gifford R M (1979a) Carbon dioxide and plant growth under water and light stress: implications for balancing the global carbon budget. Search 10:316-318.

Gifford R M (1979b) Carbon storage by the biosphere. Pages 167-181 *in* Carbon dioxide and climate: Australian research. G. I. Pearman, ed. Australian Academy of Science, Canberra.

Gifford R M (1979c) Growth and yield of CO_2-enriched wheat under water-limited conditions. Aust. J. Plant Physiol. 6:367-378.

Gifford R M (1982) Global photosynthesis in relation to our food and energy needs. Pages 459-495 *in* Photosynthesis: development, carbon metabolism and plant productivity. Vol. II. Govindjee, ed. Academic Press, New York.

Gifford R M (1987) Barriers to increasing crop productivity by genetic improvement in photosynthesis. Pages 377-384 *in* Progress in photosynthesis research. Vol. IV. J. Biggins, ed. R. I. Martinus Nijhoff, The Hague.

Gifford R M, Lambers H, Morison J I L (1985) Respiration of crop species under CO_2 enrichment. Physiol. Plant. 63:351-356.

Gifford R M, Morison J I L (1985) Photosynthesis, growth and water use of a C_4 grass stand at high CO_2 concentration. Photosyn. Res. 7:69-76.

Goudriaan J, de Ruiter H E (1983) Plant growth in response to CO_2 enrichment of two levels of nitrogen and phosphorus supply. I. Dry matter, leaf area and development. Neth. J. Agric. Sci. 31:157-169.

Jordan D B, Ogren W L (1984) The CO_2 specificity of ribulose 1, 5-bisphosphate carboxylate/oxygenase. Plants 161:308-313.

Kimball B A (1983) Carbon dioxide and agricultural yield: an assemblage and analysis of 770 prior observations. United States Department of Agriculture, Agricultural Research Service, Water Conservation Laboratory Report 14, Phoenix, Arizona. 71 p.

Lieth H (1972) Modelling the primary productivity of the world. Nat. Resour. 8:5-10.

Lincoln D E, Sionit N, Strain B R (1984) Growth and feeding response of *Pseudoplusia includens* (Lepidoptera:Noctuidae) to host plants grown in controlled carbon dioxide atmospheres. Environ. Entomol. 13:1530-1537.

Marc J, Gifford R M (1983) Floral initiation in wheat, sunflower and sorghum under carbon dioxide enrichment. Can. J. Bot. 62:9-14.
Maxon-Smith J (1977) Selection for response to CO_2-enrichment in glasshouse lettuce. Hortic. Res. 17:15-22.
Monteith J L (1977) Climate and the efficiency of crop production in Britain. Phil. Trans. R. Soc. Lond. B 218:277-294.
Monteith J L (1981) Climatic variation and the growth of crops. Q. J. Roy. Meteorol. Soc. 197:749-774.
Morison J I L, Gifford R M (1984) Plant growth and water use with limited water supply in high CO_2 concentrations. II. Plant dry weight, partitioning and water use efficiency. Aust. J. Plant Physiol. 11:375-384.
Nix H A (1975) Australian climate and its effect on grain yield and quality. Pages 183-226 *in* Australian field crops. Vol. I. Wheat and other temperate cereals. A. Lazenby and E. M. Matheson, eds. Angus and Robertson, Sydney.
Penning de Vries F W T, Van Keulen H, Van Diepen C A, Noy I G A M, Goudriaan J (1989) Simulated yields of wheat and rice in current weather and when ambient CO_2 has doubled. Pages 347-357 *in* Climate and food security. International Rice Research Institute, P.O. Box 933, Manila, Philippines.
Reynolds J F, Acock B (1985) Modeling approaches for evaluating vegetation responses to carbon dioxide concentration. Pages 33-51 *in* Direct effects of increasing carbon dioxide on vegetation. B. R. Strain and J. D. Cure, eds. United States Department of Energy, DOE/ER-0238, Washington, D.C., USA.
Silsbury J H, Stevens R (1984) Growth efficiency of *Trifolium subterraneum* at high (CO_2). Pages 133-136 *in* Advances in photosynthesis research. Vol. IV. C. Sybesma, ed. Martinus Nijhoff/Dr. W. Junk, The Hague.
Sionit N, Hellmers H, Strain B R (1980) Growth and yield of wheat under CO_2 enrichment and water stress. Crop Sci. 20:687-690.
Sionit N, Strain B R, Beckford H A (1981) Environmental controls on the growth and yield of okra. I. Effect of temperature and of CO_2 enrichment at cool temperature. Crop Sci. 21:885-888.
St. Omer L, Horvath S M (1983) Elevated carbon dioxide concentration and whole plant senescence. Ecology 64:1311-1314.
Tolbert N E (1983) Carbon metabolism. Pages 21-64 *in* CO_2 and plants: the response of plants to rising levels of atmospheric carbon dioxide. E. R. Lemon, ed. American Association for the Advancement of Science Selected Symposium 84, Washington, D.C., USA.
Warrick R A, Gifford R M (1986) CO_2, climate change and agriculture: assessing the response of food crops to the direct effects of increased CO_2 and climate change. Pages 393-473 *in* The greenhouse effect on climate change and ecosystem (SCOPE 29). B. Bolin, B R Döös, J. Jäger, and R. A. Warrick, eds. John Wiley and Sons, Chichester, UK.
White T C R (1984) The abundance of invertebrate herbivores in relation to the availability of nitrogen in stressed food plants. Oecologia 63:90-105.
Wong S C (1979) Elevated atmospheric partial pressure of CO_2 and plant growth. I. Interactions of nitrogen nutrition and photosynthetic capacity in C_3 and C_4 species. Oecologia (Berl.) 44:68-74.

Notes

Author's address: R. M. Gifford, Commonwealth Scientific and Industrial Research, Division of Plant Industry, GPO Box 1600, Canberra ACT 2601, Australia.
Citation information: International Rice Research Institute (1989) Climate and food security. P.O. Box 933, Manila, Philippines.

Development of crop germplasm with improved tolerance for environmental and biotic stresses

J. P. Srivastava

Incorporating genetic resistance into plants can help them tolerate biotic and abiotic stresses which limit their productivity. Coupled with better agronomic practices, this can improve not only productivity, but also yield stability. Characterization of the target environment is important in this process, as is the heritability of plant characteristics. Empirical and analytical approaches are discussed, with special reference to barley, the predominant cereal crop in Mediterranean countries where environmental stresses limit food crop production. Breeding strategies at the International Center for Agricultural Research in the Dry Areas include 1) identifying and studying parental lines tolerant of specific stresses, 2) developing crosses that incorporate desired traits, 3) exposing early-generation populations to the environments where they will be grown, 4) using modified bulk procedures, and 5) multienvironment testing through a network of national programs. Useful genetic variability in wild progenitors and primitive forms of barley and wheat, as well as local land races, has been identified. Barley and wheat cultivars with improved yield stability that yield satisfactorily under stress and respond to favorable conditions with substantially higher yields have been developed.

Biological yield is the outcome of genetic, environmental, and cultural interactions. Estimates of the importance of environmental stresses on crop production have indicated that crop production is rarely unaffected by such environmental stresses as limited water supply, high or low temperatures, and mineral stresses alone or in combination. These seriously reduce productivity, as do biotic stresses from weeds, diseases, and insects (Duke 1977).

Walker (1975) suggests that overall average crop loss due to insects is 14%, to diseases 12%, and to weeds 9% (Table 1). This suggests that huge rewards will repay research efforts for biotic tolerance.

Stresses can be dramatic: complete crop loss during anthesis due to frost or the effects of a hot drying wind in North Africa during grain filling. Small-scale farmers and those who raise crops on marginal or remote lands are particularly vulnerable to environmental factors (Table 2). Crop varieties more tolerant of adverse conditions would be of great value to those farmers. The benefit would be even higher if new varieties were accompanied by modifications in cropping systems and cultural practices to minimize the effects of climate variability.

Table 1. Biological cost of weeds, diseases, and insects (Walker 1975).

Crop	World production (million t)	Percent loss of yield potential		
		Weeds	Diseases	Insects
Barley	178	8.8	8.3	3.9
Maize	487	13.1	9.6	12.9
Millet and sorghum	107	17.9	10.3	9.5
Rice	468	10.6	9.0	27.5
Wheat	510	9.8	9.5	5.4

Table 2. Approximate distribution (%) of wheat area according to moisture environment in major production regions of the developing world (Byerlee and Winkelmann 1980).

Region	Moisture environment		
	Adequate rainfed, irrigated	Semiarid	Inadequate rainfed
Middle East/North Africa	14	34	51
South Asia	73	4	23
East Asia	25	39	37
Latin America	9	49	43
All developing countries	34	28	37

Drought, heat, and cold predominate in the rainfed areas of West Asia and the Mediterranean basin. Drought may occur alone or in combination with the thermal stresses, or a thermal stress may occur at different stages of crop development. Stress patterns also may differ in different climatic zones (ICARDA 1985). Even when a crop is grown in an optimum-moisture environment, occasional periods of drought during the growing season may reduce crop production. Using technologies such as irrigation, fertilization, and tillage to alleviate the effects of climatic stresses is limited or not possible in the drought-prone areas of the developing world. Genetic manipulation of plants to improve yield and yield stability under climatic stresses is important.

Definition of the environment

Before improved germplasm for stress environments can be developed, the level and frequency of specific stresses at different crop growth stages must be defined. A breeding program will be more effective if genetic material is developed to meet the specific needs of the target environment.

Given that yield is the integrated product of genotypic expression in a given environment, the definition of a stress environment will affect the breeding strategies chosen. If a target environment has a production potential of 3 t/ha or more, selection in stress-free conditions can be efficient. However, if environmental stresses limit yield potential to less than 3 t/ha, then it would be more efficient to select plants directly in that environment.

Germplasm has been classified into that destined for wet, moderate, or dry environments; that destined for warm or cool environments; and that for short or long duration and for low or high elevations. Information on the prevalence of soil toxicity, such as aluminum toxicity, has been used to classify genotypes for problem soils. Disease or insect incidence has been used to group sites and genotypes. Several statistical procedures have been used to group environments, with varying degrees of success. A great deal still needs to be added to our characterizations of different production environments.

Breeding for tolerance for environmental stresses

The emphasis in breeding programs has been on maximizing yield potential and disease resistance (Marshall 1985). Few plant breeders have concentrated on developing germplasm to minimize the effects of drought, cold, heat, and salinity.

Crop growth environments can be modified through the use of sound agronomic principles, crop rotations, and fertilizers. Improvements in water-use efficiency have been achieved by these practices. However, the availability of varieties with specific phenological adaptations that better match the most favorable growth conditions, with genetic tolerance for environmental stresses, would lead to further yield increases and production stability.

Breeding strategies used at ICARDA

The International Center for Agricultural Research in the Dry Areas (ICARDA) uses the following breeding strategies:

- Identifying and studying parental lines tolerant of specific stresses,
- Developing crosses with the desired traits,
- Exposing early-generation populations to the environments in which they will be grown,
- Using modified bulk procedures,
- Doing multienvironment testing through a network of national programs,
- Using land races, primitive forms, and wild progenitors.

Using these strategies, barley and wheat lines with improved yield stability, that yield satisfactorily in stress conditions and respond to favorable conditions with substantially higher yields, have been developed (ICARDA 1984, 1985, 1986).

The development of effective screening procedures to select for traits related to stress resistance requires preliminary research by meteorologists, physiologists, biochemists, pathologists, entomologists, and breeders. It is important to identify the major stresses that limit crop yields in the target environment, including the frequency of a particular stress and its timing in relation to crop development. With this information, a partial crop ideotype can be developed that matches crop phenology to the target environment.

The next step is to identify traits that could be beneficial in the target environment, and to establish that selection for the trait(s) and recombination with other desirable characters are feasible. Once this is established, reliable procedures and laboratory or "hot spot sites" should be used to screen potential germplasm and to develop parental lines with known sources of resistance. Determining the heritability of the trait and its relationship to yield performance in the target environment is important. Knowledge of the heritability-yield relationship allows the breeder to develop selection strategies for desirable characters (Clarke 1985).

No one set of breeding strategies is likely to suit all arid or semiarid production areas. It is desirable to evaluate breeding procedures for each situation. Ceccarelli (1986), dealing with barley improvement for low-rainfall areas, discussed two main issues: 1) the ideal environment for maximizing the efficiency of selection, and 2) whether direct selection for grain yield is still the only practical avenue, or whether analytical approaches based on morphological, physiological, and biochemical characters can be used.

Selecting the screening site

It is not easy to choose a single appropriate environment for screening and selecting genotypes for diverse conditions. Some breeders suggest that evaluation and selection should be done where there is adequate moisture and fertility, even when the successful lines are to be grown in moisture- and nutrient-deficient environments (Blum 1982, 1985; Frey 1984; Medersky and Jeffers 1973). This practice can be useful for selecting drought-tolerant genotypes if genotypes × environment interactions are small, or for selecting resistance to such stresses as diseases and insects, or where anatomical, morphological, or physiological responses of plants are clearly associated with resistance. On the other hand, a number of studies (Aufhammer et al 1959; Boyer and McPherson 1975; Buddenhagen 1983; Burton 1964; Ceccarelli et al 1985; Ceccarelli 1986, 1987; Hageman et al 1976; Hurd 1968, 1971; Johnson 1980; Johnson et al 1968; Richards 1982; Srivastava et al 1983) report the superior efficiency of direct selection under stress conditions.

Little is known about the plant characteristics and mechanisms of response to drought stress (Boyer 1982). However, the significant role of genotype × environment interaction is a common feature of rainfed agriculture. Our results indicate that selection under low-fertility, dry farming conditions was more successful in identifying drought-tolerant genotypes than selection in a high-fertility and high-moisture environment (Nachit and Ketata 1986, Srivastava et al 1983). Ceccarelli (1986), working with barley at different sites representing a range of drought stress, reported greater efficiency of selecting under stress (Fig. 1). He concluded that if yield level of around 3 t/ha with moderate rainfall and fertility is the target (stress environment), it is likely that lines selected for 5 t/ha yields with optimum rainfall and fertility (favorable environment) may maintain their superiority in moderate crop-growing conditions. However, at higher levels of stress, with the 1-2 t/ha production level that represents large expanses of barley-growing areas, these genotypes will not perform well.

It may be relatively difficult to select for high-yielding lines in drought-stressed environments, but it is far more difficult to select for drought-tolerant genotypes under stress-free environments. It is difficult to visualize how genotypes that will

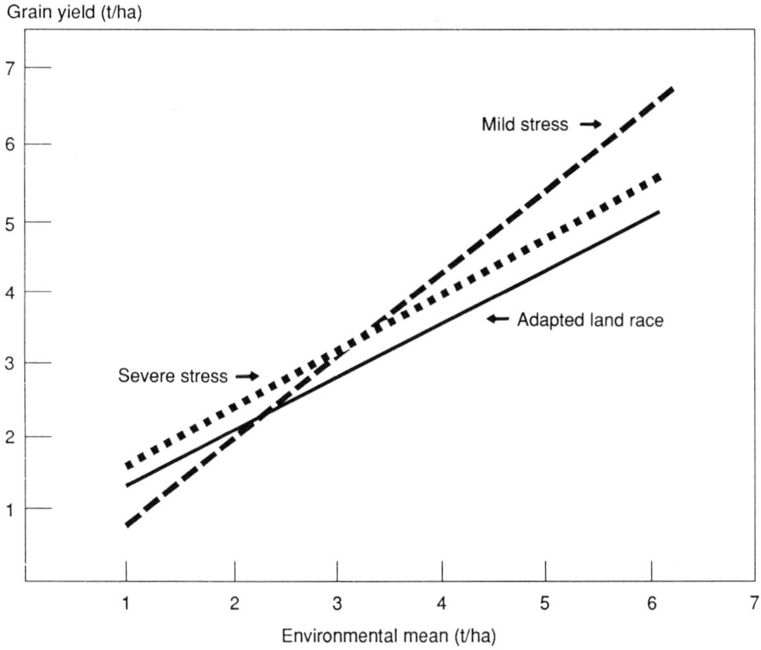

1. Barley grain yield at different levels of stress (measured as yield potential): an adapted land race, bulk families selected under severe stress, and bulk families selected under mild stress. Based on Ceccarelli (1986).

respond successfully to drought stress by means of several interactive traits would be detected without the stress. A more logical approach would be to determine the range of stress that may occur in the majority of years, and to evaluate the germplasm derived from carefully planned crosses, at sites selected to represent those environments, even in early generations.

The major problem in selecting under field stress conditions is the inherently large year-to-year variability. Choosing appropriate breeding methods and suitable experimental designs to increase the efficiency of selection in such environments is important.

The empirical vs the analytical approach

Two main, complementary approaches have been proposed for varietal improvement in highly stressed environments. The first, and most common, is empirical: this approach relies on grain yield as the main selection criterion, since yield integrates all the components in a stress-limited environment. Using this approach, progress has been made in increasing average yields, but yield increases have been modest and unreliable.

Smith (1985) suggested using a modified bulk breeding method in place of the pedigree method of selection to develop improved germplasm for rainfed areas. It identifies yield potential and, to some extent, yield stability early in the breeding program. With the method, those subpopulations reaching the advanced testing stage will have heterogeneity for a number of traits. The most promising lines are rogued or reselected for uniformity. A finished cultivar will intentionally be somewhat heterogeneous, although it will have acceptable uniform field appearance. This will contribute to yield stability in farmers' fields. Several modifications of the bulk breeding procedure are being used by breeders (ICARDA 1985, 1986).

The analytical approach is likely to be more useful for areas with relatively severe drought stress. It relies on different adaptive mechanisms of a plant in a highly stressed environment, on the assumption that breeding and selecting for those adaptations will contribute to growth and yield under stress (Acevedo 1985). The analytical approach links plant and crop physiology with breeding efforts in an attempt to hasten the slow plant improvement usually observed in highly stressed environments (Richards 1982, 1985).

Drought tolerance mechanisms

Mechanisms used by plants to minimize the effects of drought can be classified as escape, avoidance, and tolerance (Levitt 1972, Turner 1979). Timing plant developmental events so that the effects of environmental stresses are avoided is a major mechanism by which plants adapt. The avoidance mechanisms to drought inevitably reduce growth. It is possible to adapt crop plants to arid regions, but with reduced yield potential.

Schulz (1986) listed the following mechanisms through which plants endure arid and semiarid environments:

- drought escape
- drought tolerance
  desiccation tolerance
  turgor maintenance
  reduction of water loss
  maintenance of water uptake
  improved CO_2 uptake and water-use efficiency.

Ludlow and Muchow (1986) listed the following traits needed for drought resistance:

- development plasticity,
- root hydraulic resistance,
- leaf area maintenance,
- low lethal water status,
- leaf movement,
- high temperature tolerance,
- transpiration efficiency,
- rooting depth density,
- early vigor,
- osmotic adjustment,
- reduced stomatal conductance,
- leaf reflectance,
- epidermal conductance.

Ludlow and Muchow (1986) developed a list of priority traits for sorghum and cowpea in the four target situations to improve productivity. Other metabolic traits also have been reported to be associated with stress tolerance (proline accumulation, abscisic acid level, betaine accumulation), but their contribution has not been conclusively proved.

Results obtained thus far from screening for drought tolerance traits have not been fully applied in developing drought-resistant lines.

It is generally held that breeding for stress tolerance, particularly when it is associated with unpredictability, is more complex and difficult than breeding for more favorable conditions. Despite these difficulties, genetic improvement for yield in stressed environments is possible. Srivastava (1986) reported barley yields of more than 2.0 t/ha in the 200- to 300-mm rainfall zone in on-farm trials in Syria (the farmers' average was around 0.6 t/ha). Smith (1985) reported that in Oklahoma, current average dryland wheat yield is 2.0 t/ha, from about 1.0 t/ha in the mid-1950s. Half of the increase is due to plant breeding, the other half comes from improved agronomic practices. There are several other examples of improved yield and stability of production in moisture-limited areas.

Multilocation testing

Multilocation testing is an integral part of variety development procedure. The tests are replicated over years and sites representing a range of stress levels. The yield trials provide valuable information about the performance of entries in the range of environmental and biotic stresses found at geographically and environmentally diverse testing sites. Yield trials are supplemented by specific nurseries (e.g. disease screening nurseries, insect screening nurseries, cold tolerance nurseries, heat tolerance nurseries) grown at sites where a particular stress is endemic. A system of on-farm trials to verify the productive capacity of a genotype supplements on-station trials. Such trials should reliably predict the performance of new cultivars under farmers' management systems. International centers have developed very successful International Nursery system networks for such multilocation testing, with the cooperation of national scientists.

Entries used in such tests usually are fixed lines. This has been a major limitation. A multilocation selection system that could be used in early generations would take advantage of greater genetic variability. A modified bulk breeding system, in which individual plants are selected from potentially superior bulk populations identified by performance across several sites, could be used.

Use of nonconventional germplasm

In many arid and semiarid regions where crops are grown at the low-input technology and production level of traditional farming systems and where climatic stresses are the major factors limiting productivity, locally adapted land races and selections from land races are still considered the most dependable genetic material.

Long-term progress in plant breeding depends on providing an adequate reserve of genetic stocks: diverse parents with genetic variability for various desirable traits. Crops geographically or environmentally differentiated into locally adapted ecotypes, land races, or varieties may possess traits for tolerance for the prevalent stresses in the habitat. Through recombination, such germplasm has contributed significantly to the development of varieties with specific adaptation and desired characters. Several breeders have used wild progenitors to introduce some traits (mostly disease and insect resistance) into improved varieties. The use of land races or wild forms assumes greater importance when the objective of the plant breeder is to develop germplasm with improved tolerance for environmental stresses and resistance to insect pests and diseases.

The importance of land races and wild species has been emphasized in improving yield and production stability in marginal areas where drought, cold, heat, and salinity, alone or in combination, are always present (Ceccarelli 1986, Ceccarelli and Mekni 1985, Chang 1985, Nachit and Ketata 1986, Srivastava and Damania 1989, Srivastava and Winslow 1985, Weltzien and Srivastava 1981). Weltzien (1982) evaluated 280 progenies collected at 28 sites in Syria and Jordan and found a large genetic diversity, both in agronomic traits and disease resistance between and within collection sites. Van Leur et al (1986) tested the same collection with local Syrian strains of four important barley pathogens, and found high variability in disease resistance (Table 3). Ceccarelli (1986) identified a pureline selection from this collection, Tadmor, which has outperformed local check varieties in moderate to low crop-growing areas in Syria (Table 4). These and other results indicate that land

Table 3. Percentage of highly resistant (HR) and moderately resistant (MR) entries in a collection of 280 barley land races from Syria and Jordan.

Disease	Resistant entries (%)	
	Highly resistant	Moderately resistant
Yellow rust	1.1	6.4
Powdery mildew	1.8	44.2
Scald	0.7	35.7
Covered smut	31.1	32.1

Highly resistant: < 0.5% severity leaf infestation for yellow rust or head infection for covered smut. < 1 in 0-9 scoring scale for powdery mildew and scald. Moderately resistant: > 0.5 and < 3% severity leaf infestation for yellow rust or head infection for covered smut. > 1 and < 3 in 0-9 scoring scale for powdery mildew and scald. Results based on two years' data (except for covered smut) and two replications/entry (3 replications for scald).

Table 4. Grain yield (t/ha) of Tadmor, a pure line selected from a local cultivar of barley, in sites receiving less than 250 mm of annual rainfall.

Variety	Grain yield (t/ha)	
	1984-85, Bouider	1985-86, Mean of 5 sites
Tadmor	2.4 a	1.8 a
Local cultivar	1.6 b	1.6 b

Values followed by the same letter are not significantly different ($P<0.05$).

races are a source of readily available genetic diversity which can be used in short-term strategies in breeding for tolerance for climatic stresses.

Varieties selected for high yields and suitability for mechanized farming are homogeneous pure lines. Land races are heterogeneous for several traits and possess plasticity that provides buffering mechanisms aimed at adaptability and survival under stress. These features should be further utilized in developing cultivars for harsh and unpredictable crop-growing regions.

Local durum wheat (*Triticum durum*) land races that predominate in rainfed areas in West Asia and North Africa usually are adapted to prevailing dry environments. A number of varieties have been developed recently by crossing these locally adapted land races with early-maturing, high-yielding lines (Nachit and Ketata 1986, Srivastava and Damania 1989). Tahir (1986) evaluated the *T. dicoccoides* collection from Syria and reported considerable genetic variability for several traits (Table 5). This wild species has several desirable traits, and its versatility as a germplasm resource in breeding for stress tolerance is emphasized. Considerable genetic variability also has been found in *Hordeum spontaneum* collections (Ceccarelli and Grando 1987; they are being used in transferring desirable characters to barley cultivars. However, utilization of wild progenitors through hybridization with cultivated forms is difficult, because of genetic barriers and undesirable linkages.

Resistance to salt and mineral stresses

Soil salinity limits crop productivity in many arid and semiarid regions. Considerable variability exists for salt tolerance among species. Several salt-tolerant barley and wheat genotypes have been identified by screening germplasm in the laboratory and in the field (Srivastava et al 1983). Norlyn et al (1980) reported salt-resistant selections grown in dune sand irrigated with seawater that yielded as much as 1.5 t/ha. Although resistance to salinity usually is referred to as salt tolerance, in barley it is due primarily to avoidance mechanisms (e.g. salt exclusion and salt dilution) (Levitt 1972).

Mineral stress may result from excess amounts of various minerals in the soil solution. Aluminum toxicity retards root growth, reducing plant growth and yield. Genes for aluminum resistance have been identified in barley and wheat and have

Table 5. Evaluation of 15 *Triticum dicoccoides* lines collected from Syria.

Character	Value	
	Minimum	Maximum
Protein (%)	20.4	25.0
Plant height (cm)	80	120
Days to heading	159	179
Yellow rust resistance	Susceptible-resistant	
Spike length (cm)	7.6	14.0
Frost tolerance	All lines tolerant	
1,000-kernel wt (g)	10.5	26.2

been successfully incorporated into locally adapted varieties. Development of resistance to toxic levels of other minerals (e.g. copper, iron, boron, manganese) may be possible, but has received little attention from plant breeders. If large areas suffer yield reduction due to excess or deficiency of mineral nutrients, plant breeding programs to develop germplasm with tolerance for mineral stresses may be undertaken.

Winter hardiness

Winter wheat, rye, and triticale appear to be more winter hardy than barley. Breeders have sought to improve winter hardiness, to allow crops to be grown without danger of winterkill. Several laboratory and field screening techniques have been used (Cloutier and Siminovitch 1983, Evans and McGill 1963, Fejer and Schwartzbach 1978, Nachit 1983, Shevtsov et al 1978) to screen germplasm and segregating populations for low-temperature tolerance. Several genes identified and incorporated in wheat and barley have improved cold tolerance or resistance. Frost damage still remains a threat. Despite increased understanding of the physiology of frost injury, little progress has been made in breeding for improved frost tolerance. Frost avoidance, by delaying anthesis until after the frost period, is still the main strategy. Breeding programs to improve frost resistance by using nonconventional germplasm are recommended.

Breeding for insect and disease resistance

Pests often limit crop production, affecting both yield and grain quality. Pest situations are dynamic and change with different farming systems and patterns of varietal use. Experience has shown that attempts to eradicate well-established pests are expensive and usually ineffective. The goal of an integrated pest management program should be to limit pest populations to low, stable, equilibrium densities that cause minimal economic loss.

Plant resistance is often the most satisfactory method for combating plant insects. It is incorporated into the plant with no direct cost to the grower, is compatible with other control methods, and poses no environmental hazards. One limitation to using plant resistance is that it is often specific to a single pest. However, genes may be combined to gain multiple resistance. This is not easy and takes time. Another drawback to host plant resistance is the evolution of virulent biotypes that can overcome sources of resistance. For example, Hessian fly and aphid biotypes have evolved for wheat, in which resistance is conferred by a single gene.

Resistance due to morphological characters, such as solid stem in sawfly-resistant wheats and leaf pubescence in cereal leaf beetle-resistant varieties, provides relatively long-term control (Starks and Webster 1985). Progress has been made in the development of cultivars resistant to some insect pests, although total immunity to an insect is rarely, if ever, possible.

Crop diseases caused by a variety of pathogens (fungi, bacteria, viruses) can be controlled to some extent by host plant resistance; the literature contains many examples of successes and failures. Epidemics depend on initial infection and the epidemic constant. The challenge to plant breeders is to reduce the epidemic constant as far as possible, or at least to a level where crop losses are reduced to an economically acceptable level (Simmonds 1981).

Genetic control of resistance may be exerted by major genes or by polygenes. Major gene resistances are usually pathogen-specific; polygenic resistances are pathotype nonspecific. Selection of the breeding strategy depends largely on the resistance sources available, the number of loci involved and their mode of inheritance, and whether short-term or long-term protection is desired. Although long-term resistance should be the main goal of resistance breeding programs, the urgent need to develop resistant varieties, coupled with ease and a high success rate, has led many breeders toward using major genes for short-term protection (Sharp 1985). Development of cultivars possessing two or more resistance genes is common. The use of cultivars possessing single genes from different sources of resistance has been recommended, but its use has been limited due to practical difficulties.

Resistance genes from other species or genera and multiline cultivars or cultivar mixtures have been used with varied success. Some breeding programs use recurrent selection procedures to pyramid genes for resistance to diseases, insects, and even abiotic stresses in adapted populations. Both major genes and polygenes for resistance may be accumulated.

In contrast to traditional agriculture, modern agriculture has promoted the wide-scale use of monoculture farming and the use of pureline varieties possessing a few major genes for resistance to insects and diseases. This has increased the vulnerability of crops to pest outbreaks. The southern corn leaf blight and wheat rust epidemics are illustrations.

Breeding strategies are increasingly being directed toward the development of horizontal or generalized resistance. Methods of handling horizontal resistance in a breeding program are difficult and slow. More attention and resources are needed to improve understanding and use of durable generalized resistance. The methods of handling disease and insect resistance in a breeding program depend on the pest's biology. Several excellent reviews treat breeding methods for insect pests and pathogens.

References cited

Acevedo E (1985) Physiological basis for improving winter cereals. Annual report 1985. International Center for Agricultural Research in the Dry Areas, Aleppo, Syria. 45 p.

Aufhammer G, Fischbeck G, Grebner H (1959) Testing drought resistance in spring cereals. Z. Acker. Pflanzenbau. 110:117-134.

Blum A (1982) Evidence for genetic variability in drought resistance and its implication in plant breeding. Pages 53-68 *in* Drought resistance in crops with emphasis on rice. International Rice Research Institute, P.O. Box 933, Manila, Philippines.

Blum A (1985) Breeding crop varieties for stress environments. CRC Critical Rev. Plant Sci. 2(3):199-238.

Boyer J S (1982) Plant productivity and environment. Science 218:443-448.

Boyer J S, McPherson (1975) Physiology of water deficits in cereal crops. Adv. Agron. 27:1-23.

Buddenhagen I W (1983) Breeding strategies for stress and disease resistance in developing countries. Ann. Rev. Phytopathol. 27:385-409.

Burton G W (1964) The geneticist role in improving water use efficiency by crops. Pages vi-121 *in* Research on water. Special Publication Serial 4. American Society of Agronomy, Madison, Wisconsin, USA.
Byerlee D, Winkelmann D F (1980) Accelerating wheat production in semi-arid developing regions. Pages 328-362 *in* Proceedings of the 3rd International Wheat Conference. Madrid, Spain. University of Nebraska, Lincoln, USA.
Ceccarelli S (1986) Breeding strategies to improve barley yield and stability in drought prone environments. Paper presented at the Meeting on Drought Tolerance in Plants: Genetic and Physiological Aspects, 20-23 Oct 1986, Amalfi, Italy.
Ceccarelli S (1987) Tolerance to climatic stresses. Pages 689-702 *in* Proceedings of the 5th barley genetic symposium, 6-12 Oct 1986. Osaka, Japan.
Ceccarelli S, Grando S (1987) Diversity for morphological and agronomic characters in *Hordeum vulgare* ssp. *spontaneum* C. Koch. Genet. Agraria 41:131-142.
Ceccarelli S, Mekni M S (1985) Barley breeding for areas receiving less than 250 mm annual rainfall. Rachis 4(2):3-9.
Ceccarelli S, Nachit N, Ferara G O, Mekni M S, Tahir M, Van Leur J A G, Srivastava J P (1985) Breeding strategies for improving cereal yield and stability under drought. Paper presented at the Seminar on Improving Winter Cereals for Moisture Limiting Environments, 27-31 Oct 1985, International Center for Agricultural Research in the Dry Areas/Conseil National de Recherches du Canada, Capri, Italy.
Chang T T (1985) Principles of genetic conservation. Iowa State J. Res. 59:325-348.
Clarke J M (1985) Use of physiological, morphological, and metabolic traits in breeding programs to improve drought resistance of cereals. Paper presented at the Seminar on Improving Winter Cereals for Moisture Limiting Environments, 27-31 Oct 1985, International Center for Agricultural Research in the Dry Areas/Conseil National de Recherches du Canada, Capri, Italy.
Cloutier Y, Siminovitch D (1983) L'induction de la resistance au froid pour la secheresse: une methode rapide et economique pour tester et selectionner des cultivars resistants de cereales d'hiver. Can. J. Plant Sci. 63:573-580.
Duke J A (1977) The quest for tolerant germplasm. Pages 1-61 *in* Crop tolerance to suboptimal land conditions. American Society of Agronomy, Crop Science Society of America, and Soil Science Society of America, Madison, Wisconsin, USA.
Evans K H, McGill D P (1963) Response of some barley varieties to hardening and freezing during sprouting. Agron. J. 55:147-149.
Fejer S O, Schwartzbach E (1978) Hardiness of North American barley and other cereals as measured by electrical impedance at the coleoptile stage. Z. Pflanzenzuecht. 80:162-165.
Frey R L (1984) Adaptation reaction of oat strains selected under stress and non-stress environmental conditions. Crop Sci. 4:55-58.
Hageman R H, Leng E R, Dudley J W (1976) A biochemical approach to corn breeding. Adv. Agron. 19:45-76.
Hurd E A (1968) Growth of roots of seven varieties of spring wheat at high and low moisture levels. Agron. J. 60:201-205.
Hurd E A (1971) Can we breed for drought resistance? Pages 77-78 *in* Drought injury and resistance in crops. Special Publication No. 2. K. L. Larson and J. D. Eastin, eds. Crop Science Society of America, Madison, Wisconsin.
ICARDA—International Center for Agricultural Research in the Dry Areas (1984) Annual report 1983. Aleppo, Syria.
ICARDA—International Center for Agricultural Research in the Dry Areas (1985) Annual report 1984. Aleppo, Syria.
ICARDA—International Center for Agricultural Research in the Dry Areas (1986) Annual report 1985. Aleppo, Syria.
Johnson D A (1980) Improvement of perennial herbaceous plant for drought-stress western rangelands. Pages 419-433 *in* Adaptation of plants to water and high temperature stress. N. C. Turner and P. J. Kramer, eds. Wiley, New York.
Johnson D A, Shafer S L, Schmidt J W (1968) Regression analysis of general adaptation in hard red winter wheat (*T. aestivum* L.). Crop Sci. 8:187-191.
Levitt J (1972) Responses of plants to environmental stresses. Academic Press, New York.
Ludlow M M, Muchow R C (1986) Critical evaluation of the possibilities for modifying crops for higher production per unit of precipitation. Paper presented at the Meeting on Research on Drought Problems in the Arid and Semi-Arid Tropics, 17-20 Nov 1986, International Crops Research Institute for Semi-Arid Tropics, Hyderabad, India.

Marshall D R (1985) Australian plant breeding strategies for rainfed areas. Paper presented at the Seminar on Improving Winter Cereals for Moisture Limiting Environments, 27-31 Oct 1985, International Center for Agricultural Research in the Dry Areas/CNR, Capri, Italy.

Medersky H J, Jeffers D L (1973) Yield response of soybean varieties grown at two soil moisture stress levels. Agron. J. 65:410-412.

Nachit M M (1983) Use of planting dates to select stress tolerant and yield stable genotypes for the rainfed Mediterranean environment. Rachis 3:15-17.

Nachit M M, Ketata H (1986) Breeding strategy for improving durum wheat in Mediterranean rainfed conditions. *In* Proceedings of the international wheat conference, Rabat, Morocco, May 2-9. (in press)

Norlyn J C, Epstein E, Dale W R, Raloh W K, David B K, Cunningham A, Anne F W (1980) Saline culture of crops: a genetic approach. Science 210:399-404.

Richards R (1982) Breeding and selecting for drought resistance in wheat. Pages 303-316 *in* Drought resistance in crops with emphasis on rice. International Rice Research Institute, P.O. Box 933, Manila, Philippines.

Richards R (1985) Breeding temperate cereals for dry areas based on a physiological framework. Paper presented at the Seminar on Improving Winter Cereals for Moisture Limiting Environments, 27-31 Oct 1985, International Center for Agricultural Research in the Dry Areas/CNR, Capri, Italy.

Schulz E D (1986) Mechanisms of adaptation of non-cultivated arid zone plants: useful lessons for agriculture? Paper presented at the Meeting on Research on Drought Problems in the Arid and Semi-Arid Tropics, 17-20 Nov 1986, International Crops Research Institute for Semi-Arid Tropics, Hyderabad, India.

Sharp E L (1985) Breeding for pest resistance. Pages 313-333 *in* Barley. D. C. Rasmusson, ed. American Society of Agronomy, Madison, Wisconsin, USA.

Shevtsov V M, Gruntsev Y A, Tenditanaya O M, Polukhina P K, Serking M N (1978) Improving winter hardiness in barley by chemical mutagenesis. Plant. Breed. Abstr. 50:5072.

Simmonds N W (1981) Principles of crop improvement. Longman Group Ltd., New York.

Smith E L (1985) A review of plant breeding strategies for rainfed areas. Paper presented at the Seminar on Improving Winter Cereals for Moisture Limiting Environments, 27-31 Oct 1985, International Center for Agricultural Research in the Dry Areas/CNR, Capri, Italy.

Srivastava J P (1986) ICARDA annual report 1986. Aleppo, Syria. 229 p.

Srivastava J P, Damania A B (1989) Use of collection in cereal improvement in the semi-arid areas. Pages 88-104 *in* The use of plant genetic resources. Cambridge University Press, U.K.

Srivastava J P, Jana S, Gautam P L, Niks R E (1983) Parallel selection: an approach to increase grain yield and stability. Pages 725-733 *in* Proceedings of the 6th international wheat genetics symposium. Plant Germplasm Institute, Faculty of Agriculture, Kyoto, Japan.

Srivastava J P, Winslow M D (1985) Improving wheat and barley production in moisture-limiting areas. Rachis 4(1):2-8.

Starks K J, Webster J A (1985) Insects and related pests. Pages 335-360 *in* American Society of Agronomy, Crop Science Society of America, and Soil Science Society of America, Madison, Wisconsin.

Tahir M (1986) Evaluation and utilization of *T. turgidum* L var. *dicoccoides* for improvement at durum wheat. *In* Proceedings of the international wheat conference, Rabat, Morocco, May 2-5. (in press)

Turner N C (1979) Drought resistance and adaptation to water deficits in crop plants. Pages 343-372 *in* Stress physiology in crop plants. H. Mussell and R. C. Staples, eds. Wiley Interscience, New York.

Van Leur J A G, Ceccarelli S, Grando S (1986) Variability in disease resistance in barley landraces from Syria and Jordan. (in press)

Walker P T (1975) Pest control problems (pre-harvest) causing major losses in world food supplies. FAO Plant Prot. Bull. 23(3/4):70-77.

Weltzien E (1982) Barley collection and evaluation in Syria and Jordan. Plant Genet. Resour. Newsl. 52:6-7.

Weltzien H C, Srivastava J P (1981) Stress factors and barley productivity and their implications in breeding strategies. Pages 351-361 *in* Barley genetics. Editorial Subcommittee, Fourth International Barley Genetics Symposium. Edinburgh University Press, Edinburgh.

Notes

Author's address: J. P. Srivastava, International Center for Agricultural Research in the Dry Areas, P.O. Box 5466, Aleppo, Syria.

Citation information: International Rice Research Institute (1989) Climate and food security. P.O. Box 933, Manila, Philippines.

Climate and food crop production

G. O. P. Obasi

Knowledge about climate plays an important role in determining the feasibility of establishing sustainable food production systems in specific geographical regions. Natural climate variations from year to year and, in extreme cases, across decades can cause wide fluctuations in food production. Climatic change, natural or man-made, could affect food production even more extensively and conceivably could force significant changes in current global agricultural patterns. This paper examines aspects of climate and food crops.

The role of climate

Key climate parameters in plant growth

Climate is one of the key factors that determine plant growth, and hence food production. The atmosphere, and hence the climate, are directly involved in the basic production of plant matter through mass and energy exchanges. Oxygen and carbon dioxide (CO_2) are exchanged between the atmosphere and plants. Rain provides water for plant roots through the soil. Solar energy transmitted through the atmosphere is absorbed by plants. The energy from solar radiation acts on the CO_2 and water in the plant to produce carbohydrates and oxygen, through the process called photosynthesis.

It has been found that gross photosynthesis largely depends on the amount of incoming solar radiation, and only very little on temperature. Crops grow best in regions where solar energy is sufficient and water supply adequate. But because solar energy is measured only at a few locations, in practice, air temperature is often used to estimate potential growth of vegetation. Air temperature, in growing degree-days, provides a rough estimate of solar energy input and is a measurement readily available nearly everywhere. Dry matter production is more directly related to net photosynthesis (gross photosynthesis minus respiration). Respiration increases directly with temperature to burn the dry matter produced in photosynthesis. High temperatures, therefore, result in lower overall dry matter production. Plants grow where the accumulated growing degree-days are sufficient and where neither too cold nor too hot temperatures injure the plants.

The amount of water available to the plant depends not only on rainfall but also, critically, on the amount of evaporation and transpiration. Evapotranspiration in turn depends on water vapor pressure, temperature, wind speed, and solar radiation, all of which can be readily measured or estimated.

The effect of the atmosphere on plant growth can thus be determined from observed or estimated atmospheric variables. The range of these variables in observed climates is large and their effect on plants substantial.

Global vegetation patterns

Several investigators have established relationships between the key climate parameters (temperature, rainfall, evapotranspiration) and vegetation. For example, the familiar Koeppen classification of global vegetation is based on temperature and rainfall. The more modern Holdridge Life-Zone Classification System attempts to predict the vegetation of a region from temperature, annual precipitation, and potential evapotranspiration ratio (which is assumed to be dependent on biotemperature and annual precipitation). The Holdridge system classifies vegetation zones by temperature and precipitation and indicates latitudinal belts and altitudinal belts, which often are associated with specific temperatures and rainfall (Fig. 1). A critical temperature line is based on the occurrence of killing frost. No system relating only vegetation to climate variables is completely satisfactory in all respects, but the Holdridge system, as well as other systems, permits the development of a broad generalized picture of global vegetation (Fig. 2).

Agroclimatic zoning

In examining climate and food production, we are interested not only in vegetation patterns but, in particular, in the part of a crop that can be used for food—the grain

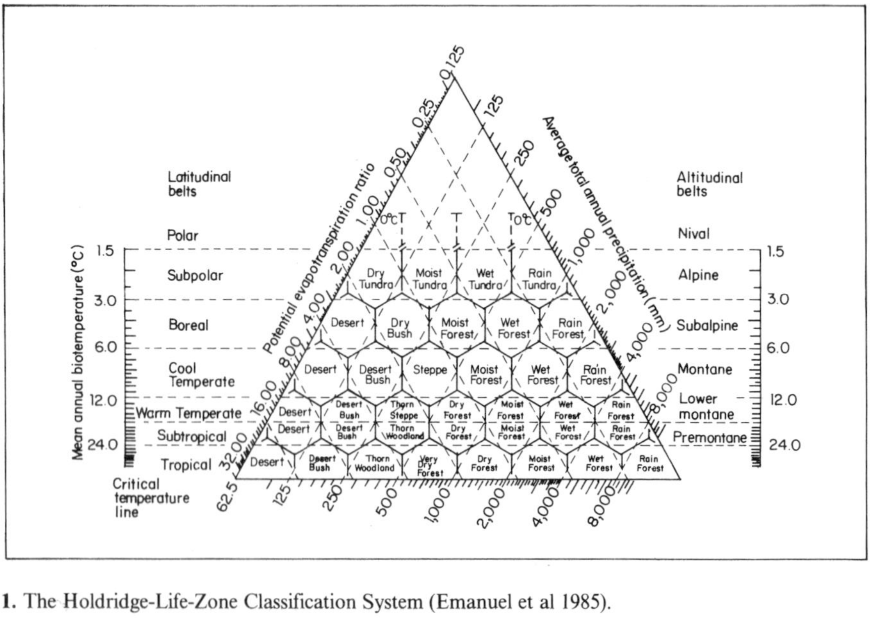

1. The Holdridge-Life-Zone Classification System (Emanuel et al 1985).

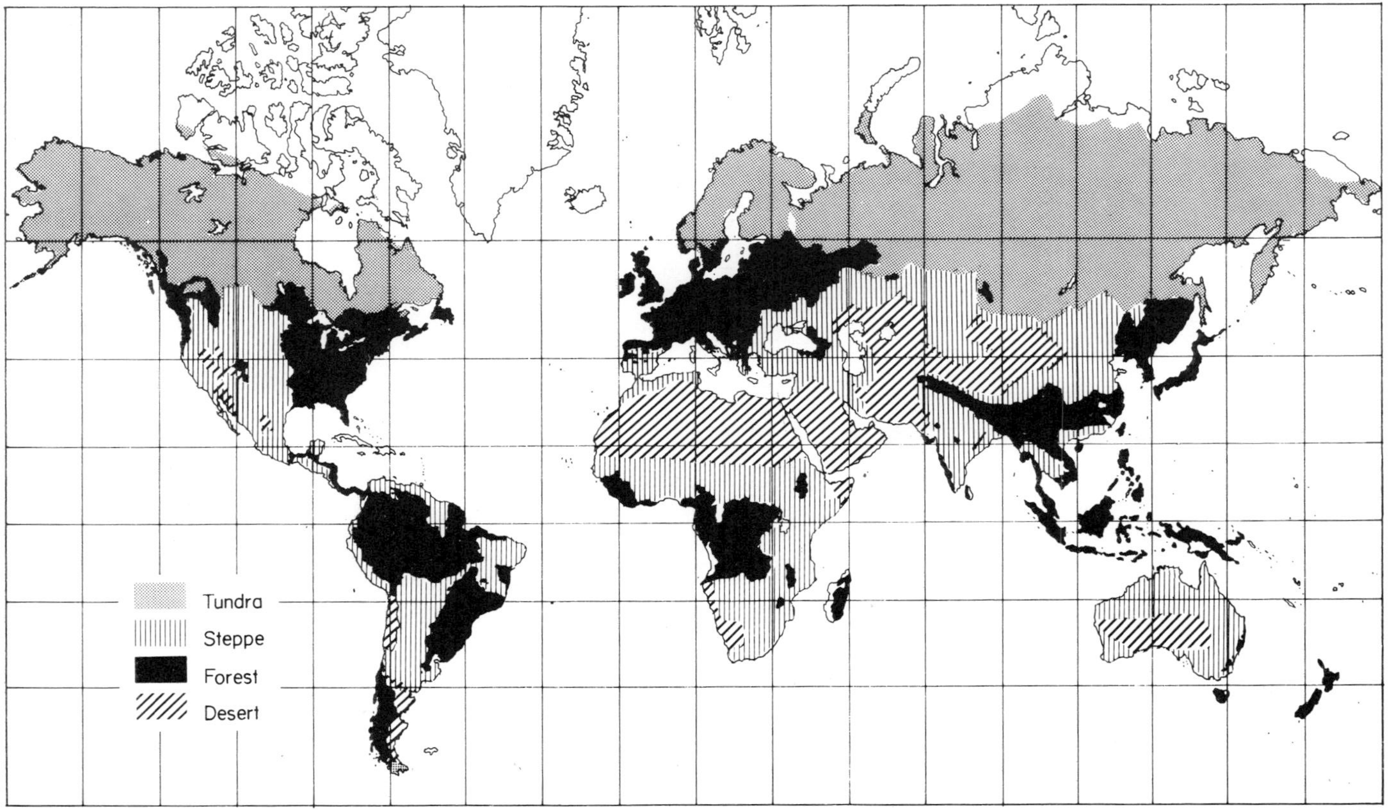

2. Simplified world map based on the Holdridge Classification (base case). The resolution is 0.5° latitude x 0.5° longitude and the extent is from 30 °N to 60 °S. Greenland is not classified (Aitof's Equal Area Projection) (Emanuel et al 1985). Precipitation differences are not shown.

of cereals or the fruits of trees. Agrometeorologists have developed techniques to determine which crops are best suited to a given region, to estimate the frequency of crop failures, and to plan long-range sustainability of the land resources. Several different agroclimatic zoning techniques of varying complexity have been developed. An example of agroclimatic zoning for sorghum and maize suitability in Ethiopia is shown in Figure 3.

Global food production systems

The actual food grown in different areas of the world depends on many factors. The major global crops are shown in conjunction with the Koeppen classification in Figure 4. Clearly, climate plays an important role in determining where major crops are grown.

Natural climate variations

Our discussion so far has dealt with mean or average conditions. Natural variations in temperature, rainfall, or evapotranspiration also affect vegetation, and hence food production.

Frosts

Each year, some regions in the world experience killing frosts during their growing seasons. These frosts may be highly localized and have only a limited effect. Or they may be very extensive and wipe out much of a crop (i.e. coffee in Brazil, the orange crop in North Africa, fruit in Canada), causing losses of several million dollars. Winterkill of winter wheat crops occurs in the absence of snow cover during very cold spells in the USSR and other cold climates. Such frosts are usually short-term weather events, but climatic change may increase their frequency. Meteorologists in Canada and other countries subject to cold spells have developed maps of the length of the frost-free period (Fig. 5) which aid in planning growing seasons and food crops.

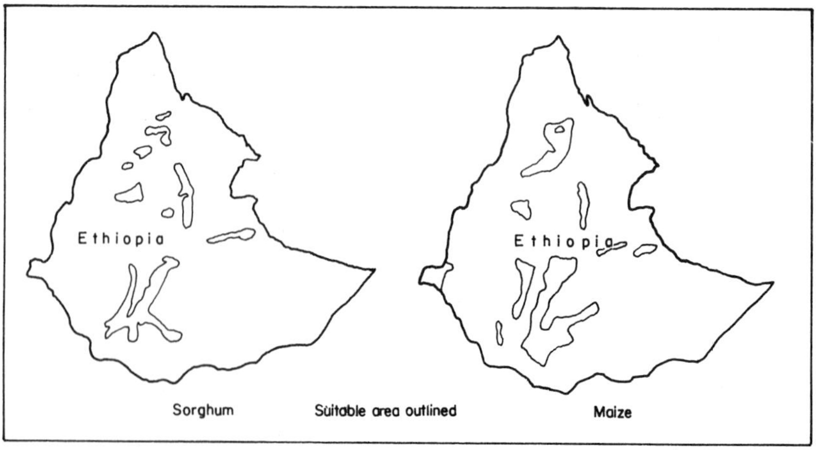

3. Ethiopia - Preliminary climatic suitability maps for sorghum and maize (Williams 1985).

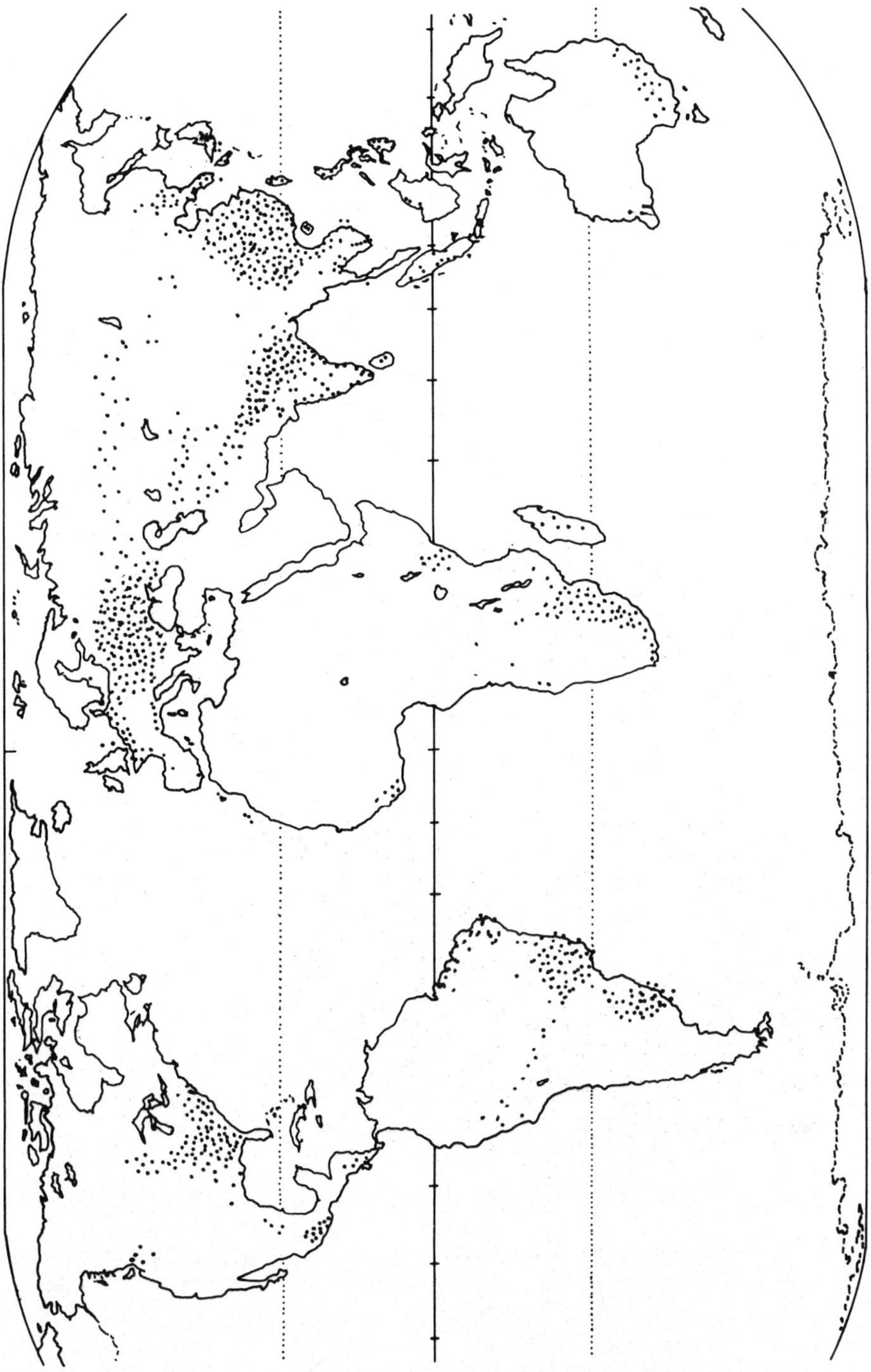

4. Production of major world food crops (simplified from National Geographic Atlas of the World).

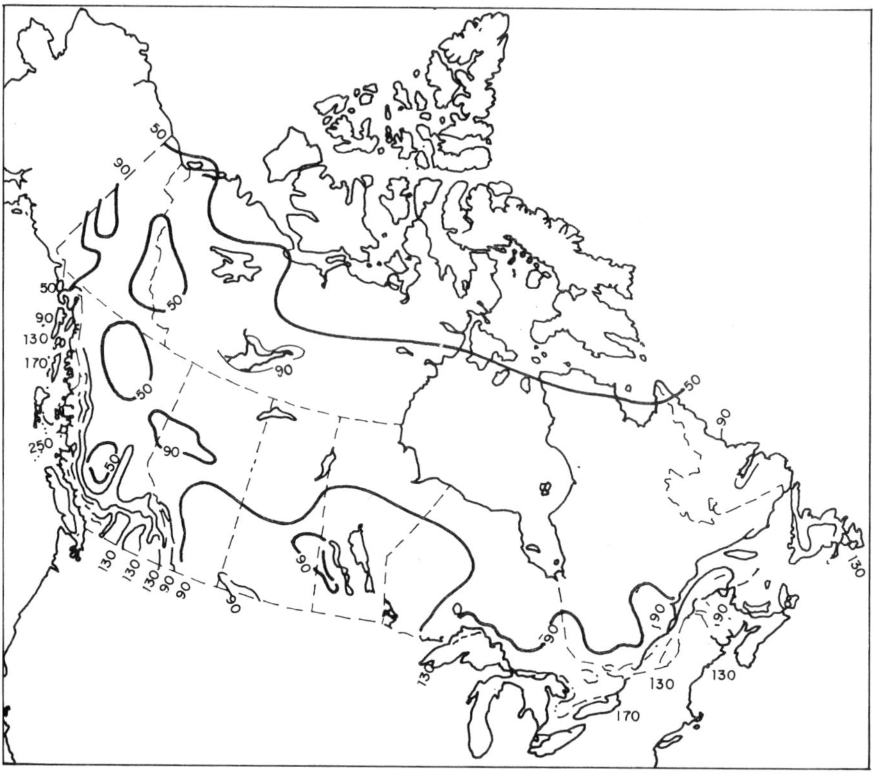

5. Length of frost-free period (days) in Canada, 1941-70 (Hare and Thomas 1974).

Heat waves

Some crops are susceptible to extremely high temperatures over extended periods (i.e. a heat wave in the maize belt of the U.S.). When such a heat wave is combined with a dry period, crop losses can become serious, reducing the availability of animal feed in the U.S. and in many places throughout the world that import maize from the U.S.

Severe weather

Violent weather such as tornadoes and typhoons can cause extensive damage to crops. Floods can wipe out crops, especially during the early growing season when floods uproot newly planted crops. Such events usually result from short-term weather, but large-scale variations in atmospheric flows may increase the frequency of such extreme weather and could be classified as a climate variation. For example, the El Niño of 1982-83 significantly increased the frequency of floods in Ecuador and northwest Peru, adversely affecting the food production systems in those countries.

Common variations in rainfall

Many of the variations in crop yields stem from variations in rainfall, and hence in water availability from place to place or across a growing season. In regions of normally adequate rainfall, a dry year or even a poor distribution of rainfall within a growing season, such as no rain during the flowering or grain filling periods, can markedly decrease crop yields. Year-to-year variations in monsoon rains in India and other parts of Asia have large consequences—the crops affected feed many hundreds of millions of people. Even in regions of abundant rainfall, such as in Zaire, interannual variations in rainfall can reduce a rainy season from 5-6 mo to 2-3 mo and decrease crop yields for that season.

In arid and semiarid zones, dry spells at critical times within a growing season are common. Such dry periods can mean the complete failure of a crop or, more commonly, a large reduction in crop yields, creating a shortage of food or a shortage of cash to buy food. Rainfall deficiencies can be pronounced over large areas for a significant period—the dreaded drought. Many droughts last for a year or less, other droughts caused by longer term large-scale variations in atmospheric flows run their course in 3-4 yr (i.e. the Australian and Brazilian droughts of recent years). Often, agricultural production during such droughts is greatly decreased.

Prolonged droughts

Sometimes droughts are prolonged over many years, as in the Sahel from 1910 to 1914, in the early 1940s, and from 1968 to 1985. Such lengthy droughts have devastating effects on sustained food production and may be associated with extensive famines. Although meteorologists know the general causes of droughts, at present neither the onset nor the duration of droughts can be predicted accurately. Studies are under way within the World Climate Research Programme to better understand climate variations, to permit more accurate predictions of droughts and other significant climatic events. But this is a difficult scientific problem that may take several years to solve.

In the meantime, meteorologists can aid food production by extending the availability of existing climate information, especially during adverse events such as droughts. The effectiveness of appropriate use of existing climate information in increasing crop production in adverse situations has been clearly demonstrated.

For example, probability distribution of rainfall amounts and timing can provide planners with information on the beginning and end of the rainy season and on the likely amount of water available during the growing season. Meteorological services also can provide information on current and near-past weather conditions which affect daily farming operations such as sowing, weeding, thinning, ridging to protect from strong winds, application of fertilizers, irrigation, and protection against crop pests and diseases. Such information has been used effectively in pilot projects in Mali and Burkina Faso to increase crop yields 20-30%, even during drought years.

Climate is an important natural resource which should be tapped now to maintain and even increase food production.

Climate and desertification

Desertification is often defined as the process by which the land loses its ability to sustain vegetation—hence, no food crops can be grown. Although drought is not the only cause of desertification, and many times not even the main cause, it is often associated with desertification, either in initiating the process or in exacerbating man-made causes such as overgrazing. Extended droughts as a leading contributor to desertification are one of the most serious climatic variations in terms of effect on food production.

Better use of existing climate information may help farmers and agriculturists prevent desertification. The World Meteorological Organization (WMO) has initiated a roving seminar project to educate meteorologists, agriculturists, and water managers in the use of meteorological information to prevent wind and water erosion, helping prevent desertification.

Natural and man-made climatic changes

While climate variations occur somewhere in the world every year, climatic changes (defined as significant changes in the mean values of atmospheric variables or in variability) occur much less often. Climatic changes can be either natural (as it occurred in the Ice Age) or man-made (e.g. from burning fossil fuels, which produce CO_2). Some scientists have speculated that the extended drought in the Sahel was the result of a climatic change caused by mankind's activities. In particular, overgrazing was theorized to have changed the albedo characteristics of the earth's surface, which resulted in downward vertical motions in the atmosphere, thereby decreasing rainfall. The vastly improved rains of 1985 and 1986 in many areas in the Sahel would seem to confirm other scientists' view, that the Sahel drought was a natural climate variation and not a true climatic change.

Natural climatic changes

Natural climatic changes can greatly affect food production. For example, a reduction in global mean temperature of 2 °C may shorten the growing season in the midlatitudes by several weeks. An increase in climate variability could bring more frequent frosts and heat waves, both of which would adversely affect food production.

Meteorologists and archaeologists have identified several situations where climatic changes are hypothesized to have dramatically affected food production, even forcing changes in the location and life-styles of large populations. Prominent among these examples are the disappearance of the Mycenean civilization in ancient Greece and of a North American Indian society in the Great Plains about 800 yr ago, both perhaps due to droughts. Some climatic changes produce beneficial results for agriculture, as in the settlement of Greenland. But most of the climatic changes recorded have been deleterious because they interfered with the growing of long-established crops.

The adverse effects on food crops in Europe during the Little Ice Age are an example of significant damage to crops across many years, until the climate improved. In some cases, climatic changes may have induced permanent ecological damage, preventing the growing of sustainable food crops. Such historical examples of the effects of climatic change are to some extent speculative, but the arguments certainly are plausible. More research is needed to establish definitively the effects of past climatic changes on food production.

Man-made effects

Mankind now has the capability to influence climate locally, regionally, and globally. Man-made effects on urban climate and on local air pollution are well known, but those local effects usually have relatively little impact on food production. Regional effects, such as transboundary acid rain, may have serious impacts on forests and other vegetation. Impacts on food production are not yet well known, but evidence is accumulating that both air and water pollution indeed affect food crop yields, at least in localized areas where crops are especially sensitive.

By far the most important potential effects on food production caused by mankind are the global climatic changes that may result from increasing CO_2 and other greenhouse gases. A joint WMO/United Nations Environment Programme/International Council of Scientific Unions International Conference on the Role of Carbon Dioxide and other Greenhouse Gases was held in Villach, Austria, in October 1985. The conference concluded, "It is now believed that in the first half of the next century, a rise of global mean temperature could occur which is greater than any in man's history."

Specifically, the general circulation models show increases of the global mean equilibrium surface temperature for an equivalent doubling of atmospheric CO_2 (by the combined effects of all greenhouse gases) of between 1.5 and 4.5 °C. The thermal inertia of oceans will delay reaching mean equilibrium temperatures by some decades, but current releases of CO_2 and other greenhouse gases make it very likely that such temperature increases will indeed occur during the next century. Regional climatic changes have not been modeled with great confidence, but the models so far show that warming will be greater in higher latitudes during late autumn and winter than in the tropics, annual mean runoff may increase in high latitudes, and summer dryness may become more frequent in middle latitudes in the Northern Hemisphere.

Some investigators have speculated that these predictions portend less rainfall in the great grain-producing areas of North America, the USSR, and China, and hence large decreases in global food supplies. Others have noted that large temperature increases in higher latitudes may lengthen growing seasons in Canada and the USSR, and hence increase food production in those regions. Different models produce different regional patterns of precipitation and temperature. Currently we cannot assert with great confidence that any particular pattern will in fact occur and result in specific effects on agricultural productivity.

In the tropics, temperature increases are expected to be smaller than the global average, but the effects on ecosystems could be large. Potential evapotranspiration will probably increase, making sustained food production even more difficult than it is now in arid and semiarid tropics. An increase of rainfall in the moist tropics probably will not offset the decrease in production in the semiarid tropics. A global warming of 1.5-4.5 °C would lead to a sea level rise of 20-140 cm, affecting food production in coastal areas, and the lives of the many millions of people who reside there.

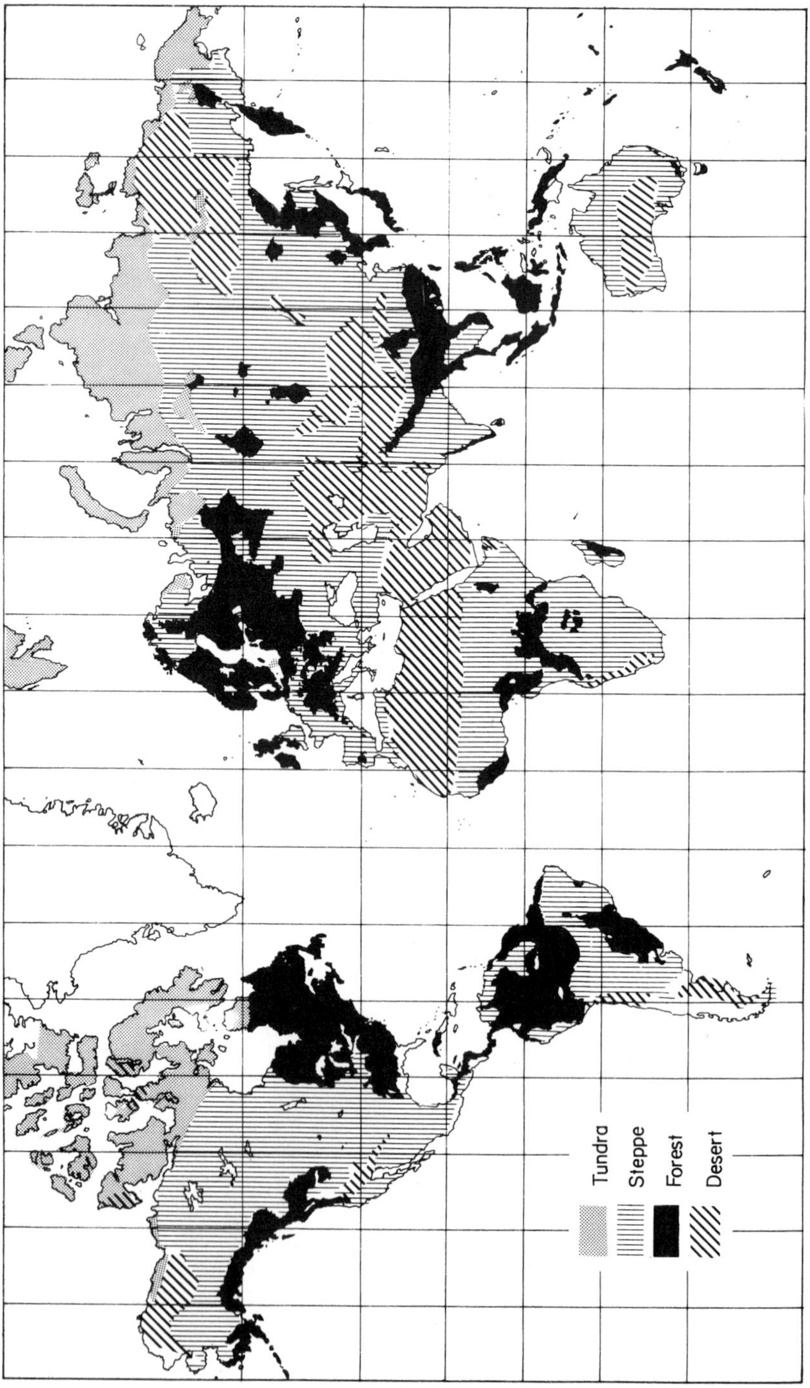

6. Simplified world map based on the Holdridge Classification with biotemperature increased to reflect climate simulated under elevated atmospheric CO_2 concentration (Emanuel et al 1985). Precipitation differences are not shown.

Based on the evidence of the effects of past climatic changes, there is little doubt that a climatic change of the magnitude estimated would have profound effects on global ecosystems, agriculture, water resources, and sea levels, and hence on sustainable food production systems. Figure 6 shows a map of the Holdridge system for world vegetation with a doubling of CO_2. Note the differences in vegetation patterns in some areas from those in Figure 2.

Studies on specific food crops in different locations show that climatic changes resulting from a doubling of CO_2 may either decrease or increase crop productivity, depending mainly on changes in temperature and precipitation and the crop in question. For Alberta, Canada, it has been estimated that crop productivity could increase more than 50% under the assumption of no change in precipitation when the direct effects of CO_2 are included. Even larger increases may occur in higher latitudes in Alberta, where the temperature increases are likely to be greater. Other locations are predicted to suffer substantial decreases in food crops. Special concern must be expressed about the possibility that large decreases in crop productivity may occur in the great food-growing areas of the world, such as China, North America, and the USSR.

The Villach Conference concluded that, despite the major uncertainties that still exist in predictions of regional temperature and precipitation patterns, enough is known now about the greenhouse gases issue to begin to explore alternative policies and adjustments. Such studies should include investigations on the sensitivity of the global agricultural resource base to climatic changes induced by greenhouse gases, the direct effects of increases in CO_2 and other greenhouse gases, and the possible combinations of these effects. Studies should be made on the historical interactions among atmosphere, oceans, cryosphere, and ecosystems, including agricultural systems. The economic and social impacts of sea level rises on agricultural systems should be studied. Policymaking procedures should be analyzed against the various kinds of risks from a greenhouse warming, so that governments and international organizations will be prepared to initiate appropriate actions when needed.

Conclusion and recommendations

Clearly, climate affects plants, and hence crops, and hence food production. Climate variations often result in substantial fluctuations in food production, some of which may last for several years and, in the worst cases, lead to famine. Climatic changes, natural or man-made, conceivably could significantly alter current global patterns of agriculture and food production, and possibly decrease markedly global food supplies. Without any doubt, climate is a significant factor in food production and in food security.

What can the meteorologist do now to help? First, research into the mechanisms that determine climate should continue, aiming at better capability to predict climatic variations and changes. At the same time, interdisciplinary research should begin on the impacts of climatic changes and on the associated policy questions. Finally, while this research is under way, meteorologists should work with agriculturists to better use available climate information in both planning and in daily farm operations to maintain or increase crop food production.

References cited

Emanuel W R, Shugart H H, Stevenson M P (1985) Climatic change and the broad-scale distribution of terrestrial ecosystem complexes. Climatic Change 7(1):29:54.

Hare F K, Thomas M K (1974) Climate Canada. J. Wiley and Co., Toronto, Canada.

National Geographic Society (1981) Atlas of the world. Washington, D.C. p. 56-57.

Williams G V D (1985) Classifying Ethiopia's climate and analyzing and mapping its agroclimatic resources. WMO Contract No. 14.553/A/CNS).

Notes

Author's address: G. O. P. Obasi, World Meteorological Organization, Geneva, Switzerland.

Citation information: International Rice Research Institute (1989) Climate and food security. P.O. Box 933, Manila, Philippines.

Irrigation as a factor in food security

D. F. Peterson

Of the 225 million ha of irrigated land in the world, 165 million ha are in developing countries—66% of that is in South and Southeast Asia, 17% in the Middle East and North Africa. Worldwide, during 1965-75, irrigation grew at about 5%/yr. But growth dropped sharply during 1975-85, probably because of rising capital costs of development and shrinking prices for agricultural commodities. Concurrently, the surge in yield called the green revolution stagnated. Serious policy, management, and technological problems also exist. Adequate maintenance has fallen by the wayside. Case studies of the effects and prospects of irrigation in enhancing food security are reported for South and Southeast Asia and for Africa, excluding the Mediterranean countries. Descriptions of geographical variation of climate related to irrigation are included. For South and Southeast Asia, a significant linear correlation between the increase in national cereal production and the increase in irrigated area exists. For Africa, irrigation development in general has been slow and sporadic, although the potential for developing arable rainfed areas in several countries is large. Development of surface reservoirs will be increasingly difficult. To ensure reliable supplies, groundwater (where available) can be used more effectively; possibilities for utilizing more modern technologies also are increasing.

While data on irrigated areas are often inconsistent because of differences in reporting, about 225 million ha—15%—of the world's 1,500 million ha of cultivated land are irrigated. This small fraction produces an estimated 37% of the total crop value. If cultivated land with surface and subsurface drainage and flood control are included, the contribution to total crop value rises to about 50%.

Nearly 75% of the world's irrigated land is in the developing countries (China, 21%; South and Southeast Asia, 28%; Middle East including North Africa, 12%; Africa, 5%; Latin America, 7%). World irrigation development generally has been accelerating, growing at 5%/yr during 1965-75, but dropping to 1.8%/yr during 1975-85 (Framji et al 1981; Horning 1986; Rangeley 1983, 1986).

The main reasons for the decline in growth rate are rapidly rising investment costs and declining prices for agricultural products. This slowing trend has had less impact where high-value crops are dominant, but the recent growth rate in developing countries may have been as low as 1.5% (ADB 1986, Horning 1986).

Problems with waterlogging and salinity seem to be increasing, even though they have received increased attention. But financing for solving waterlogging problems has been more difficult to negotiate than financing for new irrigation (Framji 1984).

Irrigation and new seeds and fertilizer have been the major technological factors contributing to increased national food security in South and Southeast Asia and the Middle East; development of all three technologies has lagged seriously in Africa. Expanding irrigation promises to be increasingly difficult.

However, yields from currently irrigated lands still remain far below potential. Probably the principal constraint is farm-level economics, often associated with pricing controls and crop quotas (Carruthers et al 1985). The main technological constraints are the unreliability of water service and mismatched irrigation scheduling and crop water requirements. The technological problems are partly managerial, partly inadequate construction and poorly maintained facilities. Fortunately, attention to these problems has increased during the last few years.

A South and Southeast Asia case history

A study of South and Southeast Asia, extending from Indonesia on the southeast to Afghanistan on the northwest (Fig. 1), examined development related to irrigated agriculture during roughly the last two decades.

Asian agriculture has three unique features: heavy monsoon rainfall in most areas, rice as the single major crop, and the predominance of subsistence farming (average farm size is 1 ha or less) (ADB 1986). Upland cropping systems are more dominant than rice in parts of India, Pakistan, and Afghanistan.

Because excess surface water limits increased production, the term irrigation implied water control that includes the removal of excess water. That hazard is

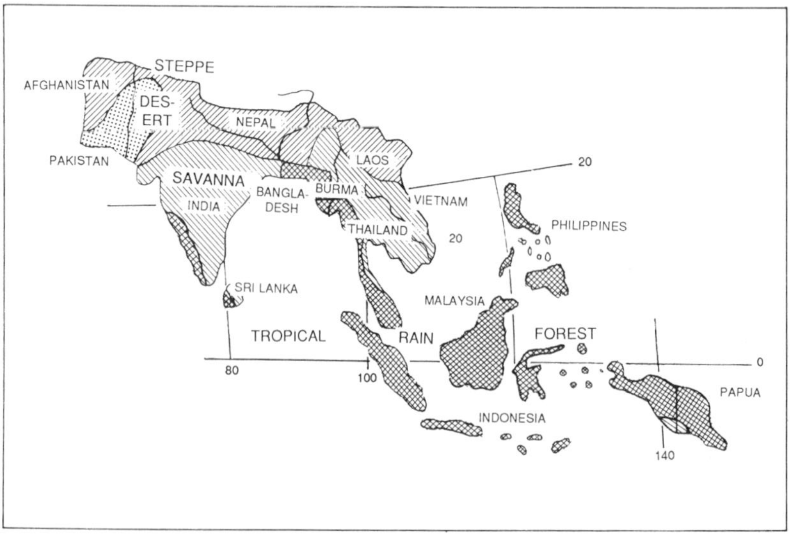

1. Climatic zones of South and Southeast Asia.

accentuated by typhoon damage, such as that which occurs in the Philippines and Bangladesh, and periodic flooding, such as that which occurs along the Indus and Ganges Rivers.

Progress in agricultural development

Rice accounts for about two-thirds of the cereal production in the region. In 1978-80, wheat constituted 16%, maize 7%, and coarse grains 10% of production (ADB 1986). Overall per capita growth in personal income from 1965 to 1984 averaged 3%/yr; 4% for the Southeast Asian countries of Indonesia, Malaysia, and the Philippines and 2.3% for the South Asian countries of Afghanistan, Bangladesh, Burma, India, Nepal, Pakistan, and Sri Lanka.

Total annual cereal production has increased 72%, from a 191 million t average 1966-68 to 330 million t average 1981-83. Net annual imports of cereals decreased from 12 million t to 9 million t. Although population increased 41%, from 950 million to 1,340 million, daily caloric intake increased 7%, from 2,110 to 2,260 calories, and protein 6%, from 51 to 54 g.

While the arable area increased 7%, about 17 million ha, the irrigated area grew 44%, from 52 million to 75 million ha, raising the proportion of arable land that is irrigated from 21 to 28%. Technical innovations, such as the introduction of high-yielding varieties and increased use of chemical fertilizers (from 2.4 million to 11.5 million t/yr); more appropriate pricing policies; and improved institutional services, also contributed to the Asian agricultural achievement.

Climate

The climatic zones of South and Southeast Asia are shown in Figure 1. The dominant weather characteristic is the strength and persistence of the southwest monsoon. The monsoon produces an average 2,000 mm/yr rainfall for most countries in the region (ADB 1986). Except for Afghanistan and some northern areas of India and Pakistan, the region is frost-free. Most of the countries near the equator are in the tropical rain forest climate zone.

Annual climate indicators for selected stations, sequenced by decreasing east longitude, are given in Table 1. The annual monsoon cycle of extremely wet summers and extremely dry winter periods (which generally become more pronounced with increasing latitude) is apparent. The exception is the northeast (winter) monsoon, which waters Southeast India and part of Sri Lanka. Mannar, Sri Lanka, experiences both monsoons. In addition to the annual cycle, climate also may vary spatially (e.g. compare Moulmein and Mandalay in Table 1).

Median monthly precipitation and estimated evapotranspiration for six selected South and Southeast Asia sites are shown in Figure 2. At Bengkulu, Indonesia, seasonal irrigation is probably not needed; at Dengpasar, the need for summer season irrigation is clearly evident. Even with the very high annual precipitation at Moulmein, irrigation would be needed during the winter season. At Pune, only such dryland crops as sorghum, millet, and some lentils can be grown without irrigation, and the risk of drought is high. At Karachi, crops cannot be grown without irrigation.

Table 1. Climate descriptors at selected sites in South and Southeast Asia.

Station	Longitude	Latitude	Mean temp (°C)	Mean precip. (mm)	Annual ETP[a] (mm)	Very dry months[b]
Merauke, Indonesia	140°24'E	8°30'S	26	1,375	1,750	Jul-Nov
Manokwari, Indonesia	134°18'E	0°54'S	26	2,539	1,611	
Dengpasar, Indonesia	115°26'E	8°54'S	27	1,477	1,654	Apr, Jun-Sep, Nov
Bengkulu, Indonesia	102°12'E	3°54'S	26	3,719	1,594	
Manila, Philippines	121° 0'E	14°36'N	27	2,115	1,473	Jan-Apr
Chiang Mai, Thailand	99° 0'E	18°45'N	26	1,266	1,626	Jan-Apr, Nov-Dec
Moulmein, Burma	97°36'E	16°30'N	27	4,801	1,639	Jan-Apr, Nov-Dec
Mandalay, Burma	96° 6'E	22° 0'N	27	865	1,803	Jan-Apr, Nov-Dec
Dhaka, Bangladesh	90°24'E	23°48'N	26	2,148	1,663	Jan-Feb, Nov-Dec
Colombo, Sri Lanka	79°54'E	6°54'N	27	2,296	1,925	Feb
Mannar, Sri Lanka	79°54'E	9° 0'N	28	962	2,060	Jan-Mar, May-Sep
Mangalore, India	74°54'E	12°54'N	27	3,416	1,924	Jan-Apr, Nov-Dec
Lahore, Pakistan	74°24'E	31°36'N	24	511	1,494	Jan-Jun, Sep-Dec
Pune, India	73°54'E	18°30'N	25	704	1,888	Jan-May, Nov-Dec
Karachi, Pakistan	67°12'E	24°54'N	26	199	1,743	Jan-Dec
MazariSharif, Afghanistan	67°12'E	36°42'N	18	188	1,459	May-Nov

Irrigation and national food-grain production

Cereal production and area irrigated increased in each of the 15 countries in the region between 1966-68 and 1981-83 (Fig. 3). The linear correlation coefficient is significant with $r^2 = 0.95$. Omitting the outlying countries Burma and Indonesia, where technological advances were delayed, improves the correlation, $r^2 = 0.99$; $y = 5.19 \times -0.15$.

For the region, annual cereal production increased about 5.2 t for each additional irrigated hectare. But this does not mean that 5.2 t/ha is the yield in the newly irrigated areas: that averages only about 2 t/ha. The period 1966-83 also reflects some of the large gains in yield resulting from the green revolution. We cannot assume that the same rate of increase will continue.

The effect of irrigation within concurrent changes in technology is unknown. Given ample water supply and responsive cultivars, the 9 million t/yr increase in use of chemical fertilizer could account for 30-50% of the 139 million t/yr increase. Seckler (1986), in seeking an explanation for the recent stagnation of cereal production in Pakistan following a period of rapid growth during the 1970s, concluded, "94% of the variation in wheat production can be accounted for by the total consumption of fertilizer in the rabi season. Ninety-five percent of the variance in fertilizer use in the rabi season is accounted for by the variance in total irrigated area."

Rainfed area actually decreased, and increased productivity from rainfed lands appears to be nil. A linear regression of increases in rainfed areas by country with increases in production essentially gave results opposite to those for irrigated areas, but with slightly weaker correlations.

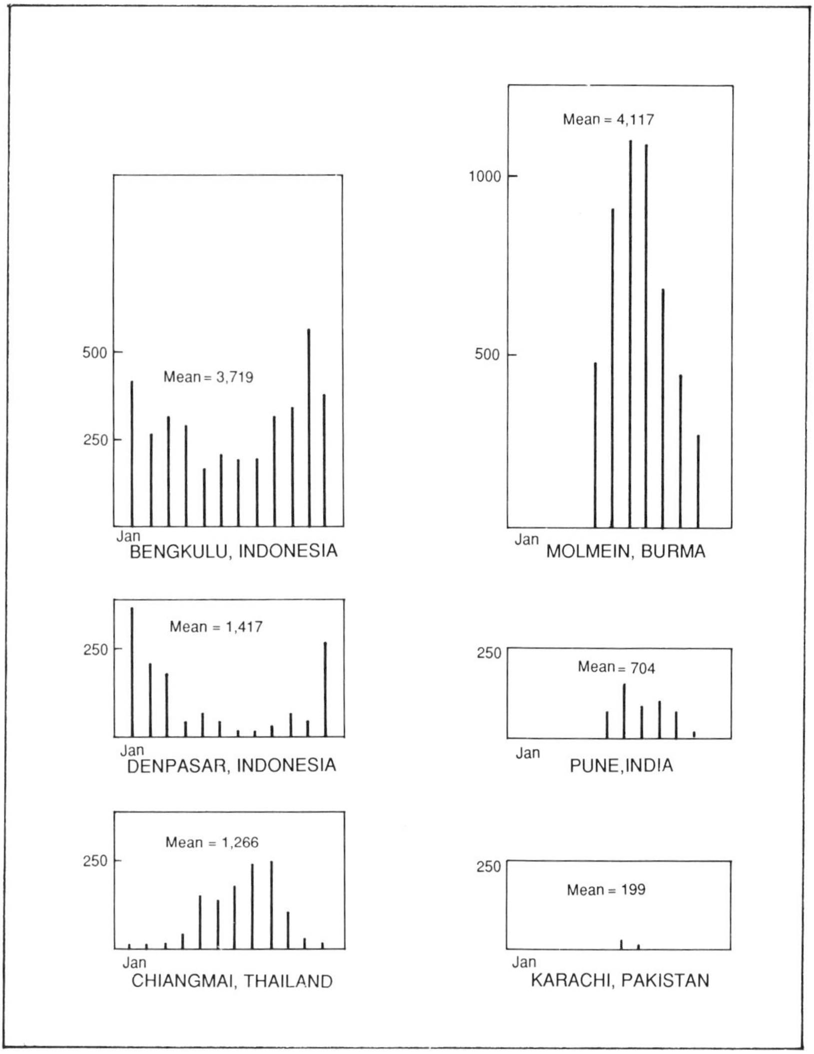

2. Average monthly precipitation (mm) at six selected South and Southeast Asia sites.

Nutrition. There was no significant linear correlation between per capita irrigated area and caloric intake in 1966 and 1982 (Table 2). Nevertheless, some observations can be made. During the period, average caloric intake for eight Southeast Asian countries increased from 2,049 to 2,316 cal/d. At the same time, the spread among the countries increased, with the standard deviation changing from 142 cal/d in 1966 to 242 cal/d in 1982. Uniformly substantial gains (300-500 cal/d) were made in Burma, Indonesia, Malaysia, Philippines, and Sri Lanka, with a more modest but still impressive increase of 110 cal/d for Thailand. The situation remained stagnant in Bangladesh and Nepal, the two countries with the lowest irrigation indices.

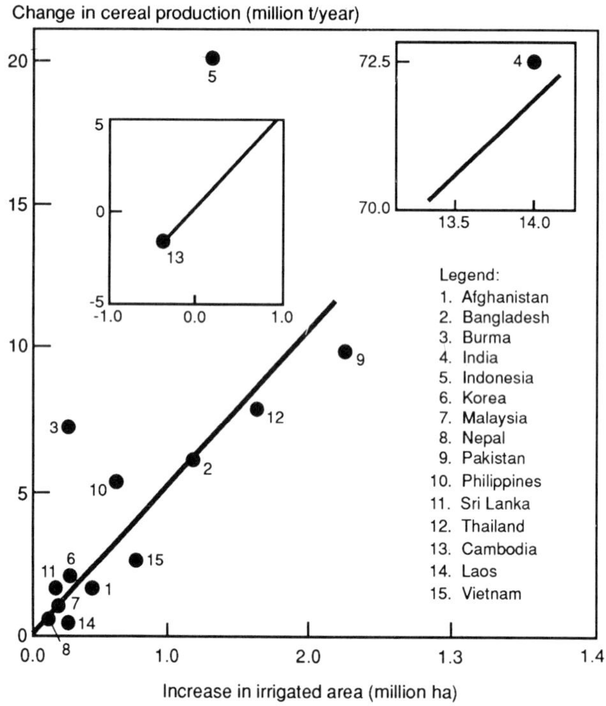

3. Average annual increase in cereal production and increase in irrigated area, 1966-83, South and Southeast Asia.

Land resources. The per capita index of irrigated land is much larger for the more arid southwestern Asia (Afghanistan, India, Pakistan, Table 2). For example, the Pakistan index for 1982 was 158 ha/1,000 persons, in contrast to an average 30.8 ha/1,000 persons for the southeastern countries. Indices of arable land and rainfed land also vary widely (Table 3). These are dramatically lower for the countries in the rain forest climate zone. For the five countries primarily in the rain forest zone, the arable land index averages 92 ha/1,000 persons compared to 301 ha/1,000 persons for the five countries with substantial savanna or arid areas. These striking contrasts emphasize the fact that water is the critical resource, even in South and Southeast Asia. Irrigation is necessary to expand multiple-season cropping at most sites.

Outlook

The changing economic situation will doubtless be the dominant factor affecting agricultural production and growth of irrigation in South and Southeast Asia during the remainder of this century. The rising debt service burden, stagnation in international trade, depressed commodity prices, and worsening terms of trade are to constrain the level of growth experienced over the last two decades. International rice prices have fallen in real terms over the past 12 yr. The projected price for rice for 1995 is still significantly below prices prevailing between 1966 and 1985. Low prices

Table 2. Irrigated area per capita and average caloric intake, Southwest and Southeast Asia.

Country	Irrigation index (ha/1,000 persons)		Caloric intake (cal/capita per d)	
	1966	1982	1966	1982
Afghanistan	199.5	150.0	2,120	2,280
Bangladesh	10.00	18.66	1,960	1,920
Burma	30.60	28.10	2,020	2,460
India	54.04	55.01	2,150	2,050
Indonesia	38.71	33.79	1,920	2,380
Korea	30.39	28.83	2,280	2,980
Malaysia	25.13	25.52	2,310	2,680
Nepal	10.18	14.93	2,020	2,020
Pakistan	236.4	157.6	2,190	2,280
Philippines	22.60	25.68	1,890	2,390
Sri Lanka	25.50	33.25	2,080	2,380
Thailand	55.25	65.85	2,190	2,300

Table 3. Arable and rainfed areas, South and Southeast Asia.

Country	Arable area (million ha)		Rainfed area (million ha)		Index (1982) (ha/1,000 persons)	
	1968	1982	1968	1982	Arable	Rainfed
Afghanistan	7.85	7.92	5.87	5.26	461	297
Bangladesh	8.86	8.91	8.23	7.11	92	74
Burma	9.91	9.67	9.14	8.63	260	232
Cambodia	2.86	2.87	2.76	2.78	NA	NA
India	158.69	165.17	132.03	125.11	224	170
Indonesia	12.61	14.27	8.43	8.82	88	55
Korea, Rep. of	2.19	2.05	1.17	0.88	50	22
Laos, FDR	0.83	0.87	0.82	0.88	NA	NA
Malaysia	0.88	1.02	0.64	0.63	67	42
Nepal	1.81	2.82	1.71	2.09	155	136
Pakistan	20.18	19.94	7.69	5.25	219	56
Philippines	7.11	7.78	6.38	6.41	146	120
Sri Lanka	0.79	1.05	0.40	0.53	67	35
Thailand	11.41	17.31	9.64	13.97	341	275
Vietnam	5.19	5.67	4.21	3.97	NA	NA
Total	251.35	267.32	199.12	192.32		

will benefit consumers and will not affect subsistence farmers, but will adversely affect incentives and use of fertilizer by market farmers, and with rapidly rising costs of development, affect the economic viability of irrigation projects.

There is still scope for increased productivity and increased multiple cropping. This will require more efficient irrigation management than that generally existing now, and in most cases will require rehabilitation and modernization of existing facilities. Thus, the emphasis of international lending agencies is expected to move toward rehabilitation and modernization rather than to development of new areas. Concern for improved management by a group of international donors, encouraged by the Consultative Group for International Agricultural Research (CGIAR), led to the establishment in 1984 of the International Irrigation Management Institute in Sri Lanka.

A study completed in 1984 by the International Food Policy Research Institute and the International Rice Research Institute, made at the request of ADB, projected an annual deficit in food grains for the region of 41 million t by 2000 A.D. (Table 4). Three-fourths of the regional deficit is generated by the two most developed countries, Taiwan and Korea, but serious deficits are also projected for most of the other countries. The sum of the national deficits totals 66.2 million t. Surpluses totaling 25.5 million t, achieved by only four countries, would be available for international trade or security storage at home.

Although irrigation has been the key to achieving national food-grain security throughout the region and will continue its critical role, a score of new and complex problems, some of them extending beyond the irrigation sector, will have to be faced over the next decade or so, if food stability is to be maintained or improved.

The Africa case

Climate

Africa is the world's driest continent. Annual runoff is estimated at 139 mm, compared to 252 mm for Asia and 445 mm for South America, the wettest continent. Africa's largest river, the Congo, is third largest in the world. Its runoff is nearly 15 times that of the Nile. Best estimates of the annual flow of principal African rivers, in billion m^3/yr, are Congo, 1,245; Zambezi, 250; Niger, 215; Nile, 84; Senegal, 70; Volta, 37. For comparison, that of the Ganges is 490 m^3/yr (Framji et al 1981, 1983; L'Vovitch 1975).

Table 4. Food grain projection for the year 2000 (million t).

Country	Production	Surplus/deficit	Total food grains (% of demand)
Afghanistan	5.81	- 1.86	-24
Bangladesh	33.74	- 6.02	-15
Burma	19.63	2.14	12
India	260.90	8.66	3
Indochina	16.14	- 8.99	-36
Indonesia	54.82	- 7.87	-13
Korea, Rep. of	11.75	-13.75	-54
Malaysia	3.26	- 3.46	-49
Nepal	5.49	- 0.47	- 8
Pakistan	43.40	7.13	20
Philippines	17.84	- 3.37	-16
Sri Lanka	3.94	- 0.37	- 9
Taiwan	4.87	-16.85	-78
Thailand	34.32	7.79	29
Other	–	- 2.98	–
Total	515.91	-40.27	- 7

Except for a narrow band of subtropical dry climate along the northeast Mediterranean, North Africa is dominated by the great Sahara Desert (Fig. 4), which extends through nearly 20° latitude and covers almost half the continent's land mass. South of the Sahara, the climate changes rapidly through the Sahelian steppe and savanna zones to the tropical rain forest of the lower Congo equatorial belt in West Africa. The savanna belt extends much more broadly in East Africa, occupying an area almost equal in size to the Sahara Desert and extending to the steppe areas of southern Africa. Annual mean precipitation varies through Central West Africa and through Central Africa (Fig. 5, Table 5).

The extensive steppe and savanna land resources provide a basis for Africa's great potential in rainfed agriculture, but these areas lie in the eastern, central, and southern portions of the continent. The most critical area is the Sahelian-Sudanese band immediately south of the Sahara, where the climate changes rapidly with latitude and wide annual, and periodic time variations occur. This situation has been aggravated by increases in population and the restriction of nomadic pastoralism. Nevertheless, two of the continent's major rivers, the Niger and the Senegal, flow through this zone.

Egypt is unique. Here, precipitation is virtually zero, but the Nile River provides an adequate water supply for the rich soils of the Nile Valley and Delta. But these

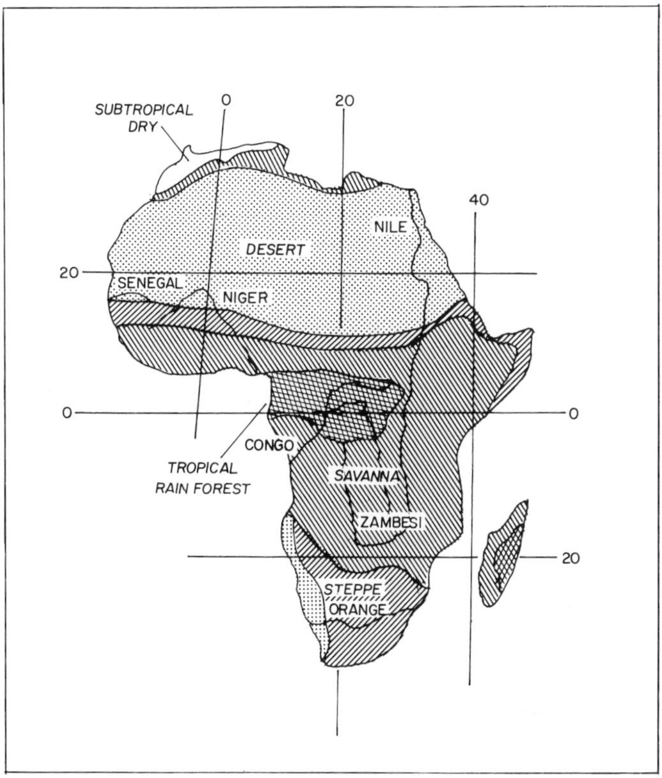

4. Climatic zones of Africa.

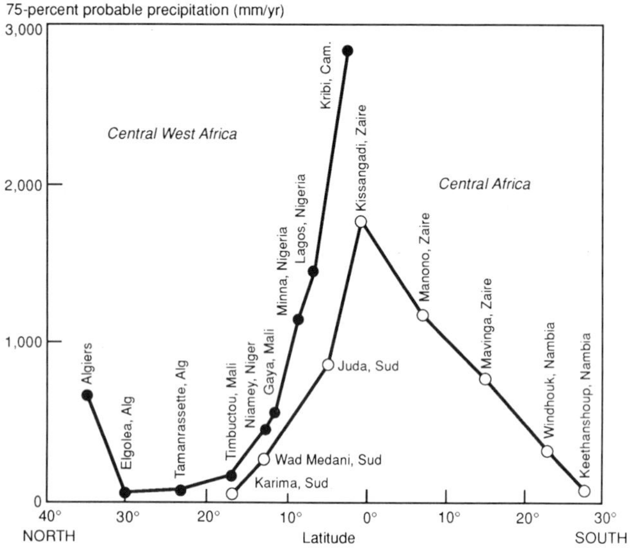

5. Variation of precipitation with latitude at selected sites in Africa.

soils are strictly limited in area. The critical task is to increase productivity on lands that have already reached relatively high yield levels.

Irrigation in Africa

There are about 9 million ha of irrigated land in Africa (Abernathy and Brethery 1986), in striking contrast to Asia. Land irrigated in Indonesia alone equals 60% of all land irrigated in Africa. Irrigation is also unevenly distributed. Only five countries—Egypt, Sudan, Madagascar, Nigeria, and Morocco—irrigate in excess of 500,000 ha/yr. They account for 6.4 million ha of the total for the continent. Ten countries irrigate between 100,000 and 500,000 ha, 23 irrigate less than 20,000 ha each. Much of present irrigated areas, particularly in Sudan and Madagascar, produces commercial crops for export. Deducting the 2.7 million ha in Sudan and Madagascar and the 3.7 million ha in the Mediterranean countries leaves only 2.4 million ha that could be considered to contribute the food security of the 450 million people who live in continental Africa. Irrigation development in most of Africa has largely failed.

Abernathy and Brethery (1986) identified 835 million ha of potential arable rainfed area, compared to 142 million ha utilized in 1982. They estimated that with the low levels of technology that might be implemented between now and 2000 A.D., this potential theoretically could support a population of 1,028 million. Sudan, Nigeria, Tanzania, Mozambique, Zaire, Zambia, Angola, and Central Africa, with a total of 524 million ha, have large rainfed potential but contain only 38% of the continent's population.

Table 5. Climate descriptors at selected African stations.

Station	Latitude	Longitude	Mean temp (°C)	Mean precip. (mm)	Annual ETP (mm)	Very dry months
Algiers, Algeria	36°42′N	3°18′E	17	762	1,237	May-Sep
El Golea, Algeria	30°24′N	2°54′E	22	39	1,803	Jan-Dec
Tamanyassett, Algeria	22°48′N	5°30′E	22	61	1,834	Jan-Dec
Timbuctou, Mali	16°48′N	2°00′W	29	196	2,075	Jan-Jun, Sep-Dec
Niamey, Niger	16°30′N	3°06′E	29	577	2,238	Jan-May, Oct-Dec
Gaya, Mali	11°54′N	3°30′E	28	741		Jan-Apr, Oct-Dec
Lagos, Nigeria	6°30′N	3°24′E	27	1,740	1,626	Jan-Feb, Dec
Kribi, Cameroon	3°00′N	9°54′E	26	2,997		
Karima, Sudan	18°36′N	31°54′E	28	37	2,298	Jan-Dec
Wad Medani, Sudan	14°24′N	33°30′E	27	358	2,150	Jan-Jun, Oct-Dec
Juba, Sudan	4°54′N	31°42′E	26	977	1,824	Jan-Feb, Nov-Dec
Kissangani, Zaire	0°30′N	25°12′E	25	1,883	1,433	
Manono, Zaire	7°18′S	27°54′E	25	1,213	1,576	May-Aug
Mavinga, Angola	15°48′S	20°24′E	21	845		Apr-Oct
Windhouk, Nambia	22°20′S	17°30′E	18	423	1,650	May-Dec
Keethanshoup, Nambia	26°36′S	18°06′E	21	157	1,781	Feb-Dec

Table 6. Ranges of food deficits and surpluses compared to estimated potential production, 2000 A.D., listed in order of critical deficit or higher surplus.

Deficit		Surplus	
50-90%	0-50%	0-100%	>100%
Rwanda	Egypt	Sierra Leone	Liberia
Kenya	Morocco	Sao Tome	Madagascar
Niger	Burkina Faso	Sudan	Ivory Coast
Libya	Swaziland	Guinea	Angola
Somalia	Senegal	Mozambique	Zambia
Burundi	Malawi	Chad	Cameroon
Comoros	Mali	Guinea Bissau	Zaire
Mauritania	Zimbabwe		Equatorial Guinea
Algeria	Gambia		Central Africa
Mauritius	Togo		Congo
Lesotho	Tanzania		Gabon
Tunisia	Ghana		
Ethiopia	Benin		
Nigeria			
Uganda			
Botswana			

Abernathy and Brethery calculated the percentage of the projected 2000 A.D. population in each country that theoretically could be supported by rainfed agriculture. With some modifications of the data to allow for irrigation already in place, the results are shown in Table 6. The tremendous obstacles to developing additional rainfed agriculture in Africa are well known; the table primarily identifies where the likely problems and opportunities are.

The large rainfed potential in some countries suggests that food security could be achieved by large-scale migration. This almost certain will not happen soon. Population is increasing rapidly, and it will nearly double by 2000 A.D. Greatly accelerated irrigation development will be necessary to avoid serious food security problems in more than half the African countries. The most critical area is the Sahel. The terrible consequences of the now chronic Sahelian famine are well known, but irrigation development has not responded to this problem.

Rangeley (1986) sums up the irony of this failure: "It is a sobering thought that with all the publicity we have had on the drought in Sahel Africa, the total deficit of the Sahel countries was no more than about 3 million t, and if we could only develop irrigation on the scale that it is developed even on relatively small projects in India, Pakistan, and the Nile Valley, if we could develop similar things in, let us say Mali, in the inland delta of the Niger or places like Chad, this would be fully adequate to meet the whole of that deficit and to solve the problems for a long time."

Many have commented on the bleak outlook for mobilizing irrigation in Africa. World Bank irrigation advisor, Lemoigne (1986) reflects the pessimism: "Most of the Bank's irrigation schemes in sub-Saharan Africa have not been successful, with one or two notable exceptions like the Rahad Project in Sudan. This is mainly because of high capital and running costs associated with low productivity and aggravated by institutional issues, unrealistic exchange rates, and trade policies. Although some of these problems are being solved, irrigation in sub-Saharan Africa is likely to remain only marginally justified, particularly when compared with conditions in other parts of the world, particularly Asia, and it will be a long time before irrigated crop production can make an impact on the total pattern of food production in Africa."

Many possible reasons for Africa's lack of success vs Asia's success have been advanced. There are significant geological differences. Lack of physical and social infrastructure, relative sparsity of populations, and high economic costs with low returns are suggested as causes. All are valid, but the more critical difficulties relate to socioeconomic conditions more than they do to technical ones. Perhaps the best general explanation was stated by J. Kellor (Utah State University, 1987, pers. comm.): "A key to being able to afford large-scale technical irrigation systems is the ability to construct within indigenous facilities and means." Africa is not at this stage of development.

Actually there is little immediate need for large-scale irrigation systems in Africa. Some small-scale indigenous systems have existed for centuries, but by and large relatively little attention has been given them. Part of the problem may be that development interests are preoccupied with large, expensive, and complex projects, at the expense of more attention to smaller-scale, more manageable ones. African systems suffer high development costs and sagging market prices (the latter, ironically, partly because of the very success of irrigation in Asia). But this situation may be temporary.

Irrigation and the poor

The Asia case demonstrates that, with irrigation, most countries can close the gap between economic demand and food grain national production. India's September

1985 food grain reserve was 29 million t. This does not mean that food security has been achieved; millions of very poor people are still undernourished, with neither access to subsistence supplies or enough income to buy basic nutritional requirements.

The situation is far worse in many African countries. The main enemy of food security is widespread poverty, either from lack of resources for subsistence farming, or low income, or both. In today's world, this is more critical than the ability of farmers to produce more. Chambers (1986) proposes that irrigation be judged on the basis of its contribution to livelihood, which he defines to include income, assets, and well-being, rather than only its contribution to production.

Chambers argues that irrigation has greatly benefited laborers and resource-poor farmers. The benefits include "amount, stability and spread of income and employment; (in) reduced vulnerability to impoverishment; and (in) a better quality of life." Both quantitative and nonquantitative data are cited. For example, an irrigated village and a nonirrigated village in West Bengal were compared. Casual laborers in the irrigated village averaged 298 d/yr, compared to 135 d in the nonirrigated.

Chambers also compared migration in several villages. The contrasts between irrigated and nonirrigated villages were great. In irrigated villages, people no longer had to go away to seek work during dead periods. Access to services, especially education (particularly for girls), was vastly improved. Chambers concludes: "As a weapon against poverty, irrigation has been undersold. Where feasible, well-implemented irrigation development is probably the single most promising short- and long-term means of reducing poverty."

Water supplies

Surface reservoirs

Surface reservoirs are used primarily to store water during the wet season for use during the following dry season. Large reservoirs like Aswan or Lake Mead, which can even out annual or cyclical rainfall fluctuations, are relatively rare. Reservoir storage is limited by available, economically developable sites. These are becoming increasingly scarce. There are geographical differences also. For example, the U.S. has about three times the reservoir storage potential of India. Increasing environmental concern about inundating productive lands and archaeological and religious artifacts, displacing people, increasing potential for waterborne diseases, and interference with wildlife will rule out many future developments. The ability of new surface reservoirs to alleviate irrigation water supply problems associated with climate variations will be, at best, modest.

One unexploited potential is the use of small reservoirs with capacity for overnight to several days water supply. Located along the alignments of long canals, they can even out varying demand resulting from day-to-day fluctuations in weather and inherent operational characteristics of canal systems. Returns from saved water and more reliable deliveries could be very large if this practice were more widely followed.

Groundwater

Groundwater reservoirs play a similar role, with the added advantage that they are more readily accessible to individual or small groups of farmers. Groundwater supplies about one-third of the world's irrigated area, but that area produces more than half the value of all irrigated crops. Wells do not need the large management infrastructure typical of even modest-sized canal systems. Groundwater can be particularly useful in conjunction with canals; pumps can be turned on when canal supplies are deficient. Groundwater development also has the additional benefit of helping to control waterlogging.

In India, the green revolution succeeded only where groundwater supplies were readily available (Vohra 1975). Studies in Pakistan by Lowdermilk (1986) document the value of this strategy (Table 7).

Publicly constructed and managed groundwater schemes have not been very successful. Most groundwater irrigation systems have been privately developed, and this pattern can be expected to continue. In many respects, groundwater has been the neglected poor relation of canal irrigation. It has received relatively little attention from international development agencies. The innovative research and development on groundwater utilization needed in developing countries include more effective programs to provide credit for development, substantial improvements in technology, reliable sources of power for pumping, better farmer information on the use of limited groundwater resources, improved information on the occurrence and potential of groundwater resources, and management of aquifers. Technology for tubewells serving less than 5 ha is virtually nonexistent; efficiencies of pumping equipment in typical operating conditions are in the 25-50% range (Chambers 1986).

Availability and quality of aquifers vary widely. In India, they vary from the deep porous sediments of the Ganges basin, which can support tubewells producing around 0.1 m^3/s, to the shallow weathered layer overlying the hard rock aquifers of the Deccan, where large-diameter shallow dug wells yield 10,000-20,000 m^3 water/yr The scarce dug-well supplies are nevertheless precious, particularly for perennial crops. Feasibility studies by Sawant and Mulik (1985) estimated that higher-technology trickle irrigation would yield an internal rate of return 76% higher than traditional technology.

In theory, aquifers could be used to even out annual and even multiannual water supply fluctuations by drawing them down during dry years and replenishing them

Table 7. Average per-hectare returns for farms under different water control situations (1975-76 prices), Pakistan.

Degree of water control	Net farm income per ha (Rs)		
	Wheat	Rice	Cotton
No control (tail farms)	745	343	–8
No control (head farms)	958	435	110
Fair control (public wells)	951	619	no cotton
Good control (private wells)	1368	950	1072

during the plentiful ones. This would require strong and persistent management in the face of large future uncertainties. It seems highly unlikely, however, that the necessary institutions could be put in place and maintained.

Annual variations in production

Availability of water storage, either in surface or groundwater reservoirs, can be expected to improve the relative stability of annual production. But it cannot substitute for carry-over food grain storage or international trade. Annual and cyclical variations will always occur. For example, the Philippines first achieved national self-sufficiency in rice in 1969-70. During 1971-74, disease, flood, drought, and the oil price increase caused a production drop, back to a deficient condition. That prevailed until 1977 (ADB 1986). The 1971-73 accumulated deficit of 1.06 million t represented 30% of the consumption level at that time. Surpluses since 1976-77 total 1.29 million t, more than enough to make up for the 1971-73 deficit. The 1976-77 to 1979-80 and 1981 surpluses also were more than adequate to cover deficits in 1980-81, 1982-83, and 1983-84.

Irrigation technology options

Over the centuries, the traditional method of irrigation has been surface flooding. Canals, stream diversions, and tanks have been constructed to serve that purpose. As long as these remained small, they were mostly built and operated by their villager owners. With the advent of larger facilities, the responsibility for constructing, operating, managing, and maintaining main canal systems became the responsibility of public agencies. The optimal arrangement would be to merge the two systems, with the water users responsible for the tertiary facilities. But this has seldom happened. Public agencies do not have the resources nor the organization to manage the tertiary infrastructure; farmers have not done so, nor have they been encouraged or assisted to do so. Greater efforts to increase farmer participation are now widely advocated (Keller 1985).

Many existing systems are crude and inefficient. Improved or modernized techniques and structures would improve both system operation and farmer participation. There are dangerous pitfalls in accommodating different social and individual values and inappropriate choices of technology. Nevertheless, as Keller notes, "Contrary to conventional wisdom, modern high technology irrigation techniques are not only feasible, but may be preferable in a variety of circumstances in Third World countries for both recently commercialized and traditional agriculture. There exists the promise of increased employment opportunities, lower per unit-area development costs, lower organizational costs to be borne by small-scale cultivators, a curtailed bureaucratic system, increased flexibility for farmers to respond to market prices, and even decreased competition between farmers for limited water supplies."

In developing countries, technological improvements could be made in on-farm management, farm-level and canal structures and controls, pumps and wells, operational techniques for scheduling the distribution of water from canals to farms, and overall system operation. Only brief outlines and examples are included here.

On-farm management

Leveling fields so that the water can be applied evenly not only saves water, it also increases yields. Farmers have done this with hoe and bullock for centuries. Modern power grading equipment lightens the burden and does a better job. Field layout and stream size are very important.

There are many options for applying water with sprinklers. Sprinklers reduce the need for land leveling, reduce the amount of water needed, and make it easier to maintain desirable moisture levels in light or sandy soils. Trickle irrigation is even more efficient, but there are technical difficulties involving filtration and costs may be higher. Improved water application reduces the hazards of waterlogging and salinity.

Many parts of the developing world have the sophistication and know-how to apply many of these higher technology systems. There have been notable failures, but there are also successes. A cash-flow economy at farm level is required to maintain sprinkler and trickle systems. These systems can range from largely manual to completely automated ones, where the farmer simply turns the valve (as at Doukkala, Morocco).

Farm and canal structures

Many innovations, ranging from simple plastic siphons for farm use to automatic control gates, will maintain constant water levels downstream. Other structures include control gates, measuring devices, turnout gates, bridges, and various types of lining. Improved structures are gradually but persistently being adapted and adopted where they are effective and where operators have the knowledge and maintenance support to use them effectively.

Canals can be operated in several modes: continuous flow, rotational scheduling, or ad hoc. The structures should be chosen to support the particular philosophy of operation adapted to the needs of the farmers and the peculiar nature of the water supply. The use of pipe conduits instead of open ditches has many attractions. With widespread availability of polyethylene materials, the slow adoption of closed conduits is difficult to explain. There is large scope for innovation in designing equipment for maintaining canals and watercourses, especially for small channels. Chemical weedicides and biological control using herbivorous fish (grass carp) are being tried for weed control in both developed and developing countries.

Pumps and wells

Pumping devices such as Persian wheels, either human or animal powered, evolved early in irrigation history. A broad choice of pump technology is available, but some adaptation is still necessary, especially for small-scale use. The same is true for well technology.

Water distribution

Most large canal systems have not been successful in delivering water to fields at the time and in the amounts needed during crop growth. This not only reduces yields, but discourages farmers from risking the extra costs of fertilizer, labor, and other inputs. Although much of the problem can be traced to inadequate design and maintenance, much also can be done to improve scheduling (Chambers 1984). Manual approaches can be used, but the process is slow, laborious, and inefficient. Computer modeling of both crop water requirements and the hydraulic performance of the canals is much more effective. Microcomputer programs are now available, and are being used in Sri Lanka and Thailand.

With the large capacity of microcomputers for processing data, canal operators learn system characteristics much more rapidly and thoroughly than they possibly could manually. Rehabilitation can be more efficient because system shortcomings are identified and better understood.

Improving overall system operation

On large canal systems, daily estimates of water needs are telephoned by section operators to district headquarters. There they are assembled and instructions are issued for setting gates. The complex, interacting effects of many gate changes on canal and river hydraulics can only be estimated, on the basis of experience. For long canals, several days or a week is required for water released at the head to reach downstream service areas. Needs often change before the release arrives (for example, if there are rains) and valuable water is wasted.

Hydraulic modeling is only one example of a research-type technique that uses automatic sensing, microwave transmission, and computer modeling, both deterministic and stochastic, to reduce operational losses and improve system reliability. Other examples include optimal reservoir operation and prediction of runoff from watersheds. Egypt has begun a comprehensive program and reservoir optimization models are in place in Pakistan.

Summary

Advanced irrigation technology can help improve productivity on farms, increase the reliability of water deliveries, reduce water losses, and lead to more efficient maintenance and rehabilitation. A shift to new technology could be very profitable. For example, sprinklers could irrigate twice as much area as a traditional surface system at a lower development cost per hectare (Keller 1985). Although they must be adopted with great caution, carefully selected and field-tested innovations could significantly increase economic and technical efficiency.

References cited

Abernathy C L, Brethery D (1986) An Africa strategy for IIMI. International Irrigation Management Institute, Digana Village via Kandy, Sri Lanka. 65 p.

ADB—Asian Development Bank (1986) Review of bank operations in the irrigation sector 1966-85. Manila, Philippines. 92 p.

Carruthers I, Peabody N S III, Bishop A A, LeBaron A D, Mehra R, Oad R, Peterson D F, Wood D H (1985) Irrigation pricing and management. U.S. Agency for International Development, Devres Inc., Washington, D.C. 106 p.

Chambers R (1984) Improving canal irrigation management; no need to wait. Ford Foundation, New Delhi. 16 p.

Chambers R (1986) Irrigation against rural poverty. Institute of Development Studies, University of Sussex, Brighton, U.K. 40 p.
Framji K K (1984) Past and likely future developments (by 2000 AD) in irrigation, drainage and flood control measures in developing countries. Int. Comm. Irrig. Drainage Bull. 33-2:1-58.
Framji K K, Garg B C, Luthra D S L (1981) Irrigation and drainage in the world. Vol. 1. International Commission on Irrigation and Drainage, New Delhi, India. 491 p.
Framji K K, Garg B C, Luthra D S L (1983) Irrigation and drainage in the world. Vol. 3. International Commission on Irrigation and Drainage, New Delhi, India. 406 p.
Hargreaves G H (1977) World water for agriculture. Utah State University, Logan, Utah, USA. 177 p.
Hargreaves G H, Samani Z A (1986) World water for agriculture. International Irrigation Center, Utah State University, Logan, Utah, USA. 617 p.
Horning H M (1986) Irrigation development and water charges. Expert consultation on irrigation water charges. Food and Agriculture Organization, Rome, Italy. 18 p.
Keller J (1985) Worldwide view of irrigation technology in world food supplies. Pages 305-311 *in* Water and water policy in world food supplies. W. R. Jordan, ed. Texas A&M University Press, College Station, Texas, USA.
Lemoigne G J M (1986) World bank involvement in irrigation and drainage. Int. Comm. Irrig. Drainage Bull. 35-1:1-7, 24.
Lowdermilk M K (1986) Private tubewell water control and anarchy on public irrigation systems. United States Agency for International Development, New Delhi. 14 p. (unpubl.)
L'Vovitch M I (1975) World water balance. Pages 41-55 *in* Selected works in water resources. A. K. Biswas, ed. International Water Resources Association, Champaign, Illinois, USA.
Rangeley W R (1983) Irrigation-current trends and a future perspective. World Bank, Washington, D.C.
Rangeley W R (1986) Address. 5th Afro-Asian regional conference. Int. Comm. Irrig. Drainage Bull. 25-2:33.
Sawant M M, Mulik B R (1985) Dug well irrigation with high technology in Maharashtra - a case study. Pages 104-115 *in* Irrigation sector strategy review India/USAID. WMS Rep. 35. J. Keller, E. W. Coward, and S. R. Daines, eds. Utah State University, Logan, Utah, USA.
Seckler D (1986) Growth and stagnation in Pakistan agriculture. Report to USAID/Pakistan. Islamabad.
Vohra B B (1975) Land and water management problems in India. Training Vol. 8. Department of Personnel and Administrative Reform, Cabinet Secretariat, New Delhi. 170 p.

Notes

Author's address: D. F. Peterson, Agricultural and Irrigation Engineering Department, Utah State University, Logan, UT 84322-4105, USA.

Citation information: International Rice Research Institute (1989) Climate and food security. P.O. Box 933, Manila, Philippines.

Cropping systems strategies for coping with climatic fluctuations

S. M. Virmani

Adaptable cropping systems are one of the best ways of coping with intraseasonal climate variability and soil fertility problems. By fine tuning the time of crop establishment and by introducing intercropping and sequential cropping systems, ICRISAT has been able to harvest more than 4 t food/ha per year in a two-crop system on Vertisols under dryland management at Hyderabad, India (annual rainfall 800 mm; mean annual temperature 25.8 °C; climate semiarid tropical). In marginal areas (e.g. Hisar, India—annual rainfall 446 mm; mean annual temperature 25.1 °C), allocation of area proportionate to the long-term occurrences of different types of seasons has been an efficient way to diversify crops and stabilize agricultural production. A set of agroclimatic analyses that could define the recommendation domain for transferring improved cropping systems technologies from research stations to farmers' fields are provided.

Attention to climate variability and change during the last decade has aimed at developing alternative agricultural strategies to cope with variability. In the semiarid tropics (the geographical area of concern to the International Crops Research Institute for the Semi-Arid Tropics [ICRISAT]), rainfall is the climate factor that varies the most from year to year. Droughts can occur in low-rainfall years, and floods can occur in high-rainfall years, or even during short but intense rainy spells in low-rainfall years. Years with a low rainfall may be followed by similar years or by years with high rainfall.

As to climate change, the question has been whether there is a definite diminishing trend in rainfall due to desertification, particularly in sub-Saharan Africa and in parts of India. The climatology research group at ICRISAT has paid particular attention to this question, and has found no evidence in available rainfall data to suggest that rainfall has diminished in any way in India. But it has found a definite diminishing trend in total annual rainfall and in length of the rainy season in sub-Saharan Africa.

Other researchers (Tyson et al 1975) have observed that the summer rainfall in southern Africa exhibited a 20-yr quasi-oscillation in annual totals. In eastern Botswana, the common rainfall pattern is 10 yr of below-average rainfall followed by 10 of above-average rainfall (Tyson 1979).

Three types of annual rainfall variation can be found in the semiarid tropics: 1) random year-to-year variations in rainfall; 2) a trend of diminishing annual rainfall; and 3) oscillation in annual rainfall over a decade or two. It is important to examine how farmers can adjust to such climatic variability, so that agriculture planners can devise effective strategies to cope with variable crop production. Should a sorghum farmer, for example, plan his cropping strategy differently each year, reconsidering the crop or cultivar, planting dates, target yields, and associated production practices? Or would a cropping system that has built-in flexibility to cope with climate variations be more advantageous?

Results of a long-term experiment conducted on an operational scale to counter rainfall variables by adopting improved land and water management practices and using a flexible cropping systems strategy are presented here. Cropping systems strategies are discussed for three types of rainfall variability: random, diminishing, and oscillatory.

Cropping systems for different rainfall patterns

Strategies for random variations

To define a rainfall environment with random variations, we analyzed the rainfall data of Hyderabad, India (17°27′ N, 78°28′ E, elev. 545 m). Table 1 shows the mean values of total annual rainfall and seasonal rainfall for the period 1901-80. The rainy season is taken as the period of the monsoon, during which more than 80% of the total annual precipitation is normally recorded. The range, standard deviation, coefficient of variation, and amount of annual rainfall for 30-yr periods from 1901 compared with the overall 1901-80 data are within acceptable norms, with an overall average rainfall of 780 mm and a CV of 28%. The seasonal rainfall figures also do not show any abnormal trends (Table 1). No trends or cycles of annual rainfall occurred.

In such a situation, it is important to define the within-year variability of rainfall and the length and characteristics of the growing season, so that cropping systems

Table 1. Annual and seasonal rainfall statistics for Hyderabad, India (1901-80).

Period	Range	Mean	SD	CV (%)
	Annual rainfall, calendar year (mm)			
1901-80	1015	780	216	28
1901-30	944	782	247	32
1911-40	944	775	239	31
1921-50	686	737	149	20
1931-60	686	771	157	20
1941-70	735	792	165	21
1951-80	968	798	211	26
	Seasonal rainfall, June to October (mm)			
1901-80	910	665	202	30
1901-30	880	661	237	36
1911-40	880	640	222	35
1921-50	506	616	137	22
1931-60	723	658	161	24
1941-70	723	695	159	23
1951-80	797	695	186	27

can be proposed that will reduce the intraseasonal impacts of climate variability. At Hyderabad, the length of the growing season varies from 15 to 30 wk with variations in rainfall and soil moisture storage capacity (Table 2). For example, in soils having medium available water-storage capacity (150 mm in the soil profile), a crop with a 19-wk growing season is likely to have adequate moisture 3 of 4 yr, while a crop having a 24-wk growing season is likely to have adequate moisture only once in 4 yr. For such a situation, intercropping a short-duration sorghum (105-110 d) with a long-duration pigeonpea (150-180 d) yielded a land equivalent ratio of 1.66, compared to when the crops were grown separately (Natarajan and Willey 1980). Where rainfall variability is random, intercropping not only increases crop yields but also provides stability.

Rao and Willey (1981) analyzed the results of 89 sorghum/pigeonpea intercrop experiments conducted in diverse environments. They found that on average, intercropping yielded the equivalent of 90% of the sole sorghum yield and about 52% of the sole pigeonpea yield. The failure of intercropping to return a specified income was far less frequent than it was for sole cropping.

Proportionate cropping can help decrease the risk of loss and increase overall productivity. In proportionate cropping, land area is allocated to crops of different growing durations on the basis of long-term probabilities of soil moisture. In research conducted at the Haryana Agricultural University, Hisar, India, allocating 40% of the land to guar (120-d duration), 40% to pearl millet (70-d duration), and 20% to mungbean (50-d duration) allows a farmer to harvest all three crops in good rainfall years and at least two crops in all but severe drought years.

Crop rotation also is used to increase production, particularly in semiarid West Africa. In low-intensity cropping, the system may include a fallow period to build up soil water and nutrients to counter climatic variability in the years when crops are raised. Charreau (1974) recommended a 4-yr rotation: 1 yr fallow or green manure plowed under, 1 yr groundnut, 1 yr cereal, and 1 yr groundnut or cowpea. In intensive cropping where fertilizers are used, the fallow period is eliminated and a cereal crop raised instead (Charreau and Nicou 1971).

Table 2. Length of the growing season (wk)[a] for three soil conditions,[b] Hyderabad, India.

Rainfall probability	Growing-season length (wk) Available water-storage capacity		
	Low (50 mm)	Medium (150 mm)	High (300 mm)
Mean	18	21	26
75%	15	19	23
25%	20	24	30

[a]From seed-germinating rains (25 Jun) to end of season [time when profile moisture reduces EA/PE (Ratio of actual evapotranspiration to potential evapotranspiration) to 0.5]. [b]Low: shallow Alfisol; medium: shallow to medium-deep Vertisols; high: deep Vertisol.

Strategies for diminishing rainfall

The sub-Saharan zone in West Africa is characterized by high evapotranspiration rates, low and variable seasonal rainfall, and sandy soils. Average rainfall barely meets the climatic water demand. Any negative change in the amount of rainfall in this region could have serious consequences for increased and stable crop production. To quantify changes in the rainfall of the region, we studied precipitation records for a number of sites. The precipitation pattern of Dakar (Yoff), Senegal, has varied considerably over the past 30 yr (Virmani and Singh 1985) (Fig. 1). Since 1955, annual rainfall has shown a diminishing trend: the number of years of less that average rainfall increased during 1960-75, compared to 1947-60. This observation is confirmed by the plot of 5-yr moving average data shown in Figure 2.

To evaluate the agricultural significance of this trend, the probabilities of weekly rainfall for 1947-55, 1956-65, and 1966-75 were calculated (Table 3). The probability of having a 10- to 12-wk growing season decreased from 80% during the 1950s to 40% for 1966-75.

Virmani and Singh also found that, apart from the reduction in the amount of seasonal rainfall, the onset of the sowing season has shifted from late June to mid-to-late July. M. K. V. Sivakumar (ICRISAT Sahelian Center, 1986, pers. comm.) reported that growing season length is related to the beginning of the rains.

These preliminary studies for the Sahelian zone show a definite trend of diminishing rainfall. Growing season and total actual rainfall have both decreased. The impact of climatic change can be minimized by introducing short-duration cultivars of commonly grown crops, such as groundnut or sorghum, or by increasing the proportion of drought-resistant short-duration crops (e.g. millet) in the cropping systems. These cropping systems interventions could help increase and stabilize crop

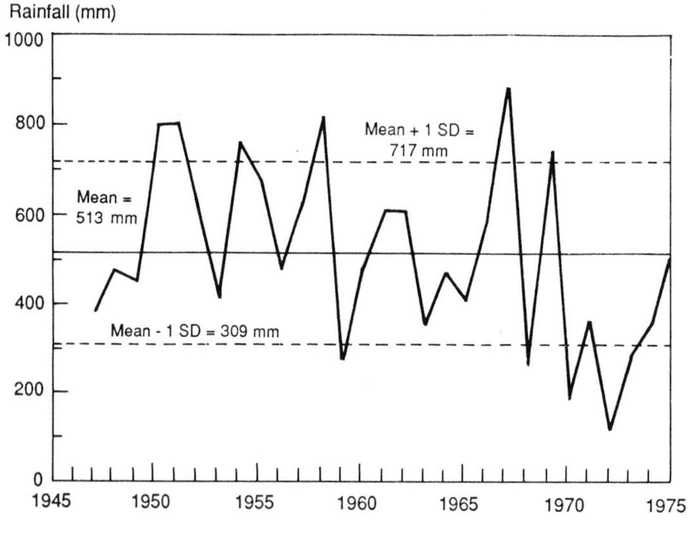

1. Annual rainfall trend at Dakar, Senegal, (14° 44' N, 17° 38' W), 1947-75.

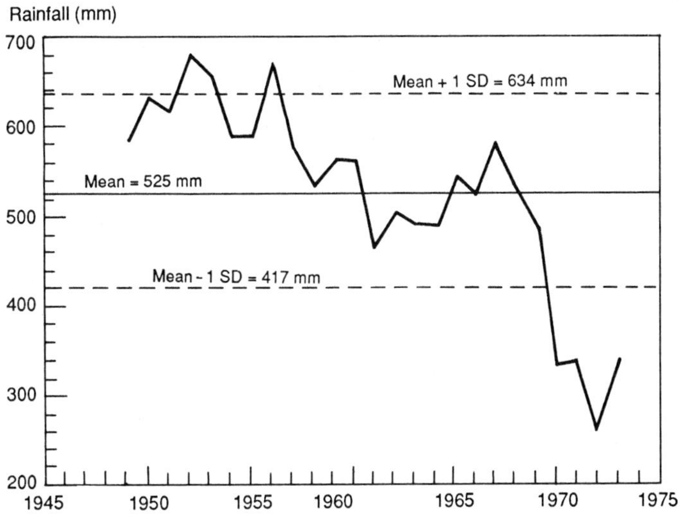

2. Five-year moving average of annual rainfall at Dakar, Senegal, 1947-75.

Table 3. Probabilities of weekly rainfall (R/PE≥0.33) in Dakar (Yoff) Senegal.

Period	Weeks (no.) rainfall probability exceeded 70%	Probability of a 10-12 wk growing season (%)
1947-55	10	80
1956-65	7	60
1966-75	6	40

production. Groundnut - millet intercropping can be recommended in areas where sole groundnut has been grown.

Sivakumar suggests that the date of onset of sowing rains could be a useful guide in crop planning for this region. If the rains start early (or in time) in a given site, cultivars of millet and other crop species recommended for a medium growing season can be grown in that site. If the onset of rains is delayed 10 d beyond the average date, short-duration cultivars may be grown to ensure yield stability.

Strategies for rainfall oscillations

Little systematic work has been done on the agricultural management of rainfall oscillations. Du Pisani (1984) studied the most recent (1963-81) full oscillation of rainfall of five weather stations (Glen, Cedara, Bethlehem, Potchefstroom, Roddeplaat) in the maize-producing area of the Republic of South Africa. Maize yields were modeled from potential evapotranspiration and moisture stress units. He found that dry and wet cycles have a little effect on optimum planting dates for maize. Long-term climatic data could be used to determine optimum planting dates.

Du Pisani's studies, however, show that dry and wet periods could markedly alter target yields, depending on soil type and rainfall regime. On soils with high water-holding capacity, target yields differed no more than 10%. On soils with a low moisture-holding capacity, however, target yields differed significantly between wet and dry periods.

Improving productivity

Most cropping in the semiarid tropics will continue to be under rainfed conditions. Current yields are low and production is unstable because of aberrant weather and poor management of soil fertility and rainwater. The semiarid tropics, with 13% of the world's land and 15% of the world's people, produces only 11% of its own food.

Research at ICRISAT during the last decade clearly shows that the introduction of improved cropping systems provides flexibility in dealing with climatic variability. However, high and stable yields can be obtained only when efficient cropping systems are practiced where improved land and water management techniques also have been applied. In Table 4, yields of an intercrop of maize or sorghum and pigeonpea and a sequential crop of maize followed by chickpea on a Vertisol are reported for the crop years 1976-77 to 1984-85, under both improved soil and water management technologies and under traditional practices. Rainfall during the experimental period varied from 550 mm to 1,100 mm (CV 26%). In most years, the improved management systems yielded more than 4 t grain/ha, with a CV of 11%. This indicates that a fair degree of weatherproofing is possible by incorporating high-yielding varieties of the crop and good agronomic management. Yields under

Table 4. Grain yields under improved and traditional technologies on Vertisols at ICRISAT Center in 9 successive years.[a]

Year	Cropping period rainfall (mm)	Grain yield (t/ha): Improved system: double cropping[b] — Sorghum or maize + Pigeonpea or chickpea	Grain yield (t/ha): Traditional system: single crop — Sorghum	or Chickpea
1976-77	708	3.9	0.4	0.5
1977-78	616	4.3	0.4	0.9
1978-79	1089	3.4	0.6	0.5
1979-80	715	3.5	0.5	0.5
1980-81	751	4.5	0.6	0.6
1981-82	1073	4.2	0.6	1.0
1982-83	667	4.4	0.6	1.2
1983-84	1045	4.8	0.8	0.5
1984-85	546	4.4	0.7	1.2
Mean	801	4.2	0.6	0.8
SD	209	0.440	0.1	0.3
CV (%)	26	11	24	42

[a]Average rainfall for Hyderabad (29 km away from ICRISAT Center) based on 1901-84 data is 784 mm with a CV of 27%. [b]Sequential or intercropping system.

the traditional management were about 600-700 kg/ha, with a CV of 24% for sorghum and 42% for chickpea.

Food production can be increased in the dryland semiarid tropics by adopting improved cropping systems and related soil, crop, and water management technology. Average yields for the major rainfed cereals and grain legumes in India range from 300 to 800 kg/ha. These can easily be doubled with improved technologies. The situation in the Sahel and other regions of Africa is less promising. According to the Organization for Economic Cooperation and Development (1976), improved technologies, and particularly improved seeds, often are not available. The institutions and infrastructure necessary for successful diffusion of new technologies are less developed. Limited evidence shows that in this region, the differences between average yields, best farmer practices, and potential yields appear to be much narrower than they are in India (Swindale et al 1981).

Summary

The most variable element of the semiarid tropical climate is rainfall. Three types of rainfall variabilities are encountered: random year-to-year and within-year rainfall variability; variability due to a diminishing trend of annual and seasonal rainfall; and variability due to oscillating rainfall, characterized by a series of wet years followed by a number of dry years. Random variability is observed in the Indian subcontinent. Evidence seems to indicate that the rainfall of Sahelian West Africa is showing a diminishing trend. The summer rainfall area of southern Africa exhibits an oscillating pattern within a period of about 20 yr.

To cope with the random variability of rainfall, intercropping has been suggested. For regions where a declining trend of rainfall is evident, short-duration, drought-resistant crops may be introduced and climate-responsive crop planning adopted. In areas where rainfall oscillations are observed, regions with deeper soils may be cropped in the drier years without much loss of crop production.

Work at ICRISAT has shown that improved cropping systems can cope best with rainfall variability when they are applied simultaneously with an efficient set of land and water management techniques.

References cited

Charreau C (1974) Soils of the tropical dry and dry-wet climatic areas of West Africa and their land use and management. Agronomy Mimeo 74-26. Department of Agronomy, Cornell University, Ithaca, New York.

Charreau C, Nicou R (1971) L'amélioration du profil cultural dans les sols sableux et sabloargileux de la zone tropicale sèche Ouest Africa et ses incidences agronomiques. L'Agric. Trop. 26(2):209-255.

Du Pisani A L (1985) Importance of rainfall oscillations in agricultural planning. S. Afr. Tydskr. Plant Grond 2:31-34.

Natarajan M, Willey R W (1980) Sorghum-pigeonpea intercropping and the effects of plant population density. 1. Growth and yield. J. Agric. Sci. (Cambridge) 95:51-58.

Nicou R, Charreau C (1980) Mechanical impedance related to land preparation as a constraint to food production in the tropics (with special reference to fine sandy soils in West Africa). Pages 311-388 *in* Priorities for alleviating soil-related constraints to food production in the tropics. International Rice Research Institute, P.O. Box 933, Manila, Philippines.

Organization for Economic Cooperation and Development (1976) Analysis and synthesis of long term development strategies for the Sahel. Paris, France.
Rao M R, Willey R W (1981) Stability of performance of a pigeonpea/sorghum intercrop system. Pages 306-317 *in* Proceedings of the international workshop on intercropping, 10-13 Jan 1979, Hyderabad, India. R. W. Willey and C. Garver, eds. International Crops Research Institute for the Semi-Arid Tropics, Patancheru, Andhra Pradesh, India.
Swindale L D, Virmani S M, Sivakumar M V K (1981) Climatic variability and crop yields in the semi-arid tropics. Pages 139-160 *in* Food-climate interactions. W. Bach, J. Pankrath, and S. H. Schneider, eds. D. Reidel Publishing Company, Dordrecht, The Netherlands.
Tyson P D (1979) Southern African rainfall: past, present, and future. *In* Symposium on drought in Botswana. Clark University Press, Worcester, Massachusetts.
Tyson P D, Dyer T G J, Mametse M N (1975) Secular changes in South African rainfall. Q. J. Roy. Meteorol. Soc. 101:817-833.
Virmani S M, Singh P (1985) Agroclimatological characteristics of the groundnut growing regions in the semi-arid tropics. Pages 35-45 *in* Proceedings of the international symposium on agrometeorology of groundnut, 21-26 August 1985, Niamey, Niger. International Crops Research Institute for the Semi-Arid Tropics, Patancheru, Andhra Pradesh, India.

Notes

Author's address: S. M. Virmani, Resource Management Program, International Crops Research Institute for the Semi-Arid Tropics (ICRISAT), ICRISAT Patancheru P.O., Andhra Pradesh 502324, India.
Citation information: International Rice Research Institute (1989) Climate and food security. P.O. Box 933, Manila, Philippines.

Some strategies to cope with drought in the Sahelian zone

M. V. K. Sivakumar

Per capita food production in the Sahelian zone of West Africa over the last two decades was negative. Substantial progress is needed during the next decade. Efforts to cope with drought must aim at understanding the entire production system—climate, management, genotypes. Droughts in the Sahel are characterized by extreme variability in time and space, persistence, and variable lengths of growing season. Aspects of current management systems and genotype selection that directly or indirectly aggravate effects of rainfall deficits are highlighted. Strategies to cope with drought, starting with an analysis of climate data, are illustrated. Simple management techniques, such as tillage, application of fertilizer, and intercropping, are an effective short-term strategy. A long-term strategy is to breed stable, high-yielding, drought-resistant cultivars.

The food production crisis in sub-Saharan Africa has been the focus of intense discussion in many national and international forums because the geographical regions and the populations involved are large and the time scale for solving the crisis is short. Thirty million people in Africa were affected by drought in 1985 (Timberlake 1985). Food deficits in sub-Saharan Africa alone are projected to be 27-34 million t by 1990 (IFPRI 1977).

About 90% of the population in sub-Saharan Africa depend on subsistence agriculture for survival, and agriculture is the main contributor to the gross domestic product (GDP) (Mudahar 1986). In the sorghum- and millet-growing countries of West Africa, the average population growth rate from 1970 to 1982 was 2.8%, while per capita food production was negative. Substantial progress in agricultural production needs to be made rapidly if the negative trends are to be reversed.

This paper reviews some strategies for coping with drought in the Sahelian zone. Pearl millet is used as the specific example because this important cereal crop is grown all over the Sahelian zone, from the extreme desert in Mauritania on the edge of the Sahara to higher rainfall regions in Burkina Faso and Nigeria.

Complexity of drought

Drought, often cited as the constraint to increasing the agricultural productivity in the Sahelian zone, is believed to result from lack of rainfall. This, at best, is an extremely simplistic view of a complex problem. In a recent report, Farmer and

Wigley (1985) said, "Although many of the meteorological features associated with drought conditions are now well-documented, the underlying causes of both the present drought and earlier droughts are unknown. It is clear that the causes of drought in Africa are complex and almost certainly not attributable to any single factor."

Drought is not a single constraint, but a complex of constraints. Crop performance in drought conditions has a genetic component and a management component (Jordan and Sullivan 1982), with complex interactions between the two components. A thorough understanding of the entire production system, including climate, management, and genotype, is a prerequisite to effective management of crop production systems in drought conditions. Insufficient rainfall creates conditions wherein crop growth suffers, but other important aspects of the current management system and genotype selection directly or indirectly aggravate the negative effects of rainfall deficiency.

Climatic causes of drought

Extreme rainfall variability. In the Sahelian zone, droughts are set off by rainfall deviations that fall far below already low and undependable rainfall. Temporal and spatial variations in rainfall are large, and in the marginal areas, agricultural systems are vulnerable to large deviations. The impact of temporal variations depends on the scale at which they are considered (these variations increase as one moves from an annual to a daily scale). Examples of temporal variation on different scales have been described in detail (Sivakumar 1987a). Spatial variability in rainfall is equally important where large variations in rainfall occur over short distances and at critical stages of crop growth.

Persistence of deficiencies. An important feature of Sahelian rainfall is the magnitude and extent of the rainfall deviations. Below-normal rainfall can persist for 10-20 yr. At Tillabery, Niger, the period between 1966 and 1984 has been consistently dry; the rainfall deviation in 1984 was 58% below the mean (Fig. 1). Rainfall fluctuations also are associated with a preferred geographic pattern. In Burkina Faso reduction in mean annual rainfall after 1969 (Fig. 2) (Sivakumar 1987a) was the geographical mean pattern. Rainfall isohyets were displaced farther south after 1969, showing the geographical extent of the drought.

Erratic growing seasons. Potential evapotranspiration (PET), or water demand, is usually high in the Sahel due to consistently high air temperatures and radiation load. Hence, the length of growing season, which is a balance between the water supply and demand, depends on the rainfall. But rainfall is erratic. Length of growing season for Niamey, computed for an 80-yr rainfall record, is shown in Figure 3. The average is 94 d. But rainfall below average since 1969 has resulted in far below average growing-season lengths. In poor-rainfall years with short growing seasons, crop failures result from a mismatch between water availability and crop phenology.

High soil temperatures. Environmental conditions during crop establishment are usually harsh, because the sowing rains follow a long, hot dry season. Although one or two showers facilitate sowing, soil moisture evaporates quickly. If a period of dry, clear weather follows, soil surface temperatures increase rapidly, up to 55 °C. Under these conditions, pearl millet seedling death is common, leading to plant

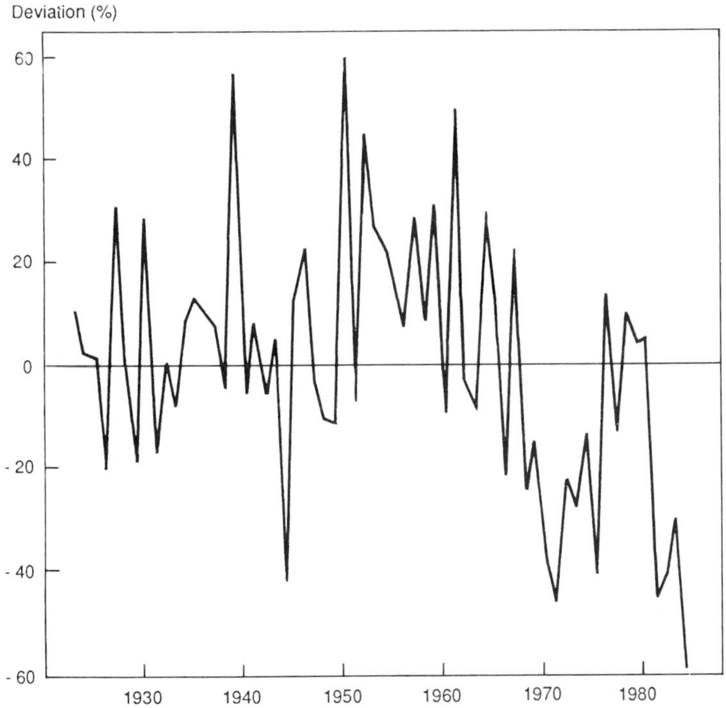

1. Rainfall deviation at Tillabery, Niger, 1923-84.

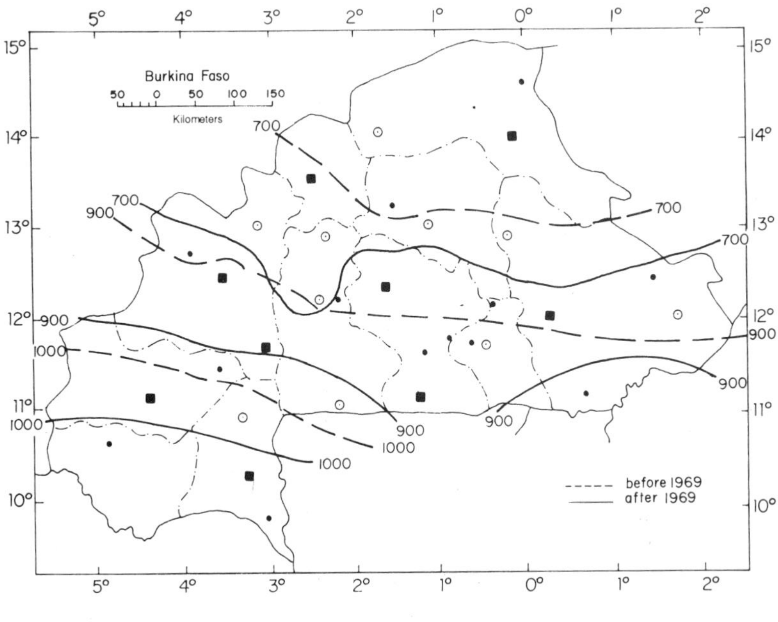

2. Deplacement of rainfall isohyets in Burkina Faso before and after 1969.

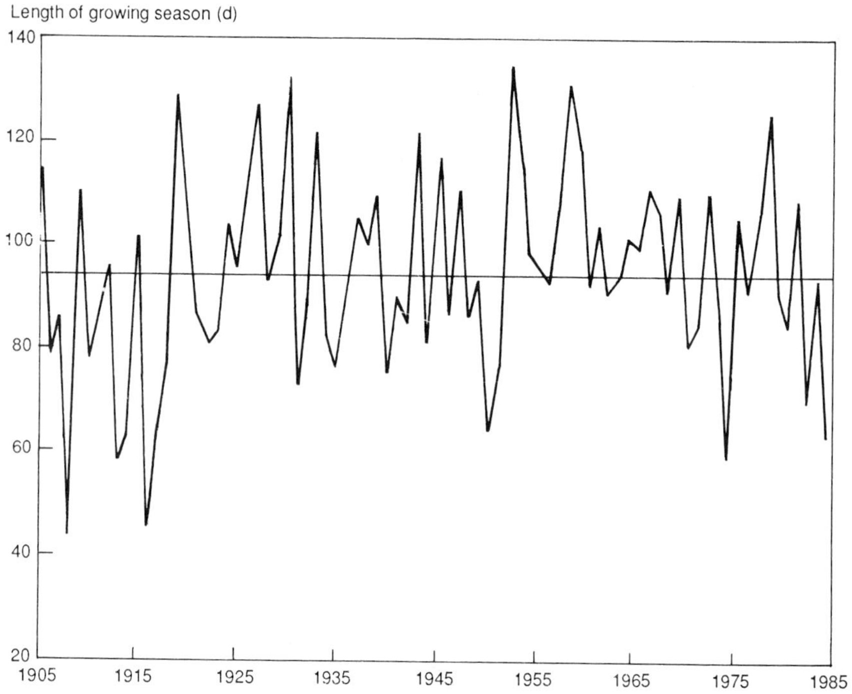

3. Variation in the length of the growing season at Niamey, Niger, 1905-85.

stands much below the recommended 10,000 hills/ha. In field surveys in the Niamey district, hill populations declined from a mean of 4,900/ha at 5 d after sowing (DAS) to 2,300/ha at 12 DAS, with stands failing completely in nearly half of the fields (Soman et al 1986). In those fields, soil temperatures exceeded 50 °C at midday. Such conditions often force the farmers to replant their fields 2-3 times a year.

Nonclimatic causes of drought

Limited, untimely cultural practices. Two soil types in the Sahelian zone occupy 40.2 million ha, 61% of the total area (Fig. 4) (Sivakumar 1986).

Arenosols are coarse-textured soils containing more than 65% sand and less than 18% clay (Swindale 1982). They have low moisture-holding capacity, which imposes a severe drought risk when extended dry periods occur during the crop season.

Luvisols are characterized by clay content and bulk densities that increase with depth, with low cation exchange capacities, hydraulic conductivities, infiltration rates, and available moisture (Perrier 1986). During the rainy season, the soil surface forms a hard crust. Reduced infiltration and increased runoff cause substantial moisture losses. Charreau (1972) showed that as much as 32% of the annual rainfall could be lost as runoff on a cultivated, well-tilled soil, and as high as 60% could be lost on bare soil. Perrier (1986) estimates surface runoff losses from any given area in the Sahel to vary from 40 to 80% of annual rainfall.

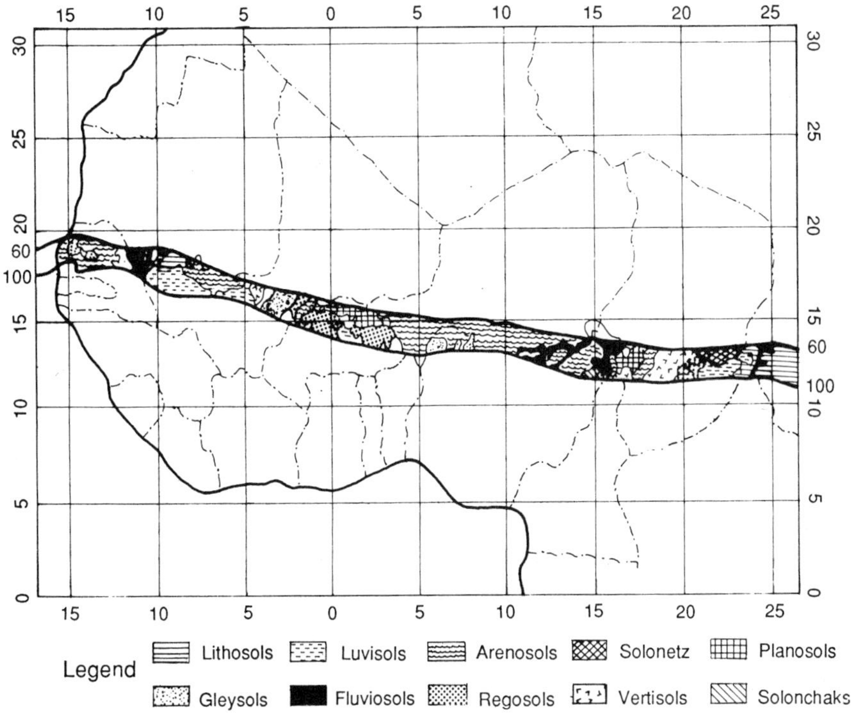

4. Major soil types in the southern Sahelian zone of West Africa.

It should be obvious that appropriate soil and water management practices are crucial for efficient use of the limited and variable rainfall. However, soil management practices are virtually unknown at the farm level. Traditionally, human labor has been the major source of power, and tillage was minimal. In a major part of the millet-growing region of the Sahel, animal traction is not used for preparatory cultivation, and the soils are seldom plowed (Spencer and Sivakumar 1987). Matlon (1985) estimates that less than 15% of the farmers use animal traction and less than 5% of the sorghum/millet area is plowed before planting. On soils that are hard and crusty, no tillage results in low infiltration rates and high runoff.

Other prevailing farm practices for millet production also are not conducive to efficient water use. Millet is sown in hills spaced 45 × 45 cm to 100 × 100 cm apart. Spacings of 100 × 200 cm or even 200 × 200 cm are not uncommon. Traditionally, farmers sow at low densities—about 5,000 hills/ha. Under these conditions, water-use efficiencies are low: a large proportion of the water is lost through evaporation from the soil.

Poor soil fertility management. Arenosols in the Sahelian zone are low in organic matter, nitrogen, and phosphorus. Soils in Niger are very sandy, with the sand fraction usually exceeding 92% (Table 1). Organic matter and cation exchange capacity are low and the soils are acidic.

Table 1. Physical and chemical characteristics of sandy soils in four villages of Niger.

Characteristic	Hamdalaye	Tounga	Mai Gamji	Magaria
Sand (%)	94.5	92.0	94.7	96.6
Silt (%)	0.6	1.4	1.3	0.6
Clay (%)	4.7	6.1	3.6	2.2
pH	5.9	6.6	5.6	6.2
Organic matter (%)	0.22	0.51	0.11	0.33
Cation exchange capacity (meq/100 g)	1.0	2.2	1.5	1.3

Source: Dr. A. Bationo, ICRISAT Sahelian Center, Niamey, Niger, pers. comm.

Traditional soil management was based on alternating arable and fallow phases to allow nutrient replenishment and organic matter buildup. Because the crops grown had low yield potential, dry matter yields and nutrient uptake were also low. Such yields could be sustained for substantial period. When crop yields declined to unacceptable levels, overcropped fields were abandoned and new ones opened (shifting cultivation), leaving fields under natural fallow to restore their soil fertility. This was possible because of the low populations.

Increasing population pressure has reduced land availability, reducing the ratios of length of fallows to cropping years to the point where shifting cultivation is losing its effectiveness. As a result, soil fertility is decreasing in many areas (Sanchez and Buol 1975). Without added manure or fertilizers, crop yields have declined (Sivakumar, this volume).

Use of fertilizers in the sorghum- and millet-growing regions during 1979-81 averaged only 5 kg/ha and was virtually negligible in some countries (Table 2). Where the fertilizer was used, it was applied mostly on export crops (Mudahar 1986).

The absence of a system of crop rotations or regular and optimum fertilizer applications to restore soil fertility led Penning de Vries and Djiteye (1982) to conclude that poor soil fertility rather than water supply is the major constraint to increased production.

Lack of early crop varieties. Lack of improved, suitable varieties is recognized as a major biological constraint (Matlon 1985, Spencer 1985, Stoop et al 1982). Given the rainfall variability and short growing season, it is apparent that short-duration cultivars are to be preferred. Short duration offers several farm management options (intercropping, relay cropping) that would reduce the risk of drought in late planting.

The choice of currently available improved pearl millet varieties is limited. In Burkina Faso, a promising variety IRAT S-10 was recommended for the dry northern zone (Labeyrie 1977). But it was not adopted by farmers because it had a long duration (Stoop et al 1982).

Lack of pest and disease management strategies. One constraint that limits realization of the potential of existing millet varieties and land races is the lack of

Table 2. Average fertilizer use (1979-81) in sorghum- and millet-producing countries of West Africa.[a]

Country	Fertilizer use (kg/ha)
Benin	1.3
Burkina Faso	2.9
Cameroon	5.3
Chad	0.5
Gambia	15.2
Ghana	7.4
Guinea	1.1
Guinea Bissau	0.8
Ivory Coast	13.3
Mali	5.7
Mauritania	7.2
Niger	0.8
Nigeria	5.4
Senegal	4.6
Togo	2.5
Av	5.0

[a]Compiled from various sources by Mudahar (1986).

effective pest and disease management. No effective control measures are available for the two major insect pests of pearl millet in the region: stem borer (*Acigona ignefusalis*, Hmps.) and earhead caterpillar (*Raghuva albipunctella* De Joannis). These two pests are not known to exist elsewhere (ICRISAT 1984). Downy mildew, ergot, and smut, major diseases that cause significant economic losses, have not received much research attention so far.

Some strategies to cope with drought

The crises that droughts create in the Sahel could be avoided with accurate forecasting. In a report on climatic trends for tropical Africa, Farmer and Wigley (1985) concede that forecasting is difficult at present. Still, they predict a continuation of the low rainfall levels of the 1970s and 1980s as more likely than a return to the wetter conditions of earlier decades. A range of strategies are needed to deal with this situation.

Reducing risk

Exploiting the environment. Effective and stable soil and crop management practices in the drought-prone Sahel can only be developed with an understanding of the environment and its variability.

In a recent analysis (Sivakumar 1987b), we reported a highly significant relationship between the date of the onset of rains and the length of the growing season for several sites in the southern Sahelian zone. This analysis points to possibilities for assessing the potential length of the growing season from the date of the onset of rains.

In field tests at the ICRISAT Sahelian Center, Sivakumar (1989) showed that by tailoring management tactics to weather conditions in years with early onset of rains in the Sahelian zone, it is possible to establish a second crop of cowpea for hay after a first crop of millet. If the rains are delayed 10 d beyond the calculated average date of onset, short-duration cultivars that will mature early may be grown. The objective is to minimize the effects of drought by making efficient use of scarce rainfall in a drought year, but to maximize production in good years by exploiting the longer growing season. In disaster planning, delayed rains signal the need for early-maturing cultivars, since traditional and improved cultivars of medium season length are likely to give poor yields.

Stewart (1987) recommends using a combination of date of onset of rains and the first 30-day rainfall to make early season adjustments of plant populations and fertilizer rates. These recommendations need to be field-tested, but the concept appears promising given the already established relationship between onset of rains and length of growing season.

Intercropping. Another strategy adopted by farmers to reduce risks due to climatic variability is intercropping. Steiner (1984) estimates that 80% of the cultivated area in the West African tropics is intercropped. In the Sahelian zone, millet/cowpea is the most prevalent intercropping system. Recent reviews suggest that yields in traditional millet/cowpea intercropping could be increased by improved agronomic management (Fussell and Serafini 1985, Ntare et al 1987).

Planting and harvest schedules, crop densities and spacing, soil fertility, and varieties with different durations were found to be the important factors determining the performance of a millet/cowpea intercrop. Manipulation of one or more of these components led to substantial yield increases (Table 3). In traditional combinations,

Table 3. Reported yield advantage (%) by agronomic manipulation in intercropping.[a]

Component	Yield advantage (%)
Fertilizer	
40 kg N	49
Cultivars	
Local millet + local cowpea	28
Improved millet + local cowpea	38
Local millet + improved cowpea	40
Improved millet + improved cowpea	69
Date of planting of cowpea relative to millet	
Same day as millet	35
6 d after millet	24
25 d after millet	22
Time of harvesting of cowpea	
40 d after planting	102
60 d after planting	146
80 d after planting	94
End of season	35

[a]Compiled from Ntare et al (1987).

a local cultivar of millet is intercropped with a long-duration, photoperiod-sensitive, local cowpea that flowers at the end of the rains. When the rains end early, the local cowpea often produces little or no grain. Substituting an improved cultivar for either of the crops gave similar yield increases, but the maximum advantage was with improved cultivars for both crops.

Another management tool that can maximize intercropping advantages is to adjust the cowpea planting date relative to the development stage of millet and to the probable length of the rainy season (Ntare et al 1987). Early planting of cowpea with, or shortly after, millet gave good cowpea growth and maximized the advantages of the association. Competition between cowpea and millet for moisture and nutrients during early growth should be considered. When millet and cowpea were planted simultaneously, harvesting cowpea early for hay was another option that helped stabilize millet yields (Table 4).

Intercropping short- and long-duration cultivars is currently receiving increased attention in reducing drought risk.

Increasing water-use efficiency. In semiarid regions, efficient water use is the key to yield and yield stability. Regardless of our ability to forecast droughts, no one can dispute the need to maximize the use of rainwater. In a recent review on the water-use efficiency of crops in the semiarid tropics, Gregory (1987) defined water-use efficiency (WUE) as

$$\mathrm{WUE} = \frac{\mathrm{N/T}}{1 + \mathrm{E/T}} \qquad (1)$$

where N where N is dry matter yield,
T is transpiration, and
E is evaporation.

Since runoff (R) and drainage (D) are also substantial components of total water balance in the semiarid regions, WUE was expressed as

$$\mathrm{WUE} = \frac{\mathrm{N/T}}{1 + (\mathrm{E} + \mathrm{R} + \mathrm{D})/\mathrm{T}} \qquad (2)$$

Because N/T is constant for a given saturation deficit and crop, Gregory (1987) concluded that to produce dry matter with the greatest WUE, either the total

Table 4. Effects of nitrogen, phosphorus, and potassium fertilizer on water use (WU), grain yield (Y), and water-use efficiency (WUE) for pearl millet grown at 3 sites in Niger during 1985 rainy season.

Site	Rainfall (mm)	Treatment	WU (mm)	Y (t/ha)	WUE (kg/ha per mm)
Sadore	543	Fertilizer	382	1.57	4.14
		No fertilizer	373	0.46	1.24
Dosso	583	Fertilizer	400	1.70	4.25
		No fertilizer	381	0.78	2.04
Bengou	711	Fertilizer	476	2.23	4.68
		No fertilizer	467	1.44	3.08

amount of water available to the crop should be increased or T should be maximized with respect to all other losses.

Research at the ICRISAT Sahelian Center (ISC) and by ICRISAT/SAFGRAD (Semi-Arid Food Grain Research and Development Project) in Burkina Faso over the last 4 yr showed that the denominator term in equation 2 could be manipulated by simple cultural practices to enhance WUE.

Tillage. The major soil types in the Sahelian zone (Fig. 4) have a high propensity to compaction and hardening during the dry season (Nicou and Charreau 1985). In these soils, which have limited water availability and poor fertility, deep root establishment is essential for plant growth. Cultivation promotes better crop establishment and root growth through increased porosity, reduced bulk density, improved infiltration, enhanced soil water availability, and water conservation (Klaij and Hoogmoed 1987, Nicou 1977). Chopart (1983) found that plowing doubled the dry weight of millet roots in the first 50 d of growth. Nicou and Charreau (1985) reported significant yield advantages due to tillage for several crops.

On the sandy soils at the ISC, the beneficial effects of cultivation have been attributed primarily to enhanced rooting (Klaij and Hoogmoed 1987). Interrow cultivation during the rainy season helps control weeds; at the end of the rainy season, it kills weeds—saving precious soil moisture for a subsequent crop (Dancette and Nicou 1974).

Different tillage methods have been tested in the Sahelian zone. For sandy soils, ridging has reduced wind erosion and increased effective plant populations (Klaij and Hoogmoed 1987). Tied ridges or microcatchment basins increase water storage on the surface and trap runoff, which increases water infiltration and storage in the soil profile (Boa 1966). Perrier (1986) showed significant yield advantages with tied ridging on Alfisols in Burkina Faso. On-farm trials in the central plateau region of Burkina Faso also confirmed significant yield increases (Ohm et al 1985).

Mulching. In the Sahelian zone, millet is traditionally grown in wide rows. Water loss through soil evaporation is a significant component of total evapotranspiration (ET). In recent investigation at ISC, evaporation losses were as high as 40% of total ET. Mulching is an effective means of reducing evaporation losses. It also improves infiltration by absorbing the impact of wind-driven rain, by increasing termite or biological activity, and by improving soil organic matter status. In trials in Burkina Faso, use of mulches in the traditional flat cultivation system was superior to in-place water harvesting methods such as tied ridges (Perrier 1986). Mulching with crop residues helped reduce the aluminum and hydrogen saturation of the exchange complex, a major problem on the acidic soils in the Sahelian region (Bationo et al 1987). Despite the much demonstrated beneficial effects of mulching, the availability of crop residues for mulching is often a problem, because of their use as cattle feed.

Interaction of ridging and mulching. On structurally poor soils, a combination of ridging and mulching has been reported to be an effective management strategy. Experimental results on Arenosols and Luvisols showed considerable yield advantages, because adding plant residues in the tied ridges improved infiltration and water storage capacity (Klaij and Hoogmoed 1987, Perrier 1986). In Burkina Faso, sorghum cultivar E 35-1 with tied ridges yielded 2.2 t/ha; with mulch added, yield improved to 3.3 t/ha (Perrier 1986).

Restore and maintain soil fertility

Several studies have shown that lack of phosphorus is a major constraint to crop growth in semiarid West Africa (Hauck 1966, Jones and Wild 1975, Pichot and Roche 1972). Applying as little as 8.8 kg P/ha can double millet yields. But even this small quantity of fertilizer may be expensive for the resource-poor farmers in the Sahelian zone. Direct application of ground indigenous phosphate rock is an alternative to using imported commercial phosphate fertilizers. In field trials in Niger, two sources of indigenous phosphate rock were compared with commercial single superphosphate (SSP). The less reactive PARC-W phosphate rock was 48% as agronomically effective as SSP, the more reactive Tahoua rock was 76% as effective as SSP (Bationo et al 1987).

Another way to utilize unreactive phosphate rock is to increase available phosphorus by chemical conversion to a partially acidulated phosphate rock (PAPR). In field tests in Niger, PAPR acidulated to a level of 50% was agronomically as effective as triple superphosphate (Bationo et al 1987). Niger has phosphate reserves large enough to meet the country's phosphate requirement, with a positive, reasonably high internal rate of return (Mudahar 1986). Use of PAPR also confers an added advantage: it supplies sulfur, an important plant nutrient, and has a long-term residual effect.

Nitrogen, especially a split application, has also been shown to give substantial yield increases (Bationo et al 1985).

An important consequence of the use of fertilizers is increased water-use efficiency. Early vigorous growth builds up a larger canopy, which shades the ground early in the season when a significant proportion of the water is lost through soil evaporation, and helps in effective and efficient use of the scant rainfall. Studies at three sites in Niger showed substantial increases in WUE with the use of fertilizer (Table 4).

Fertilizer application also helps ensure better plant stands. Klaij and Hoogmoed (1987) showed that hill survival in fertilized plots was 74%, against 43% in control plots.

Combine technologies

The management strategies discussed demonstrate their individual value. There is accumulating evidence that a combination of these inputs offers much greater advantages in the Sahelian region. On sandy soils, tillage alone did not result in much yield increase, but a combination of tillage and fertilizer gave a 300% yield increase (Klaij and Hoogmoed 1987). Soil tillage combined with fertilizers and residues improved the long-term productivity despite rainfall variability in West Africa (Pieri 1985). This point merits consideration because cost-benefit analysis of the inputs is often done on the basis of field trials conducted over only 1-3 yr. The advantages of using these inputs to improve long-term productivity are often ignored.

Breed improved cultivars

Developing stable, high-yielding, drought-resistant cultivars for the Sahelian zone is a long-term, complex task, given the climatic and soil variability under which the cultivars are to be grown. Analysis of the long records of daily rainfall available for many sites could help us understand some aspects related to crop breeding. As much attention needs to be paid to variability as to the averages themselves (Table 5).

Too often in the past, "improved" varieties and genetic material were brought into the Sahelian region from other regions of the world, in hopes of achieving "quantum leaps" in crop yields. That has not happened, because the genetic material was not well adapted to the Sahelian environment. Improved varieties bred in situ have performed remarkably well. With adequate soil fertility at ISC, improved variety CIVT consistently outyielded the local variety for 4 yr (Table 6). Significantly, the maximum yield advantage (78%) was achieved in 1984, a year with the lowest annual rainfall on record. These data point to the advantages of increased varietal development and testing in the Sahelian region itself.

For varietal improvement in drought-prone regions, information on the probabilities of dry spells is often more important than rainfall totals. Knowledge of the relative susceptibility of millet to drought spells during the three important growth stages—emergence to panicle initiation (GS1), panicle initiation to flowering (GS2), and flowering to physiological maturity (GS3)—may provide breeders with useful criteria in breeding for drought resistance. Assuming the date of the beginning of rains each year as the date of sowing, we have computed the length of dry spells (or days to next day with rainfall greater than a threshold value) at different probability levels for consecutive 10-d periods. For two selected locations in the Sahelian zone (Fig. 5), dry spells in the GS1 phase were longer than those during GS2. At Hambori, the length of dry spells is progressively longer from 75 DAS and at Niamey, from 90 DAS. These data could be used as a guide for the growth durations of target

Table 5. Variability in annual rainfall and growing season characteristics pooled over 30 sites in the Sahelian zone.

Parameter	Av	Range	Av SD
Annual rainfall (mm)	480	330-640	125
Onset of rains	23 Jun	10 Jun-4 Jul	19 d
End of rains	12 Sep	6 Sep-16 Sep	9 d
Growing-season length (d)	82	66-99	22

Table 6. Yields of improved millet variety CIVT and local variety Sadore local (SL) grown at the ICRISAT Sahelian Center, Sadore, Niger, 1982-86.[a]

Year	Sowing date	Harvest date	Rainfall from sowing to harvest (mm)	Yield (t/ha)		Yield increase for CIVT (%)
				CIVT	SL	
1982	1 Jul	12 Oct	372	1.69	1.21	40
1984	1 Jun	14 Sep	213	1.12	0.63	78
1985	18 Jun	21 Sep	536	2.35	1.90	24
1986	29 May	10 Sep	499	2.26	1.95	16

[a]Source: Dr. K. Anand Kumar, ICRISAT Sahelian Center, Niamey, Niger, pers. comm.

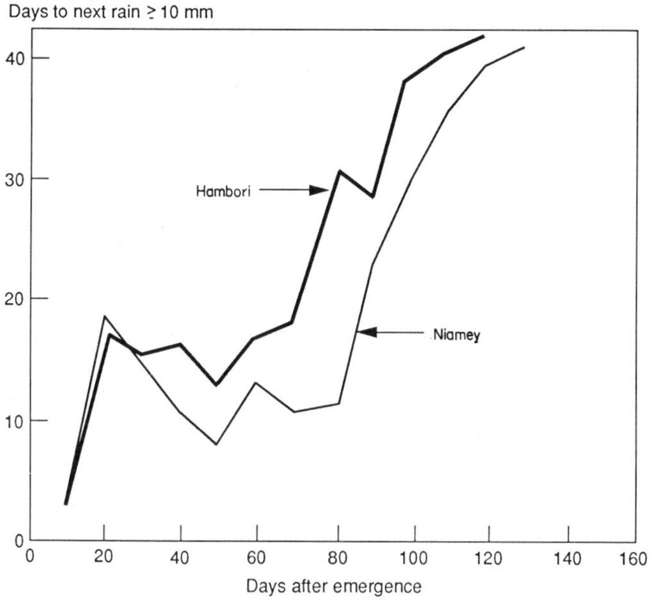

5. Days to next rain ≥10 mm at 90% probability for two sites in the southern Sahelian zone.

varieties. At Hambori and Niamey, breeding strategies should be oriented toward growth durations of 80-90 d.

This information is important because many varieties presently available mature in 100-110 d and are exposed to drought risk during grain filling. Development of slightly earlier varieties would mean that the new lines could escape late-season drought. This is necessarily a long-term strategy, but some of the improved cultivars available from ISC now show promise and yield stability. ICRISAT's millet varieties (such as ITMV 8303) yielded more than 1.3 t/ha in 1984, which recorded the lowest rainfall in the century.

The ability of plants to withstand high temperatures is an important character, especially during germination, emergence, and plant establishment. Sivakumar (1987a) analyzed the frequency distribution of air temperatures and showed that mean maximum temperatures could exceed 40 °C at time of sowing and that absolute temperatures could be much higher. High soil temperatures at this time lead to poor plant stands. Research is under way at ISC to identify millet and sorghum genotypes that germinate and emerge under high soil temperatures (> 55 °C at the surface, 3 d after a rain).

Even with the limited rainfall in the Sahelian zone, millet cultivars can yield total dry matter of more than 8 t/ha. Low harvest index (usually less than 0.20) of existing cultivars is responsible for the low grain yields. A long-term research strategy is to alter the harvest index in favor of higher grain yields.

Incorporation of resistance to insect pests and diseases is a long-term research strategy that is likely to play a significant role in efforts to bring stability to present production systems.

References cited

Bationo A, Christianson C B, Mokwunye A (1987) Soil fertility management of the millet producing sandy soils of Sahelian West Africa: the Niger experience. Paper presented at the Workshop on Soil, Water and Crop/Livestock Management Systems for Rainfed Agriculture in the Sudano-Sahelian Zone, 11-17 Jan 1987, Niamey, Niger.

Bationo A, Mokwunye A, Baanante C A (1985) Agronomic and economic evaluation of alternative phosphate fertilizer sources in Niger. Pages 110-122 *in* Appropriate technologies for farmers in semi-arid West Africa. H. W. Ohm and J. G. Nagy, eds. Purdue University, Indiana, USA.

Boa W (1966) Equipment and methods for tied-ridge cultivation. Farm Power and Machinery Informal Working Bulletin No. 28. Food and Agriculture Organization, Rome, Italy.

Charreau C (1972) Problemes poses par l'utilisation agricole des sols tropicaux par des cultures annuelles. Agron. Trop. 27:905-929.

Chopart J L (1983) Etude du systeme racinaire du mil (*Pennisetum typhoides*) dans un sol sableux du Senegal. Agron. Trop. 38:37-46.

Dancette C, Nicou R (1974) Economie de l'eau dans les sols sableux au Senegal. Institute for Research in Tropical Agriculture/Centre National de Recherches Agronomiques, Bambey, Senegal. (mimeo.)

Farmer G, Wigley T M (1985) Climatic trends for tropical Africa. Climatic Research Unit, School of Environmental Sciences, University of East Anglia, Norwich, England. 136 p.

Fussell L K, Serafini P G (1985) Crop associations in the semi-arid tropics of West Africa: research strategies, past and future. Pages 218-235 *in* Appropriate technologies for farmers in semi-arid West Africa. H. W. Ohm and J. G. Nagy, eds. Purdue University, Indiana, USA.

Gregory P J (1987) Water use efficiency of crops in the semi-arid tropics. Paper presented at the Workshop on Soil, Water and Crop/Livestock Management Systems for Rainfed Agriculture in the Sudano-Sahelian Zone, 11-17 Jan 1987, Niamey, Niger.

Hauck F W (1966) Fertilizer needs and effectiveness with tropical crops in West Africa. FFHC Bulletin. Food and Agriculture Organization, Rome, Italy.

ICRISAT—International Crops Research Institute for Semi-Arid Tropics (1984) Entomology. Pages 31-37 *in* ICRISAT (International Crops Research Institute for the Semi-Arid Tropics) Sahelian Center annual report 1983. Niamey, Niger.

IFPRI—International Food Policy Research Institute (1977) Food needs of developing countries: projections and consumption to 1990. Research Report No. 3. Washington, D.C., USA.

Jones M J, Wild A (1975) Soils of the West African savanna: the maintenance and improvement of their fertility. Technical Communication No. 55. Commonwealth Agricultural Bureaux, Farnham Royal, UK. 246 p.

Jordan W R, Sullivan C Y (1982) Reaction and resistance of grain sorghum to heat and drought. Pages 131-142 *in* Sorghum in the eighties: proceedings of the international symposium on sorghum, 2-7 Nov 1981. International Crops Research Institute for the Semi-Arid Tropics, Patancheru, Andhra Pradesh, India.

Klaij M C, Hoogmoed W (1987) Crop response to tillage practices in a Sahelian soil. Paper presented at the Workshop on Soil, Water and Crop/Livestock Management Systems for Rainfed Agriculture in the Sudano-Sahelian Zone, 11-17 Jan 1987, Niamey, Niger.

Labeyrie (1977) L'IRAT et l'amelioration du sorgho grain en Haute Volta. Agron. Trop. 32:287-292.

Matlon P J (1985) A critical review of objectives, methods and progress to date in sorghum and millet improvement: a case study of ICRISAT/Burkina Faso. Pages 154-179 *in* Appropriate technologies for farmers in semi-arid West Africa. H. W. Ohm and J. G. Nagy, eds. Purdue University, Indiana, USA.

Mudahar M S (1986) Fertilizer problems and policies in sub-Saharan Africa. Pages 1-32 *in* Management of nitrogen and phosphorus fertilizers in sub-Saharan Africa. A. Mokwunye and P. L. G. Vlek, eds. Martinus Nijhoff Publishers, Dordrecht, The Netherlands.

Nicou R (1977) Le travail du sol dans les terres exondees du Senegal, motivations, contraintes. Institute for Research in Tropical Agriculture/Centre National de Recherches Agronomiques, Bambey, Senegal. 50 p. (mimeo.)

Nicou R, Charreau C (1985) Soil tillage and water conservation in semi-arid West Africa. Pages 9-32 *in* Appropriate technologies for farmers in semi-arid West Africa. H. W. Ohm and J. G. Nagy, eds. Purdue University, Indiana, USA.

Ntare B R, Serafini P G, Fussell L K (1987) Recent developments in millet/cowpea cropping systems for low rainfall areas of the Sudano-Sahelian zone of West Africa. Paper presented at the Workshop on Soil, Water and Crop/Livestock Management Systems for Rainfed Agriculture in the Sudano-Sahelian Zone, 11-17 Jan 1987, Niamey, Niger.

Ohm H W, Nagy J G, Sawadogo S (1985) Complementary effects of tied ridging and fertilization with cultivation by hand and donkey and ox traction. Pages 61-73 *in* Appropriate technologies for farmers in semi-arid West Africa. H. W. Ohm and J. G. Nagy, eds. Purdue University, Indiana, USA.

Penning de Vries F W T, Djiteye M A (1982) La productivite des paturages Saheliens - une etude des sols, des vegetations et de l'exploitation de cette ressource naturelle. Agric. Res. Rep. (Versl. Landbouwk. Onderz.) 918. 525 p.

Perrier E R (1986) Adaptation of water management practices to rainfed agriculture on Alfisols in the Sahel. Paper presented at the OAU/STRC-SAFGRAD International Drought Symposium, 17-23 May 1986, Nairobi, Kenya.

Pichot J, Roche P (1972) Le phosphore dans les sols tropicaux. Agron. Trop. 27:939-965.

Pieri C (1985) Food crop fertilization and soil fertility: the IRAT experience. Pages 74-109 *in* Appropriate technologies for farmers in semi-arid West Africa. H. W. Ohm and J. G. Nagy, eds. Purdue University, Indiana, USA.

Sanchez P A, Buol S W (1975) Soils of the tropics and world food crisis. Science 188:598-603.

Sivakumar M V K (1986) Soil-climatic zonation for West African semi-arid tropics - implications for millet improvement research. Paper presented at the Regional Pearl Millet Workshop, 7-10 Sep 1986, Niamey, Niger.

Sivakumar M V K (1987a) Agroclimatic aspects of rainfed agriculture in the Sudano-Sahelian zone. Paper presented at the Workshop on Soil, Water and Crop/Livestock Management Systems for Rainfed Agriculture in the Sudano-Sahelian Zone, 11-17 Jan 1987, Niamey, Niger.

Sivakumar M V K (1987b) Predicting rainy season potential from the onset of rains in the Sahelian and Sudanian climatic zones of West Africa. Agric. For. Meteorol. 42:295-305.

Sivakumar M V K (1989) Climate vulnerability of sorghum and millet. Pages 187-192 *in* Climate and food security. International Rice Research Institute, P.O. Box 933, Manila, Philippines.

Sivakumar M V K (1989) Exploiting rainy potential from the onset of rains in the Sahelian zone of West Africa. Agric. For. Meteorol. 43 (in press)

Soman P, Stomph T J, Bidinger F R, Fussell L K (1986) Improvement in stand establishment in pearl millet. Paper presented at the OAU/STRC-SAFGRAD International Drought Symposium, 17-23 May 1986, Nairobi, Kenya.

Spencer D S C (1985) A research strategy to develop appropriate agricultural technologies for small farmer development in Sub-saharan Africa. Pages 308-325 *in* Appropriate technologies for farmers in semi-arid West Africa. H. W. Ohm and J. G. Nagy, eds. Purdue University, Indiana, USA.

Spencer D S C, Sivakumar MVK (1987) Pearl millet in African agriculture. Pages 19-31 *in* Proceedings of the international pearl millet workshop, 7-11 April 1986, ICRISAT Center, India Patancheru AP 502 324, India. ICRISAT.

Steiner K G (1984) Intercropping in tropical smallholder agriculture with special reference to West Africa. German Agency for Technical Cooperation, Eschborn. 304 p.

Stewart J I (1987) Potential for response farming in sub-Saharan Africa. Paper presented at the Workshop on Soil, Water and Crop/Livestock Management Systems for Rainfed Agriculture in the Sudano-Sahelian Zone, 11-17 Jan 1987, Niamey, Niger.

Stoop W A, Pattanayak C M, Matlon P J, Root W R (1982) A strategy to raise the productivity of subsistence farming systems in the West African semi-arid tropics. Pages 519-526 *in* Sorghum in the eighties: proceedings of the international symposium on sorghum, 2-7 Nov 1981. International Crops Research Institute for Semi-Arid Tropics, Patancheru, Andhra Pradesh, India.

Swindale L D (1982) Distribution and use of arable soils in the semi-arid tropics. Pages 67-100 *in* Managing soil resources, plenary session papers. Transactions of the 12th International Congress of Soil Science. Indian Society of Soil Science, New Delhi.

Timberlake L (1985) Africa in crisis. The causes, cures of environmental bankruptcy. Earthscan International Institute for Environment and Development, London, UK.

Notes

Author's address: M. V. K. Sivakumar, International Crops Research Institute for the Semi-Arid Tropics, B.P. 12404, Niamey, Niger.

Citation information: International Rice Research Institute (1989) Climate and food security. P.O. Box 933, Manila, Philippines.

Climatic variation and food production in Jiangsu, China

Gao Liangzhi, Juan Fang, and Li Bingbai

Jiangsu, located in the transition zone between subtropical and warm temperate climates, is one of the main food-producing provinces in China. The region has substantial climatic variability, with strong influences of cold from Siberia and high pressure from the Pacific Ocean. In the 30 yr 1951-80, the coefficients of variation were 28-31% for annual rainfall, 2.5-4% for temperature, and 9-12% for sunshine. Food crop yields have been increasing. In 1980-84, rice yields averaged 5.7 t/ha, wheat yields 3.6 t/ha, and maize yields 4 t/ha. Crop yield variabilities were much smaller in 1971-80 than they were in 1951-70. Some important measures for maintaining high and stable food production are building irrigation networks, developing new cropping systems, improving varieties, applying agrometeorology research results to grain production, and undertaking a policy of family responsibility.

Jiangsu, with a population of 70 million, is the most economically developed province and one of the largest food-producing areas in China. Since 1980, annual grain production has been more than 30 million t, which meets the food, feed, and industry raw material requirements of the province and provides 1-1.5 million t/yr for export to other provinces.

The province is located between 30°57′ and 35°07′ N, in the transitional zone between warm temperate and subtropical climates. Its landscape is 69% alluvial plains, 17% water surface, and 14% hills and mountains. Primary food crops produced are rice, wheat, maize, and sweet potato (Li and Gao 1986).

The agroclimate

The agroclimate in Jiangsu differs from north to south. Total sunshine during the growing season (mean daily temperature ≥0 °C) ranges from 1,876 to 2,237 h/yr. Annual total solar radiation ranges from 4,522 to 5,275 MJ/m^2. Both these factors decrease with decreasing latitude because of greater cloud cover in the south. Mean annual temperature is 13-16 °C, with frost-free 210-245 days. Annual accumulated temperatures are 4,900-5,000 degree days in the north, 5,100-5,500 in the central zone, and 5,500-5,850 in the south. This means different cropping patterns and crop varieties are needed for different areas.

Annual precipitation is 800-1,500 mm, increasing gradually from the north to the south and from inland to the eastern coast. Spring rainfall is plentiful in the south, averaging more than 300 mm, 35% of it during the growing season. Rainfall patterns in the north are different: spring rainfall is usually less than 180 mm, summer rainfall is as high as 400-500 mm, with almost half concentrated during the growing season.

The climates of Jiangsu Province are, to a large extent, controlled by the East Asia monsoon. Cold currents from Siberia in the winter and high pressure from the Pacific Ocean in the summer also substantially affect the climate. The high variability in occurrences and in their strength results in frequent cold waves, droughts, and typhoons. Because no big mountain range bars the winter and summer monsoons, temperature and rainfall vary widely from year to year. There can be 20-30 times more river water from the upper region in 1 yr than in another.

The result is meteorological stresses, such as drought, waterlogging, typhoons, low and high temperatures, dry hot wind, and hailstorms, that occur frequently and affect agricultural production considerably. Meteorological data for the 30 yr 1951-80 show a coefficient of variation (CV) as high as 28-31% for annual rainfall, 2.5-4.0% for annual temperature, and 9-12% for total sunshine duration. The compound CV index (mean of 3 CVs) was 13.2-15.0%. During this period, the probability of drought and waterlogging reached 33%. Typhoons swept the province an average of 3.3 times/yr.

Increased grain production

Since 1949, the rate of increase in grain production has exceeded the rate of population increase. From 1949 to 1984, grain production increased from 7.5 million t to 31.3 million t a year, a 4.1% annual growth rate. The population growth rate during the same period was only 1.7%. Per capita grain stores grew from 270 to 503 kg, ensuring food requirements.

Land-use capacity and labor productivity also have increased. Grain yield has grown from 1.2 t/ha in 1952 to 4.9 t/ha in 1984 (rice, 5.7 t/ha; wheat, 3.6 t/ha; maize, 4.0 t/ha), a 304% increase (Fig. 1). At the same time, food production per labor year increased from 614.5 to 1,293.5 kg.

More and more grain has been sold on the market. On the average, 2.9 million t of grain were sold per year in the 1950s, 3.2 million t in the 1960s, and 5.7 million t in the 1970s. Jiangsu ranks eighth in the country in area of cultivated land and second in grain output (Jiangsu Agricultural Regional Division Committee 1985).

Stability of grain yield under variable climate

Because of climate variability and the many adverse weather events, grain production in the 1950s and 1960s was highly variable. Since 1970, food production has tended to be more stable. The 30-yr meteorological yields (yield deviation caused by meteorological factors) of rice and wheat were analyzed statistically at Nanjing and Xuzhou, using perpendicular polynomial regression to reduce the time tendency (Table 1). The variabilities of rice yields in the whole province and of wheat yield in the north were considerably smaller in 1971-80 than in 1951-70.

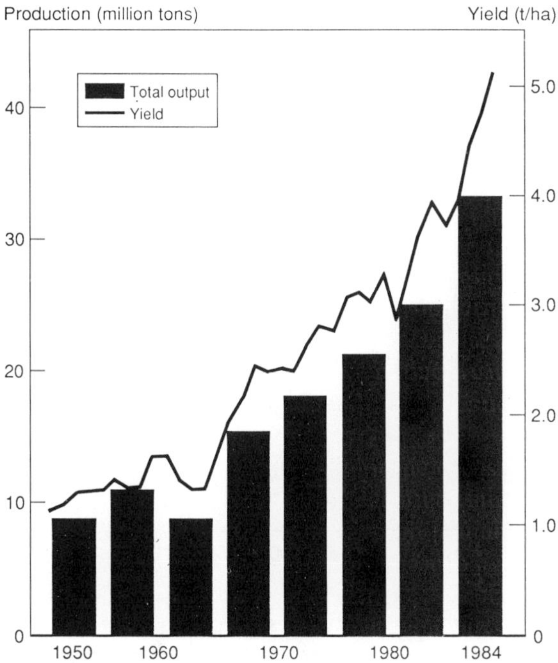

1. Grain production and yield per unit area in Jiangsu Province, China.

Table 1. Rice grain yield and coefficient of variation (CV) in different regions.

Region	Rice						Wheat					
	1951-60		1961-70		1971-80		1951-60		1961-70		1971-80	
	Yield (t/ha)	CV (%)	Yield (t/ha)	CV (%)	Yield (t/ha)	CV (%)	Yield (t/ha)	CV (%)	Yield (t/ha)	CV (%)	Yield (t/ha)	CV (%)
Nanjing (South Jiangsu)	2.4	13.1	3.5	17.6	4.4	8.3	0.8	19.4	1.4	12.6	2.3	15.9
Shuzhou (South Jiangsu)	3.7	9.1	4.8	17.1	4.5	8.6	1.0	18.1	1.6	12.6	3.0	22.7
Xuzhou (North Jiangsu)	0.9	19.6	2.2	21.6	3.5	10.0	0.8	13.2	0.8	17.7	2.1	12.4
Huaiyin (North Jiangsu)	1.5	13.0	2.0	14.4	3.6	9.2	0.7	22.5	0.8	19.2	1.8	14.7

The relationship between the meteorological yields of rice and wheat and climatic factors from 1951 to 1980 was analyzed by stepwise regression. The results at Nanjing and Xuzhou are shown in Table 2. The limiting climatic factor for rice in

Table 2. The statistical models for stepwise regression of meteorological yields of rice and wheat at Xuzhou and Nanjing.[a]

Site		Rice
Nanjing (South Jiangsu)	$Y = -64.7+15.1X_1-0.4X_2$ (F = 8.2** n = 30 SD = 47.7)	Y = meteorological yield X_1 = mean sunshine duration in Jul-Aug X_2 = rainfall in Sep-Oct
Xuzhou (North Jiangsu)	$Y = -597.7+25.4X_1+0.93X_2$ (F = 4.8** n = 30 SD = 39.8)	X_1 = mean temp in May-Jun X_2 = mean temp in Aug-Sep
Site		**Wheat**
Nanjing	$Y = 6.98+21.6X_1-0.4X_2-0.3X_3$ (F = 6.98** n = 30 SD = 34.8)	X_1 = mean temp during winter. X_2 = rainfall during winter. X_3 = rainfall during filling period.
Xuzhou	$Y = -25.1+13.5X_1+0.4X_2-5.2X_3 +1.6X_4$ (F = 4.4** n = 30 SD = 19.7)	X_1 = mean temp during winter. X_2 = rainfall during winter. X_3 = mean temp during filling period. X_4 = mean sunshine duration during filling period.

[a]** = significant at 0.01 level, n = sample size, SD = standard deviation.

the north is primarily temperature. In the south, constraints are primarily sunshine and precipitation during the middle and late parts of the year.

In north Jiangsu, temperature and rainfall in winter and sunshine during grain filling have a positive effect on wheat yields, temperature during grain filling has a negative effect. In the south, temperature during winter has a positive effect on wheat yields, rainfall during winter and grain filling has a negative effect.

Measures to achieve high and stable grain production

Despite considerable climatic variability, remarkably high and stable grain yields have been achieved through the following measures:

- Irrigation networks. Since the founding of the People's Republic of China, the people of Jiangsu have constructed 3,500 km of dikes and dams, dredged and cleaned 420 main rivers, and built 1,100 reservoirs. The ability to combat disasters has been greatly strengthened. Adverse weather events such as flood, drought, and waterlogging are almost under control. In the north, the rice area has been expanded from 133,000 to 667,000 ha as a result of installations for transferring water from the Yangtze River. Since 1949, Jiangsu residents have conquered 4 serious floods, 6 droughts, and 11 waterloggings. The area affected by natural disaster apparently has decreased, which should ensure rapid development in agriculture.
- New cropping systems with increased multiple cropping index (MCI). With an increase in land use for alternate purposes, the area for agriculture decreased 0.7%/yr. To compensate, farmers developed intensive cultivation and improved cropping systems. They raised the MCI to 185% in 1985. In the 1960s and 1970s, a triple-cropping system of wheat - rice - rice established in

the south played an important role in increased grain production (1.1-1.5 t/ha higher than the rice - wheat system).

In Central Jiangsu, the original cropping pattern was single-crop rice. Between 1965 and 1970, farmers changed to double cropping (rice - wheat) in 270,000 ha and grain production increased rapidly. Since 1980, the central area has become the second biggest grain production region in Jiangsu Province, marketing more than 1.5 million t of grain a year.

In the north, maize, soybean, and sweet potato had been the main food crops, with either one crop/yr or three crops/2 yr. Yield per hectare was 30-40% lower than the provincial average. Since the 1960s, irrigation has expanded and irrigated fields now total 0.45 million ha. Double cropping rice + wheat and maize + wheat is practiced and the MCI has been raised from 130 to 170%. In the 20 yr 1955-75, grain production doubled; by 1985, it had reached 11.4 million t/yr. North Jiangsu is now the largest grain-producing region in the province.

- Improved crop varieties. Traditional varieties with lower productivity are being continually replaced by improved varieties. Rice varieties have been changed six times and wheat varieties five times. Each time a variety was changed, average yield increased 10%. In the early and mid-1960s, high-yielding rice variety Nongken 58 was widely adopted, and total rice yields increased from 3.8 to 5.2 t/ha. In the mid- and late 1970s, hybrid rice was introduced. In 1985, the area planted to hybrid rice was 780,000 ha, with a yield of 7.8 t/ha. In the 1980s, several new rice varieties with high yield potential, high resistance to diseases, and better grain quality were introduced. Yields averaged 6.8-7.5 t/ha, with a maximum of 9.8 t/ha.

  For wheat, several highly disease-resistant and early-maturing varieties have been developed. Wheat yields have increased from 1.2 to 3.8 t/ha. Short-duration maize varieties with black spot resistance are widely grown. They can be sown in both spring and summer, with yields of more than 6 t/ha.
- Agrometeorology research applied to grain production. The Department of Agrometeorology in the Jiangsu Academy of Agricultural Sciences (JAAS), established in 1953, conducts research on the agrometeorological problems affecting grain production that plays an important role in ensuring high and stable grain production. It has investigated excess-soil-moisture injury to wheat in south Jiangsu and proposed preventive measures. Wheat should produce one or two tillers before winter as well as escape spring frost injury; the agrometeorological team established optimum wheat sowing dates (Zhang et al 1980). Single- and double-crop late rice have late heading and can be injured by low temperatures; JAAS agrometeorology scientists established safe heading dates for rice (Gao et al 1983). Using a newly developed photothermo model for rice growth duration (Gao et al 1982), they analyzed the safe growing seasons for different rice varieties.
- New agricultural policies. Since 1978, the family responsibility system has been practiced in Jiangsu. Profit from increased grain production now belongs to the farmer and his family. During the same period, the government

raised grain prices. These policies stimulated grain production. In 1983, total grain production reached 30 million t, with per capita grain production exceeding 500 kg for the first time. Many technical service systems—crop protection, seed supply, irrigation, agricultural mechanization—have increased the application of science and technology to grain production. In recent years, surplus labor has been transferred from farm production to rural industry and commerce. Farmlands are gradually being consolidated and worked by fewer farmers. With the development of mechanized farming and the improvement of economic efficiency, grain production should increase still more.

References cited

Gao L Z, Jin Z Q, Li L (1982) Meteor-ecological models for the growth and development duration of various rice varietal types and their application in China [in Chinese]. Agric. Meteorol. 3(2):1-8.

Gao L Z, Li L, Guo P (1983) An investigation of the growth season and climatic ecology of cultivated system of rice in China [in Chinese]. Agric. Meteorol. 4(1):50-55.

Jiangsu Agricultural Regional Division Committee (1985) A comprehensive regional division for crop farming [in Chinese]. Nanjing, China. 95 p.

Li L, Gao L Z (1986) A climatic-ecological regionalization for cropping systems in Jiangsu province [in Chinese]. Farming Cultivation 1-2:19-24.

Zhang L Z, Guo P, Gao L Z (1980) Studies on the climatic ecology models and the optimum sowing dates for wheat and barley [in Chinese]. Agric. Meteorol. 1(3):15-21.

Notes

Authors' address: Gao Liangzhi, Juan Fang, and Li Bingbai, Department of Agrometeorology, Academy of Agricultural Sciences, Nanjing, 210014 China.

Citation information: International Rice Research Institute (1989) Climate and food security. P.O. Box 933, Manila, Philippines.

Climatic variability and agronomic management in Mediterranean barley - livestock farming systems

M. J. Jones and H. C. Harris

The demand for livestock products in the Middle East and North Africa is increasing rapidly. The major sources are sheep and goats, many of which are found in dry areas, subject to wide variability in rainfall and therefore feed supply. This paper considers a livestock production system in northern Syria, its built-in buffers against variability, and current pressures on these buffers from increasing population. Decreased access to natural grazing is increasing the dependence on annually sown fodder crops, principally barley. It has been shown that the use of fertilizers and the replacement of alternate years of bare fallow with a fodder legume can greatly increase productivity, but this is likely to be coupled with an increasing variability in annual carrying capacity. Already many dry-area farmers buffer their flocks against local shortages by supplementary feeding, but few countries can afford large imports of feed grains. Any intensification of livestock production in dry areas through improved agronomic practices will therefore require increased support from wetter and irrigated areas, in the form of crop residues and by-products and probably bulk feed production and large fodder banks.

A recent survey of the Middle East and North Africa noted that agricultural policy in the region has so far concentrated on increasing production of the main food staple, wheat (Khaldi 1984). However, livestock products also are an important component of the diet, and demand is increasing rapidly. As a result, the gap between supply and demand of livestock feed is expected to widen, to become the largest production deficit by the year 2000.

We may anticipate that fluctuations arising from climatic variability will be imposed on these general trends. The production of livestock feed, and of livestock, is subject to the vagaries of rainfall and temperature no less than the production of crops for human food. A large proportion of the region's livestock products are derived from small ruminants (sheep and goats). In turn, a large proportion of those sheep and goats are found in the driest areas, where stock raising constitutes the predominant farming enterprise. It is those areas that experience the widest climatic variability.

In the northern Syria livestock production system, built-in buffers against climatic variability are being subjected to increasing stresses through increasing population pressure. There is potential for certain innovations in agronomic management to increase and stabilize production, with some wider implications.

Barley - livestock systems of the Mediterranean region

A broad view of the agricultural systems of North Africa and West Asia (ignoring local variations imposed by mountains and irrigation) shows a transition from predominantly arable farming in wetter areas to predominantly livestock-based systems in drier areas. At one extreme, we find intensive production of wheat, grain legumes, vegetables, and fruit; at the other, steppe land grazing by nomadic flocks of small ruminants. We may reasonably assume that given the technology available, these systems represent the most stable agricultural adaptations to the amount and variability of rainfall.

At the drier end of the rainfall spectrum, adjacent to the steppe in Syria and elsewhere, is a zone of sheep production systems based largely on annually sown feed crops. Relatively little is known about barley - livestock systems, how they function, and their potential. National research institutions have directed most of their research efforts to the wetter areas, with their potential for substantial yield gains in the wheat and food legumes staples. Our understanding derives from 10 yr interdisciplinary research by the International Center for Agricultural Research in the Dry Areas (ICARDA) in northern Syria. We believe, but have yet to confirm, that the farming system there has many elements in common with other barley - livestock systems across the region.

The common elements almost certainly include a preponderance of poor farmers, a dependence on common grazing lands to provide feed for part of the year, and an increasing trend toward overgrazing and degradation. Many, perhaps all, of these systems are no longer self-sustaining, but rely on the availability of alternative feed sources, in good as well as in bad seasons. In Syria, these are either crop by-products, such as cottonseed cake, sugar beet pulp, and wheat bran, or crop residues in higher rainfall and irrigated areas. In Egypt, substantial quantities of fodder and crop residues are exported from the Nile Delta to support sheep in the northwest coast region. In Tunisia, hay (oat and vetch) is produced in wetter areas to meet the seasonal demand for feed in drier areas. In well-watered areas, especially in Egypt, there is probably direct competition for resources between food and feedstuff production. That can only be exacerbated by climatic variability.

Livestock systems in Northern Syria

Current trends

In northern Syria, sheep production depends strongly on annually sown feeds. The major crop is barley, with small areas of fodder legumes. In some years, the barley may be lightly grazed as a spring pasture, but its main purpose is to provide grain and straw for winter feed. Any stubble is grazed during the summer (Fig. 1). Natural grazing and other feed sources are also utilized, and many farmers grow small areas of subsistence crops, primarily wheat and lentils. The core of the system is barley.

Such a system appears to combine those elements of arable and livestock farming that maximize semiarid rainfed production at an acceptable level of reliability. Under low rainfall and on predominantly calcareous soil, often shallow

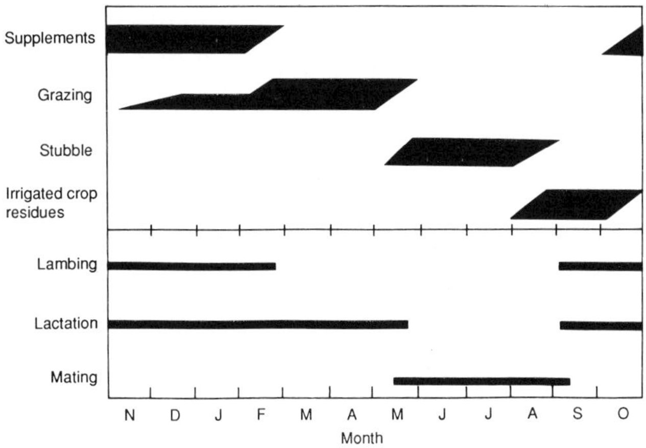

1. Typical sheep feeding and breeding cycles in northern Syria (Jaubert and Oglah 1985).

and droughty, barley outyields other cereals. Fallows in alternate years help to buffer its productivity. Greater buffering is provided by the livestock component, through alternative feed sources, and by the option to sell and purchase animals. A substantial element of nomadism remains. Although some flocks stay near the home village all year, grazing marginal lands, fallows, and crop stubble, others are moved to where feed is available—grazing the steppe in spring and early summer, and, especially in dry years, grazing on crop residues in higher rainfall or irrigated areas. Stored barley grain and straw are the winter staples, but supplementary feeds, such as cottonseed cake and sugar beet pulp, may be purchased as needed (Table 1). Large urban centers provide a ready market for meat and dairy product surpluses.

There is evidence that this system has changed a great deal over the last 20-30 yr (Jaubert and Oglah 1985). With competition for the increasingly overgrazed steppe, the numbers of sedentary animals have increased and seminomadic flocks are

Table 1. Contribution of feedstuff to sheep in Bueda/Breda subarea, Aleppo Province, northern Syria during winter 1983-84, as percentages of metabolizable energy (Jaubert and Oglah 1985).

Feedstuff	Nov-Dec	Jan	Feb	Mar	Apr	Mean, Nov to mid-Feb
Barley straw	33.4	29.3	29.5	24.9	7	31.6
Barley grain	47.7	48.7	46.3	50.5	59.0	48.8
Wheat straw	2.0	–	–	–	–	0.5
Lathyrus straw	2.4	5.3	0.3	4.7	–	2.9
Lathyrus grain	–	0.4	1.4	–	–	0.7
Wheat bran	3.5	5.3	8.5	3.8	41.0	4.7
Cottonseed cake	3.8	5.8	10.1	8.2	–	6.5
Cottonseed hulls	5.6	2.2	–	–	–	1.3
Sugar beet pulp	1.6	3.1	3.9	7.9	–	3.0

spending more time in the villages. New settlements have proliferated and plowing has encroached farther and farther into the shallow soils on the hillsides, reducing the area of local marginal grazing. The resulting need to increase on-farm production of feed has increased the area of barley, at the expense of wheat and fallow. The reduction in fallowing has meant, in turn, that many fields now grow barley continuously, with declining yields. Some relief is provided by the greater availability of commercial feeds, but population pressure is intensifying.

The original system had considerable resilience to climatic variability. The challenge to the agronomist and livestock specialist now is to accommodate the pressures and changes that are still occurring to raise production without losing the original resilience.

As agronomists, we concentrate here on crop and soil factors in the system—keeping the goal of animal production firmly in view. Other work at ICARDA encompasses improving marginal grazing, selecting fodder and pasture legume species, and studying sheep nutrition.

Fallows and rotations

When barley is grown continuously, yields decrease. This has been ascribed variously to declining fertility, a buildup of pathogens, or allelopathic effects. On current evidence, no firm conclusion can be drawn. Some alleviation appears to be possible, through the use of fertilizer. On the other hand, fallows are presumed to protect fertility and to act against allelopathic effects and pathogen buildup. Certainly, they afford a minor grazing opportunity and enhance the yield of the subsequent barley crop. As pressures mount, however, fallowing is widely perceived to be an unproductive use of land.

The alternative to continuous barley cropping is some form of rotation. Given the requirement for sheep feed, most ICARDA work has concentrated on a barley - fodder legume cycle, with vetch, peas, or lathyrus as the legume. Results of two rotation trials comparing barley - barley, barley - fallow, and barley - vetch are presented in Table 2 (following Thomson 1986b, they are given as "sheep equivalents": the number of standard ewes that could be supported per year by the metabolizable energy value of the crops produced).

- Barley - fallow, although the least productive rotation (only half the area is cropped in any 1 yr), had the lowest variability over the 4 yr.
- Barley - vetch was slightly more productive than barley - barley and much less variable, but was more expensive to produce.
- Applying fertilizer every second year increased the productivity of all three rotations, with annually fertilized barley - barley the most productive at no increased cost per sheep equivalent.
- All barley - barley rotations showed high year-to-year variability in productivity.

Results from water balance studies in an older rotation trial at Breda help to explain these findings. Although much of the water accumulated in fallow soil during the winter is lost to evaporation during the summer, after a wet year some remains for use by the following crop. But after a dry year, such as 1983-84, little remains (Fig. 2). Where barley is grown continuously, there is no carry-over of water

Table 2. Four-year means of productivity in 2-yr rotations at 2 sites[a] in northern Syria, expressed in terms of sheep equivalents (theoretical carrying capacity).

Mineral fertilizer[b]	Rotation[c]	Sheep equivalents/ha						Cost (sheep equiv.)	
		Breda			Tel Hadya				
		Mean	SD±	CV %	Mean	SD±	CV %	Breda	Tel Hadya
–	B-V	4.00	1.14	28.4	7.83	1.38	17.6	260	155
	B-F	2.87	0.85	29.5	5.14	0.64	12.4	152	120
	B-B	3.87	1.50	38.7	6.21	3.17	50.7	149	126
+	B-V	6.40	2.24	34.9	9.07	2.43	26.6	203	160
	B-F	4.50	1.40	31.1	6.79	1.21	17.8	143	130
	B-B	5.51	2.67	48.4	8.67	3.71	42.9	148	130
++	B-B	6.88	2.92	42.4	10.95	4.28	39.1	150	130

[a]Breda, 35° 55′N, 37° 10′E; Tel Hadya, 35° 55′N, 36° 55′E.

Seasonal rainfall (mm):	1982-83	1983-84	1984-85	1985-86	4-yr mean
Breda	279	204	277	218	244
Tel Hadya	322	230	373	316	310

[b]– = none; + = 20 kg N/ha and 26 kg P/ha supplied to barley once every second year; ++ = same fertilizer rates supplied to barley annually. (All N rates doubled at Tel Hadya.) [c]B = barley, V = vetch, F = fallow. 1 sheep equivalent is defined as 4.2 GJ of metabolizable energy (Thomson, 1986b). Values given are based on grain and straw yields of barley and hay yields of vetch. (Alternative calculations based on crude protein contents give a similar, although not identical, pattern of values.) Costs are based on locally surveyed costs, to the farmer, of cultivations, fertilizers, seed, and harvesting.

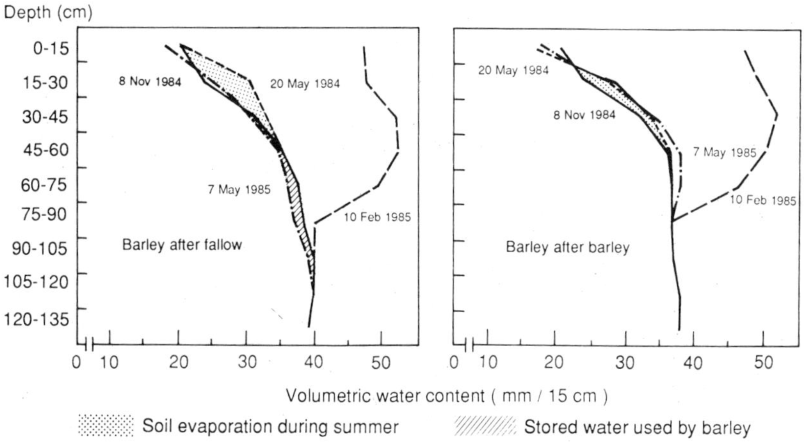

2. Soil water profiles under barley at four points in the crop cycle in a wet year following a dry year. Breda, northern Syria.

to the next crop, even after a wet year (Fig. 3). The relative yield stability of the barley - fallow rotation can, therefore, be attributed to the buffering effect of fallow-stored water.

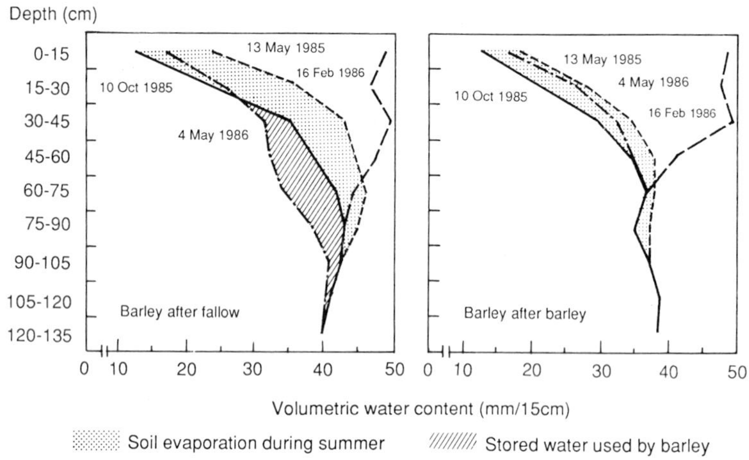

3. Soil water profiles under barley at four points in the crop cycle in a dry year following a wet year. Breda, northern Syria.

Vetch, cut early as a hay crop, uses less water than barley and acts more like a fallow. After a wet year, the vetch soil retains some available water for the succeeding barley crop (Fig. 4). However, higher production costs of the barley - vetch rotation arise from the cost of harvesting vetch as hay (Table 2). A better economic option is probably to grow the legume to maturity, harvesting some seed for future planting before grazing the rest in situ as standing hay. But it must be recognized that greater water use by the vetch would in this case eliminate any chance of buffering for the next barley crop.

A third option is green-stage grazing of vetch. If available immediately after sheep are weaned, this could provide high-quality feed for milking ewes, for yogurt and cheese production (important cash commodities in the system), or for fattening lambs. Either seems economically viable, but fattening lambs appears to offer greater returns (Nordblom and Thomson 1986, Thomson 1986a). Again, seed supply could be a problem, but part of the crop area might be saved for seed production. As far as soil water storage is concerned, green-stage grazing would offer the benefits of a hay crop without the costs of hay making.

However, this form of utilization has the important requirement of timeliness. It might be the most vulnerable to variability in the weather. If a delayed start to the rainfall season and/or a period of exceptionally low temperatures restricted early growth, there would be insufficient feed at a vital time. This highlights what is probably the main issue regarding fodder legumes in this farming system: It is essential to balance utilization with respect to both the economics and the animal feeding cycle.

Fertilizer use

Most of the soils in the barley - livestock zone of northern Syria are deficient in phosphate, and many are also low in available nitrogen. Of 32 fertilizer trials in farmers' fields 1984-85 and 1985-86, 26 showed significant yield response to

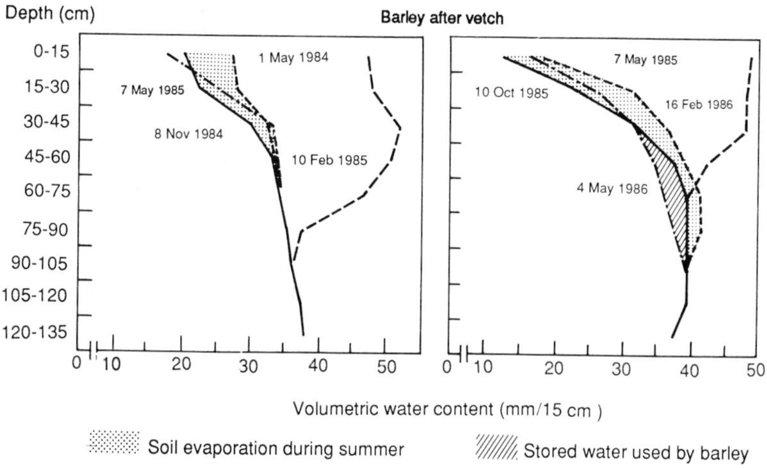

4. Soil water profiles under barley following a vetch hay crop in a wet year following a dry year (left) and in a dry year following a wet year (right).

phosphorus and 15 to nitrogen. At current prices, the increases in feed production achieved were highly profitable. But what of stability? In one sense, the use of fertilizer reduced site-to-site variability (Table 3). Particularly where N and P were used together, any increase in the standard deviation was proportionately less than the yield increase, so that percentage of variability (CV) decreased.

However, this is rather misleading. The major determinant of yield over the 32 trials was rainfall. A linear regression on total rainfall (range: 147-339 mm) accounted for 54% of the variability in mean dry matter production. Fertilizer—at least P—was effective over the whole rainfall range, but response to P increased with increasing rainfall (Fig. 5). One may view this as meaning that fertilizer efficiency is increased by rainfall or that rainfall efficiency is increased by fertilizer. Either way, in a variable rainfall environment, fertilizer use increases year-to-year yield variability (Table 4).

As the r-squared values for each of the 16 fertilizer treatments show (Fig. 5), fertilizer use increased the relative dependence of yield on rainfall. This is because

Table 3. Summary of 32 on-farm barley fertilizer trials in northern Syria: means, standard deviations (SD), and coefficients of variation (CVs) of total dry matter production (t/ha).

Treatment[a]	Dry matter				SD±				CV (%)			
	P0	P1	P2	P3	P0	P1	P2	P3	P0	P1	P2	P3
N0	3.0	3.4	3.7	3.9	1.3	1.4	1.4	1.7	43.7	39.7	38.9	43.6
N1	3.2	3.9	4.4	4.4	1.4	1.6	1.7	1.5	44.2	40.9	39.0	34.4
N2	3.4	4.1	4.5	4.8	1.5	1.6	1.5	1.6	43.1	38.7	32.9	33.9
N3	3.4	4.2	4.7	5.2	1.5	1.6	1.6	1.7	45.1	39.4	33.9	33.6

[a]N0 = 0 kg N, N1 = 20 kg N/ha, N2 = 40 kg N/ha, N3 = 60 kg N/ha. P0 = 0 kg P_2O_5/ha, P1 = 30 kg P_2O_5/ha, P2 = 60 kg P_2O_5/ha, P3 = 90 kg P_2O_5/ha.

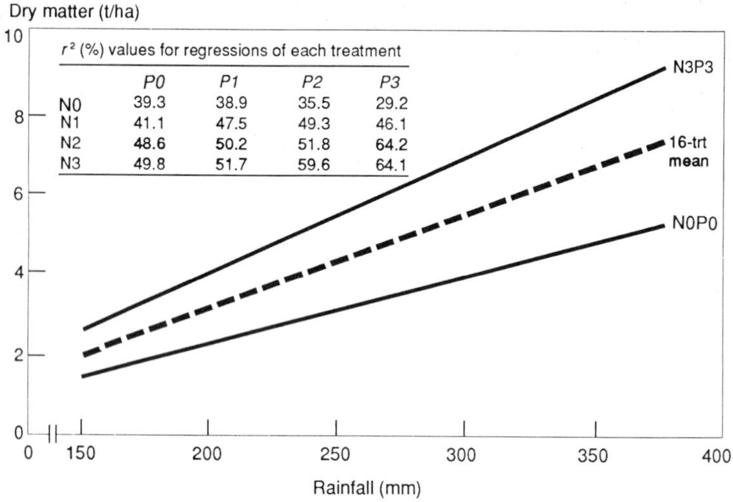

5. Regressions of barley dry matter response to seasonal rainfall under different fertilizer regimes, from 32 trials in northern Syria.

Table 4. Calculated values for the highest rate of application (N3P3) and zero-fertilizer control, at the limits of the experimental rainfall range.

Fertilizer[a]	Predicted t dry matter/ha at 150 mm rainfall	Predicted t dry matter/ha at 350 mm rainfall	Difference
N0P0	1.60	4.91	3.31
N3P3	2.78	8.34	5.56
Difference	1.18	3.43	

[a] N0P0 = 0 kg N, 0 kg P, N3P3 = 60 kg N, 90 kg P_2O_5/ha.

fertilizer greatly reduced the effect of the next most important site factors, available nitrogen and phosphate initially present in the soil. It may seem a truism to say that, the more successful one is at controlling the controllable factors (in this case, nutrient supply), the more dependent production becomes, albeit at a higher level, on uncontrollable factors such as rainfall. Evans (1986), in his summary of the proceedings of a workshop on cereal yield variability, wrote "variability tends to fall as agronomic control of the environment becomes more complete, as in the case of wheat in Western Europe and the Punjab." Unfortunately, this does not apply in more marginal agricultural situations. There, the most important feature of the environment is beyond control.

Other factors

Other agronomic practices which might be used to improve yield in more favorable environments offer little promise for the barley - livestock system. Variable timing of the start of the rains (Cooper et al, this volume) largely determines sowing date.

Some farmers practice dry sowing, but this brings the added risk of seed loss when there is a false start to the season. This risk is minimized by farmers' use of sowing methods that place seeds over a wide range of depths. Shallow-sown seeds may germinate early, but seeds at greater depth (to 10-12 cm) remain viable and can germinate later, when there is heavy rain. Even though plant populations are reduced, yield potential probably is not reduced much.

Minimum tillage is already practiced, and there appears to be little scope for yield improvement by modifying tillage methods. Retention of stubble, which in other places is used to improve the water balance of land under rainfed crops, is scarcely relevant to a system where the feed value of stubble is so high.

Discussion

We are concerned with the consequences of further increases in population on the production stability of an agricultural system under perennially low and irregular rainfall. Over the last 30 yr, the system has evolved considerably. But scope for further change in the same direction appears limited. Little marginal land remains that could, or should, be plowed. While fallowing might be further reduced, farmers are aware of the problem of declining yields.

Our results suggest that productivity can still be increased. Fertilizer, little used so far in this zone, could increase barley grain and straw yields 25-50%. Replacing fallow with fodder legumes or well-fertilized continuous barley could provide similar increments. Work remains to be done on the best way to utilize legumes and to control the adverse effects of continuous barley cropping, but the potential to increase substantially current yields of feed and forage on farmers' fields appears assured. What is almost equally assured, however, is an accompanying increase in the rainfall-driven variation in annual productivity. And the gradual decline in the buffering of the system provided by natural grazing (both local marginal land and steppe pasture) seems certain to continue as well.

Sandford (1982), in considering the effects of variable rainfall on livestock numbers, drew attention to the spectrum of stocking-rate strategies, from highly conservative to opportunistic. Although he was writing about rangeland, his ideas have some relevance to northern Syria. A conservative strategy maintains a relatively low and constant animal population through good years and bad, an opportunistic one varies that population according to the current availability of feed. Which is favored in any particular situation is determined by many factors, not least the stockowner's priorities (subsistence, wealth accumulation, commercial turnover). Ease of stock disposal in droughts and replacement afterward clearly have an important bearing.

In the barley - livestock system of northern Syria, commercial sheep production overlays a strong subsistence element. Stocking rates tend toward the opportunistic, since the local market of meat is large and can, with some reduction in price, absorb a higher offtake during droughts. Also, sheep reproduction rates are relatively high. Any trend toward an increased dependence on higher but relatively more variable supply of planted feeds, fodders, and pastures seems likely to encourage this strategy, perhaps to a point of instability. Sheep equivalents of the annually fertilized, continuous barley rotation in our trials ranged from 6.1 to 15.3/ha at Tel Hadya and from 4.7 to 11.8/ha at Breda. The 4 yr during the trials were not exceptional.

It seems, therefore, that if the productivity of the system is to be stabilized at a higher level, an increased burden of buffering must fall on supplementary feeds. Khaldi (1984) fears these feeds will come from increasing imports of coarse grains. To avoid that, efforts must be made to increase the utilization of agricultural residues from irrigated and higher rainfall areas. A number of organizations that process and market such residues as sugar beet pulp and cottonseed cake already exist in Syria, but we suspect that the potential is much greater than that currently exploited. Burning of straw in wheat lands appears to be a common practice. There may also be a case for increased bulk feed production in wetter areas. In combination with a high-volume fodder bank, this could go a long way to iron out dry-zone fluctuations.

However, for the dry-zone farmer, higher production will be at the price of increased dependence on other agricultural producers.

References cited

Cooper P J M, Harris H C, Goebel W (1989) Climatic variability and food production in West Asia and North Africa. Pages 69-83 *in* Climate and food security. International Rice Research Institute. P.O. Box 933, Manila, Philippines.

Evans L T (1986) Yield variability in cereals: concluding assessment. Pages 1-12 *in* Summary proceedings of a workshop on cereal yield variability. P. B. R. Hazell, ed. International Food Policy Research Institute, Washington, D.C., USA.

Jaubert R, Oglah M (1985) Farming systems management in the Bueda/Breda subarea, 1983/84. Research report, Farming Systems Program. International Center for Agricultural Research in the Dry Areas, Aleppo, Syria.

Khaldi N (1984) Evolving food gaps in the Middle East/North Africa: prospects and policy implications. Research Report 47. International Food Policy Research Institute, Washington, D.C., USA.

Nordblom T L, Thomson E F (1986) A whole-farm model based on experimental flocks and crop rotations at Tel Hadya, North-West Syria. Research report, Farming Systems Program. International Center for Agricultural Research in the Dry Areas, Aleppo, Syria.

Sandford S (1982) Pastoral strategies and desertification: opportunism and conservatism in dry lands. Pages 61-80 *in* Desertification and development: dryland ecology in social perspective. Academic Press, London, UK.

Thomson E F (1986a) Livestock management and nutrition. Pages 305-324 *in* ICARDA annual report, 1985. International Center for Agricultural Research in the Dry Areas, Aleppo, Syria.

Thomson E F (1986b) Feeding systems and sheep husbandry in the barley belt of Syria. Paper presented at Tropentag 1986: Bedeutung der kleinen Wiederkauer zur Nutzung marginaler Standorte den Tropen und Subtropen. Justus-Liebig Universitat Giesen, Federal Republic of Germany.

Notes

Authors' address: M. J. Jones and H. C. Harris, Farming Systems Program, International Center for Agricultural Research in the Dry Areas, P.O. Box 5466, Aleppo, Syria.

Citation information: International Rice Research Institute (1989) Climate and food security. P.O. Box 933, Manila, Philippines.

Food security and food stocks: some managerial issues

N. C. B. Nath

Major changes in the world food situation have changed Asian food imperatives, from the dependence on superabundant world stocks of the 1960s to national self-sufficiency and some participation in the world markets now. In this context, food security has taken a number of forms, from holding large national stocks to regional cooperation. The issues involved in managing these large stocks, cost efficiencies, operational structure, and, more fundamentally, the concept of stockholding are discussed. The usual distinction between privately held stocks and publicly owned stocks does not illuminate the complexity of the behavior of stockholders, from consumers to those responsible for state holdings. Policymakers need to consider ways of harnessing the fractionated inventory to minimize social distress. In practice, major stockholding is done by public entities—perhaps because it is felt that the private sector is not dependable. A more efficient alternative might be integrated responsibilities, coupled with modern technology.

Management of food stocks is an integral part of food security systems. The United Nations Food and Agriculture Organization (FAO) International Undertaking on World Food Security reads:

> "To ensure the availability at all times of adequate world supplies of basic foodstuffs, primarily cereals, so as to avoid acute food shortages in the event of widespread crop failures or natural disasters: to sustain a steady expansion of production and reduce fluctuations in production and prices."

Food stock management is a complex concept, with multiple objectives and a program covering all areas of food supply: production, storage, distribution, and pricing. It has a developmental aspect and a market stabilization concern.

An adequate supply of foodstuffs is a naive phrase: it contains within itself conflicting stocking concepts—from commercial stocks, which could help price stabilization, to earmarked stocks for maintaining a public distribution system, to emergency reserves that can be drawn on to minimize human distress when crops fail.

International support to these different concepts of food stock holding has varied. The food exporting economies do not take kindly to international food price stabilization measures; emergency reserves receive greater conceptual support.

In this paper we look at stockholding concepts and public enterprise in the international food business. We suggest that the time has come in at least some developing countries to reexamine organizational requirements to ensure food policy objectives. We have drawn on consulting experience at the Economic and Social Commission for Asia and the Pacific and in India in formulating the issues.

Current systems have historical justification, and any dilution of public presence is likely to stir emotional responses. But it is necessary to raise this issue, even if only to ultimately reject it.

Asian food imperatives

The 1982 World Development Report identifies the sharp increases in the prices of rice, wheat, and fertilizer between 1972 and 1974 as a dramatic manifestation of a world food crisis. What occurred was a combination of factors: a reduction in world stock levels; anxiety about land availability in the traditional exporting countries; emergence of new power buyers; and a growing feeling that buyers from developing countries were becoming increasingly vulnerable in the international markets.

Asian food imperatives have changed. During the 1960s food needs were imported from what seemed to be superabundant American stocks. Food was drawn from an assured source through aid and trade. The 1970s were years of a conscious thrust toward national food self-sufficiency. In the 1980s, the region has achieved near self-sufficiency and is poised to become a food exporter in its own right, definitely in rice and possibly in wheat. Two classes of food exporters have emerged: those who are consistent (e.g. Thailand and Burma) and those who sell surpluses (e.g. India and Pakistan). Both need assured offtake and stable markets. Their concerns are different but interrelated.

Thus, the scenario of the 1980s and possibly of the 1990s is qualitatively different from what it was earlier. A different approach is needed toward holding food stocks and toward the concept of food security.

Rationale for holding stocks

According to FAO, a desirable minimum of global stocks for world food security is 17-18% of world consumption. Recently world levels have reached this level (in 1981, for instance). But total world level is not enough if the stocks are concentrated in a few exporting countries where they may not be accessible because of price and logistics. Thus, the rationale for holding stocks in the importing countries was developed.

Even if stocks are held nearer to consumption centers, they can achieve their policy objectives only if there is a cost-efficient infrastructure to manage storage and delivery. Such facilities do exist in most South Asian countries, but the efficiency of the structure and the cost of their delivery systems need to be reexamined.

Concept of fractionated stocks

The literature has many references to the use of stocks as a useful supply management device for greater price stability. There are many learned papers on the optimum level of stocks that need to be held in the world markets for orderly trading. These depend on the nature of the market and its structure. Concepts of insurance and regional cooperation have been used to work out cost-efficient alternatives.

Stocking (i.e. withdrawal from current offer) is a useful aggregate concept. But for incisive policymaking, it may be too broad. Perhaps there are more distinctions to be considered than the usual privately owned vs government stocks. In developing country markets, with their mix of state intervention and private trade, their alternating surpluses and deficits, and the unstated feeling that abundance is not a permanent state, stockholding has many nuances and varies by a number of criteria: ownership, objective, how acquired, disposal method, decisionmaking process, and the setting.

In the Asian economies, food stocks are acquired and held by many sectors—consumers, processors, traders, speculators, producers, and government. Barring legal restrictions (and sometimes in spite of them), all of the sectors hold and dispose of stocks to suit their convenience. The size of stocks held varies with holding capacity.

The objectives for holding stocks can vary—from providing a feeling of security for one's family, to profit-making, to avowed social policy. Social policy can range from market control and stabilization to support programs for the less privileged. When the government is the owner, it may have a number of objectives and many eventual uses for its supplies.

Some stocks are acquired at market prices (e.g. imports) and some at lower procurement prices, either voluntarily or as a compulsory reflection of market rigidities.

Disposal has similar variety. Sale could be part of an allocation system, at fixed prices; it could be market intervention by the government; it could be a normal commercial sale between traders or from government to business.

Decisionmaking styles also vary, reflecting the owner and the objective of holding stock.

The components of government-owned stocks and privately owned stocks, therefore, are not homogeneous. They may at times move in contrary directions. Even government stocks, which are theoretically one stock that can be switched from one purpose to another, are often earmarked for different purposes—a powerful if unseen boundary.

In policy, we need to make a distinction between policy intention and policy effect. The pattern of stocking and destocking in the farm sector is of specific interest. Smaller farmers tend to depend more on market supplies than on their own production. The policy implication of this behavior is that a substantial number of producers, not to mention agricultural labor, are as vulnerable to market vagaries as are urban consumers.

What emerges is a concept of fractionated stocks. The stocks, while usually treated as one category (at best a two-part composite of private and government stocks) in reality consist of a number of elements, each of which has an identifiable utility function (e.g. the consumer holding stocks for security or the government acquiring stocks because of a fall in prices). The elements in total constitute the stocks held in a country.

The configuration has three characteristics:

1. The elements do not move together.
2. The elements are an aggregate that affects market behavior.
3. Even small additions or subtractions to individual holdings affect price behavior significantly.

   It is this marginal effect that has significant policy implications.

Stock management

In the food grains business, the Food Corporation of India (FCI) is one of the megacorporations among Indian public enterprises, directly employing 80,000 persons and carrying about 25 million t of stocks. It is perhaps the largest stockholding operation in South Asia.

Operationally, it uses the facilities of other public enterprises, such as central and state warehousing cooperatives and private trade. Over the years, its management has had to face recurring industrial unrest due to its size and number of employees. It has to act at the behest of the Centre as well as the states, while linking with railways, road contractors, and several external agencies. FCI has learned to handle large surpluses at short notice and is familiar with the problems of procurement, movement, and storage.

This all adds up to a complex commercial juggernaut, and has led to inefficiencies, which have been criticized both in the Parliament and outside.

It is not our intention to magnify the criticisms. In our view, inefficiencies are not the cause but a symptom of a suboptimally managed system.

The links in the FCI distribution chain are many, and are differently motivated. Producer - FCI purchase center - State agency purchase center - Cooperative societies - FCI storage depot (transit) - FCI depot in the secondary market - State Government storage agency - Fair Price shops - Private trade.

The obvious question is, “Can such a complex series of activities involving so many agencies be made into a well-orchestrated, efficient machine?”

Possible policy alternatives

Historically, a private trader in India is regarded as an unscrupulous appropriator of market rent. Public policy from time to time has sought to whittle this role down. While the extreme step, nationalization of the food grains trade, has failed, the private sector remains a nonrisk taking agent, and an instrument of the public enterprise distribution system. Could private trade take a larger risk taking role, necessarily under socially acceptable surveillance?

For example:

- Is it possible to think of private companies holding FCI stocks more as a principal than as an agent? Such participation would perhaps diffuse the pressure of lobbies, which invariably get exercised about public enterprise.
- Could private industry on behalf of FCI help with technology, procurement, storage, and delivery, for a stated quid pro quo?
- Could a joint sector activity be developed as a logistical purchase-to-storage support for food grains?
- Would it be possible to consider brand-name marketing of food stocks, which would assure the consumer quality, weight, and price (criteria not fully satisfied in the current public distribution system)?

The problem of improving public sector stockholding organizations such as FCI is crucial, in view of the expected increase in the demand for holding stock of food grains.

Our studies of FCI indicate the need for more industry-oriented management. We realize such a step could be considered radical, but we feel that some debate on this issue would be beneficial to developing countries examining the management of stocks as the way to food security.

Notes

Author's address: N. C. B. Nath, National Council of Applied Economic Research, Institute of Economic Growth and Development, New Delhi, India.

Citation information: International Rice Research Institute (1989) Climate and food security. P.O. Box 933, Manila, Philippines.

Food security in the changing global climate

S. K. Sinha, N. H. Rao, and M. S. Swaminathan

Food security is a cause for concern in the best of times. Even under present climatic conditions, food grain requirements in all regions of the world will rise as populations increase. Except in Africa and South America, there is no possibility of more land opening up for cultivation. Most of the increase in agricultural production will have to come from increased crop yield. For most crops, no major improvements in productivity have occurred in the past two decades, and sustaining a growth rate in food production higher than the growth rate in population while providing for national food security are major challenges. The international community would have to rise high above regional and national interests to overcome problems in the global food system. How future agricultural production and food systems will be affected by global climatic change depends on the magnitude of the change in specific regions. But geographical variations in the projected climatic changes are uncertain. Considered in isolation, the enhanced levels of CO_2 in the atmosphere appear to promote production in some crops, if adequate inputs are available. However, most projections are based only on studies of vegetative growth. A warming of 2 °C promotes sterility in rice, reducing yield 25% or more. A similar rise in temperature could greatly affect wheat. A detailed regional analysis is needed to project the effects of climatic change on agricultural production.

The Food and Agriculture Organization of the United Nations (FAO) defines food security as physical and economic access to food by all people at all times. Swaminathan (1986) has pleaded that this concept be enlarged into one of "nutritional security." Only access to balanced nutrition and safe drinking water can ensure that all children have an opportunity for full expression of their innate genetic potential for physical and mental development. Today most developed and some developing countries, like China and India, have marketable surpluses of food grains. The widespread hunger prevailing in many nations is not due to nonavailability of food in the market, but to inadequate purchasing power among the rural and urban poor. Inadequate purchasing power in turn is due to insufficient opportunities for gainful employment. The famines of jobs and purchasing power are the primary causes of famines of food in poor households.

Crop husbandry, animal husbandry, fisheries, and forestry are the major sources of employment and income in the rural areas of most developing countries. In this context, agriculture assumes a more significant role than being just the source of food in the development of national and global food and nutrition security systems. In predominantly agricultural countries, the importation of food to compensate for inadequate national attention to agricultural development has the additional consequence of increasing rural unemployment. Thus, food security must be viewed in the interactive context of food production, job creation, and income generation.

An additional issue of overriding importance, if we are to ensure that today's progress is not at the expense of tomorrow's prospects, is conservation of the ecological base for sustained agricultural production. The various issues relating to the sustainable production of food for the growing population have been dealt with by the Panel on Food Security and Environment of the World Commission on Environment and Development (WCED) (Food 2000, 1987).

Although the problems we face in promoting sustainable nutrition security are staggering, we must be prepared to face the challenges of the future, particularly in relation to probable changes in climate, including changes in precipitation and temperature induced by increasing concentrations of CO_2 and other industrial gases in the atmosphere. These changes will have visible impact in about 25 yr. Whatever the magnitude of the changes, it would be prudent to make the scientific investment necessary to face different climate scenarios. We should maximize the advantages of favorable weather and minimize the adverse impacts of unfavorable weather on human, animal, and plant populations. Oceans and inland waters may not be able to provide more than 5% of total food needs; soil-based cultivation is the mainstay of our food and nutrition security system. But land is a shrinking resource for agriculture: we must produce more and more food from less and less land and water in the decades ahead.

We discuss here some of the major problems and possibilities associated with food security and the projected changes in climate. We deal mainly with implications for the production of cereals, because these are central to food security. We consider the relevant issues under two scenarios.

Scenario I: Food security in the current climate

Scenario II: Climate change and food security

Food security in the current climate

Population growth and food production

The important consideration for food security is whether the food production growth rate will remain higher than the population growth rate. Between 1970 and 1982, world population grew at 1.8%/yr while cereal production, which constitutes 94% of total grain production, grew at 2.3%/yr. On a global scale, food production outstripped population growth by 0.5%. Food grain production in 1986 was 1,942 million t for a population of 4,915 million people. This corresponds to about 395 kg food grain per person on earth. But there were regional disparities, to the extent that near famine conditions occurred in many parts of the world. Hunger existed amid plenty; food production did not provide food security to everyone.

A United Nations study (World Commission on Environment and Development 1987) has projected population size and growth rates for 1985 to 2000 and 2000 to 2025. The growth rate is likely to decline to 1.6% between 1985 and 2000 and to 1.2% between 2000 and 2025. The world population is projected to be 6.1 billion in 2000 and 8.2 billion in 2025 (Table 1). Based on per capita consumption in the recent past, Sanderson (1984) estimated per capita grain consumption in 2000 A.D. and the food grain requirements in various regions of the world (Table 1). The global requirement for food, feed, and industrial use grains in 2025 would be about 3,050 million t, assuming no significant changes in per capita grain consumption.

The WCED report (Food 2000, 1987) pointed out that food imports are not the answer to increasing populations in developing countries: importing food leads to growing crops with export potential. In predominantly agricultural countries, imported food also results in unemployment. Coupled with this is the low price of exported farm commodities in developed countries. This has resulted in the increasing indebtedness of developing countries, with several undesirable ecological and political consequences.

Assuming no significant change in food consumption patterns, the projected additional demand for food grains in 2025 over that in 1986 would be 330 million t in Africa, 130 million t in South America, 582 million t in Asia, 73 million t in Europe, and 16 million t in the USSR. If individual regions are to be self-sufficient in food grains, these projections lead to the following questions:

1. What changes in productivity and cultivated areas will be needed to grow the additional food grains?
2. Can the regions requiring additional food grains produce them?

Increase in area and productivity

Food production in different regions must increase substantially if each region is to meet its own requirements. Production in Africa would have to increase almost 400% and in Latin America 200% to meet the projected demand. In Asia, the required increase would be 75%; in the USSR and Europe, increases of 25-82% would be needed.

In Asia and Europe, 83 and 88%, respectively, of potentially arable land is already being cultivated (Table 2). Yield increases are almost the only means of realizing projected food grain demands. In Africa and South America, where only

Table 1. Food grain requirement of different regions of the world in 2025.

Region	Population (billion)	Average per capita consumption (kg)	Food grain requirement (million t)
Africa	1.62	257	416
South America	0.78	296	231
Asia	4.54	300	1362
North America	0.35	885	310
Europe	0.52	700	364
USSR	0.37	983	364
Oceania	0.04	578	23
World	8.22	373	3070

Table 2. Additional land requirement to meet cereal demand in 2025.

Region	Additional land requirement (million ha)				
	Potential arable	Cultivated land 1986	Cultivated for cereals 1986	To be cultivated for cereals 2025	Additional for cereals
Africa	734	158	74	277	203
Asia	627	519	307	340	33
Australia and New Zealand	153	32	16	16	–
Europe	174	154	71	78	7
North America	465	239	103	103	–
South America	681	77	39	93	54
USSR	356	227	111	146	35
World	3190	1406	721	1053	332

Table 3. Changes needed in productivity of cereals and corresponding land requirements to maintain adequate supplies in different regions.

Region	1986		2025	
	Land area (million ha)	Productivity (kg/ha)	Land area (million ha)	Productivity (kg/ha)
Africa	73.5	1175	277	1500
Asia	307.0	2534	340	4000
Europe	70.5	4137	77.4	4700
South America	39.0	1991	92.4	2500
USSR	111.0	1796	277.5	2500
World	721.0	2587	1053	2915

22 and 11%, respectively, of the potentially arable land is under cultivation, increase in area could be the major means of additional food grain production. The USSR will have to increase both yield and area under cultivation to meet the requirement.

In regions where land is available, an appropriate strategy that balances increases in productivity with increases in land area needs to be worked out. The relative emphasis on each component will vary in different regions. One feasible combination of area and productivity is given in Table 3. The rates of growth required to attain the needed levels of production are given in Table 4.

Globally, arable land is estimated at 3,190 million ha, of which 1,406 million ha are already cultivated. Cereals account for 721 million ha, about 50% of total cultivated land. Assuming no change in this pattern, 2,102 million ha would need to be cultivated by 2025, 65.8% of the arable land against the present 44%.

Factors limiting agricultural production

Energy. Agricultural operations include land preparation, use of appropriate crop cultivars, application of fertilizers and pesticides, and water management. All these

Table 4. Projected rates of growth of area, productivity, and production to meet the food grain demand of different regions of the world.

Region[a]	Growth rate (%)		
	Area	Productivity	Production
Africa	3.50	0.06	4.1
Asia	0.02	1.20	1.4
Europe	0.02	0.03	0.05
South America	2.20	0.08	1.50
World	1.00	0.003	1.30

[a]It is assumed that North America and Oceania would maintain present area, productivity, and production.

operations require energy. Historically, draft animals provided a major source of energy (Pimentel and Pimentel 1979). When mechanization provided a more efficient means of farm operations, draft energy was gradually replaced by fossil fuels. This has led to a gradual decrease in numbers of farm families, with the result that less than 5% of the total population in developed countries today are engaged in agriculture. However, in developing countries, agriculture continues to be the major occupation of the majority of the population (FAO 1987).

In developed countries, the main agricultural operations and inputs changed very little before the early 1960s. Since then, high-yielding, short-stature cultivars have replaced conventional, locally adapted cultivars. The new cultivars of wheat and rice require more fertilizer and better pest management to achieve their yield potential. With the higher productivity of these cultivars, mechanization became imperative. Commercial energy input almost equalled energy output (food value of grain) (Slesser 1986). Estimating the energy input to the whole agricultural system, from land preparation to processed foods, then more commercial energy was being spent than solar energy harvested through crops. It could be said that in developed countries, fossil fuels ultimately serve as food.

In developing countries, agricultural production practices range from traditional to transitional to modern. Traditional agriculture is based on the use of local cultivars, almost no inputs of fertilizer and pesticides, minimal water management, and draft energy. Transitional agriculture includes improved seeds, fertilizer, pesticides, and water management, but mostly draft and human energy. Modern agriculture is mostly a replica of industrialized country agriculture, but some practices, such as weeding and harvesting, are still manual. In developing countries, more than 50% of the population continue to work on farms.

In projecting the food grain production needed to meet the additional requirements of each region, it is important to consider the technological level of agriculture and the source of energy in the region. Agriculture in Asia and Africa is being transformed, with many countries moving from traditional to transitional production systems. However, draft energy continues to be the mainstay of agricultural production. In Africa and South America, the additional land required cannot possibly be brought under cultivation with draft energy alone. Commercial energy will be needed. Sinha (1986) estimated the amount of oil needed to produce 2,412 million t of grain (the world demand in 2000) based on U.S. commercial energy-intensive and on Indian transitional agricultural production. For each ton of grain (wheat, rice, coarse grains), the U.S. technology used 0.110 t oil, the Indian technology 0.038 t (Table 5).

The implications to food security should be viewed in the context of the availability of fossil fuels. If the food grain-deficit countries have to import grain from the U.S. or any other country, their costs will rise. But the indigenous resources of these nations may not be adequate to produce the required food grain.

Decreasing fertilizer response. Fertilizers and chemical control agents are important components of farm inputs. FAO attributed 55% of the 1965-76 increase in yield in developing countries to fertilizers (Food 2000 1987). This implies that fertilizer consumption in developing countries will have to rise substantially in the future. However, the fertilizer response ratio is likely to decrease with increasing fertilizer use (as has happened over the last five decades) (Table 6).

Limitations on genetic improvement of crops. A breakthrough in rice and wheat improvement was responsible for increased agricultural production worldwide, and for the green revolution in developing countries. Scientists, administrators, planners, politicians, and the public have great faith in the possibility of developing new cultivars with even greater yield potential. However, the maximum yields of rough rice and wheat in experimental trials at the International Rice Research

Table 5. Projected world grain requirement and estimated oil needed for its production in 2000 A.D. (Sinha 1986).

	Grain requirement (million t)	Oil requirement (million t)	
		U.S. technology	Indian technology
Wheat	601.3	51.1	10.5
Coarse grain	1186.6	123.6	44.5
Rice	633.6	89.7	36.9
Total	2412.5	264.4	91.9

Table 6. World grain production and fertilizer use, 1934-38 to 1979-81.

Period	World grain production[a] (million t)	Increment (million t)	World fertilizer use (million t)	Increment (million t)	Incremental grain-fertilizer response ratio
1934-38	651		10		
1948-52	710	59	14	4	14.8
1959-61	848	138	26	12	11.5
1969-71	1165	317	64	38	8.3
1979-81	1451	286	113	49	5.8

Source: Food 2000 (1987).
[a]Annual average for period.

Institute and the International Maize and Wheat Improvement Center (CIMMYT) were reached in the later part of the 1960s (Table 7). The situation in other cereals is no different. The yield of some recently developed rice cultivars is higher on a per day basis. Nevertheless, it is reasonably clear that we may have to wait for another scientific breakthrough, for another major advance in yield potential in cereal crops.

Degradation of resource base. The urgent need to meet the food demand of growing populations and industrialization has led to degradation of the agricultural resource base on almost every continent. Often the poor are held responsible for environmental degradation, including the decrease in agricultural resources. Statements at public hearings of the WCED sum up the situation beautifully.

Geoffrey Bruce of the Canadian International Development Agency said, "Small farmers are held responsible for environmental destruction as if they had a choice of resources to depend on for their livelihood, when they really don't. In the context of basic survival, today's needs tend to overshadow consideration for the environmental future. It is poverty that is responsible for the destruction of natural resources, not the poor."

Adolfo Mascarenhas of the International Union for Conservation of Nature and Natural Resources highlighted the African situation succinctly. "There are many contradictions in agricultural development. The blind imitation of models developed under different circumstances will have to give way to the realities and conditions existing in Africa. Large areas of virgin land have been opened up for export crops whose prices keep declining. This is not in the interest of developing countries.

"There are so many problems to be overcome that we forget that every problem is an opportunity to do something positive. This is an opportunity for us to think of conservation and environment in a broad educational context. In doing so, we will be able to capture the next generation and demonstrate the wonder and the benefits of the world around them."

The need to open up new land for cultivation, timber, and fuel wood has caused extensive deforestation in different parts of the world. This has adversely influenced water conservation and led to soil erosion, silting, and flooding, particularly in mountainous regions. Many river basins in India are experiencing the impact of deforestation. While there is no conclusive evidence linking afforestation with

Table 7. Maximum yields of the best varieties of rice (rough rice) in 1966-67 and 1986, and of wheat in 1968-69 and 1983-84.

Year	Variety	Yield (t/ha)			
		Dry season	Wet season	Bread wheat	Durum wheat
1966-67	IR8	10.1	7.5		
1968-69				9.3	8.5
1984				8.4	8.9
1986	IR36	6.5	4.1		
	IR28	8.2	3.4		

Sources: Annual reports of the International Rice Research Institute and the International Maize and Wheat Improvement Center.

precipitation, the loss of animal and plant genetic resources is evident in many parts of the world.

Measures to be adopted. The main objective of future research and extension should be to narrow the gap between maximum and average yields of crops. It will be important to analyze the contribution of different industrial inputs and environmental factors to assess the realizable potential of genetically superior cultivars. Actual realization of the potential will be governed by the technologies adopted in relation to three factors:

- land and water management,
- crop management, and
- postharvest management.

Food 2000 (1987) discusses these aspects in detail.

Climatic change and food security

Food production in any given year is affected most directly by the levels of the critical climate variables (temperature, radiation, precipitation) during the year. The stability of available food supplies is governed by the interannual variability of these variables. Access to food supplies in different regions of the world is determined by the share of food production, the role of cereals in the diet, and the various political and market forces that act upon the global food security system. Climate anomalies during the 1970s caused relatively small fluctuations in world cereal supplies. But they occurred at a time when the use of cereal as livestock feed was increasing. The food shortages were particularly severe in the Soviet Union, and its large grain purchases led to dramatic fluctuations in world cereal prices. The disastrous effects all this had on the world food security system are now well documented (Garcia 1981).

Climate fluctuations of the kind witnessed in the 1970s lie within the variability of the present climate. They could have been anticipated by prudent societies, if the climatic record had been monitored. In addition to the normal variability in climate, there is increasing evidence of a change in atmospheric optical properties as a result of buildup of CO_2 and other "greenhouse gases." It is also clear that this buildup will continue. It is expected that in the long run, this will result in a climatic change.

Projections of climatic change

Mathematical models of the potential climatic impact of a change in greenhouse gases concentrations have been developed by various groups. The models attempt to predict changes in critical climatic variables with a doubling of CO_2 concentration. While there is little agreement between various models about the specific magnitudes of the changes at the regional level during the next 50-100 yr, there is considerable agreement on the nature of such changes at the global level. They may be summarized as follows:

- The lower atmosphere will be warmed and the stratosphere cooled.
- The average annual global warming will be 1.5-4 °C. This is much greater than any natural climatic change. The rise in temperature in general will be greater

in the Northern Hemisphere than in the Southern, and will increase (by a factor of 2 or 3) poleward.

- The temperature rise will be greater (by about 50%) in winter than summer. Consequently, we may expect the production of winter season crops to be more affected than summer crops.
- Freeze-free periods will increase in higher latitudes and larger areas may be brought under cultivation, if soil conditions are suitable. The increase in the freeze-free period will depend on the current duration of this period (e.g. a 1 °C rise in temperature increases an 80-d freeze-free period by about 20 d but a 120- to 130-d period by only 6 d).
- General warming will be accompanied by a weakening of temperature differences between the equator and the poles. That will affect the atmosphere's general circulation, which can lead to longer dry periods.
- Global annual precipitation average will increase 7-11%, but regional and temporal variations are uncertain.
- The relationship between precipitation and evaporation in the lower latitudes is not likely to change. In the mid- to higher latitudes, evaporation will increase more than precipitation.
- Soil moisture conditions will be wetter in some regions of the world (35° N - 35° S), but drier in others.
- A rise in sea level is foreseen, but its magnitude and time scale are uncertain. The effects on agriculture in coastal regions could be disastrous.
- Agroclimatic zones will shift toward the poles (about 100 km per degree of warming).
- Variability of temperature and precipitation may decrease because of weaker upper air circulation. Regional variations are uncertain.
- Only changes in mean climate conditions are specified by the models. Sizable uncertainties remain about timing, intensity, and direction of specific effects.

Climate is a complex, nonlinear multiple feedback system with dominant positive feedbacks. From a cybernetic systems viewpoint, a relatively rapid forced change, such as a change in CO_2 concentration, is likely to destabilize the system. The magnitude of the destabilization tends to be proportional to the rate of change of the forcing function. Because the rate of change of CO_2 concentration is expected to be greatest between 2000 and 2060, those decades may experience chronic and severe weather variability (Markley and Hurley 1983).

One aspect of climatic change, even at the present level of CO_2 increase, is provided by Rowntree and Boulton's simulation study of 1983. It shows that, as a result of positive feedback, a hot dry spell occurring in mid-summer in Central Europe, when coupled with a weak atmospheric circulation pattern, could persist for as long as 50 d and expand into Scandinavia, Spain, and North Africa. Even in conditions of moist airflow, initial hot dry conditions lasted for up to 20 d.

Even though mean changes in global climate resulting from a doubling of the CO_2 concentration can be anticipated, the specific changes in annual variability of climate are uncertain during the period when the doubling is taking place. But in keeping with the complex, nonlinear, and multiple feedback characteristics of the climate system, it is likely that during the next 50-100 yr, the world will experience chronic and severe climatic variability.

What are the implications of climatic change to the world food security system? To answer this it is necessary to examine changes in regional food production, as a result of changes in mean climate, and in variability in food production, as a result of increased climatic variability. In the ultimate analysis, annual variability in total food production as well as regional shares of production are the determinants of food security.

Climatic change and food production

Increased CO_2 and crop yield. The climatic changes envisaged in the next century are mostly attributed to the increasing concentration of CO_2 and other gases. Since CO_2 is an essential reactant in photosynthesis to produce organic matter, it has been postulated that farmers could look forward to better harvests (Wittwer 1986). But these postulates have often been based on short-term experiments in controlled environments or glasshouses, with adequate supplies of water and plant protection measures. Rosenberg (1987) analyzed gas exchange to conclude that climatic change, at least as far as CO_2 concentrations are concerned, may prove advantageous. However, Gifford (this volume) has made a more cautious assessment of CO_2 effects by including temperature change as an additional component. The following observations are relevant in assessing the effects of climatic change, including CO_2 concentration, on crop yields:

- Highest yields in C_3 crops are obtained around a mean daily temperature of 15 °C; in C_4 crops, around 30 °C.
- The temperature optima for vegetative growth and reproductive phases are often different. An increase of temperature beyond a mean of 22 °C although it has no effect on photosynthesis, causes sterility in rice, resulting in reduced grain yield (Fig. 1). In wheat, an increase in mean temperature beyond 16 °C results in decreased grain weight and poor yield (Fig. 2). Higher temperature significantly reduces tillering, which is essential to building shoot population.
- Crops having a high growth rate at the preflowering stage usually deplete the soil moisture that is needed to complete grain development. Consequently, without irrigation, high initial growth results in poor grain yield despite high dry matter accumulation.

Gifford estimated the rise in temperature that would cancel out the advantageous effects of CO_2 fertilization. At locations ranging from 50° N in Canada to 37° S in Australia, a rise of 1.5-2.4 °C was required to cancel the advantageous effects of CO_2 on grain yield, presumably in irrigated conditions. Without irrigation, crop yields may in fact be reduced.

The optimistic predictions for agricultural production based on CO_2 fertilizer effects recently made should not lead to complacency on the question of food security. These projections have been based on individual effects of only one or two factors. Agricultural production is a complex process. The available evidence on CO_2 fertilizer effects when two or more factors are considered simultaneously is, at best, inconclusive.

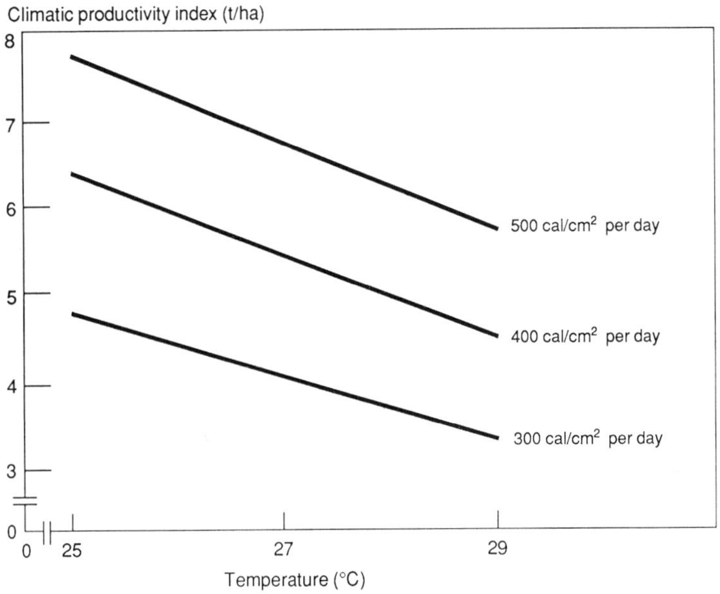

1. Effect of increasing temperature on productivity of rice at different levels of radiation (Yoshida 1978).

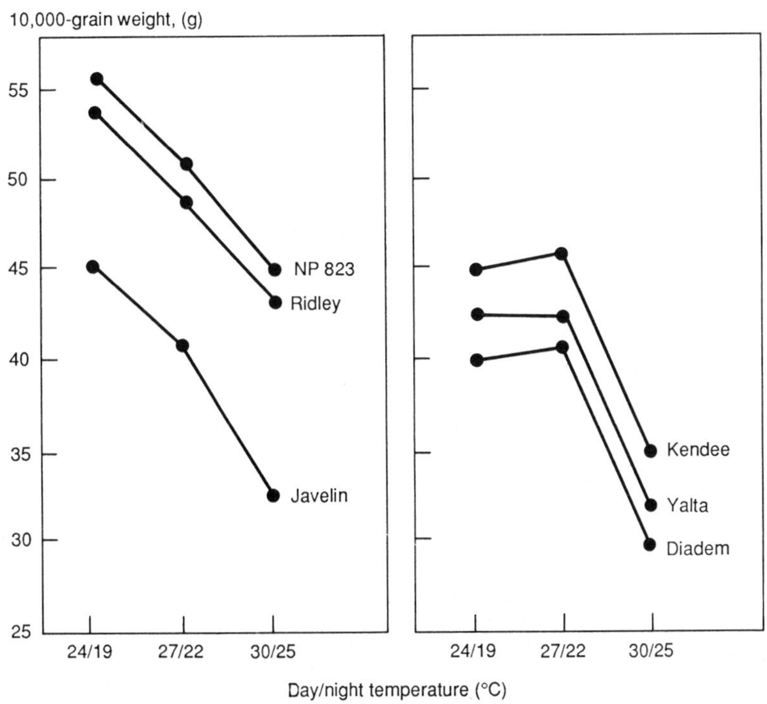

2. Effect of increasing temperature on grain development in wheat (Asana 1976).

Changes in mean temperature and precipitation. Crop production is directly affected by temperature and precipitation: both are critical variables for determining agroclimatic zones. Temperature determines the duration of a crop's growing season and controls its phenological development and water requirements. Precipitation provides water, the critical input for crop growth. The distribution of these variables during crop growth is also critical. But information on anticipated intraseasonal variations in temperature and precipitation is not available. This discussion is, therefore, limited to anticipated changes in seasonal or annual values of the variables. Direct effects of CO_2 on crop yields are excluded because available information is inconclusive.

In general, increased average annual temperature will result in shorter freeze periods than at present. As a result, large areas at the climatic margins can be brought under cultivation. But in the existing crop belts, the growing season will be shortened and productivity (crop yield) will decline. Some crops may be forced out of cultivation, to be replaced by others. To what extent losses in crop yield can be compensated for by increases in crop area is uncertain; that will be governed to a significant extent by technology. Increased precipitation will be beneficial to both productivity and expanded crop area.

Wheat, rice, and maize, the three major crops of the world, accounted for about 80% of total cereal production in 1985. The climate scenario considered here is based on temperature changes predicted by the Geophysical Fluid Dynamics Laboratory (GFDL) model for various latitude limits for the summer and winter seasons. This model predicts maximum increases (in contrast to other commonly used models, such as those of the Goddard Institute of Space Studies [GISS] and the National Center for Atmospheric Research [NCAR]) (Rosenberg 1987). Precipitation scenarios are considered indirectly, using the results presented by Kellogg and Schware (1981). Impacts on production are indicated as increases (+) or decreases (-) over current crop yields or areas. The results are presented for the countries or regions which accounted for significant shares of each crop's production in 1985.

The impacts of a changed climate on each crop's production, assuming current levels of technology, are shown in Tables 8-10. The following observations may be made.

- In the mid- to higher latitudes of developed countries, significant increases in area, accompanied by crop yield, reductions are expected. In the lower latitudes, with increasing temperature, significant areas become unsuitable for wheat, and yields decline. Increased water requirements may be anticipated in all regions in these latitudes, increasing the importance of irrigation management. A net reduction in wheat production is anticipated, with the balance of production shifting even further in favor of developed countries.
- Up to 94% of the world's rice is grown in developing countries, mostly during the summer monsoon. However, better yields are obtained in the dry season, when temperatures are lower and pest problems are fewer. Increases in precipitation would increase the area and yield of rice, but increases in temperature promote sterility that would reduce yields. Almost one-third of rice is grown in nonirrigated conditions, which may benefit from increased

Table 8. Effect of doubling of atmospheric CO_2 on area and productivity of wheat (total production in 1985 = 510 million t).

Country or region	Share of production (%)	Yield (t/ha)	2 X CO_2 climate scenario		Effects on crop production[a]	
			ΔT °C	Soil moisture conditions	Area	Yield
China	17	3.0	4	Some regions wetter	+	–
India	9	1.9	3	Wetter	–	–
USSR	16	1.6	6	Drier	++	–
Canada	5	1.7	8	Drier	++	–
USA	13	2.5	5	Drier	+	–
Western Europe	16	4.6	6	Wetter	+	–
Australia	3	1.4	2	Wetter	–	–
Developed countries	60	2.3	5-8	Drier	++	–
Developing countries	40	2.1	2-4	Wetter	+	–

[a]+ indicates positive effect, – indicates negative effect.

Table 9. Effect of doubling of atmospheric CO_2 on area and productivity of rice (total production in 1985 = 465 million t).

Country or region	Share of production (%)	Yield (t/ha)	2 X CO_2 climate scenario		Effects on crop production[a]	
			ΔT °C	Soil moisture conditions	Area	Yield
China	37	5.3	3	Wetter	+	+
India	20	2.2	3	Wetter	+	+
Indonesia	8	2.1	2	Drier	–	+
Bangladesh	5	4.1	3	Wetter	+	+
Developing countries	94	3.1	2-4	Wetter	+	+

[a]For rice, sterility effects of increasing temperature may neutralize increase in production. + indicates positive effect, – indicates negative effect.

Table 10. Effect of doubling of atmospheric CO_2 on area and productivity of maize (total production in 1986 = 490 million t).

Country or region	Share of production (%)	Yield (t/ha)	2 X CO_2 climate scenario		Effects on crop production[a]	
			ΔT °C	Soil moisture conditions	Area	Yield
USA	46	7.4	4	Drier	+	–
Western Europe	8	5.5	4	Drier	+	–
China	13	3.5	3	Wetter	+	+
Brazil	4	1.9	3	Wetter/Drier	+	+
Developed countries	65	6.0	4	Drier	+	–
Developing countries	35	2.2	3	Wetter	+	+

[a]+ indicates positive effect, – indicates negative effect.

precipitation. On balance, it would appear that developing countries will continue to be major growers of rice.

- The mid-latitudes account for more than 50% of maize production. Increased precipitation may increase the share of developing countries.

Overall, under the doubled CO_2 climate, declines in wheat production are expected, production of rice may be unaffected, and maize may increase. The existing wheat production imbalances in favor of developed countries may be further accentuated. In rice production, the share of developing countries may rise. Maize production may become more uniform across developed and developing countries.

This analysis excluded the possibility of replacement of one crop by another crop more suitable for the changed climate conditions. But this is expected to happen. For example, wheat belts may be replaced by barley, barley by maize, maize by sorghum, and so on. Where adequate irrigation facilities are available, rice may be substituted. These substitutions are likely to take place gradually, because there will be problems of adaptation.

With respect to food security, the current advantage is with the developed countries, primarily because of the rapid progress of technology, absence of population, pressure, or both. From currently available information, it is clear that these two factors will continue to be the significant determinants of food security, whether or not a climatic change occurs.

Pests and diseases. Currently, about 25% of food crops are lost to pest damage. The high productivity of crops in temperate regions can be attributed not only to improved technology, but also to limited numbers of diseases and pests; better soil health, including microflora; and better response to fertilizers. The generally warmer and more moist conditions in the changed climate, coupled with low general circulation and longer freeze-free periods, would be conducive to the proliferation of crop pests. By far the most predictable effect of climatic change is that it will significantly increase pest populations, and some of the advantages of present temperate regions may be canceled out.

Variability in food production. In the current climate, the coefficient of variation (CV) of total global cereal production in the 1970s was about 3% (Anderson et al 1987). Production variabilities in individual countries are higher. Variability is highest in semiarid areas and least in humid areas. The high variability regions also have low average crop yields. Variability tends to be less in larger countries because of risk pooling across regions and crops.

For individual crops, the global-level CV for maize and rice is about 4%; for sorghum, 6%; and for wheat, 5%. But the variabilities in crop production in individual countries are higher. For example, in India, the CV for wheat is 11%; for rice, 8%; and for sorghum, 16%. In the U.S., the CV for maize and winter wheat is 10%, and for sorghum, 11% (Hazell 1984). Six of the sources of interannual variability in national agricultural production are

1. Variability in weather,
2. Variability in areas planted to different crops,
3. Variability in yield correlations between regions and crops,

4. Production expansion into riskier regions,
5. Increased sensitivities of new technologies to weather and diseases, and
6. Variations in agricultural prices, policies, and levels of rural infrastructure.

The relative contribution of these factors to overall food production variability is different in different regions, and depends on current production variability. For example, variability in cereal production in India was a consequence of increased adoption of high yield technology and variability in weather, crop yields, areas cropped, and prices (Hazell 1984). The predominant sources of production variability in the U.S. were crop yields and yield correlations between states.

The relevant question to food security under CO_2-induced changes in climate is whether such climatic changes will increase or decrease variability in food production. Even when anticipated changes in mean climate conditions only are considered, it is clear that increased variability of production would result from sources 2-5. When increased weather variability (source 1) is included, significant increases in the variability of food supplies are foreseen.

This leads to many concerns that need to be addressed by the international community. Prominent among these are

- Perceptions of increased risk may discourage adoption of new technologies and retard agricultural growth,
- Increased instability of national and international food supplies,
- Increased frequencies of droughts and floods over larger areas,
- Increased destabilizing effects of agricultural prices on food production and consumption, and
- Risk pooling across regions and crops by diversification of crop production systems.

Climate impact assessment

Climate fluctuations affect the lives of millions of people around the world (Kates et al 1985). Whether climate change-induced effects could be discerned as a warning or an attempt to mitigate adverse effects is an important question. Parry and colleagues have developed a methodology based on climate impact assessment in marginal areas (Parry et al 1987). This is useful when the objective of studies is to evaluate sensitivity of ecosystems to climatic changes.

However, food security is concerned more with stabilizing the available food supplies. Both national and global food security depend more on the stable and productive areas than on marginal areas. For example, in India, 7 of 35 meteorological divisions are important determinants of food security (Sinha 1987).

This is shown in the impact on national food production of droughts of nearly equal magnitude in 1979-80 and 1982-83. In 1979-80, there was a 17% shortfall in grain production. However, in 1982-83, food grain production declined only by 3.7%. That year, the highly productive areas were not affected by drought (Sinha 1987). In assessing food security, it is important to study the impact of climate fluctuations on productive areas. In studying global food security, it may be prudent to conduct such studies on a selected group of nations.

Conclusions

- Projected population growth rates and the ensuing food demands, even in the current global climate, will make it difficult to provide for human sustenance and food security in the 21st century.
- The most vulnerable regions from the point of view of food security are Africa and South America. To meet the food demands of these regions, the cultivated area must be increased. Asia is next in vulnerability. Significant increases in crop productivity are required in this region, because it would be difficult to bring additional area under cultivation. Appropriate strategies for increasing agricultural production that balance increases in cultivated area with increases in crop productivity need to be designed for the various regions of the world.
- There is a near consensus that the rising concentrations of CO_2 and other greenhouse gases in the atmosphere will lead to a climatic change. This will directly affect agricultural production and food security.
- Some recent reports on the favorable impact of climatic change on agricultural production should not lead to complacency. Available evidence is inconclusive and is not based on the complex, dynamic interactions between agricultural production processes and the environment.
- Two aspects of the effects of changing climate on agricultural production and food security need to be considered: 1) changes in mean values of critical variables which will affect trends of global agricultural production and regional shares of production, and 2) increased instability of climate, which will result in greater instabilities in food supplies.
- In a $2 \times CO_2$ climate, an overall decline in production of wheat may be expected as a consequence of increases in temperature in the mid-latitudes. Production of rice may be unaffected. That of maize may increase as a result of increased area and rainfall in the lower latitudes. Food supplies of smaller nations are likely to be affected more by climatic change than those of larger nations. Larger nations have the advantage of risk pooling across regions and crops.
- Even in a changed climate, the current balance of food supplies in favor of developed nations will continue. In the future, even more than now, technology and resource management will be an even more important factor in determining this balance.
- Availability of energy, decreasing marginal response of crops to fertilizers, leveling off of the production potential of major cereals, and an eroding natural resource base will be the major factors limiting agricultural technology developments.

Implications for research and policy

Measures to improve food security by maintaining adequate food reserves and developing efficient transportation and distribution mechanisms require huge capital resources. In many small nations, planners are reluctant to commit the financial resources needed to establish such facilities. But in a changing climate, the smaller nations will be the most vulnerable from the point of view of food security. Such nations may be encouraged to group and pool the resources required for stepping up food security measures. This will also provide the advantage of risk pooling across regions and crops, making a group of nations as a whole less susceptible to climatic effects.

In the long term, adoption of food security measures will come to be viewed in the broader context of regional development. Specific studies on criteria to be adopted for forming nation groups; the crop and production technologies needed; the appropriate balance between increasing crop area and productivity; and related aspects of infrastructural development, such as transportation, distribution, and marketing networks, need to be initiated.

Ensuring food security in the future, particularly with changing climatic conditions, will require greater emphasis on land and water management, crop management, and postharvest management. Availability of energy will be a significant factor. Developments in the use of nonpolluting, renewable sources of energy will play a role in conserving the resource base of agricultural production. Any major breakthrough in this sector should be shared globally, without political, social, or economic reservations. Such a breakthrough, in addition to supporting agricultural production, would help control the greenhouse effect itself, benefiting all mankind.

Considerable additional research on the annual, seasonal, and intraseasonal variability in climate in individual regions is needed. Links between the results of this research and agricultural production should be established. The predictive power of general circulation models must be extended considerably before this can be done.

An important issue for research relates to the methodology of climate impact assessment. Current methodologies developed for impact assessment in marginal areas are not directly applicable. Such methodologies are useful as long as the goal of research is to analyze and demonstrate the sensitivities of agroecological and social systems to climatic change. In this case, the marginal areas provide the analytical advantage required for isolating the effects of climate variables. However, for providing food security, the major source of food is agricultural activity in regions with high potential for production. In these regions, it is difficult to isolate climate effects from other factors. Methodological issues relating to climate impact assessment on food production need to be critically addressed.

Specific research on the effects of climatic change is required for planning and designing agricultural production and food security systems that span several decades, such as irrigation systems or agricultural expansion in coastal areas.

Instability of food supplies results from many sources. In addition to climatic instability, these include uncertainties of agricultural policy. While effects of climatic instability are difficult to anticipate and control, instabilities resulting from food policy perspectives can be minimized. This requires, from individual nations, explicit statements of food policy objectives (efficient growth of agriculture and related sectors, income distribution through employment, subsidies, insurance, etc.), and specific programs and policies (not just projects). Many projects in tropical Africa achieved only limited success because a framework for monetary, fiscal, and trade policy did not exist.

Considerable progress has been made recently in drawing the attention of policymakers to the issues related to climatic change. But much more persuasion is needed before that interest will be translated into public policy and action. This requires developing capabilities for obtaining and analyzing information and for developing policy alternatives. Simultaneously, it must be recognized that the political will to initiate action is driven by public opinion. Informed public opinion must be mobilized through systematic communication of available research information.

References cited

Anderson J R, Hazell P B R, Evans L T (1987) Variability of cereal yields: sources of change and implications for agricultural research and policy. Food Policy 12(2):199-212.

Asana R (1976) Physiological approach to breeding of drought resistant crops. Indian Council of Agricultural Research Technical Bulletin, New Delhi, India.

FAO—Food and Agriculture Organization (1987) FAO production yearbook 1986. Rome, Italy.

Food 2000 (1987) Global policies for sustainable agriculture. Report to the World Commission on Environment and Development. Zed Books Ltd., London, UK.

Garcia R (1981) Drought and man: the 1972 case history. Vol. 1. Nature pleads not guilty. Pergamon, New York, NY, USA.

Gifford R M (1989) Exploiting the fertilizer effect of increasing atmospheric carbon dioxide. Pages 477-487 *in* Climate and food security. International Rice Research Institute, P.O. Box 933, Manila, Philippines.

Hazell P B R (1984) Sources of increased instability in Indian and US cereal production. Am. J. Agric. Econ. 66(3):302-311.

Kates R W, Ausubel J H, Berberian M, eds. (1985) Climate impact assessment. SCOPE 27. Wiley, London, UK.

Kellogg W W, Schware R (1981) Climate change and society—consequences of increasing atmospheric carbon dioxide. Westview Press, Boulder, Colorado, USA.

Markley O W, Hurley T J III (1983) A brief technology assessment of the carbon dioxide effect. Technol. Forecasting Soc. Change 23:185-202.

Parry M L, Carter T R, Konijn N T (1987) The impact of climatic variations on agriculture. Vols. 1 & 2. Reidel, Dordrecht, The Netherlands.

Pimentel D (1979) Food, energy and society. Edward Arnold Ltd, Berks, England.

Rosenberg N J (1987) Drought and climate change: for better or worse. Pages 297-316 *in* Planning for drought - toward a reduction of societal vulnerability. D. A. Wilhite and W. E. Easterling, eds. Westview Press, Boulder, Colorado, USA.

Rowntree P R, Bolten J A (1983) Effects of soil moisture anomalies over Europe in summer. Pages 447-462 *in* Variations of the global water budget. A. Street-Perrot, M. Beran, and R. A. S. Rachliffe, eds. Reidel, Dordrecht, The Netherlands.

Sanderson F H (1984) World food prospects to the year 2000. Food Policy 9(4):363-373.

Sinha S K (1986) Energy balance in agriculture: the developing world. Pages 57-84 *in* Global aspects of food production. M. S. Swaminathan and S. K. Sinha, eds. Tycooly International, London, UK.

Sinha S K (1987) The 1982-83 drought in India; magnitude and impact. Pages 37-42 *in* The societal impacts associated with the 1982-83 worldwide anomalies. M. Glantz, R. Katz, and M. Krenz, eds. Stan Grossfield/Boston Globe, USA.

Slesser M (1986) Energy balance in agriculture: the developed world. Pages 47-56 *in* Global aspects of food production. M.S. Swaminathan and S.K. Sinha, eds. Tycooly International, London, U.K.

Swaminathan M S (1986) Building national and global nutrition security systems. Pages 417-449 *in* Global aspects of food production. M. S. Swaminathan and S. K. Sinha, eds. Tycooly International, London, UK.

Wittwer S H (1986) Research and technology needs for the twenty-first century. Pages 85-116 *in* Global aspects of food production. M. S. Swaminathan and S. K. Sinha, eds. Tycooly International, London, UK.

World Commission on Environment and Development (1987) Our common future. Oxford University Press, Delhi. p. 101.

Yoshida S (1978) Tropical climate and its influence on rice. IRRI Res. Pap. Ser. 20. 25 p.

Notes

Paper is published here as it was revised for presentation at the World Conference on The Changing Atmosphere: implications for global security, Toronto, Canada, 27-30 June 1988.

Authors' address: S. K. Sinha, N. H. Rao, and M. S. Swaminathan, Indian Agricultural Research Institute, New Delhi, India.

Citation information: International Rice Research Institute (1989) Climate and food security. P.O. Box 933, Manila, Philippines.

Participants

B. L. Amla
Central Food Technological
Research Institute
Mysore
India

Roger B. Austin
Plant Breeding Institute
Maris Lane, Trumpington
Cambridge CB2 2LQ
United Kingdom

Wilfrid Bach
Center for Applied Climatology and
Environmental Studies
Department of Geography
University of Munster/R.-Koch-Str. 26,
4400 Muenster
Federal Republic of Germany

Guy Baird
United States Agency for International
Development
U.S. Embassy — New Delhi
Department of State
Washington, D.C. 20520
USA

J. M. P. Cooper
International Centre for Agricultural
Research in the Dry Areas
P.O. Box 5466
Aleppo
Syria

A. N. Chaturvedi
Tata Energy Research Institute
7 Jor Bagh, New Delhi 110-003
India

V. K. Dadhwal
Space Applications Center (ISRO)
Ahmedabad 380 053
India

P. K. Das
Centre for Atmospheric Sciences
Indian Institute of Technology
Hauz Khas, New Delhi 110-016
India

William E. Easterling
Climate and Meteorological Section
Illinois State Water Survey Division
2204 Griffith Drive
Champaign, IL 61820
USA

Jay S. Fein
Climate Dynamics Branch
National Science Foundation
Washington, D.C. 20550
USA

Sulachana Gadgil
Centre for Ecological Services
Indian Institute of Science
Bangalore
India

Liangzhi Gao
Agricultural Meteorology
Jiangsu Academy of Agricultural Sciences
Nanjing 210014
China

Roger M. Gifford
Commonwealth Scientific and Industrial
Research Organization
Division of Plant Industry
G.P.O. Box 1600
Canberra, A.C.T. 2601
Australia

Peter H. Gleick
Energy and Resources Group
University of California at Berkeley
Berkeley, CA 94720
USA

G. Golubev
GEMS/PAC
U.N. Office of Environment Programme
P.O. Box 30552
Nairobi
Kenya

Hazel C. Harris
International Centre for Agricultural
Research in the Dry Areas
P.O. Box 5466
Aleppo
Syria

N.S. Jodha
International Crops Research Institute
for Semi-Arid Tropics
Patancheru P.O.
Andhra Pradesh 502 324
India

Michael J. Jones
International Centre for Agricultural
Research in the Dry Areas
P.O. Box 5466
Aleppo
Syria

Harold E. Kauffman
International Soybean Program
College of Agriculture
University of Illinois at Urbana
Urbana, IL 61901
USA

Peter Kenmore
Food and Agriculture Organization
P.O. Box 1864
Manila
Philippines

Renu Khanna-Chopra
Water Technology Centre
Indian Agricultural Research Institute
New Delhi 110012
India

Rattan Lal
International Institute of
Tropical Agriculture
c/o Miss M. Larkin
L.W. Lambourn & Co. Ltd., Carolyn
House
26 Dingwall Road, Croydon CR9 3EE
England

T. L. Lawson
International Institute of
Tropical Agriculture
Project de Recherche, IITA/Benin
B. P. 062523
Cotonou
Republique Populaire du Benin

Li Jingxiong
Institute of Crop Breeding and
Cultivation
Chinese Academy of Agricultural
Sciences
Beijing 30
China

W. John Maunder
New Zealand Meteorological Service
P.O. Box 722
Wellington
New Zealand

Darell E. McCloud
Department of Agronomy
University of Florida
Gainsville, FL 32611
USA

M. G. K. Menon
International Council of
Scientific Unions
Science Advisor to the
Prime Minister of India
New Delhi
India

Mohanty
Indian Institute of Technology

Raphael Harun Muturi
Kenya National Environment Secretariat
P.O. Box 67839
Nairobi
Kenya

C. B. Nath
National Council of Applied Economic
Research
Institute of Economic Growth
and Development
New Delhi
India

Peter A. Oram
International Food Policy Research
Institute (IFPRI)
1776 Massachusetts Avenue, N.W.
Washington, D. C. 20036
USA

John W. Pendleton
Turner Hall
University of Illinois
Urbana, Illinois 61801
USA

Dean F. Peterson
Agriculture and Engineering Education
Utah State University
Logan, Utah 84322-4105
USA

Thomas D. Potter
World Meteorological Organization
41, Giuseppe-Motta
Case Postale No. 5
CH - 1211 Geneva 20
Switzerland

N. Ganga Prasada Rao
Marathwada Agricultural University
Parbhani 431 402
Maharashtra
India

Eugene M. Rasmusson
Cooperative Institute for
Climate Studies
Department of Meteorology
University of Maryland
College Park, MD 20742
USA

Roger B. Revelle
Program in Science, Technology
and Public Affairs
University of California at San Diego
La Jolla, CA 92093
USA

William E. Riebsame
Natural Hazards Research and
Applications Information Center
University of Colorado
Boulder, CO 80309
USA

Norman J. Rosenberg
Center for Agricultural Meteorology
and Climatology
University of Nebraska
Lincoln, NE 68583-0728
USA

Eneas Salati
Fundacao Salim Farah Maluf
Rua Tagipuru, 235 — 2o andar
CEP 01156, Sao Paulo — SP
Brazil

R. P. Sarkar
Indian Meteorological Department

J. S. Sarma
International Food Policy Research
Institute
1776 Massachusetts Avenue, N.W.
Washington, D.C. 20036
USA

D. V. Seshu
International Rice Research Institute
P.O. Box 933
Manila
Philippines

Jagadish Shukla
Center for Ocean-Land Atmosphere
Interactions
Department of Meteorology
University of Maryland
College Park, MD 20742
USA

S. K. Sinha
Water Technology Centre
Indian Agricultural Research Institute
New Delhi 110-012
India

M. V. K. Sivakumar
ICRISAT Sahelian Centre
P.O. Box 12404
Niamey
Niger

J. P. Srivastava
International Centre
for Agricultural Research in Dry Areas
P.O. Box 5466
Aleppo
Syria

M. S. Swaminathan
International Rice Research Institute
P.O. Box 933
Manila
Philippines

Kalilou Traore
Meteorologie Nationale
B.P. 237
Bamako
Republique du Mali

S. Unninayar
World Climate Data Programme
World Meteorological Organization
41, Giuseppe — Motta
Case Postale No. 5
CH - 1211 Geneva 20
Switzerland

Hermannes Van Keulen
Centrum Voor Agrobiologisch Onderzoek
[CABO] / Bornsesteeg 65
P.O. Box 14
6700 AA Wageningen
The Netherlands

B. Venkateswarlu
Plant Physiology
Directorate of Rice Research
Hyderabad 500 030
Andhra Pradesh
India

S. M. Virmani
International Crops Research Institute
for the Semi-Arid Tropics
Patancheru P.O.
Andhra Pradesh 502324 A.P.
India

Jan C. Zadoks
Department of Phytopathology
Agricultural University
Binnenhaven 9
6709 PD Wageningen
The Netherlands